E2-17

$21.05

Ed & Judy 867-1858

FINAL
MAY 13

JOHN
947-0431 Thursday

MATH 233
BUILDING
6:30 — 8:20

8 MORE CLASS
PERIODS

BARNETT
PHY. - II

TRIG
Rudowski

65,84

ARTHUR SMITH
432-7091

Applied Engineering Mechanics

By Alfred Jensen and
Harry H. Chenoweth

Applied Engineering Mechanics, Third Edition
Applied Strength of Materials, Second Edition
Statics and Strength of Materials, Second Edition

Applied Engineering Mechanics

Third Edition

by

ALFRED JENSEN

Emeritus Professor, Engineering
University of Washington, Seattle

and

HARRY H. CHENOWETH

Associate Professor, Civil Engineering
University of Washington, Seattle

McGRAW-HILL BOOK COMPANY

New York San Francisco St. Louis Düsseldorf Johannesburg
Kuala Lumpur London Mexico Montreal New Delhi
Panama Rio de Janeiro Singapore Sydney Toronto

APPLIED ENGINEERING MECHANICS

Printed in the United States of America.

Library of Congress catalog card number: 73-157481

8 9 0 DODO 8 3 2 1 0

07-032480-8

Contents

PART 1 STATICS

1 Introduction

2 Basic Principles of Statics

3 Coplanar, Parallel Force Systems

4 Coplanar, Concurrent Force Systems

5 Coplanar, Nonconcurrent Force Systems

6 Noncoplanar, Parallel Force Systems

7 Noncoplanar, Concurrent Force Systems

8 Noncoplanar, Nonconcurrent Force Systems

9 Friction

14 Kinematics of Rectilinear Motion

15 Kinetics of Rectilinear Motion

16 Curvilinear Motion

17 Kinematics of Rotation

18 Kinetics of Rotation

Preface

The object of this third edition of *Applied Engineering Mechanics* is to provide students who intend to enter the engineering profession with the basic engineering mechanics knowledge that is so essential to their education, regardless of the branch of engineering they eventually may follow.

As in the previous editions, the text material is based upon easily and commonly understood physical concepts and principles, rather than upon the more abstract mathematical relationships that often are not as well understood. For greater clarity, the text refers in a few places (especially in the appendexes) to the principles of calculus. However, all the text material is so simply stated and the mathematical formulas so carefully developed that nothing more than an understanding of high school mathematics is required. In this respect, this new edition is similar to its companion volume, "Applied Strength of Materials," also written by us.

Our primary aim has been to develop and to present the material in an easily understood manner in order to make the text suitable in content and approach both for college students in engineering and architecture and for those in junior colleges, technical institutes, and in many industrial training and Armed Services programs. Furthermore, we believe the text will be a continuing aid to the many practicing engineers and engineering aides who, in the past, have found the previous editions useful as a reference text.

The book is divided into two parts: statics and dynamics. In statics, the problems most often occuring in practice are given more than usual emphasis. Analytical and graphical solutions are presented side by side in order to show their close relationship and to encourage students to use them concurrently, one as a check upon the other. Experience has shown that graphical solutions enable a student to *visualize* force analysis and that they definitely better the students' understanding of corresponding processes in analytical solutions.

In arranging and developing the subject matter in both sections, we have proceeded very gradually from the elementary problems to those more difficult, in the honest belief that, by such gradation, students learn more quickly and easily. Our experience has been that students learn more efficiently by thoroughly mastering one step at a time and later integrating the various steps and concepts into a completed whole.

The new edition retains the best of the well-tested problems, provides new data for many other problems, and presents more than a hundred new problems. Answers are given in the text to many of the problems. Classroom instructors may, upon request to the authors, obtain a list of the answers to the remaining problems: our addresses are listed at the end of this preface.

New material in the third edition includes a "three-dimensional" approach to space frames, V-belt theory and problems, jackscrews, and a section on the coriolis acceleration. This latter concept leads to some very interesting problems.

Practical problems are used in the text wherever possible, because they stimulate the student's interest and often lie within his personal experience or observation. However, many practical problems, especially in dynamics, appear confusing and complex: when stripped of its complexities, the actual object may be reduced to a mere block, or simply "a body," and the problem is then of the so-called academic type. Such a problem, however, has definite teaching value, because it enables the student to concentrate entirely on the theory involved and to apply this in the solution, leaving any minor complexities for later study.

In the solution of analytical problems, great emphasis is placed on the *complete* free-body diagram. Students are encouraged to develop this diagram gradually as the solution progresses until, upon completion of the problem, it shows all forces or their components. Simple arithmetical summations then prove the correctness of the solution without further computation.

The content is divided into 21 chapters, each composed of several articles. Each article presents additional theory, a new concept, or a different aspect and contains one or more completely solved problems illustrating the application of some theory or concept. A number of problems follow, carefully arranged in order of difficulty. Each chapter closes with a summary of the important points and formulas covered in the chapter; this summary affords students a quick review of the essential parts. The summary, too, is followed by a series of review problems and by a number of review questions carefully arranged to test the student's grasp of the subject matter.

Despite all reasonable efforts by us, the publishers, and the printer, some errors may creep into the first printing of any book: we shall be grateful to all who report any errors they may discover; subsequent printings will benefit greatly.

ALFRED JENSEN
Emeritus Professor, Engineering
University of Washington
Seattle, Wash. 98105

HARRY H. CHENOWETH
Associate Professor of Civil Engineering
Harris Hydraulics Laboratory
Department of Civil Engineering
University of Washington
Seattle, Wash. 98105

PART 1

STATICS

CHAPTER 1

Introduction

1-1 Definition of Mechanics

The science of **mechanics** treats of motion, forces, and the effects of forces on the bodies upon which they act.

Applied engineering mechanics concerns itself mainly with applications of the principles of mechanics to the solution of problems commonly met within the field of engineering practice.

Mechanics is generally divided into two main branches of study: (1) statics and (2) dynamics. **Statics** is that branch which deals with forces and with the effects of forces acting upon rigid bodies at rest. The subject of statics, therefore, is essentially one of *force analysis;* that is, a study of force systems and of their solutions. **Dynamics** deals with motion and with the effects of forces acting on rigid bodies in motion. Dynamics is divided into two branches: (1) **kinematics,** the study of motion without consideration of the forces causing the motion, and (2) **kinetics,** the study of forces acting on rigid bodies in motion and of their effect in changing such motion.

1-2 Problems in Applied Mechanics

Engineers conceive, plan, design, and construct buildings, machines, airplanes, and countless other objects, for the comfort and use of the human race. Each of these objects serves a definite and useful purpose; behind each lies an absorbingly interesting story of engineering skill and achievement.

Having in mind a definite purpose for the object to be constructed, the engineer then conceives its appropriate form, either entirely new, or an improvement upon one already in existence. Next he must analyze and determine the forces, known and unknown, acting on the object, and the motions, if any, of its various related parts.

To do this successfully *the engineer must have a thorough knowledge and understanding of the principles of mechanics, and of their applications to his particular problem.* Having thus determined the forces and the motions involved, he may proceed with the design of the object, using available materials of suitable strength and other requisite properties. The final size and shape of the object, and of each of its separate parts, may then be expressed in blueprints, after which the object is ready for production.

From this brief analysis of the work of the engineer, we see clearly that *a knowledge of mechanics is as fundamental to success in the fields of engineering as is an understanding of the alphabet to those who would learn to read and write their own language.* The extent of his knowledge of mechanics may have an important bearing on the opportunities that will open to the student in this great field of work. Certainly without some knowledge and training in mechanics he would have little or no chance of entering the engineering profession.

1-3 Procedures in the Solution of Mechanics Problems

Successful and efficient solution of any engineering problem calls for a well-organized and logical method of attack, involving a number of steps, each of which must first be well understood and then carefully executed. Among these steps, the following five include in a general way the entire process of solving any problem:

1. Analyze carefully the given data and ascertain the known quantities and the unknown quantities to be determined.
2. Recognize all the acting forces, known and unknown.
3. Decide on a suitable type of solution to use to determine the unknown quantities.
4. Formulate the steps to be taken to complete this solution.
5. Execute these steps, using available methods of checking the results.

The necessity for checking intermediate as well as final results as the solution progresses cannot be overemphasized, and yet it is most difficult to impress this fact on students, especially during the early part of their training; too many insist on dashing on to some answer, often finding it to be wrong and then, on a recheck, discovering some foolish or careless mistake which a second glance at the proper time would have quickly revealed.

1-4 Standards of Workmanship in Problem Solution

Because of the great responsibility that attaches to the practice of engineering, high standards of workmanship are demanded by the profession. These standards call for clear and neat figures and letters, and for uncrowded and logical arrangement of all computations and diagrams. Squared paper, $8\frac{1}{2}$ by 11 in., and a straightedge for drawing diagrams are recommended. All diagrams must be complete; that is, they must show all forces, dimensions, and other items that are parts of the problem.

In order that computations may readily be checked, as they must be, all work except the simplest additions, subtractions, multiplications, and divisions must be shown. Of course, if the slide rule is used, no multiplications or divisions need be shown, but the processes must be indicated. One of the best ways of checking data and computations is to glance over them *as soon as they are completed,* to see that no mistake has been made and that the result obtained so far is *reasonable.* Often an absurd answer or a misplaced decimal point is thus quickly detected. The use of scratch paper encourages sloppy work and should, therefore, not be allowed.

In most engineering computations, a *degree of accuracy* to three significant figures is considered satisfactory. (The numbers 64,800 and 0.0648 both contain three significant figures, 6, 4, and 8.) The process of learning the subject matter and of attaining the required standards of workmanship is progressive. Students are encouraged, therefore, to file completed problems in a loose-leaf notebook which should always be available for ready reference.

CHAPTER 2

Basic Principles of Statics

2-1 Force

In his *first law of motion* Sir Isaac Newton states that a body will continue in its state of rest or of uniform motion unless acted upon by a force that changes or tends to change its state. Therefore, we may state that **force is an action that changes or tends to change the state of motion of the body upon which it acts.** In statics we are interested only in bodies at rest. When applied to bodies at rest, as in statics, this definition more appropriately is that **force is an action that changes the shape of the body upon which it acts.**

In his *third law of motion* Newton states that **to every action there is an equal and opposed reaction.** A force, then, being an action, is always opposed by an equal reaction. Therefore, *forces exist not singly, but always in pairs, equal and opposite.* In analyzing forces and their effects, however, we may consider a force singly in order to study and evaluate its effect.

A body is any object, or any part of an object, which may be considered separately. When two objects are in contact, equal and opposite forces are produced at the contacting surfaces. A ball resting on a person's hand presents an example of two equal and opposite forces: the weight of the ball pressing down upon the hand, and the hand pushing upward with a force equal to this weight, force being exerted on the two surfaces of contact.

When acted upon by forces a body is necessarily somewhat deformed; thus the relative positions of the points at which the forces are applied are changed slightly. In solid bodies, such as are normally encountered in engineering practice, these changes are so insignificant that, for the purpose of force analysis, the bodies may be considered to be "rigid" and are therefore referred to as *rigid bodies.*

2-2 Types of Forces

Force can be exerted only through the action of one physical body upon another, either in contact or at a distance. Accordingly, forces may be classified under two general headings: (1) contacting or **applied forces,** such as a *push* or a *pull* which might be produced by muscular or mechanical effort; for example, a pull on a rope, the push of steam on the piston of a steam engine, or the retarding force produced by applying the brakes on a moving automobile, and (2) noncontacting or **nonapplied forces,** such as the gravitational pull of the earth on all physical bodies, magnetic force, or inertia force such as manifests itself within a body when its state of motion is changed.

A distinction is made between the **external forces** acting on a body and the resulting **internal forces,** also called **stresses,** produced in its various parts. In Fig. 2-1, for example, the external forces acting on the truss produce internal forces in its various members. The existence of these internal forces is easily visualized and can readily be shown, for should a single member be cut the truss would collapse.

A further illustration of external and internal forces is Fig. 2-2, in which an external force P must be applied to the loose end of the rope passing over the sheave, in order to counteract the downward pull of the weight W, which is also an external force. Quite obviously, if friction on the sheave pin is disregarded, an internal force equal to the pull P is produced in the rope.

External forces are further divided into **acting and reacting forces.** Gravity forces, forces caused by wind or water pressures, and all applied forces are examples of *acting forces.* Resisting forces at the supports of a structure are

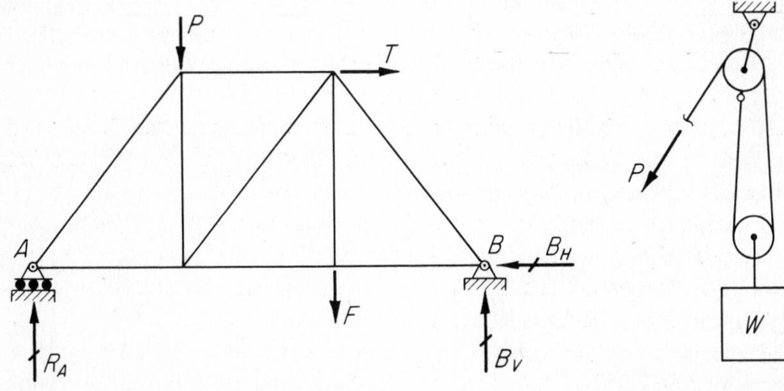

FIG. 2-1 Acting and reacting forces.

FIG. 2-2

called *reacting forces*. Therefore, of the forces on the truss in Fig. 2-1, P, T, and F are acting forces, and R_A, B_V, and B_H are reacting forces. A short diagonal bar across the shank of the arrow denoting a force indicates that it is a *reacting force.*

Since mechanical forces must be resisted by material of limited strength, they must necessarily be distributed over some area of material. In order, however, to simplify the solutions of many force systems, we regard as a **concentrated force** one which is resisted by an area of material relatively so small that, for practical purposes and with inappreciable error, it may be considered as a point. The pull in a rope or cable, the pressure of the leg of a table or chair on the floor, and the forces exerted on the rails by the wheels of a locomotive are examples of concentrated forces.

A distributed force is one which acts on an area relatively too large to be considered as a point without introducing an appreciable error. The pressure on the floor of a warehouse caused by piles of goods, the wind pressure against the side of a building, and water pressure on the bottom and against the side of a tank are examples of distributed forces.

2-3 Characteristics and Units of a Force

A force is completely described through statement of its (1) magnitude, (2) direction,[1] (3) sense, and (4) point of application. These items are called the *characteristics of the force.*

The *units* most commonly used for expressing the **magnitude** of a force are the pound, the kip (1,000 lb), and the ton (2,000 lb). The **direction** of a force is the direction of the line along which it acts, and may be expressed as vertical, horizontal, or at some angle with the vertical or horizontal. Since an applied force is transmitted from one body to another by contact, the **point of application** is the point of contact between the two bodies. A straight line extending through the point of application in the direction of the force is called its *line of action.* The **sense** of a force is indicated by the *way* it acts along its line of action, upward or downward, to the right or to the left, etc., and is generally denoted by an arrowhead. In Fig. 2-3, for example, the

[1]Some authors of texts on mechanics prefer to include the *sense* of a force within the meaning of the term *direction*. However, in most problems involving the solution of an unknown force, its direction is known but its magnitude and sense are unknown. Preferably, therefore, a distinction is drawn between direction and sense.

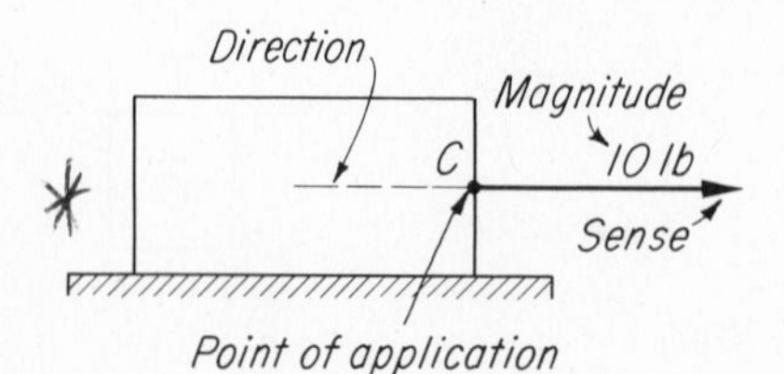

FIG. 2-3 Characteristics of a force.

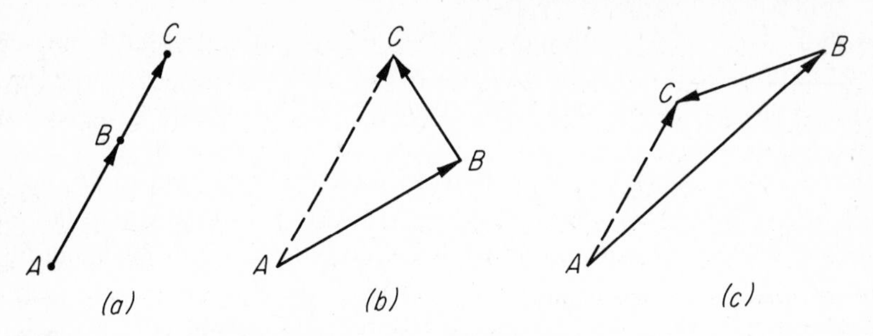

FIG. 2-4 Graphical vector addition.

force applied to the block may be described thus: a 10-lb (magnitude) force acting horizontally (direction) to the right (sense) through point C (point of application).

2-4 Vector and Scalar Quantities

All quantities, such as forces, velocities, accelerations, and moments, that possess direction and sense as well as magnitude are called *vector quantities*. The graphical representation of a vector quantity in its proper magnitude, direction, and sense is called a **vector.** Vectors may be added graphically (geometrically). For example, in Fig. 2-4 the vector AC in each of the three illustrations is the vector sum, or geometric sum, of vectors AB and BC.

Quantities that have magnitude only, such as pounds, feet, and dollars, are referred to as *scalar quantities;* these may be added algebraically, and the algebraic sum, or scalar sum, has magnitude only. Figure 2-4*a* shows that, when a number of vectors lie along the same straight line, they may be added algebraically, as scalar quantities, since the direction of the vector sum remains unchanged.

2-5 Transmissibility of Force

The horizontal beam shown in Fig. 2-5 supports two vertical loads, P and T, having a common line of action and applied midway between the end

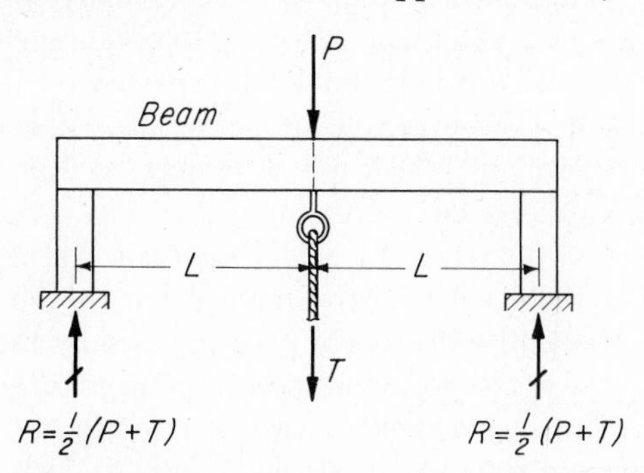

FIG. 2-5 Principle of transmissibility.

reactions. Each end reaction R then is $\frac{1}{2}(P + T)$. These end reactions will remain unchanged if T is moved upward along the common line of action and is supported on the top of the beam, or if P is moved downward and is supported by the rope.

The principle illustrated, known as the *principle of transmissibility,* is that *the point of application of an external force acting on a body may be transmitted anywhere along its line of action without changing other external forces also acting on the body.*

2-6 Types of Force Systems

To simplify their solutions, force systems are conveniently classified into six groups, the solutions of which differ according to their differing characteristics. Before these groups are named, some definitions are necessary. When all forces of a system lie in one plane, the system is said to be **coplanar.** Conversely, when they do not all lie in one plane, the system is **noncoplanar.** When the action lines of all the forces of a system intersect at a common point, the system is said to be **concurrent.** Conversely, when the action lines do not intersect at a common point, the system is **nonconcurrent.** In many force systems, all forces are **parallel.** When all forces in a parallel system act along a single line of action, illustrated by a tug of war in which two or more persons pull opposite ways on a straight rope, the system is said to be **collinear.** This system is a special case of coplanar, concurrent forces.

The six force systems then are (1) *coplanar, parallel;* (2) *coplanar, concurrent;* (3) *coplanar, nonconcurrent;* (4) *noncoplanar, parallel;* (5) *noncoplanar, concurrent;* and (6) *noncoplanar, nonconcurrent.* Each of these six systems is more fully discussed in following chapters.

2-7 Components of a Force

A force may be replaced by two or more components that will produce the same effect as the force they replace. The solution of many engineering problems is greatly simplified by reducing the force system involved to one containing only vertical and horizontal forces. To obtain such a system, forces that are neither vertical nor horizontal are replaced by their vertical and horizontal components. This principle applies to all vector quantities, forces, velocities, and accelerations. The truth of it may be illustrated as follows:

Consider a man walking due north from O in Fig. 2-6, at the rate of 3 miles per hour, across the deck of a ship which is traveling due east at the rate of 4 mph. Let these velocities be represented by the vectors OA and OB, respectively. Evidently, the *actual* path traveled by the man is directly from O toward C. The vector OC then represents his *resultant velocity,* while OA and OB represent his *component velocities,* north and east. When the desired components are *rectangular* (at right angles to each other), they may be evaluated analytically as follows:

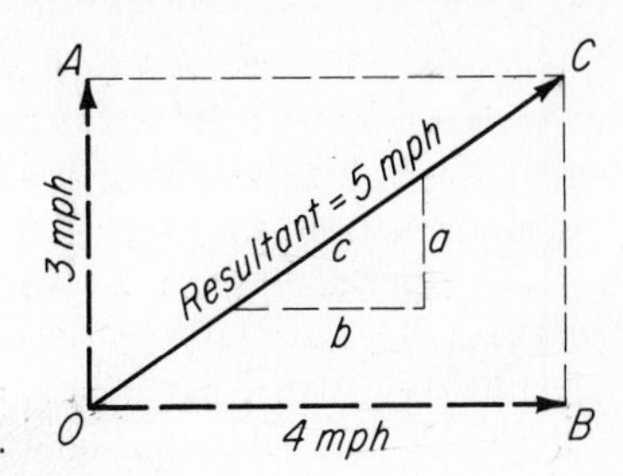

FIG. 2-6 Components of a vector.

If the angle θ defines the direction of OC, then, by trigonometry,

$$\mathbf{OA = OC} \sin \theta \qquad \text{and} \qquad \mathbf{OB = OC} \cos \theta \tag{2-1}$$

If the direction of OC is defined by the **slope triangle** abc, and vector OA is placed in the position of side BC, we have *two similar triangles* in which, by geometry, corresponding sides are proportional. That is,

$$\frac{a}{OA} = \frac{b}{OB} = \frac{c}{OC}$$

By cross-multiplication, we obtain

$$\mathbf{OA} = \frac{a}{c} \cdot \mathbf{OC} \qquad \text{and} \qquad \mathbf{OB} = \frac{b}{c} \cdot \mathbf{OC} \tag{2-1a}$$

from which it is clearly seen that $a/c = \sin \theta$ and $b/c = \cos \theta$.

Graphically, these components may readily be evaluated by laying out the diagram in Fig. 2-6 to some convenient scale.

PROBLEMS

2-1. Determine the vertical and horizontal components of each of the forces P and Q in Fig. 2-7. Check graphically (scale 1 in. = 500 lb).

Ans. $P_V = P_H = 0.707$ kip; $Q_V = 0.866$ kip, $Q_H = 0.5$ kip

2-2. In Fig. 2-8, compute the vertical and horizontal components of each of forces P and Q. Check graphically (scale 1 in. = 100 lb).

2-3. Calculate the vertical and horizontal components of each of forces P and Q in Fig. 2-9. Check graphically (scale 1 in. = 100 lb).

Ans. $P_V = 300$ lb, $P_H = 160$ lb; $Q_V = 240$ lb, $Q_H = 100$ lb

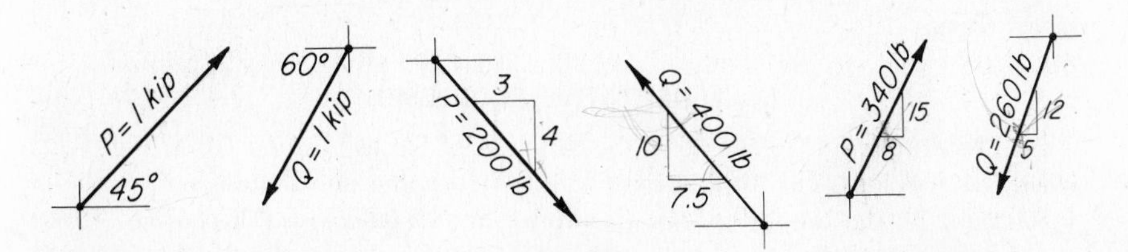

FIG. 2-7 Prob. 2-1. FIG. 2-8 Prob. 2-2. FIG. 2-9 Prob. 2-3.

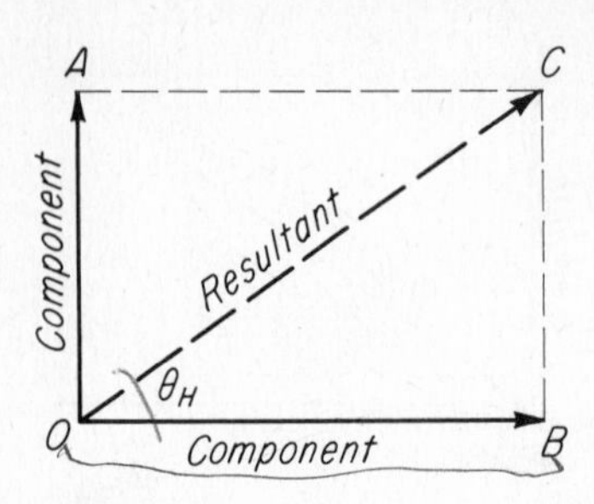

FIG. 2-10 Resultant force.

2-8 Resultant of Two Concurrent Forces

Two or more forces may be replaced by a single resultant force which will produce the same result as the forces it replaces. Clearly, then, in Fig. 2-10, if forces *OA* and *OB* are the components of *OC*, then *OC* is the *resultant* of forces *OA* and *OB*.

If the two forces are at right angles, we find by geometry that

Resultant force:

$$OC = \sqrt{\overline{OA}^2 + \overline{OB}^2} \tag{2-2}$$

When the forces are not at right angles to each other, as in Fig. 2-11*a*, they may be laid out *graphically* to scale, each with its proper magnitude, direction, and sense. The diagonal *OC* of the completed *parallelogram* then is the resultant force. The angle θ, defining the direction of the resultant force, may be scaled by a protractor, or its tangent may be scaled as follows: From *O*, along *OB*, lay off one unit of distance (or 10) to any convenient scale. The vertical distance t is then the tangent of θ, the value of which may be obtained from a table of trigonometric functions.

If, in Fig. 2-11*a*, force *OA* is placed in the position of line *BC*, a *triangle OBC* is obtained whose diagonal *OC* is the resultant force, as shown in Fig. 2-11*b*. Thus, a resultant force is obtained by the principle of vector addition, described in Art. 2-4.

The resultant of two nonrectangular forces may also be found *trigonometrically* by means of the cosine law, as in Prob. 2-4, and *algebraically* by means of summations of rectangular components, as in Prob. 2-5.

A force exactly equal and opposite to a resultant force is called an **equilibrant force.**

ILLUSTRATIVE PROBLEMS

2-4. Determine by trigonometry the resultant of forces *OA* and *OB* in Fig. 2-11, when $OA = 8$ lb, $OB = 10$ lb, and $\phi = 60°$. Determine also the angle θ.

Solution: By the law of cosines, $a^2 = b^2 + c^2 - 2\,bc \cos \alpha$. (The cosine of α is positive for values of α up to 90° and is negative for values of α between 90 and 180°.) Since $\alpha = 180° - 60° = 120°$, and $\cos 120° = -\cos(180° - 120°) = -0.5$,

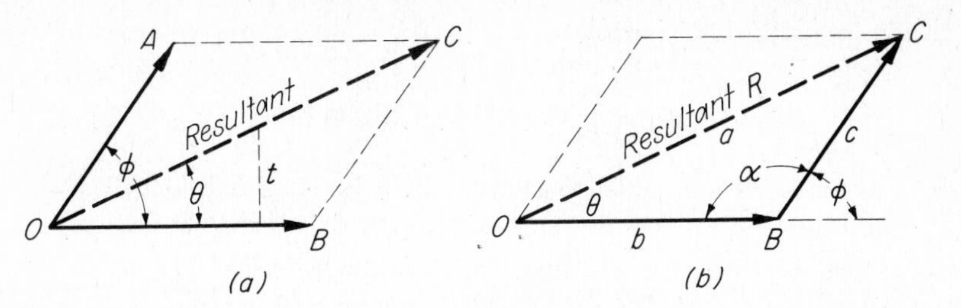

FIG. 2-11 Resultant of two concurrent forces. (*a*) Parallelogram method. (*b*) Triangle method.

we have

$$R^2 = (10)^2 + (8)^2 - (2)(10)(8)(-0.5) = 244 \qquad \text{and} \qquad R = 15.6 \text{ lb} \qquad \textit{Ans.}$$

To find θ by the law of sines,

$$\frac{\sin\theta}{c} = \frac{\sin\alpha}{R} \qquad \text{or} \qquad \sin\theta = \frac{c\sin\alpha}{R}$$

Since $c = 8$ lb, $R = 15.6$ lb, and $\sin\alpha = \sin(180° - 120°) = 0.866$, we obtain

$$\sin\theta = \frac{(8)(0.866)}{15.6} = 0.4441 \qquad \text{and} \qquad \theta = 26°22' \qquad \textit{Ans.}$$

2-5. Determine the resultant R of the two forces shown in Fig. 2-12*a*, and its angle of inclination θ_H with the horizontal.

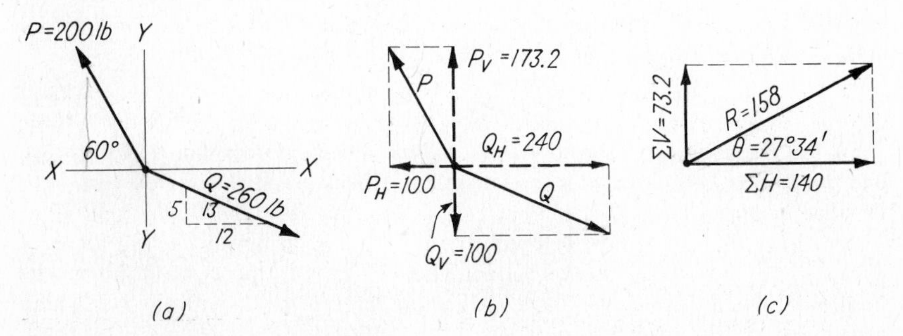

FIG. 2-12 Resultant force. Algebraic method. (*a*) Space diagram. (*b*) Component diagram. (*c*) Summation diagram.

Solution: The vertical and horizontal components of forces P and Q are computed in accordance with the methods described in Art. 2-7 and are then recorded in the component diagram. The components are

$$P_V = P\sin 60° = (200)(0.866) = 173.2 \text{ lb}$$
$$P_H = P\cos 60° = (200)(0.5) = 100 \text{ lb}$$

$$Q_V = \frac{5}{13} Q = \frac{5}{13}(260) = 100 \text{ lb}$$
$$Q_H = \frac{12}{13} Q = \frac{12}{13}(260) = 240 \text{ lb}$$

For convenience, let upward-acting forces and forces acting to the right be positive. Summing up the vertical and the horizontal components then gives

$$\Sigma V = 173.2 - 100 = 73.2 \text{ lb}$$
$$\Sigma H = 240 - 100 = 140 \text{ lb}$$

from which

$$R = \sqrt{(73.2)^2 + (140)^2} = \sqrt{5{,}358 + 19{,}600} \qquad \text{or} \qquad R = 158 \text{ lb} \qquad \textit{Ans.}$$

and

$$\tan \theta_H = \frac{\Sigma V}{\Sigma H} = \frac{73.2}{140} = 0.522 \qquad \text{or} \qquad \theta_H = 27°34' \qquad \textit{Ans.}$$

PROBLEMS

2-6. Determine the resultant of the two forces shown in Fig. 2-13 and the angle θ_H it makes with the horizontal. Check graphically (scale 1 in. = 20 lb).
Ans. $R = 50$ lb; $\theta_H = 36°52'$

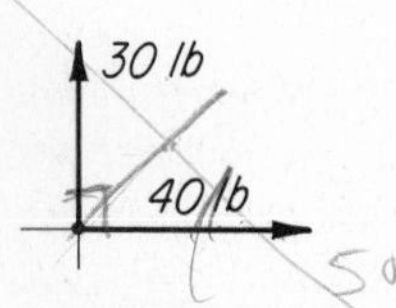

FIG. 2-13 Prob. 2-6.

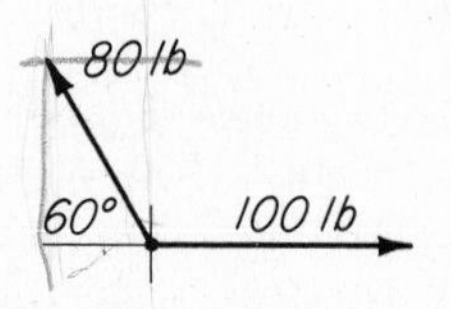

FIG. 2-14 Probs. 2-7 and 2-10.

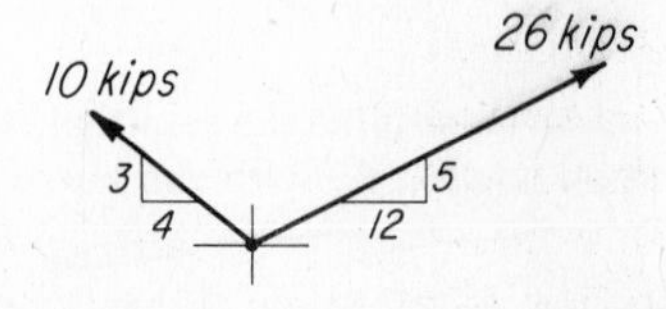

FIG. 2-15 Prob. 2-9.

2-7. By the cosine law, solve for the resultant R of the forces shown in Fig. 2-14, and the angle θ_H it makes with the horizontal. Check graphically (scale 1 in. = 40 lb). (See Prob. 2-4.) *Ans.* $R = 91.7$ lb; $\theta_H = 49°10'$

2-8. Using the cosine law, compute the resultant R of upward acting forces of 4 kips to the left at 45° to the horizontal and 7 kips to the right at 30° to the horizontal. Compute also the angle θ_H that the force R makes with the horizontal. Check graphically (scale 1 in. = 2 kips).

2-9. By algebraic summations of components, solve for the resultant R of the forces shown in Fig. 2-15. Find also its direction angle θ_H with the horizontal. Check graphically (scale 1 in. = 6 kips). (See Prob. 2-5.)
Ans. $R = 22.6$ kips; $\theta_H = 45°$

2-10. Solve Prob. 2-7 by the *component method* as shown in Prob. 2-5.

2-11. Calculate the resultant R of forces A and B shown in Fig. 2-16. Use the *component method* of Prob. 2-5. Calculate also the angle θ_H.
Ans. $R = 431$ lb; $\theta_H = 21°50'$

2-12. Determine the resultant R of forces A, B, and C, shown in Fig. 2-17, and

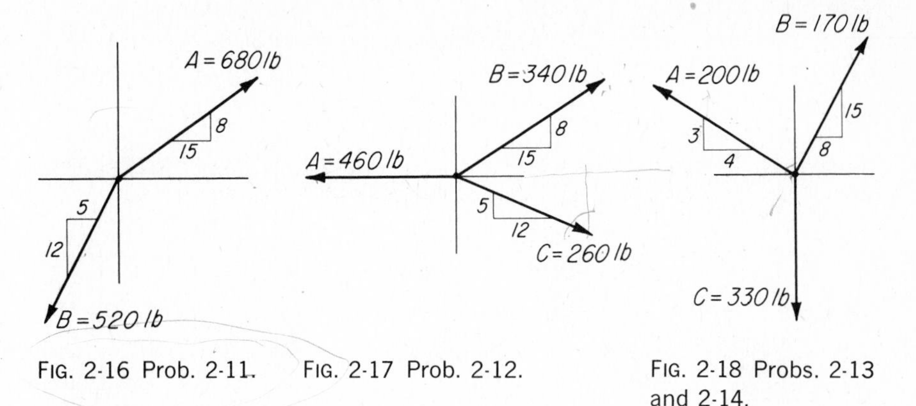

FIG. 2-16 Prob. 2-11.

FIG. 2-17 Prob. 2-12.

FIG. 2-18 Probs. 2-13 and 2-14.

the angle θ_H it makes with the horizontal. (*a*) Solve analytically using the summations of V and H components. (*b*) Solve graphically, using the method of triangles. (Force scale: 1 in. = 100 lb.)

2-13. Using an algebraic summation of components, calculate the resultant R of forces A, B, and C shown in Fig. 2-18. Find also the angle θ_H it makes with the horizontal. *Ans.* $R = 100$ lb; $\theta_H = 36°52'$

2-14. Solve Prob. 2-13 if $A = 400$ lb, $B = 340$ lb, and $C = 230$ lb.

2-9 Moment of a Force

The moment of a force is *its tendency to produce rotation of the body upon which it acts,* about some axis. **The measure of a moment is the product of the force and the perpendicular distance between the axis of rotation and the line of action of the force.** This distance is called the **moment arm.** The intersection of the axis of rotation with the plane of the force and its moment arm is called the **center of moments.** Since *the center of moments is a point,* it is often referred to as such.

These definitions are illustrated in Fig. 2-19, in which the force F, applied to the end of a wrench in an effort to tighten a nut, rotates the wrench and the nut about an axis through the center of the nut. When F is vertical, its moment arm is a, and the moment of F then is $F \cdot a$ and is clockwise. To obtain a maximum turning effort, F must be so directed that its moment arm is maximum.

Units of Moments. Since a moment is the product of force and distance, its unit is likewise the product of the respective units of force and distance. The basic unit of moment is the pound-foot (lb-ft).[1] Other units are the pound-inch (lb-in.), the kip-foot (kip-ft), and the kip-inch (kip-in.).

[1]No agreement has yet been reached among authors as to whether the unit of moment should be the pound-foot or the foot-pound. Many authors prefer to use the foot-pound as the unit of moment, work, energy, torque, and bending moment, each of which is the product of force and distance, or of their equivalents. Others prefer the pound-foot as the unit of moment and of torque to distinguish these quantities from those of work and energy.

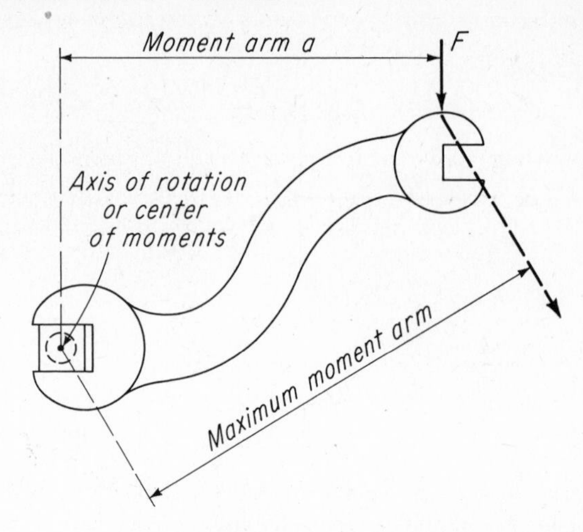

FIG. 2-19 Moment of a force.

The moment of a force has *magnitude*, and has *direction* clockwise or counterclockwise with respect to a given moment axis. To aid in visualizing its direction at a glance, the following is recommended:

In Fig. 2-20, the magnitudes and directions of the moments of each of the four forces acting on bar *AB*, about the moment center *A*, are to be determined. Imagine the bar to be *hinged* at the moment center. Then, placing a finger at the moment center, consider *all* supports and forces removed except that force whose moment is desired. *The direction in which the given force will rotate the bar about the moment center is the direction of the moment* and should be obvious at a glance.

About *A*, then, the moment of *P* is $P \cdot a$ clockwise, and the moment of R_B is $R_B \cdot L$ counterclockwise. The amounts of A_V and A_H about A must both be zero, because their moment arms are zero. Similarly, about *B* as the moment center, the moment of *P* is $P \cdot b$ counterclockwise, that of A_V is $A_V \cdot L$ clockwise, and that of A_H is zero, since its moment arm with respect to *B* is zero.

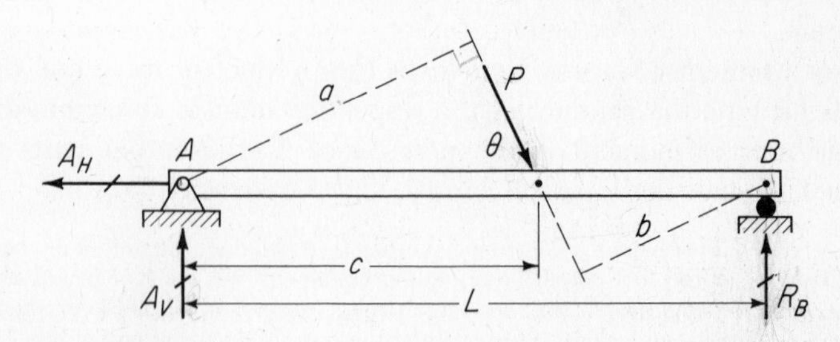

FIG. 2-20 Magnitude and direction of the moment of a force.

We may now state, as an important fact: **the moment of a force about an axis passing through its line of action is zero.**

PROBLEMS

In the following problems let counterclockwise moments be positive.

2-15. Compute the moment of force F, in Fig. 2-21, about each of points A, B, and C. *Ans.* $M_A = +60$ lb-ft↺; $M_B = +60$ lb-ft↺; $M_C = 0$

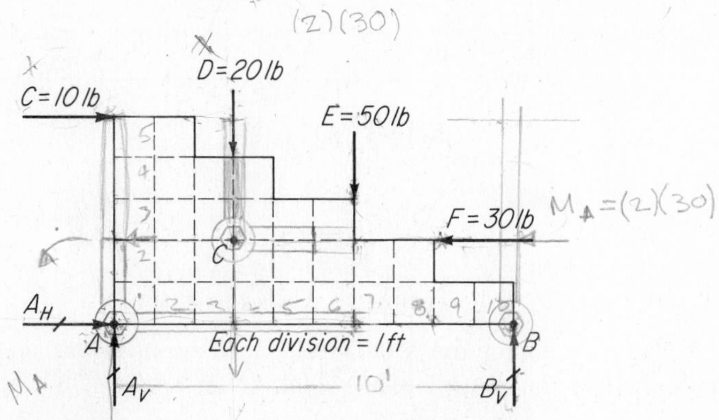

FIG. 2-21 Probs. 2-15 through 2-18.

2-16. Calculate the moment of force D, in Fig. 2-21, about each of points A, B, and C.

2-17. Find the algebraic sum of the moments of forces C and D, in Fig. 2-21, about each of points A, B, and C.

Ans. $M_A = 110$ lb-ft↻; $M_B = 90$ lb-ft↺; $M_C = 30$ lb-ft↻

2-18. If, in Fig. 2-21, the algebraic sum of the moments of forces C, D, E, F, and B_V is zero, about point A, what must be the value of force B_V?

2-10 The Principle of Moments. Varignon's Theorem

A force may be replaced by its components without changing its total effect, as is stated in Art. 2-7. Therefore, **about any point, the algebraic sum of the moments of the components of a force equals the moment of the force.** This principle is known as *Varignon's theorem.* To prove the theorem, consider the force R in Fig. 2-22, and its two components T and P, obtained by the parallelogram method. Let α (alpha), β (beta), and γ (gamma) be the angles between the axis OX and the forces T, P, and R respectively. If t, p, and r are the moment arms of forces T, P, and R, respectively, from the center of moments M, we must prove that $R \cdot r$, the moment of R, equals $T \cdot t + P \cdot p$, the algebraic sum of the moments of T and P, the components of R.

Line AB is parallel to P and is equal in length to P. Therefore, BC equals DE. Also, we see that

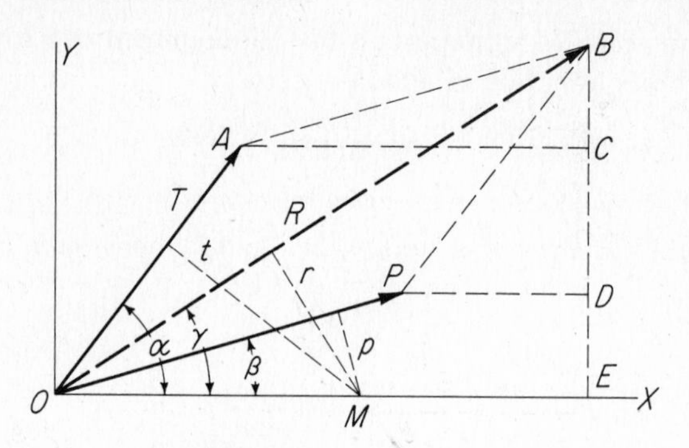

FIG. 2-22 Varignon's theorem.

$$BE = CE + BC \tag{a}$$

or

$$R \sin \gamma = T \sin \alpha + P \sin \beta \tag{b}$$

Without changing the total result, each term in this equation may be multiplied by the distance OM, which gives

$$R(OM \sin \gamma) = T(OM \sin \alpha) + P(OM \sin \beta) \tag{c}$$

However, from Fig. 2-22, $r = OM \sin \gamma$, $t = OM \sin \alpha$, and $p = OM \sin \beta$. Substituting these values of r, t, and p in Eq. (c), we obtain

$$\mathbf{R} \cdot \boldsymbol{r} = \mathbf{T} \cdot \boldsymbol{t} + \mathbf{P} \cdot \boldsymbol{p} \tag{2-3}$$

which is the required proof.

The forces of a system may be regarded as the components of their resultant force. Hence, **about any point, the moment of the resultant of a system of forces equals the algebraic sum of the moments of the separate forces.**

ILLUSTRATIVE PROBLEM

2-19. An application of Varignon's theorem is shown in Fig. 2-23. For the purpose of determining R_C, we wish to obtain the moment of force P about A as the moment center. This moment may be determined in three ways:

1. The moment is $P \cdot a$. Triangles AEB and DFB are similar. Hence

$$\frac{a}{6} = \frac{4}{5} \qquad \text{or} \qquad a = \frac{(6)(4)}{5} = 4.8 \text{ ft} \qquad \text{and} \qquad M_A = (10)(4.8) = 48 \text{ kip-ft}\circlearrowright$$

2. Replace P with its V and H components, and apply these at D. The components are

$$P_V = \frac{4}{5}(10) = 8 \text{ kips} \qquad \text{and} \qquad P_H = \frac{3}{5}(10) = 6 \text{ kips}$$

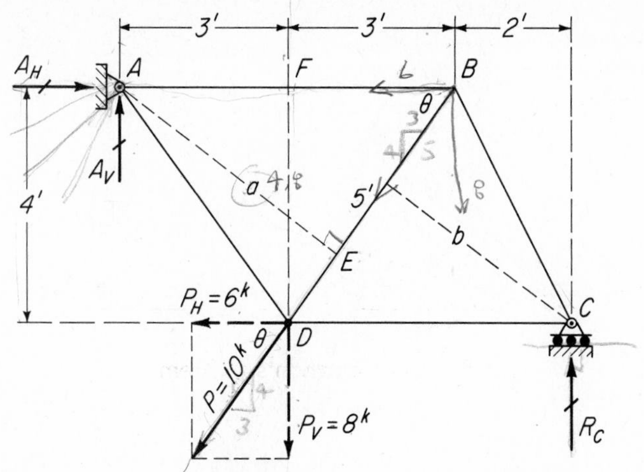

FIG. 2-23 The moment of a force equals the algebraic sum of the moments of its components.

The algebraic sum of their moments about point A then is

$$\Sigma M_A = (8)(3) + (6)(4) = 48 \text{ kip-ft}$$

In general, method 2 is preferable to method 1.

3. According to the principle of transmissibility (Art. 2-5) we may transmit the point of application of P to any other point along its line of action without changing its effect on R_C. If we apply P at B and there replace it with its components, again we obtain

$$\Sigma M_A = (8)(6) + (6)(0) = 48 \text{ kip-ft}$$

PROBLEMS

2-20. Compute the moment of force B in Fig. 2-24 about point A. (Draw brief point sketch only.)

2-21. Calculate the moment of force D in Fig. 2-24 about point A. (Draw brief point sketch only.) *Ans.* $M_A = -48$ kip-ft

2-22. Determine the algebraic sum of the moments of forces C and F in Fig. 2-24 about point A. (Draw brief point sketch only.)

2-23. In Fig. 2-24 the algebraic sum of the moments of forces B, C, D, F, and R_E about point A is zero. Compute the value of R_E. (NOTE: Calculate all components and draw complete sketch showing all forces and components. Suggested scale: 1 in. = 2 ft.) *Ans.* $R_E = 55.25$ kips

2-24. The algebraic sum of the moments about point E of all forces acting on the truss shown in Fig. 2-24 is zero. Calculate the value of A_V. (NOTE: Calculate all components and draw complete sketch of the truss showing all forces and components. Suggested scale: 1 in. = 2 ft.)

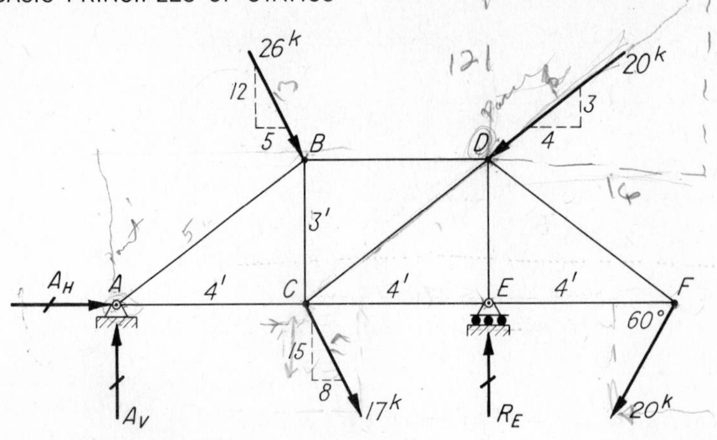

FIG. 2-24 Probs. 2-20 to 2-24.

2-11 Couples

Many special combinations of forces are encountered in engineering practice. Among them is a **couple,** which consists of **two equal, opposite, and parallel forces having separate lines of action.** The plane in which the forces lie is called the **plane of the couple,** and the perpendicular distance between their lines of action is the **arm of the couple.**

Such a couple is shown in Fig. 2-25. In this figure the two forces F are equal and opposite and might represent the push and pull on the opposite ends of a die holder cutting threads on a steel bar. Other examples of couples are the two forces applied by the hands on opposite sides of the steering wheel of an automobile and the forces applied by thumb and forefinger in turning a nut on a bolt.

A couple has the following characteristics:

1. *The resultant force of a couple is zero.*
2. *The moment of a couple is the product of one of the forces and the perpendicular distance between their lines of action.*
3. *The moment of a couple is the same for all points in the plane of the couple.*

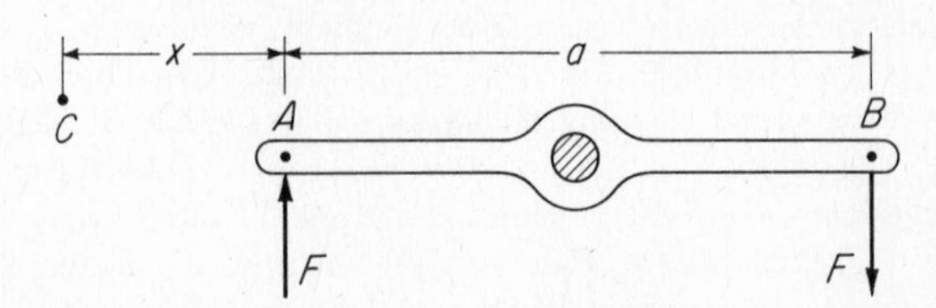

FIG. 2-25 Moment of a couple.

STEERING WHEEL
Tap + Die Wrench

(To prove this fact, we will select three different moment centers in Fig. 2-25: A, B, and a point C chosen at random. Then,

$M_A = F \cdot a$ clockwise; $M_B = F \cdot a$ clockwise
$M_C = F(x + a) - F \cdot x = F \cdot x + F \cdot a - F \cdot x = F \cdot a$ clockwise

thus proving the statement.)

4. *A couple can be balanced only by an equal and opposite couple in the same, or in a parallel plane.* (That is, a couple cannot be balanced by a single force, since the moment of a couple is constant for all points in its plane, while the moment of a force is dependent upon the position of its moment center.)

2-12 Resultant of Two Parallel Forces

In Fig. 2-26, the magnitude of the resultant R of the parallel forces A and B equals their algebraic sum, $A + B$. **The distances *a* and *b* from the action line of *R* to the action lines of forces A and *B*, respectively, are inversely proportional to the magnitudes of these forces.** That is,

$$\frac{\mathbf{A}}{b} = \frac{\mathbf{B}}{a} \tag{2-4}$$

Proof is obtained by the principle of moments (Art. 2-10), according to which the moment of R about any point equals the algebraic sum of the moments of A and B about the same point. With point c as the moment center, letting the counterclockwise moment be positive, we have

$$\mathbf{R \cdot 0 = A \cdot a - B \cdot b} \quad \text{or} \quad \mathbf{Aa = Bb}$$

from which
$$\frac{\mathbf{A}}{b} = \frac{\mathbf{B}}{a} \tag{2-4a}$$

which is the required proof.

The action line of R may be located *graphically* by the **inverse-proportion method,** as shown in Fig. 2-26. On one side of any reference line, xx, and on the action line of B, is laid off to scale the magnitude of force A; on the opposite side of xx, and on the action line of A, is laid off the magnitude

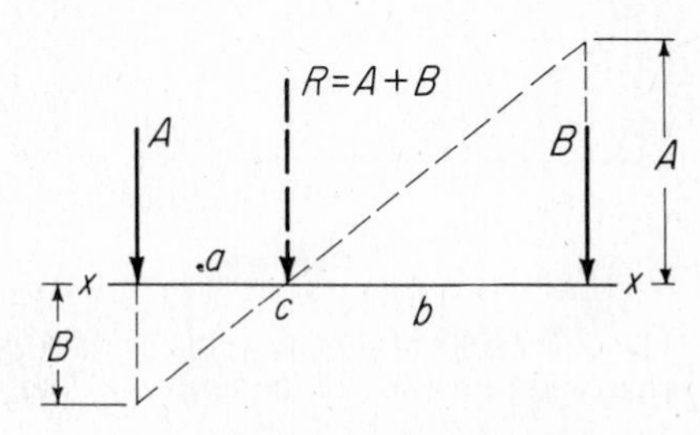

FIG. 2-26 Resultant of two parallel forces.

of B. Point c, where a diagonal line crosses xx, then lies on the action line of R. That this is true is seen by the fact that the two triangles thus formed are similar and that, therefore, $A/b = B/a$, as was stated above.

If A and B are oppositely directed, the resultant R equals the difference between A and B, and both are laid off *on the same side* of xx. The intersection c then lies to the right of B when A is less than B, and to the left of A when A is greater than B. The following convenient rules should be remembered: *When both forces have similar senses, their resultant lies between them; when they have different senses,* the resultant lies outside the space between them. It always lies *near the greater* of the two forces.

2-13 Resolution of a Force into Two Parallel Components

Suppose that we wish to resolve the force F shown in Fig. 2-27 into two parallel components, one component to be located a distance a to the left of F and the other a distance b to the right of F. The magnitudes of the components will be inversely proportional to their distances from the force F. That is,

$$\frac{\mathbf{A}}{\mathbf{b}} = \frac{\mathbf{B}}{\mathbf{a}} \qquad \textbf{(2-4}b\textbf{)}$$

Proof is obtained by the principle of moments. Taking moments about point c in Fig. 2-27a gives

$$Aa = Bb \qquad \text{from which} \qquad \frac{A}{b} = \frac{B}{a}$$

which is the required proof. The sum of components A and B must equal F.

The components A and B may be found graphically by the **inverse-proportion method** as shown in Fig. 2-27b. The force F is drawn to scale at c. A line dc is then drawn parallel to xx. The line de is drawn next. This line will divide the force F into the two required components A and B.

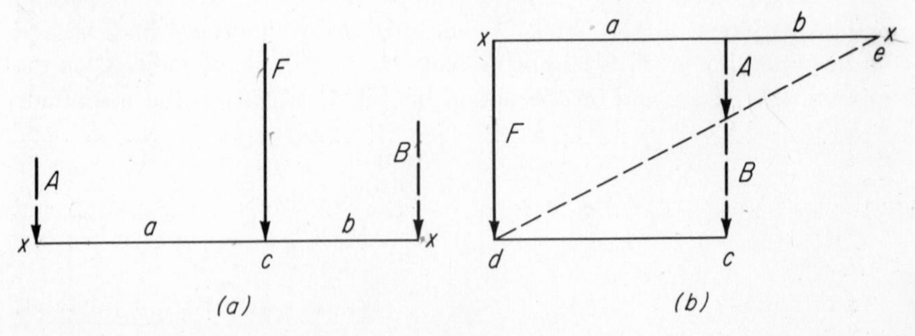

FIG. 2-27 Resolution of a force into two parallel components. (a) Force F to be resolved into two components. (b) Method of resolution.

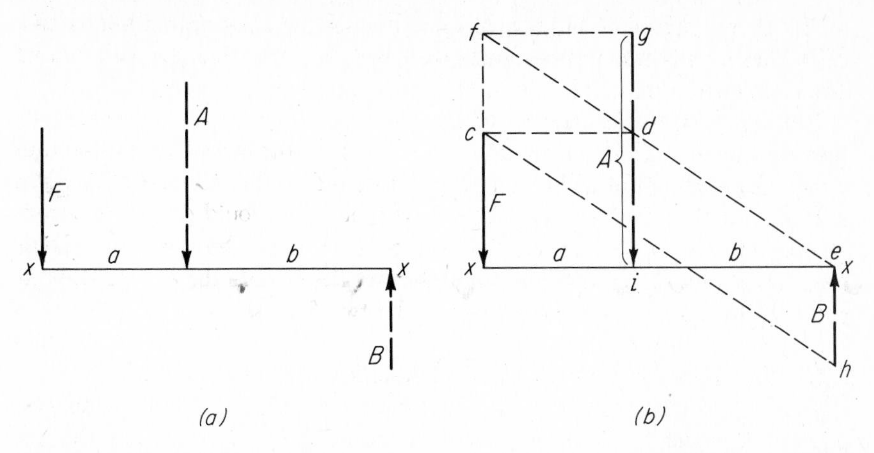

FIG. 2-28 Resolution of a force into two parallel components on the same side of the force. (*a*) Force F to be resolved into two components A and B. (*b*) Method of resolution.

If both components are to lie on the same side of the force, the graphical procedure shown in Fig. 2-28 should be followed. Draw the force F to scale. Draw cd parallel to xx. Next draw the line de and extend it to f. Then draw fg parallel to xx and ch parallel to fe. The line gi will give the magnitude of component A and eh will give the magnitude of component B. It will be noted that the component nearest the force F will be *larger* than F and that the component farthest from F will be reversed in *sense* from that of F.

PROBLEMS

In each of the following problems determine *graphically by the inverse-proportion method* the magnitude of the resultant force R, and the horizontal distance x from A to its line of action. Check analytically by moments.

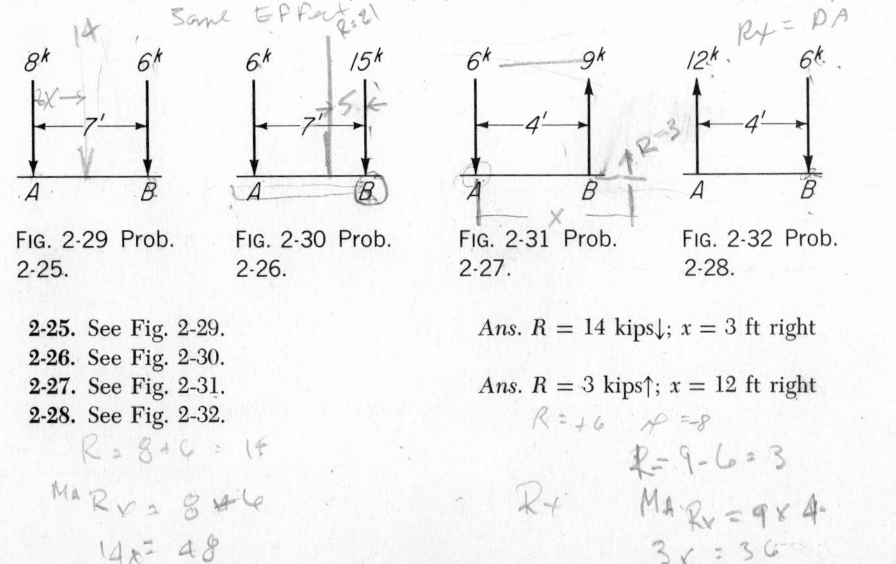

FIG. 2-29 Prob. 2-25.

FIG. 2-30 Prob. 2-26.

FIG. 2-31 Prob. 2-27.

FIG. 2-32 Prob. 2-28.

2-25. See Fig. 2-29. *Ans.* $R = 14$ kips↓; $x = 3$ ft right

2-26. See Fig. 2-30.

2-27. See Fig. 2-31. *Ans.* $R = 3$ kips↑; $x = 12$ ft right

2-28. See Fig. 2-32.

In each of the following problems determine *graphically by the inverse-proportion method* the magnitudes of the parallel components at A and B of the force shown. Check analytically by moments.

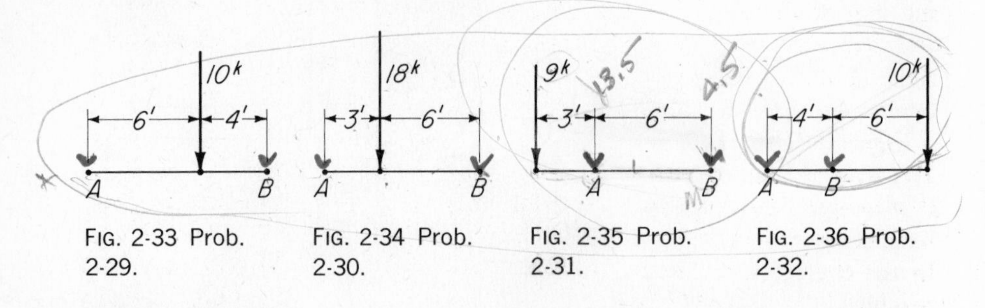

FIG. 2-33 Prob. 2-29.

FIG. 2-34 Prob. 2-30.

FIG. 2-35 Prob. 2-31.

FIG. 2-36 Prob. 2-32.

2-29. See Fig. 2-33. *Ans.* $A = 4$ kips; $B = 6$ kips

2-30. See Fig. 2-34.

2-31. See Fig. 2-35. *Ans.* $A = 13.5$ kips↓; $B = 4.5$ kips↑

2-32. See Fig. 2-36.

2-14 Equilibrium of Force Systems

Equilibrium is essentially a state of balance. Equilibrium in a force system is, then, a state of balance between opposing forces within the system, and a balance about any point of the moments of all the forces in the system. If a body at rest is acted upon by a system of forces and remains at rest, the system is *balanced* and the body is said to be in **static equilibrium.** Within the subject of statics all bodies are presumed to remain at rest.

The meaning of equilibrium may be simply illustrated as in Fig. 2-37. The simple balance shown in Fig. 2-37*a* is symmetrical about the fulcrum A, and the weights of its various parts therefore balance about A. The magnitude of an unknown weight W, placed on one scale pan, is ascertained when a

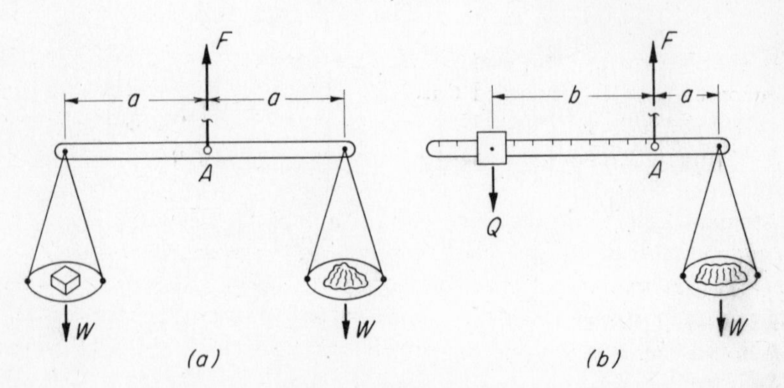

FIG. 2-37 Equilibrium of forces and of moments.

known and equal weight, placed on the opposite pan, balances the weighing arm in a horizontal position.

To have *complete equilibrium* the single upward-acting force F must equal the sum of all the downward-acting forces, which are the two weights W and the weight of the entire balance, and the tendencies of the forces to the right of A to cause clockwise rotation about A must be equaled by the tendencies of the forces on the left to cause counterclockwise rotation.

Figure 2-37*b* shows especially well a balance of moments of unequal forces. The scale balances with W and Q removed. When the unknown weight W is placed on the weighing pan, the position of Q is adjusted until the arm balances horizontally. Then, the clockwise moment $W \cdot a$ about A is equaled by the counterclockwise moment $Q \cdot b$, and a state of equilibrium has been reached.

A body at rest is in static equilibrium under the combined action of all forces acting on it. When *all* forces acting on a body are considered, we have a **complete force system.** Hence, **a complete force system is always in equilibrium.**

Force Law of Equilibrium. Equilibrium implies a *balance of opposing forces* within a system. Therefore, *in any direction,* **the algebraic sum of all forces acting upon a body in static equilibrium is zero.** This statement is known as the *force law of equilibrium.* Stating the law in symbols, we have

$$\Sigma F = 0 \tag{2-5}$$

The Greek letter Σ (sigma) means "the algebraic sum of." To be assured of *complete* force equilibrium, the algebraic sum of the forces must be shown to be zero in at least two directions, usually vertical and horizontal (V and H). We may then substitute for Eq. (2-5) the following *two equations of force equilibrium:*

$$\Sigma V = 0 \tag{2-6}$$

$$\Sigma H = 0 \tag{2-7}$$

If the directions of the summations are not vertical and horizontal, the summations may be indicated thus: $\Sigma F_x = 0$ and $\Sigma F_y = 0$, where x and y are any two rectangular axes. The principle embodied in this law is especially useful in determining the unknown forces of a *concurrent* system in equilibrium.

Moment Law of Equilibrium. Equilibrium also implies a *balance of opposing moments* of the forces within a system. Therefore, about *any* moment axis, **the algebraic sum of the moments of all forces acting upon a body in static equilibrium is zero.** This statement is known as the *moment law of equilibrium.* Stating this law in symbols, we obtain *the equation of moment equilibrium:*

$$\Sigma M = 0 \tag{2-8}$$

About *any* axis, then, the sum of the clockwise moments must be equal to the sum of the counterclockwise moments. The principle embodied in this law is especially useful in determining the unknown forces of a *nonconcurrent* system in equilibrium.

Signs of Forces and Moments. The preceding discussions have indicated that no sign convention is necessary in the summations of forces and of moments of forces in equilibrium. That is, *in any one direction*, the sum of the forces acting *one way* is simply equated to the sum of the forces acting the *opposite way*. Similarly, the sum of the clockwise moments is equated to the sum of the counterclockwise moments.

Occasionally, when dealing with groups of forces *not* in equilibrium, and in some instances of force systems *in* equilibrium, signs are conveniently given to forces and moments. When signs are so given, *forces acting upward or to the right are positive, and forces acting downward or to the left are negative.* Similarly, *counterclockwise moments are usually positive, and clockwise moments are negative.*[1]

2-15 Principles of Force Equilibrium

The following principles are extremely important, for they are very useful in the solution of many engineering problems:

The Two-force Principle. When two forces are in equilibrium, they must be equal, opposite, and collinear. Proof of this principle is embodied in Newton's third law, which states: To every action there is an equal and opposed reaction.

The Three-force Principle. When three nonparallel forces are in equilibrium, their lines of action must intersect at a common point. The proof of this principle may be deduced from the two-force principle and the principle of resultant force. Let the three forces *P*, *Q*, and *S*, acting on the body shown in Fig. 2-38, be in equilibrium. Then let *Q* and *S* be replaced by their re-

[1]The convention of designating forces acting upward and to the right as being positive is perhaps universal. In the matter of moments, however, no agreement exists among authors. But in mechanics, most authors designate the counterclockwise moment as being positive.

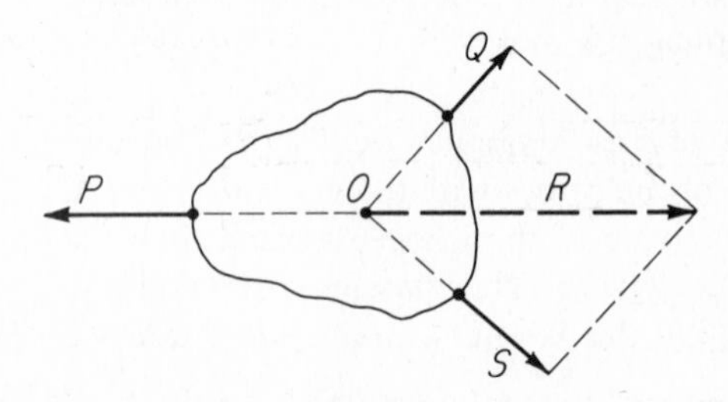

FIG. 2-38 The three-force principle.

sultant R. Now the two forces R and P are in equilibrium and hence are collinear. The intersection O of the action lines of Q and S lies on the action line of P. Hence, the three forces are concurrent; that is, their lines of action intersect at a common point.

Also, *when the three concurrent forces are in equilibrium, they must be coplanar.* This relation we deduce from the fact that the intersecting action lines of two concurrent forces form a *plane.* The action line of the third force must also lie in that plane; else that force would have an *unbalanced component* perpendicular to the plane, and the system would not be in equilibrium.

The Four-force Principle. When four forces are in equilibrium, the resultant of any two of the forces must be equal, opposite, and collinear with the resultant of the other two. Again, proof of this principle may be deduced from the two-force principle and the principle of resultant force. The four forces divide into two pairs. When each pair is replaced with a resultant, the two resultants are then in equilibrium, and hence must be equal, opposite, and collinear. This principle of two equal and opposite resultant forces applies, of course, to any number of forces in equilibrium.

2-16 Supports and Support Reactions

When a body exerts a force against a point of support, the support will *react* with a force of equal magnitude but with opposite sense. Here, then, action and reaction are equal, opposite, and collinear. To be solved by the laws of statics only, a single, rigid body may be supported at not more than two points in one plane, or at three points in space.

The truss shown in Fig. 2-39*a* is held to its support at A by a pin, presumed to be frictionless, thus allowing *rotation* of the truss in its plane but preventing *displacement.* To allow for small expansions or contractions of the truss due to changing loads or to changes in temperature, some freedom of displacement is provided at the other support B by placing that end of the truss on **rollers.** No resistance (force) is then offered to displacement parallel to the surface supporting the rollers.

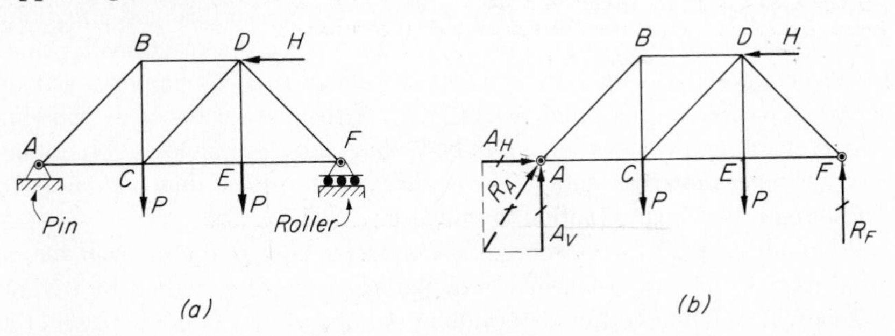

FIG. 2-39 Supports and support reactions. (*a*) Bridge truss and supports. (*b*) Supports replaced by reactions.

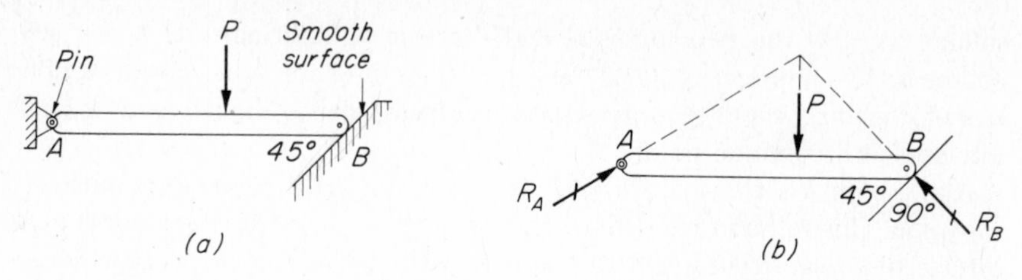

FIG. 2-40 Support reactions. (*a*) Beam with supports. (*b*) Supports replaced by reactions.

For lighter structures or single members, such as the beam shown in Fig. 2-40, an equivalant condition is provided by assuming one end to rest on an ideally *smooth surface* which likewise offers no resistance (force) to displacement parallel to itself.

Hence, **the reacting force at a roller support, or at a smooth surface, is always perpendicular to the supporting surface.** Only one reacting force exists at each support. The direction of the reaction at a pinned support is generally unknown. When so unknown, the reaction is usually replaced with its rectangular components acting in known directions, as at support A in Fig. 2-39*b*, thus simplifying analytical solution.

2-17 Free-body Diagrams

Two common engineering problems are (1) to determine the external forces, called *support reactions,* at the supports of a structure and caused by the loads it carries, and (2) to determine the internal forces, called *stresses,* which the external forces produce in the various members or parts of the structure. The first essential step in determining such unknown forces is to recognize *where* and *in which direction* they act. The following paragraphs outline an important aid in *visualizing* all these unknown factors.

For example, to determine the reactions at supports A and F of the truss in Fig. 2-39, the supports are removed and are replaced with the support reactions R_A and R_F which hold it in equilibrium. The vertical and horizontal components A_V and A_H at support A in turn replace the reaction R_A, since the direction of R_A is unknown. When so isolated from its supports, a body is said to be *free* and is called a **free body.** A diagram of the body showing *all* forces acting on it is called a **free-body diagram.** The unknown forces may now be determined because *the complete force system thus shown is in equilibrium.* A similar situation is shown in Fig. 2-40.

When all external forces acting on a body are known, the *internal forces,* or stresses, in its various members or parts may be determined by several methods. In one method of determining stresses in members of trusses, the members are cut near a joint, as at joints A and B in Fig. 2-41*a*. The joint is then isolated as a free body, and a *free-body diagram* is drawn. The external

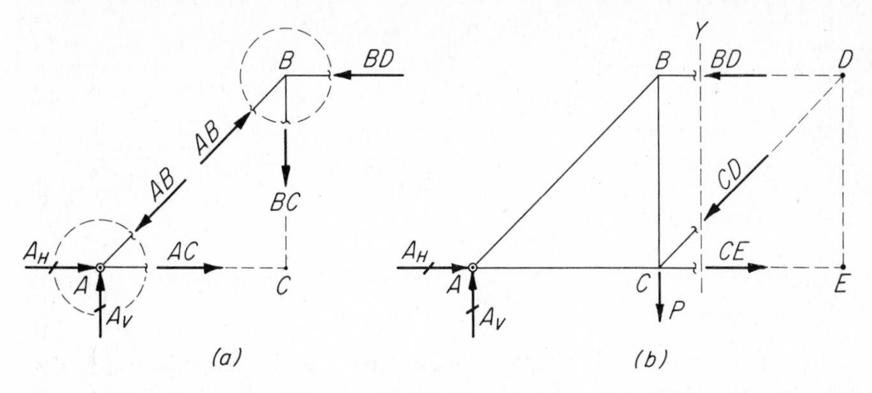

FIG. 2-41 Free-body diagrams. (*a*) A joint is isolated. (*b*) A section is isolated.

forces acting at the joint, together with the internal forces in the cut members, then form a *complete* force system which is in equilibrium, and the unknown forces may thus be determined.

In another method of determining stresses in members of trusses, a section is cut through the entire truss, as in Fig. 2-41*b*. Either of the two parts is then isolated as a free body, and a *free-body diagram* is drawn. Again a *complete* force system is formed by the external forces acting on the part isolated and the internal forces in the cut members; this system then is in equilibrium and the unknown forces may thus be determined.

From the foregoing discussion, we conclude that **a body isolated from its supports, or any part of such a body isolated from the remainder, is in static equilibrium under the influence of all forces acting on it.**

2-18 Problems in Equilibrium of Coplanar Force Systems

By use of the three equations of equilibrium, $\Sigma V = 0$, $\Sigma H = 0$, and $\Sigma M = 0$, we may determine three unknown quantities of any coplanar, nonparallel force system in equilibrium, or two unknown quantities of a parallel system in equilibrium. An unknown quantity may be (1) the magnitude and sense of a force, (2) the direction of the line of action of a force, or (3) the location of the line of action of a force, or of its point of application. The most commonly occurring problems in force analysis involve the determination of either two or three unknown magnitudes of forces together with their unknown senses. Problem 2-33 illustrates the determination, in a single problem, of all these three quantities. To emphasize methods and to simplify solutions, the weight of a structure mentioned in a problem is generally disregarded. The true sense of an unknown force can usually be visualized at a glance. If not, a probable sense is assumed. If the assumption is correct, the numerical value of the force, when solved for, will be positive; if incorrect, it will be negative.

ILLUSTRATIVE PROBLEMS

2-33. The horizontal plank shown in Fig. 2-42*a* is acted upon by forces P and Q and is supported by a frictionless pin at C. We need to determine (*a*) the magnitude of the reacting force R at support C, (*b*) the direction of its line of action, as indicated by the direction angle θ, and (*c*) the location of its point of application C in order that equilibrium may be maintained. All forces are in pounds. The weight of the plank is disregarded.

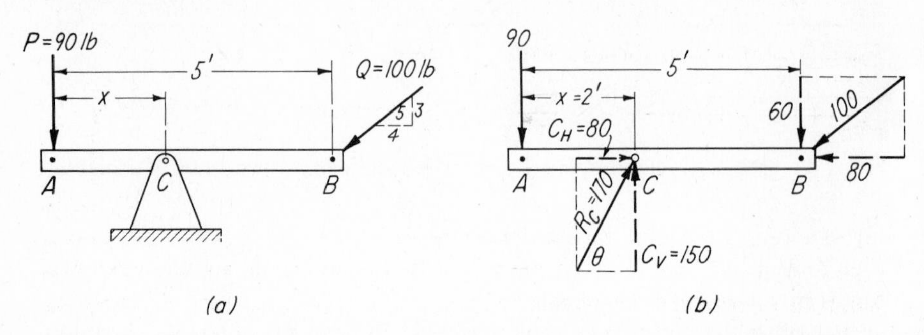

FIG. 2-42 Static equilibrium. (*a*) Space diagram. (*b*) Free-body diagram.

Solution: Our first step is to remove the support, thus isolating the plank as a free body, and then to draw the free-body diagram shown in Fig. 2-42*b*. To simplify the solution, force Q is replaced by its V and H components. By similar triangles,

$$\frac{Q_V}{3} = \frac{Q_H}{4} = \frac{100}{5} \quad \text{or} \quad Q_V = \frac{3}{5}(100) = 60 \text{ lb} \quad \text{and} \quad Q_H = \frac{4}{5}(100) = 80 \text{ lb}$$

Since the direction of R_C is unknown, we replace it with its components C_V and C_H. We now have a *complete* force system in equilibrium. C_V clearly acts upward, and C_H acts to the right. Applying the two equations of force equilibrium, we have

$$[\Sigma V = 0] \qquad C_V = 90 + 60 = 150 \text{ lb}$$

$$[\Sigma H = 0] \qquad C_H = 80 \text{ lb}$$

Then $$R_C = \sqrt{C_V^2 + C_H^2} = \sqrt{(150)^2 + (80)^2} = 170 \text{ lb} \qquad \textit{Ans.}$$

We next obtain the direction angle θ,

$$\tan\theta = \frac{150}{80} = 1.875 \quad \text{and} \quad \theta = 61°56' \qquad \textit{Ans.}$$

All forces are now known. A moment equation about A (or B) will give the location of support C required for equilibrium. When A is chosen as the center of moments, the moments of each of the two 80-lb forces become zero and are therefore eliminated from consideration. Thus

$$[\Sigma M_A = 0] \qquad 150x = (60)(5) = 300 \quad \text{and} \quad x = 2 \text{ ft} \qquad \textit{Ans.}$$

We note with interest here that this simple problem contains in its solution nearly all the basic principles of statics, namely: (*a*) force, (*b*) components of a force, (*c*) resultant force, (*d*) moment of a force, (*e*) the principle of moments, (*f*) force equilibrium, and (*g*) moment equilibrium.

2-34. The beam shown in Fig. 2-43*a* is resting freely on supports *A* and *B*. Determine the reacting forces R_A and R_B necessary at these supports for equilibrium.

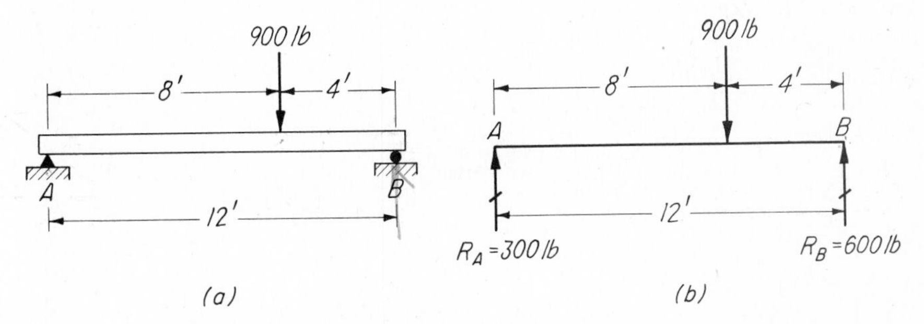

FIG. 2-43 Equilibrium of a parallel force system. (*a*) Space diagram. (*b*) Free-body diagram.

Solution: The supports are removed and are replaced by the two reacting forces, R_A and R_B, as shown in the free-body diagram (Fig. 2-43*b*). The roller at *B* rests on a horizontal surface. Hence, R_B is vertical. R_A is also vertical, because, when two of three forces in equilibrium are parallel, all must be parallel. Both reactions are clearly upward-acting forces. Then only two unknown magnitudes need be determined. A moment equation with *B* as the moment center will eliminate R_B from consideration, since its moment about *B* is zero, and R_A may be solved for. A second moment equation with *A* as the moment center will give R_B *independently*. A summation of vertical forces will furnish the required check. The equations are

$[\Sigma M_B = 0]$ $\quad 12R_A = (900)(4) = 3{,}600$ and $R_A = 300$ lb *Ans.*
$[\Sigma M_A = 0]$ $\quad 12R_B = (900)(8) = 7{,}200$ and $R_B = 600$ lb *Ans.*
$[\Sigma V = 0]$ $\quad 900 = 300 + 600 = 900$ *Check.*

2-35. In Fig. 2-44 is shown a ladder supporting a person weighing 200 lb. The ladder rests on the floor at *A* and against a smooth wall at *B*. The negligible friction at *B* may be disregarded. Compute the horizontal reaction R_B exerted by the wall at *B*, the vertical reaction component A_V exerted by the floor at *A*, and the horizontal reaction component A_H (friction or applied) necessary to prevent the bottom of the ladder from slipping to the left. Neglect weight of ladder.

Solution: Three unknown magnitudes of forces are to be determined. Their correct senses are readily deduced, and are as shown in the free-body diagram, Fig. 2-44*b*. Three simple steps solve this problem: (*a*) a moment equation about *A* eliminates the moments of A_V and A_H, and R_B may be solved for; (*b*) A_V is then found by a summation of vertical forces; and (*c*) A_H by a summation of horizontal forces. A

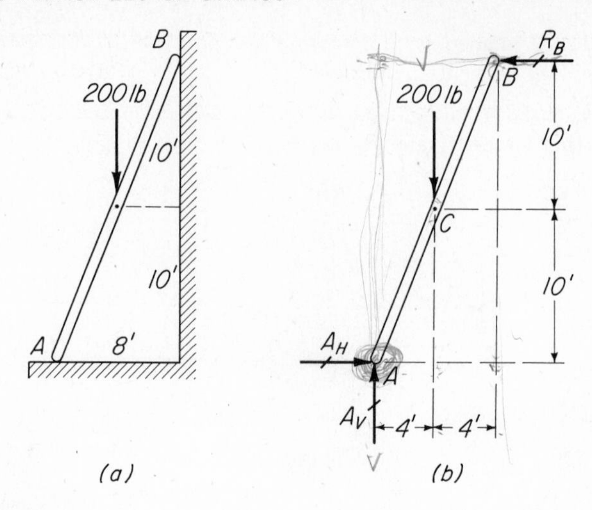

FIG. 2-44 Equilibrium of a nonconcurrent force system. (*a*) Space diagram. (*b*) Free-body diagram.

moment equation with C as moment center will then *check* the results obtained. The equations are

$[\Sigma M_A = 0]$ $\quad 20R_B = (200)(4) = 800 \quad$ and $\quad R_B = 40$ lb $\quad$ *Ans.*

$[\Sigma V = 0]$ $\quad A_V = 200$ lb. $\quad$ *Ans.*

$[\Sigma H = 0]$ $\quad A_H = 40$ lb $\quad$ *Ans.*

$[\Sigma M_C = 0]$ $\quad (200)(4) = (40)(10) + (40)(10 \quad$ or $\quad 800 = 800 \quad$ *Check.*

PROBLEMS

2-36. Compute the reacting forces R_A and R_B at the ends of the beam shown in Fig. 2-45, required to maintain equilibrium. *Ans.* $R_A = 70$ lb; $R_B = 30$ lb

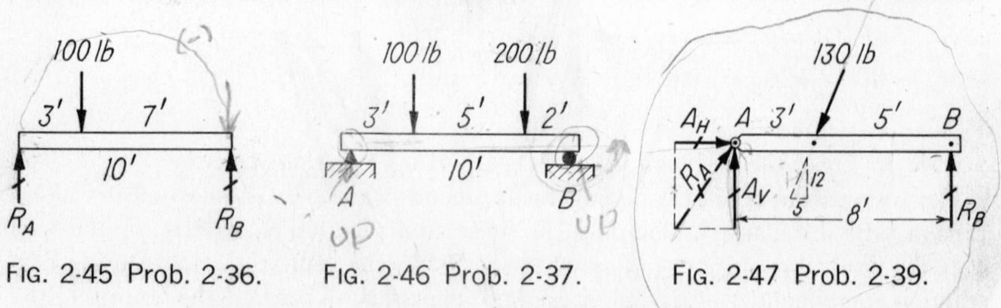

FIG. 2-45 Prob. 2-36. FIG. 2-46 Prob. 2-37. FIG. 2-47 Prob. 2-39.

2-37. The beam shown in Fig. 2-46 is resting freely on supports at A and B. Draw a free-body diagram of the beam and compute the vertical reacting forces R_A and R_B.

2-38. Calculate the reaction components shown in Fig. 2-23.

2-39. In Fig. 2-47 is shown a beam supported at A and B. Compute the reacting forces R_B, A_V, A_H, and R_A necessary for equilibrium.

Ans. $R_B = 45$ lb; $A_V = 75$ lb; $A_H = 50$ lb; $R_A = 90$ lb

G.E. 112	1-16-71	Jones, R. C.	18-2.35 (Page No.)	1/1

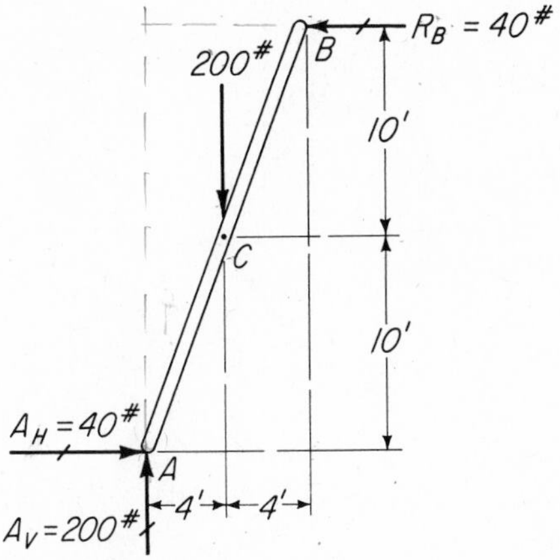

$R_B = 40^{\#}$

$200^{\#}$ B

10′

C

10′

$A_H = 40^{\#}$

A

$A_V = 200^{\#}$ 4′ 4′

Free-body diagram

Find: R_B, A_H, A_V

To find R_B, $\Sigma M_A = 0$:
$20 R_B = (200)(4) = 800$ — $R_B = 40^{\#}$ — R_B

To find A_H, $\Sigma H = 0$:
$A_H = R_B = 40^{\#}$ — $A_H = 40^{\#}$ — A_H

To find A_V, $\Sigma V = 0$:
$A_V = 200$ — $A_V = 200^{\#}$ — A_V

Check, $\Sigma M_C = 0$:
$(200)(4) = (40)(10) + (40)(10)$
$800 = 400 + 400$ — Check

SAMPLE PROBLEM SOLUTION SHEET

Paper: $8\frac{1}{2}$ by 11 in. squared.

Heading: Left to right, course number, date, name, problem number.

Problem Number: The number 18 represents the student's consecutive problem number; 2.35 is the number of the problem in the book; and the third number is the page number in the book.

Upper Right Corner: *Upper,* student's consecutive sheet number; *lower,* total number of sheets in the solution.

2-40. The frame shown in Fig. 2-48 carries a vertical load at C and is supported at A and E. Compute the reaction R_A and the reaction components E_V and E_H required for equilibrium.

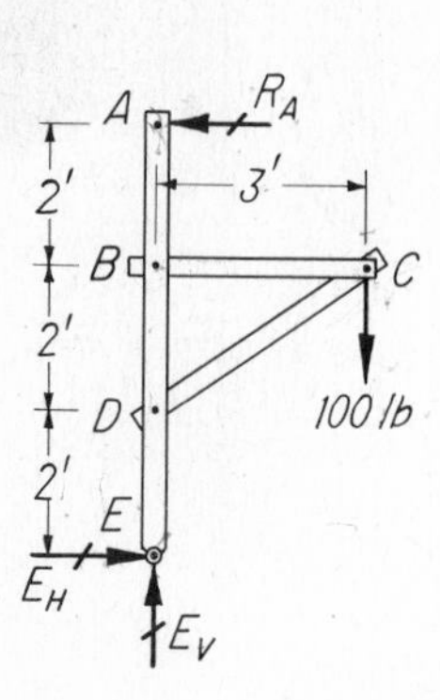

FIG. 2-48 Prob. 2-40.

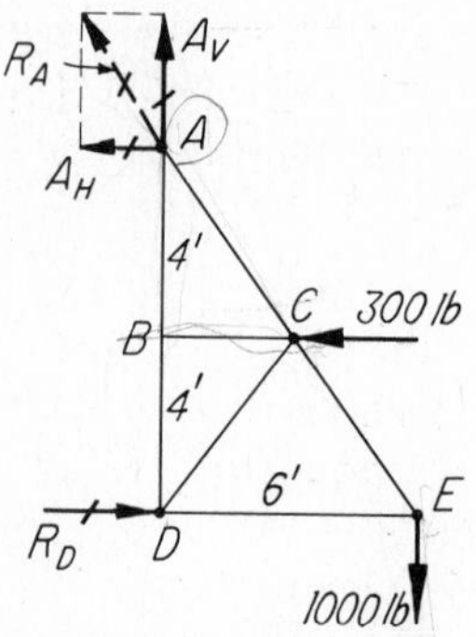

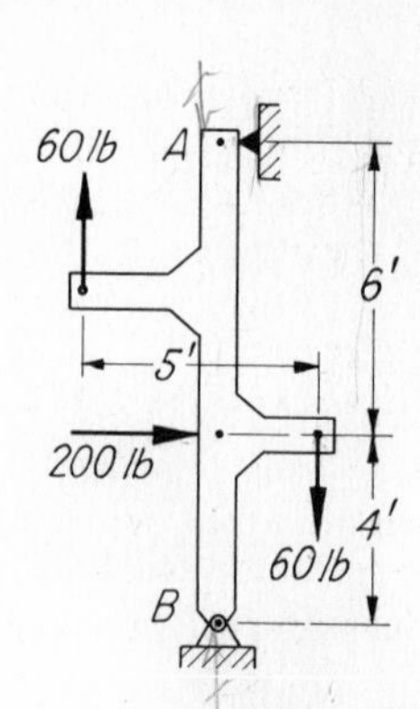

FIG. 2-50 Prob. 2-42.

FIG. 2-49 Prob. 2-41.

2-41. In Fig. 2-49 is shown a truss acted upon by forces at C and E and supported at A and D. Compute the reacting forces R_D, A_H, A_V, and R_A.

Ans. $R_D = 900$ lb; $A_H = 600$ lb; $A_V = 1{,}000$ lb; $R_A = 1{,}167$ lb

2-42. The frame shown in Fig. 2-50 is acted upon by a force and a couple as shown and is supported at A and B. No vertical force can be resisted at A. Draw a free-body diagram of the frame, and solve for the vertical and horizontal reacting forces, at A and B, required for equilibrium.

2-43. The materials hoist shown in Fig. 2-51 is held against motion by the cable tension T. The load P is 4 kips. Find the horizontal reactions on the wheels at A and B. Neglect the weight of the hoist. The hoist dimensions are $a = 3$ ft, $b = 3$ ft, and $c = 6$ ft. *Ans.* $A_H = 2$ kips←; $B_H = 2$ kips→

2-44. Solve Prob. 2-43 if $P = 6$ kips, $a = 4$ ft, $b = 2$ ft, and $c = 8$ ft.

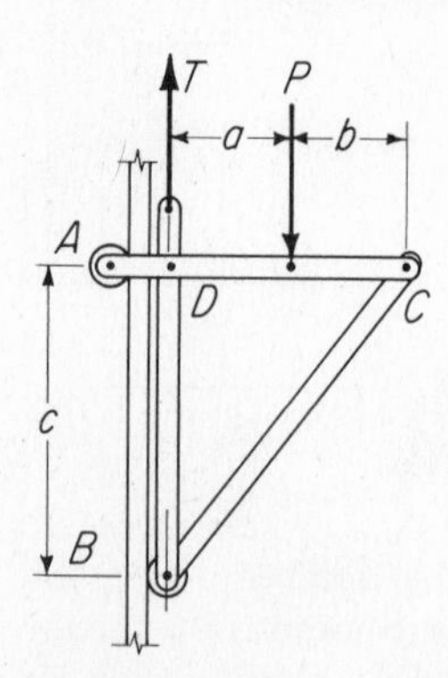

FIG. 2-51 Probs. 2-43 and 2-44.

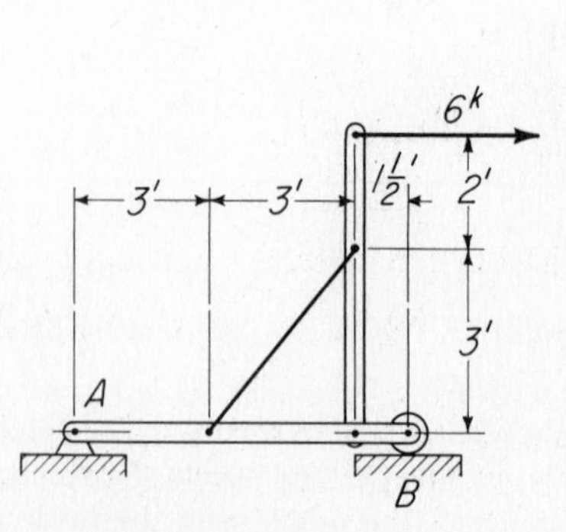

FIG. 2-52 Prob. 2-45.

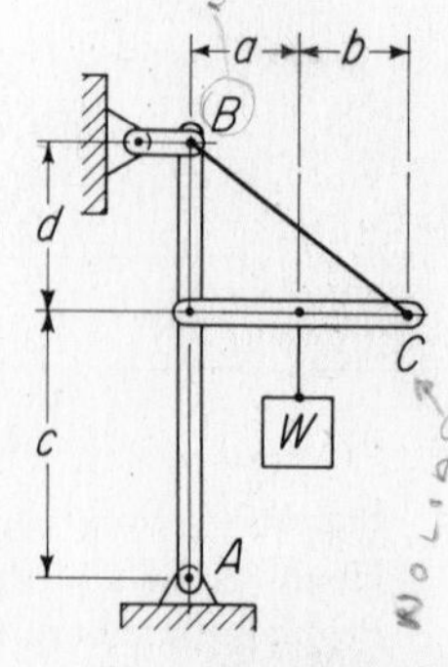

FIG. 2-53 Probs. 2-46 and 2-47.

2-45. The frame shown in Fig. 2-52 is subjected to the horizontal load of 6 kips as shown. Find the vertical reaction at B and the vertical and horizontal reactions at A.

Ans. $B_V = 4$ kips↑; $A_V = 4$ kips↓; $A_H = 6$ kips←

2-46. The frame shown in Fig. 2-53 supports a weight $W = 1{,}200$ lb. The link at B can supply only a horizontal reaction at B. Find the horizontal reaction at B and the vertical and horizontal reacting forces at A. The dimensions of the frame are $a = 2$ ft, $b = 2$ ft, $c = 5$ ft, and $d = 3$ ft. *Partial Ans.* $A_H = 300$ lb→

2-47. Solve Prob. 2-46 if $W = 4$ kips, $a = 5$ ft, $b = 3$ ft, $c = 6$ ft, and $d = 4$ ft.

2-48. A ladder AB is leaning against a smooth, vertical wall at A in Fig. 2-54 and is resting on a horizontal floor at B. A person weighing 200 lb exerts a vertical force at C; another person applies a horizontal force of 60 lb at D to help reduce the horizontal pressure against the block at B. Draw a free-body diagram of this ladder; compute the horizontal reacting force at A, and the vertical and horizontal reaction components at B, necessary to maintain equilibrium. Check by moments about E.

Ans. $R_A = 70$ lb→; $B_V = 200$ lb↑; $B_H = 10$ lb←

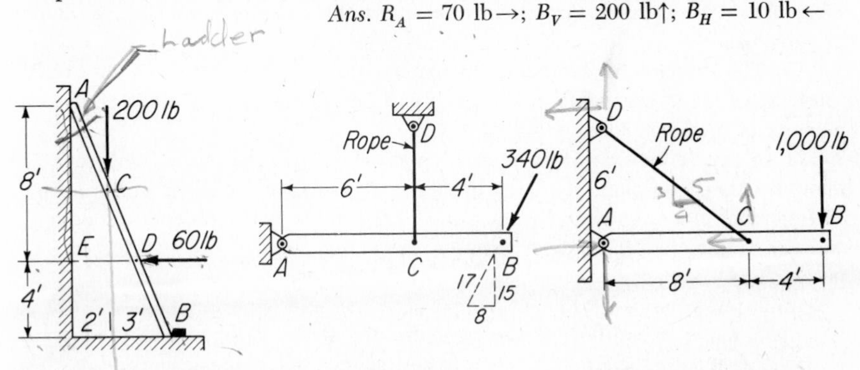

FIG. 2-54 Prob. 2-48. FIG. 2-55 Prob. 2-49. FIG. 2-56 Prob. 2-50.

2-49. The horizontal beam in Fig. 2-55 is acted upon by a single force at B and is supported at A and D as shown. Draw a free-body diagram of the beam, and compute the vertical and horizontal reacting forces at pins A and C.

2-50. Draw a free-body diagram of the horizontal member AB, shown in Fig. 2-56, and compute the vertical and horizontal reacting forces at pins A and C. Compute also the tension T in the rope CD. *Ans.* $A_V = 500$ lb↓; $A_H = 2{,}000$ lb→; $C_V = 1{,}500$ lb↑; $C_H = 2{,}000$ lb←; $T = 2{,}500$ lb

SUMMARY

(By article number)

2-1. Force is an action which changes or tends to change the state of motion of the body upon which it acts. A force is completely defined when its (1) magnitude, (2) direction, (3) sense, and (4) point of application are known. Force is a *vector quantity*.

2-4. A vector is a graphical scale representation of a vector quantity. When vectors are added graphically, the result is called their *vector sum*.

2-5. Transmissibility of a force means that its point of application may be transmitted to any other point along its line of action, without changing other *external* forces also acting on the body.

2-6. Coplanar forces all lie in one plane; **noncoplanar forces** do not all lie in one plane. The lines of action of **concurrent forces** all pass through a common point; those of **nonconcurrent forces** do not. All force systems are combinations of these four types. When all forces of a system have a common line of action they are said to be **collinear.**

2-7. The components of a force acting through its points of application may replace it without change in effect.

2-8. A resultant force will produce the same *external* effect as the forces it replaces. An **equilibrant force** is exactly equal and opposite to a resultant force.

2-9. The moment of a force is its tendency to produce rotation of the body upon which it acts and is the product of the force and the perpendicular distance between the axis of rotation and the line of action of the force. This distance is called the **moment arm.** In coplanar systems, the moment axis becomes a point, called the **center of moments.**

2-10. The Principle of Moments. Varignon's Theorem. The algebraic sum of the moments of the components of a force about any point in their plane equals the moment of the force about the same point.

2-11. A couple consists of two equal, opposite, and parallel forces with separate lines of action. The plane in which the forces lie is called the **plane of the couple.** The **moment of a couple,** *about any point in this plane,* is the product of one force and the perpendicular distance between the action lines of the two forces. This perpendicular distance is called the **arm of the couple.**

2-14. Equilibrium is a balance of opposing forces and opposing moments. For all systems of forces in equilibrium **the algebraic sum of the acting and reacting** forces is zero, or, in symbols, $\Sigma F = 0$. Likewise, **the algebraic sum of the acting and reacting moments is zero,** or $\Sigma M = 0$.

2-15. Principles of Force Equilibrium. Three principles of force equilibrium are:

The Two-force Principle. When *two forces* are in equilibrium, they must be equal, opposite, and collinear.

The Three-force Principle. When *three nonparallel forces* are in equilibrium, their lines of action intersect at a common point, and they are coplanar.

The Four-force Principle. When *four forces* are in equilibrium, the resultant of any two of the forces must be equal, opposite, and collinear with the resultant of the other two.

2-16. The reaction of a **roller support,** or at a **smooth surface,** is always perpendicular to the supporting surface. The direction of the reaction at a pinned support is generally unknown.

REVIEW PROBLEMS

2-51. Determine the V and H *components* of a 4-kip force acting upward and to the right at an angle of 70° with the horizontal. Check graphically (scale, 1 in. = 1 kip). *Ans.* $V = 3.76$ kips↑; $H = 1.37$ kips→

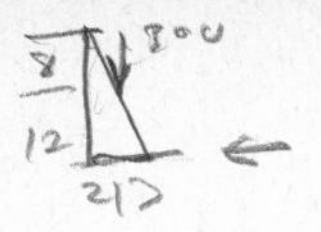

2-52. A 100-lb force acts upward and to the right with a slope of two units vertical to three units horizontal. Compute the *V* and *H components*. Check graphically (scale, 1 in. = 20 lb).

2-53. Using the *sine law*, resolve a 100-lb force acting upward and to the right at 45° with the horizontal, into two *components* *P* and *Q* also acting upward and to the right at angles of 20° and 80° with the horizontal, respectively. Check graphically (scale, 1 in. = 20 lb). *Ans.* $P = 66.3$ lb; $Q = 48.8$ lb

2-54. Using the *cosine law*, solve for the *resultant* *R* of two 10-kip forces, one acting upward and to the right at 30° with the horizontal and the other upward and to the left at 70° with the horizontal. Find also the angle θ_H the resultant makes with the horizontal. Check graphically (scale, 1 in. = 4 kips).

Ans. $R = 15.3$ kips; $\theta_H = 70°$

2-55. The right end of the beam shown in Fig. 2-57 rests on a smooth frictionless plane. Calculate the vertical and horizontal components of the reactions at supports *A* and *B*.

2-56. The truss shown in Fig. 2-58 is loaded as shown. Solve for the vertical and horizontal components of the reaction at *A* and the vertical reaction at *B*. (HINT: The solution will be considerably simplified if the horizontal and vertical components of the 25-kip load are recognized as parts of couples acting on the truss.)

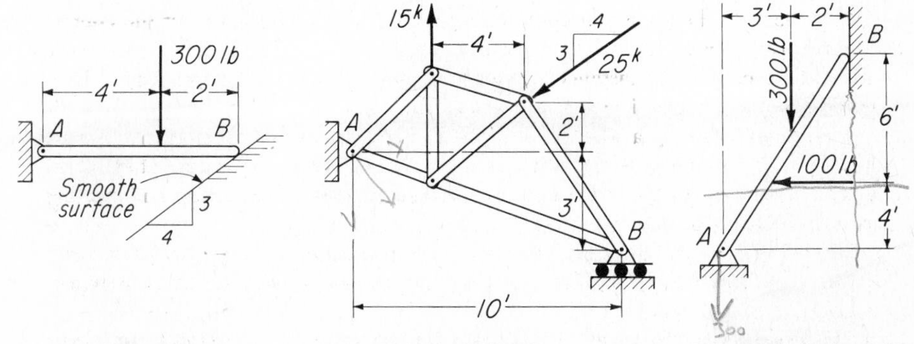

FIG. 2-57 Prob. 2-55. FIG. 2-58 Prob. 2-56. FIG. 2-59 Prob. 2-57.

2-57. The pole *AB* shown in Fig. 2-59 leans against a smooth frictionless wall at *B*. Calculate the vertical and horizontal components of the reactions at *A* and *B*.

Ans. $A_V = 300$ lb↑; $A_H = 150$ lb→; $B_H = 50$ lb←

2-58. Solve Prob. 2-54 by the algebraic-component method. Draw three diagrams similar to those shown in Fig. 2-12.

2-59. Beam *BC* in Fig. 2-60 is supported at its left end by the rope *AB*, and its right end is resting on another beam *DE*. Compute the tension *T* in the rope and the *reactions* R_D and R_E required to maintain *equilibrium*. Draw separate free-body diagrams of *BC* and *DE*. *Ans.* $T = 90$ lb↑; $R_D = 20$ lb↑; $R_E = 40$ lb↑

2-60. Draw a free-body diagram of bar *AB* in Fig. 2-61. Solve for the *V* and *H* components of the *reacting forces* at *A* and *C*, necessary for *static equilibrium*, and compute the reactions R_A and R_C. *Ans.* $R_A = 671$ lb↙; $R_C = 1{,}000$ lb↗

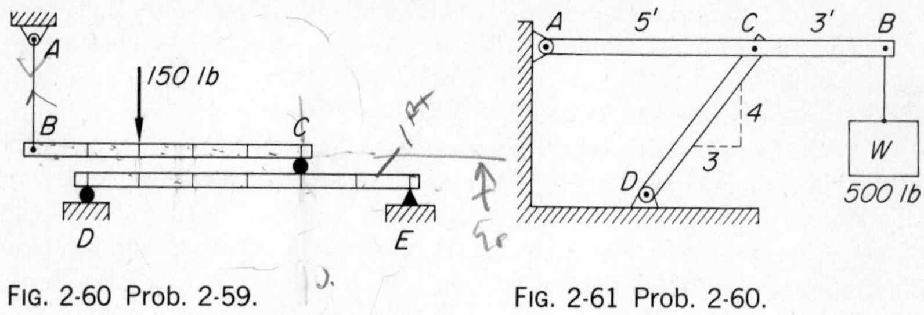

FIG. 2-60 Prob. 2-59.

FIG. 2-61 Prob. 2-60.

2-61. The weight W shown in Fig. 2-62 is 400 lb. Calculate the reactions at A and C.

2-62. The nose landing gear shown in Fig. 2-63 is subjected to a horizontal force $D_H = 1{,}400$ lb and a vertical force $D_V = 7{,}000$ lb. Calculate the magnitude and directions of the reaction at A and B. Note that the resultant reaction at B must point along the line BC.

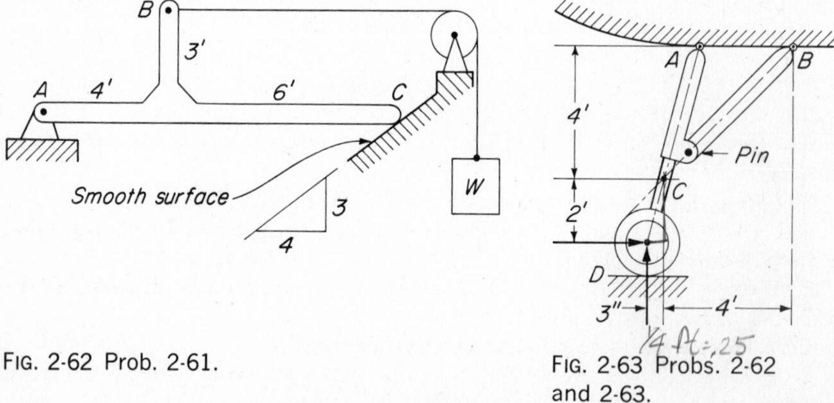

FIG. 2-62 Prob. 2-61.

FIG. 2-63 Probs. 2-62 and 2-63.

2-63. Solve Prob. 2-62 if $D_H = 0$ and $D_V = 10{,}000$ lb.

Ans. $A = 12{,}320$ lb, $\theta_H = 80°$; $B = 3{,}020$ lb, $\theta_H = 45°$

2-64. The truss shown in Fig. 2-64 is subjected to the loads shown. Calculate the vertical and horizontal components of the reactions at A and D.

2-65. The bar AB in Fig. 2-65 is supported at B as shown and is held at A by a horizontal rope passing over a sheave. Draw a free-body diagram of bar AB and compute the weight W required for *static equilibrium*, if the load P is 400 lb. Compute also reaction components B_V and B_H, and the *reaction* R_B. Neglect sheave friction tion and weight of bar.

Ans. $W = 120$ lb; $R_B = 418$ lb ↗

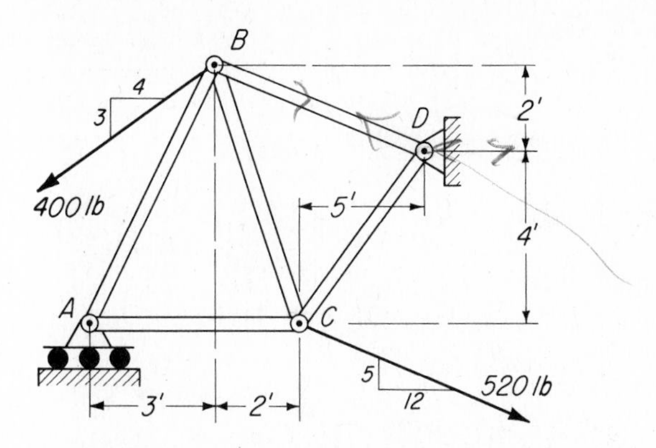

FIG. 2-64 Prob. 2-64.

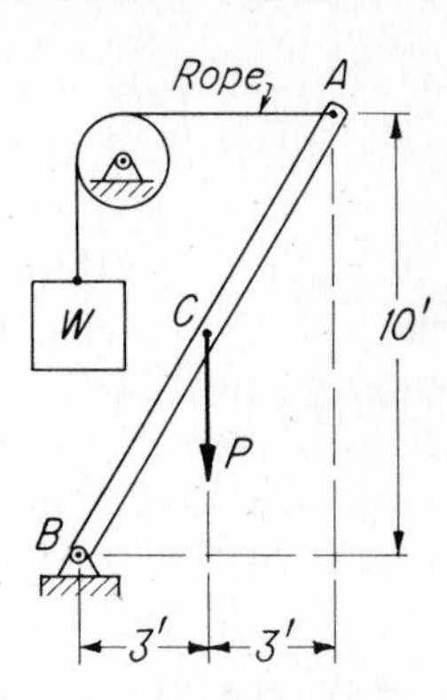

FIG. 2-65 Prob. 2-65.

REVIEW QUESTIONS

2-1. Define briefly the following terms: (*a*) mechanics, (*b*) statics, (*c*) force.

2-2. Give the four characteristics of a force.

2-3. What is meant by (*a*) a vector quantity, (*b*) a scalar quantity? Give a few examples of each.

2-4. Explain briefly the principle of transmissibility.

2-5. Name the six types of force systems.

2-6. State briefly the principle of components.

2-7. Give briefly the principle of resultants.

2-8. Define the moment of a force. By what product is the moment of a force measured?

2-9. State the principle of moments known as Varignon's theorem.

2-10. What is a couple? What is the algebraic sum of the forces of a couple? What is the moment of a couple?

2-11. State, or illustrate by a diagram, how to locate the line of action of the resultant of two parallel forces.

2-12. Explain the essential meaning of equilibrium.

2-13. What is meant by a complete force system? Is such a system always in equilibrium?

2-14. State the force law of equilibrium.

2-15. Give the moment law of equilibrium.

2-16. When dealing with force systems in equilibrium, is a sign convention necessary? If it is used, which forces and moments are generally positive?

2-17. State the two-force principle.

2-18. Define the three-force principle.

2-19. Explain the four-force principle.

2-20. Give the specific direction of the reaction at a roller support or at a smooth supporting surface.

2-21. What is meant by a free body? What is the main purpose of a free-body diagram?

CHAPTER 3

Coplanar, Parallel Force Systems

3-1 Introduction

Because of the attraction of the gravitational force of the earth upon all outside bodies, vertical forces and systems of vertical forces (parallel) are commonly encountered in engineering practice.[1] The usual problem in this, as in all other types of force systems, is (1) to determine the magnitude of a resultant force and the location of its line of action or (2) to determine the unknown reacting forces of a system in equilibrium. Both analytical and graphical solutions are available and are described in the following articles.

3-2 Resultant of Coplanar, Parallel Forces

The solutions of problems involving parallel force systems are often simplified by use of the resultant force. The *analytical method of moments* is simple, as is shown in the example below, and is generally used. Graphical methods are more involved but are useful at times when all-graphical solutions are desired, or for checking results obtained analytically. The graphical method

[1] The *gravity forces* of a system are not truly parallel; actually they are *concurrent* at the center of the earth's attraction. In such problems as normally arise in engineering practice, we do, of course, consider them to be parallel. Theoretically, a system of applied forces can be made truly parallel. The *inertia forces* acting on the various parts of a body in rectilinear motion are truly parallel. The *centrifugal forces* acting on the various parts of an automobile traveling in a circular path on a horizontal road appear *parallel* when projected onto a vertical plane, and appear *concurrent* through the axis of rotation when projected onto a horizontal plane.

of inverse proportion, described in Art. 2-12, may be applied to three or more parallel forces by combining the resultant of any two forces, obtained as in Fig. 2-26, with a third force, and by similarly combining the resultant of those three with a fourth force, etc. The method of determining a resultant force by the *graphical string-polygon method* is described in Chap. 5. Although this description refers to the determination of the resultant of a system of non-parallel forces, it applies as well, of course, to a system of parallel forces.

The magnitude of the resultant of a system of parallel forces is their algebraic sum. *Its line of action* is parallel to those of the forces of the system. *The center of a system of parallel forces* is a point through which the resultant passes. Specifically, it is the point where the action line of the resultant pierces any chosen perpendicular plane.

ILLUSTRATIVE PROBLEM

3-1. Determine the magnitude of the resultant R of the four forces shown in Fig. 3-1 and the distance x from A to its line of action.

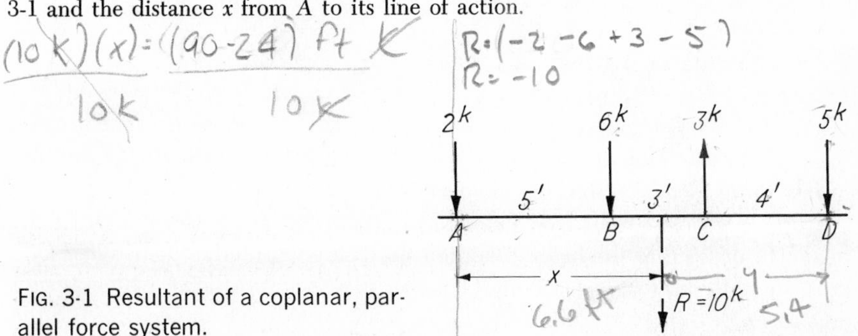

FIG. 3-1 Resultant of a coplanar, parallel force system.

Solution: The resultant of a system of parallel forces must be equal in magnitude to the *algebraic sum* of the forces of the system it replaces. Also, its moment about any point must equal the *algebraic sum* of the moments of the separate forces about that point, and its line of action is parallel to those of the forces. Then, if upward-acting forces are positive,

$$R = -2 - 6 + 3 - 5 = -10 \text{ kips}$$

Let A be the moment center to determine x. Then

$$-10x = -(6)(5) + (3)(8) - (5)(12) = -66 \qquad \text{and} \qquad x = 6.6 \text{ ft} \qquad \textit{Ans.}$$

The negative sign of R shows it to be downward-acting. The positive sign of x indicates that the action line of R lies to the right of A, as was assumed.

PROBLEMS

3-2. Compute the resultant R of the forces shown in Fig. 3-2, and the distance x from A to its line of action. *Ans.* $R = 9$ kips; $x = 7$ ft

3-3. Calculate the resultant R of the forces shown in Fig. 3-3, and the distance x from A to its line of action. *Ans.* $R = -12$ kips; $x = 4.25$ ft

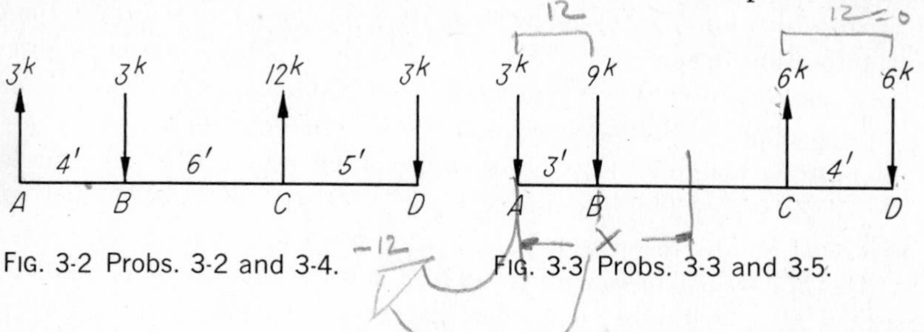

FIG. 3-2 Probs. 3-2 and 3-4.

FIG. 3-3 Probs. 3-3 and 3-5.

3-4. In Fig. 3-2 change the 3-kip force at D to 30 kips; solve for R and for distance x from A.

3-5. Change force B in Fig. 3-3 from 9 kips downward acting to 6 kips upward acting. Then solve for R and for distance x from A.

Ans. $R = 3$ kips; $x = -2$ ft (to the left of A)

3-3 Resultants of Distributed Loads

Distributed forces were defined and discussed in Art. 2-2. They are usually referred to as **distributed loads.** The pressure of warehouse contents on the floor and of the floor planking on the supporting joists or beams are examples of distributed loads. Unless clearly otherwise distributed, such loads on beams are usually considered to be uniformly distributed along the length of the beam. In order that the beam reactions may be computed, *a distributed load must be replaced by its equivalent concentrated resultant load,* which always acts through the center of gravity of the load, or through the centroid of the load area.

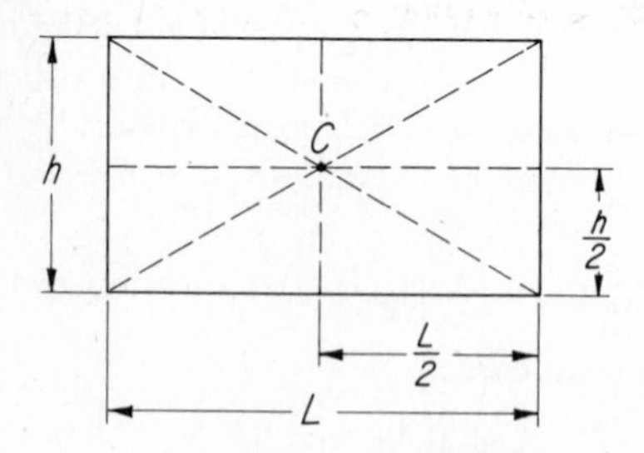

FIG. 3-4 Centroid of rectangle.

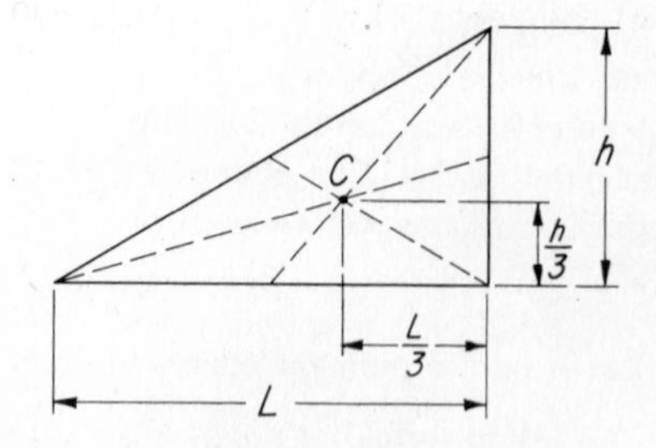

FIG. 3-5 Centroid of triangle.

By geometric construction, the centroid of a rectangular area is at the intersection C of its diagonals, as in Fig. 3-4; that of a triangular area is at the intersection C of its medians, which is always one-third of any altitude above its base, as shown in Fig. 3-5.

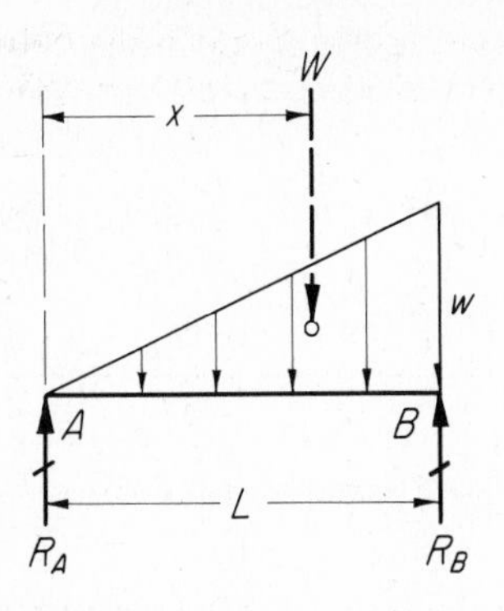

FIG. 3-6 Triangular load.

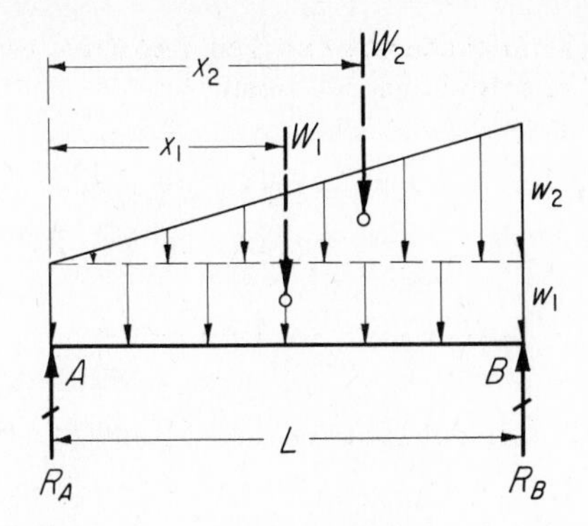

FIG. 3-7 Trapezoidal load.

Nonuniformly distributed loads may usually be reduced to either the triangular form, shown in Fig. 3-6, or the trapezoidal, as in Fig. 3-7, without appreciable error. A trapezoidal load is usually replaced with its equivalent rectangular and triangular loads, whose two resultants then have the same effect as the single resultant of the trapezoidal load.

3-4 Equilibrium of Coplanar, Parallel Force Systems

The usual problem is that of determining two unknown support reactions. Analytical solutions by moments are simplest; a graphical solution using the string and force polygons is shown in Chap. 5.

ILLUSTRATIVE PROBLEM

3-6. Solve by moments for the reactions R_A and R_B at the supports of the beam shown in Fig. 3-8.

Solution: The trapezoidal load area is divided into a rectangle and a triangle. The resultant W_1 is (12)(3) or 36 kips, and W_2 is ($\frac{1}{2}$)(12)(4) or 24 kips. When these

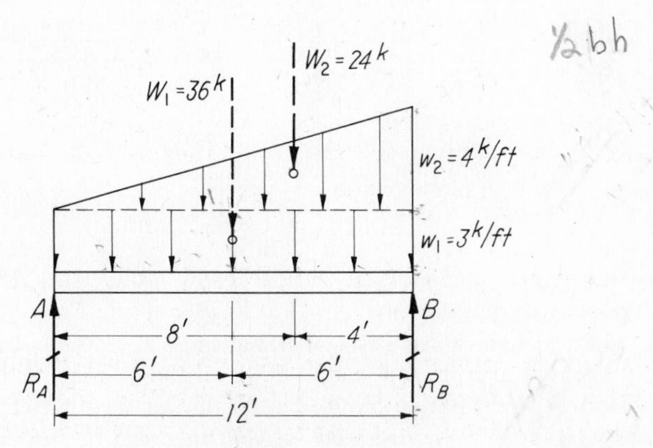

FIG. 3-8 Support reactions.

resultants are applied at their respective centroids, we find R_A by moments about B, and R_B by moments about A. A summation of vertical forces will then provide a sufficient check. The equations are

$[\Sigma M_B = 0]$	$12R_A = (4)(24) + (6)(36) = 312$	and	$R_A = 26$ kips	*Ans.*
$[\Sigma M_A = 0]$	$12R_B = (6)(36) + (8)(24) = 408$	and	$R_B = 34$ kips	*Ans.*
$[\Sigma V = 0]$	$36 + 24 = 26 + 34$ or $60 = 60$		*Check.*	

PROBLEMS

3-7. Compute the reactions R_A and R_B at supports A and B of the beam shown in Fig. 3-9.

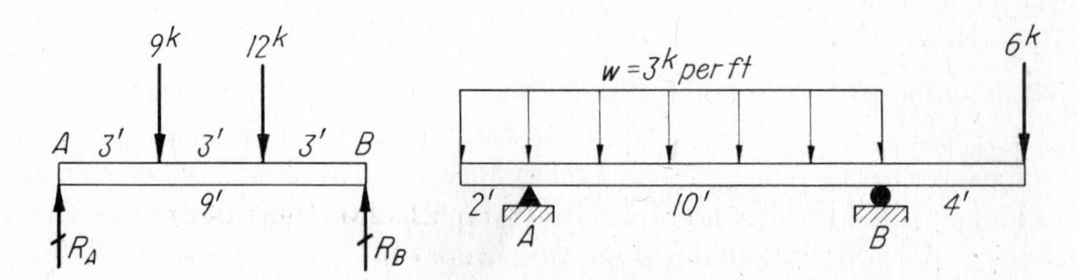

FIG. 3-9 Prob. 3-7. FIG. 3-10 Prob. 3-8.

3-8. Calculate the reacting forces at supports A and B caused by the loads on the beam shown in Fig. 3-10.

3-9. Find the reactions at support A and B of the beam shown in Fig. 3-11.
Ans. $R_A = 47.83$ kips; $R_B = 27.17$ kips

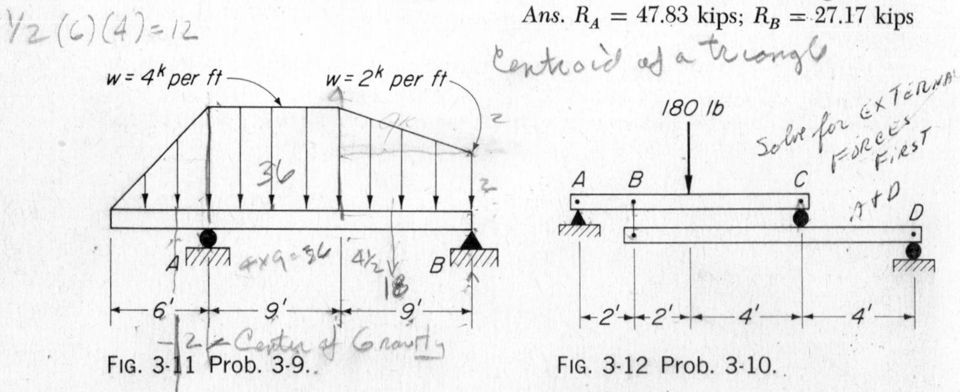

FIG. 3-11 Prob. 3-9. FIG. 3-12 Prob. 3-10.

3-10. In Fig. 3-12 compute the vertical forces acting at A, B, C, and D. (Draw three free-body diagrams.) *Ans.* $A = 120$ lb; $B = 40$ lb; $C = 100$ lb; $D = 60$ lb

3-11. In Fig. 3-13 is shown a typical arrangement of platform-scale levers, the two identical levers A and levers B and C. If, in all these levers, the ratio of distances a and b is 1 to 9, determine the load L on the platform which will be balanced by a 1-lb weight W on the right end of weighing arm C. Does it matter whether or not the load is centered on the platform? *Ans.* $L = 900$ lb; no!

3-12. Solve Prob. 3-11 when the ratios of distances a and b on levers A, B, and C are, respectively, 1 to 9, 1 to 14, and 1 to 8.

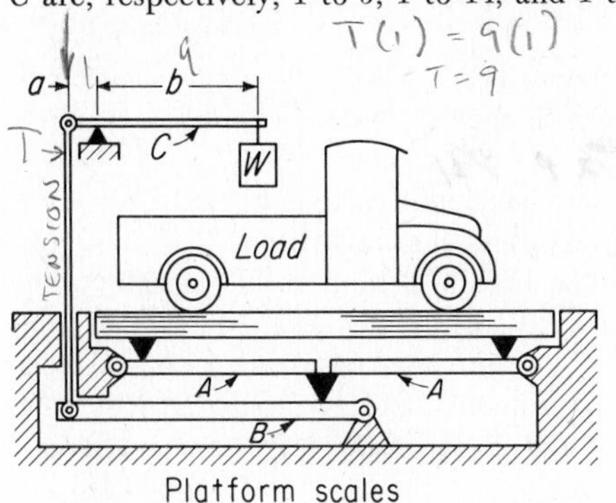

Bar 1 2 3 4 5 Link Piston W Steam 6" 24" Relief valve

FIG. 3-13 Probs. 3-11 and 3-12.

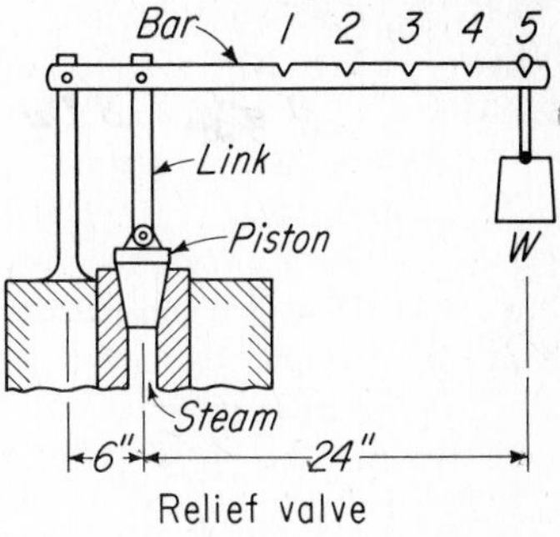

FIG. 3-14 Probs. 3-13 and 3-14.

3-13. In Fig. 3-14 is shown a simple form of relief valve. The weight of the link and the piston is 10 lb. The horizontal bar weighs 20 lb, which may be considered to be concentrated at its mid-point. If the circular surface of the bottom of the piston is 2 in. in diameter, compute the least weight W required to maintain a steam pressure of 30 psi. *Ans.* $W = 6.85$ lb

3-14. If, in Prob. 3-13, the notches in the horizontal bar are 4 in. apart and a weight of 30 lb is placed in notch 3, compute the maximum steam pressure p, in psi, under which the valve will remain closed.

3-15. A steel beam 12 ft long and weighing 45 lb per ft is supported at its ends and carries a triangular load, as in Fig. 3-6. If the rate of load w at B is 3,000 lb per ft, compute the reactions R_A and R_B. *Ans.* $R_A = 6{,}270$ lb; $R_B = 12{,}270$ lb

3-16. In Fig. 3-15 block A has two sheaves and block B has one. Determine (a) the pull P required to lift a weight W, and (b) the tension T in the rope C. Neglect

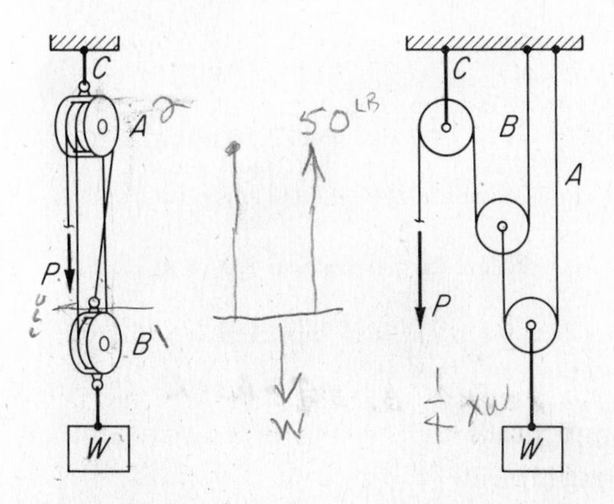

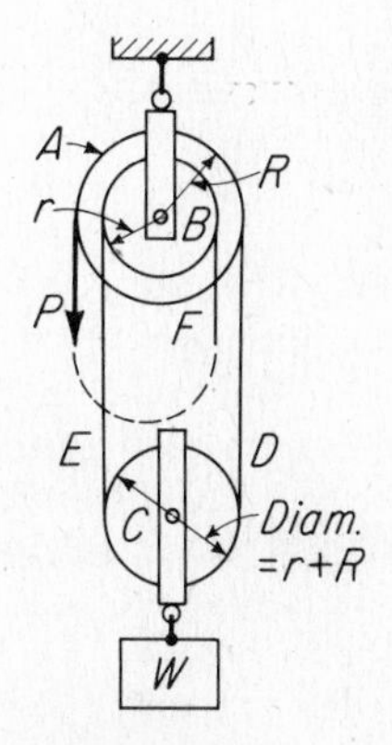

FIG. 3-15 Prob. 3-16.

FIG. 3-16 Prob. 3-17.

FIG. 3-17 Probs. 3-18 and 3-19.

friction. The ropes may be assumed to be coplanar and parallel with inappreciable error. (HINT: Cut ropes just above block B, and draw a free-body diagram.)

Ans. $P = W/3$; $T = 4W/3$

3-17. Determine (*a*) the pull P required to lift a weight W, using sheaves as arranged in Fig. 3-16 and (*b*) the tension T in each of ropes A, B, and C. Neglect friction.

3-18. In Fig. 3-17 is shown a differential chain hoist such as is commonly used in shops and garages. Two sprocket sheaves A and B of radii R and r are fastened together. A third sprocket sheave C of diameter $r + R$ supports the weight to be lifted. A continuous chain passes over sheave A, then under sheave C, and up over sheave B, returning to sheave A. A pull P on the chain raises side D and lowers side E in the ratio of R to r. There is no stress in side F of the chain. Assuming P, D, and E to be parallel, cut chains D and E, and show that, for equilibrium,

$$P = \frac{W(R - r)}{2R}.$$

3-19. In the differential chain hoist shown in Fig. 3-17, let $R = 4$ in. and $r = 3.5$ in. Neglect friction. (*a*) What pull P is required to lift a weight W of 1,000 lb? (*b*) If R is 4 in., what should r be in order that the pull P will be equal to 5 per cent of W?

Ans. (*a*) $P = 62.5$ lb; (*b*) $r = 3.6$ in.

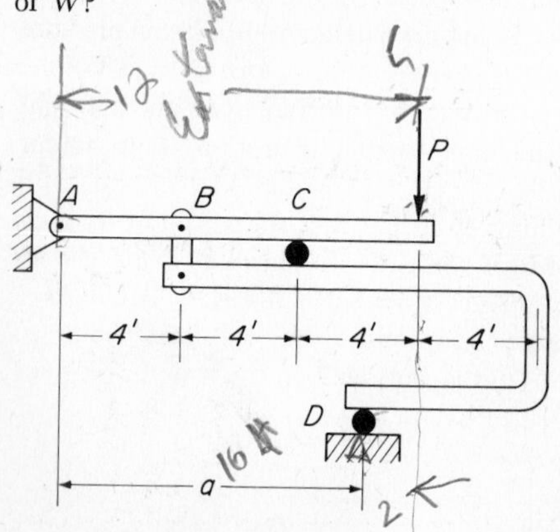

FIG. 3-18 Probs. 3-20 and 3-21.

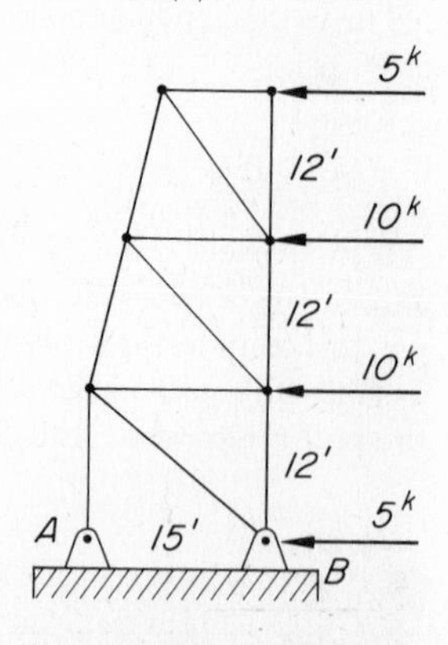

FIG. 3-19 Prob. 3-22.

3-20. The load P in Fig. 3-18 is 5 kips. Calculate the vertical forces at A, B, C, and D if a equals 10 ft.

3-21. The load P in Fig. 3-18 is 7 kips. Calculate the vertical forces at A, B, C, and D if a equals 14 ft.

Ans. $A = 1$ kip, $B = 9$ kips, $C = 15$ kips, and $D = 6$ kips

3-22. In Fig. 3-19 compute the reactions at A and B.

SUMMARY

(By article number)

3-1. Parallel force systems are common, since probably most forces encountered in nature and in engineering practice arise from gravitational attraction of bodies, thus creating parallel vertical forces.

3-2. The resultant of a parallel force system is most easily found analytically by moments. Graphical solutions may be obtained either by the inverse-proportion method or by the string-polygon method outlined in Chap. 5.

3-3. Resultants of distributed loads are always applied at the **center of gravity** of the load, which is the centroid of the load area.

3-4. The unknown quantities of a **parallel force system in equilibrium** are generally the unknown magnitudes of two *support reactions*. These are most easily found by moments.

REVIEW PROBLEMS

3-23. Find the *resultant* R of the forces shown in Fig. 3-20 and the horizontal distance x from A to its line of action. All forces are in pounds.

Ans. $R = 100$ lb; $x = 13.5$ ft

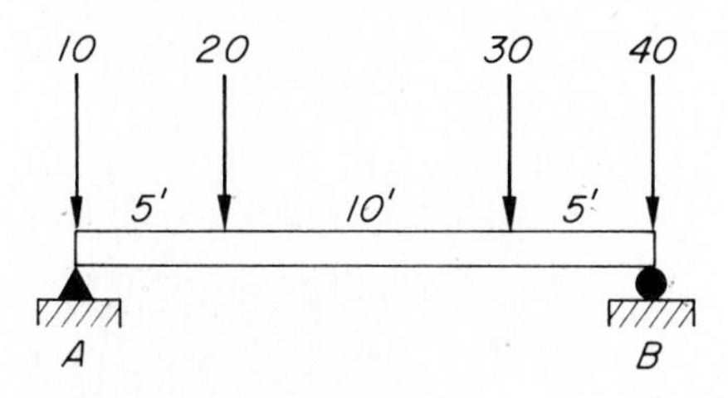

FIG. 3-20 Prob. 3-23.

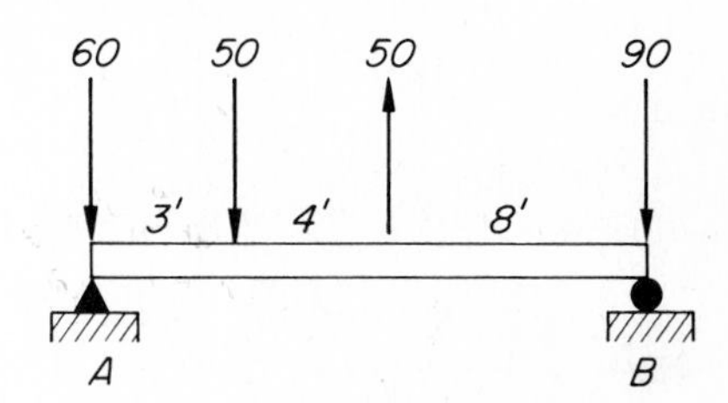

FIG. 3-21 Prob. 3-24.

3-24. In Fig. 3-21, determine the *resultant force* R, and the horizontal distance x from A to its line of action. All forces are in pounds.

3-25. Compute by moments the *reactions* R_A and R_B of the beam shown in Fig. 3-22. (HINT: See Art. 2-11.)

Ans. $R_A = 4$ kips↑; $R_B = 6$ kips↑

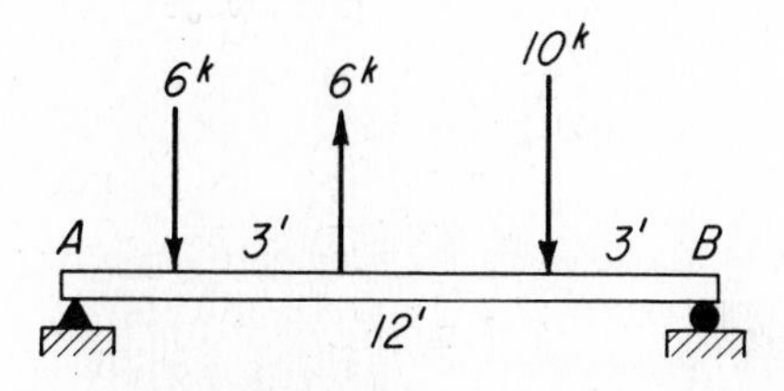

FIG. 3-22 Prob. 3-25.

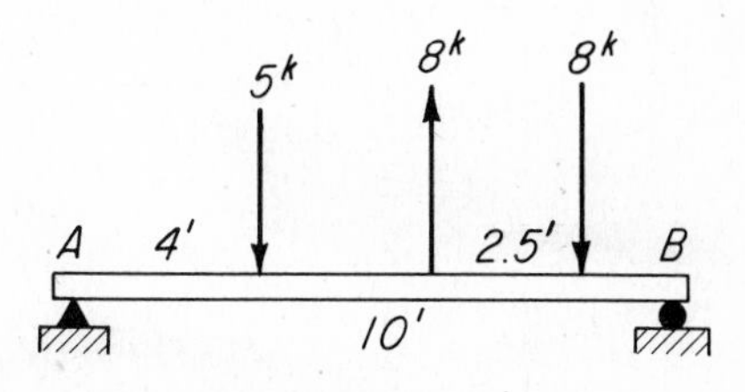

FIG. 3-23 Prob. 3-26.

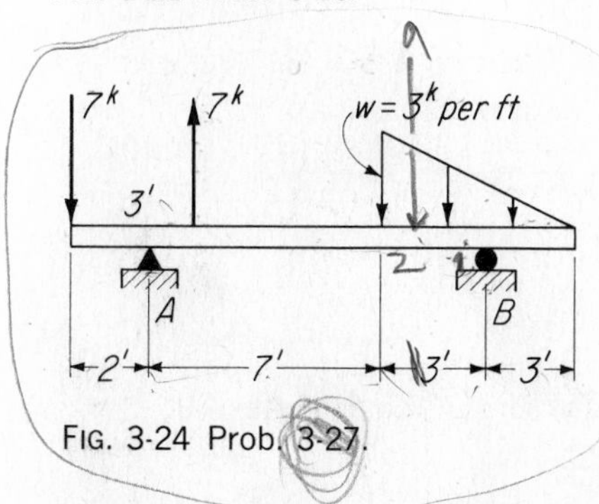

FIG. 3-24 Prob. 3-27.

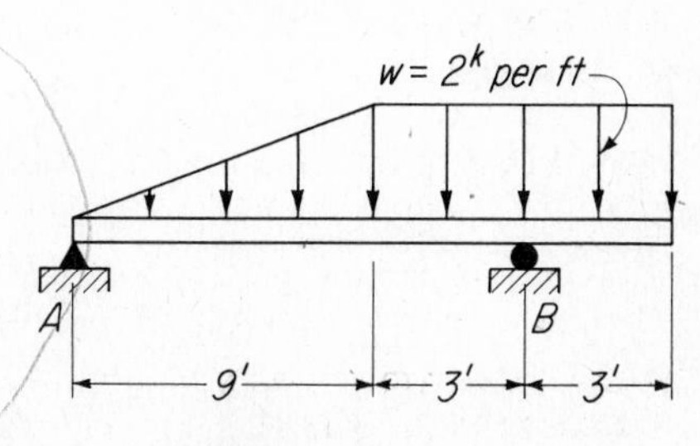

FIG. 3-25 Prob. 3-28.

3-26. Compute by moments the *reactions* R_A and R_B of the beam shown in Fig. 3-23.

3-27. Compute by moments the *reactions* R_A and R_B of the beam shown in Fig. 3-24. *Ans.* $R_A = 3$ kips↑; $R_B = 6$ kips↑

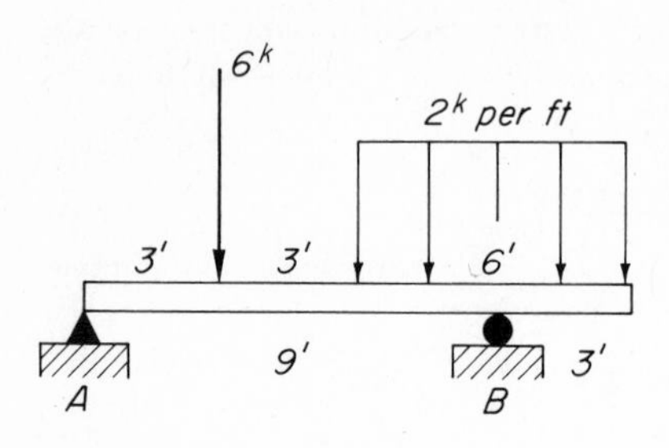

FIG. 3-26 Prob. 3-29.

FIG. 3-27 Prob. 3-30.

3-28. Compute by moments the *reactions* R_A and R_B of the beam shown in Fig. 3-25.

3-29. Compute by moments the *reactions* R_A and R_B of the beam shown in Fig. 3-26. *Ans.* $R_A = 4$ kips; $R_B = 14$ kips

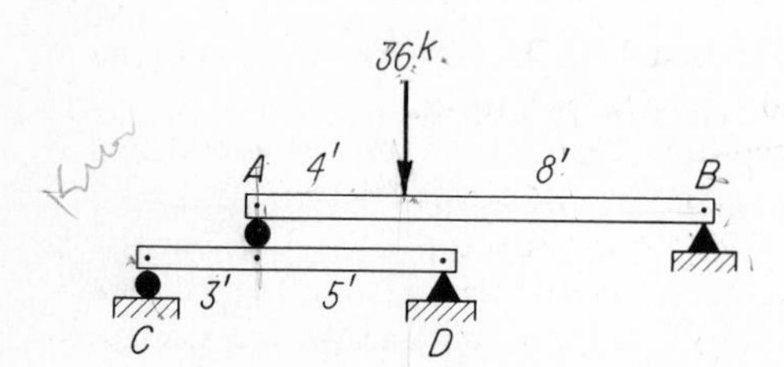

FIG. 3-28 Prob. 3-31.

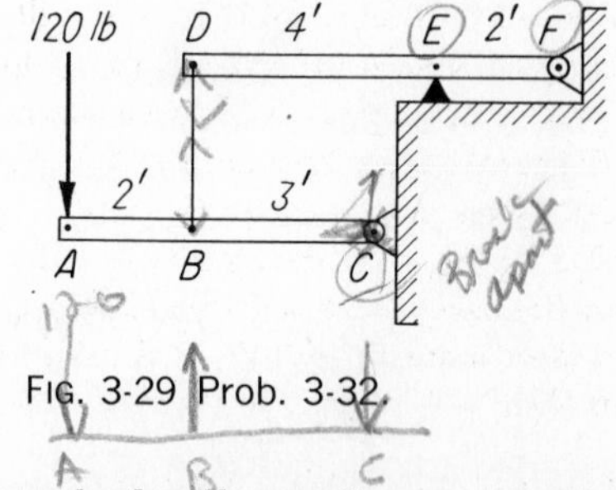

FIG. 3-29 Prob. 3-32.

3-30. The beam in Fig. 3-27 is subjected to a load as shown. Compute *reactions* R_A and R_B.

3-31. Draw separate *free-body diagrams* of beams *AB* and *CD* in Fig. 3-28; then compute and show all forces acting on each beam. *Ans.* $R_C = 15$ kips

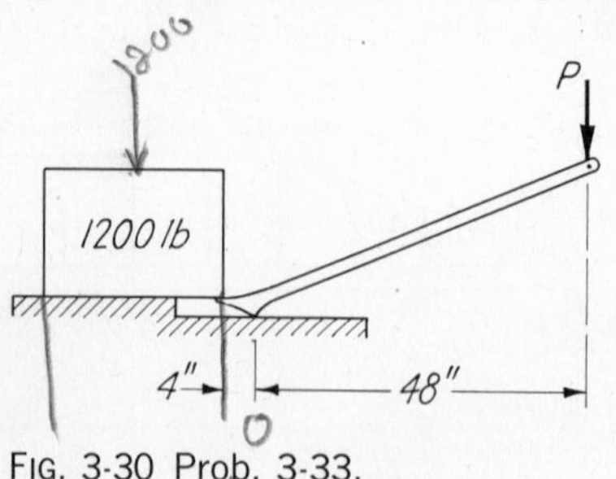

FIG. 3-30 Prob. 3-33.

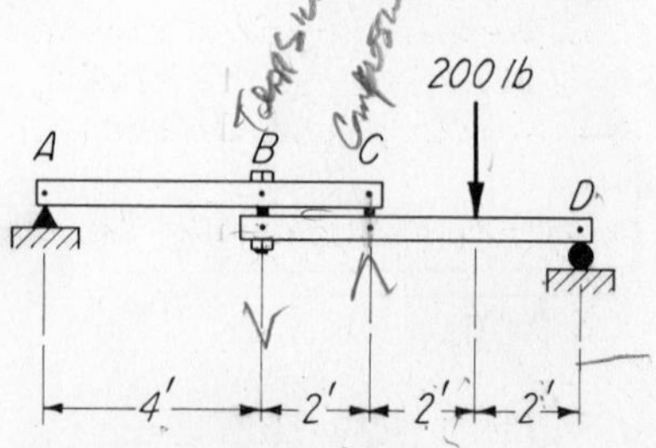

FIG. 3-31 Probs. 3-34 to 3-36.

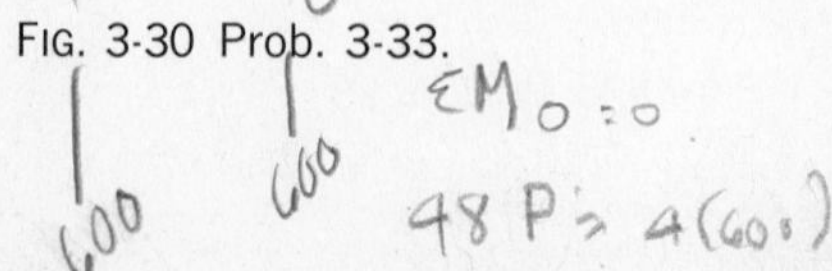

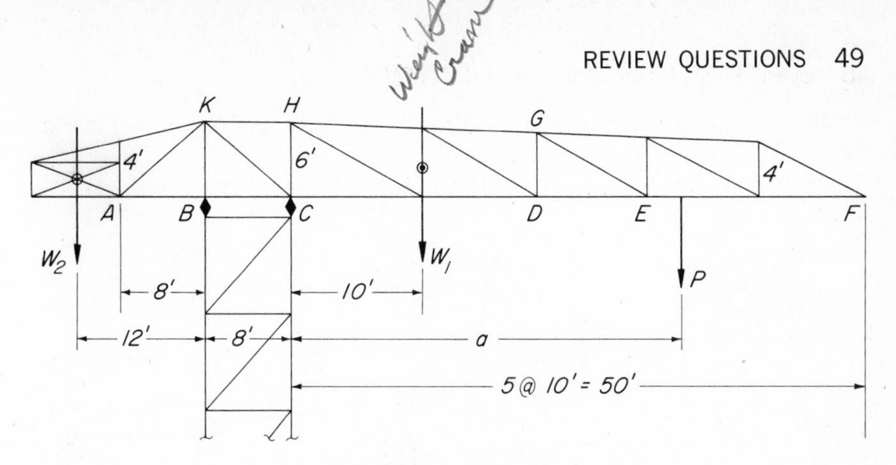

FIG. 3-32 Prob. 3-37.

3-32. In Fig. 3-29 draw separate *free-body diagrams* of the two horizontal levers; then compute and show values of all forces acting on each lever.

3-33. The solid, rectangular stone block shown in Fig. 3-30 weighs 1,200 lb. Compute the vertical force P required barely to lift the right end off the horizontal supporting surface by using the bar as shown. *Ans.* $P = 50$ lb

3-34. Two 6-ft planks are held together by a single bolt at B, as shown in Fig. 3-31. Compute the tensile force T in the bolt. Spacers are used at B and C. Show complete free-body diagrams.

3-35. If the bolt at B in Fig. 3-31 will safely stand 420 lb in tension, what maximum additional downward-acting vertical load P may be placed at C?

3-36. Add an 80-lb load to the beam shown in Fig. 3-31 midway between A and B. Calculate the forces acting at A, B, C, and D due to the two loads.

3-37. The dead weight W_1 of the revolving crane shown in Fig. 3-32, but not including the counterweight W_2, is 6 kips. (*a*) If the reaction at C is zero when the load P is zero, what is the value of W_2? (*b*) If $a = 50$ ft, the reaction at B is zero, and the counterweight W_2 is as calculated in (*a*), what is the value of P? (*c*) Repeat part (*b*) if a is limited to 20 ft.

REVIEW QUESTIONS

3-1. What is the magnitude of the resultant of a system of parallel forces? What is known about its line of action?

3-2. What is meant by the center of a system of parallel forces?

3-3. By what principle of moments may the line of action of the resultant of a system of parallel forces be located?

3-4. At what point is the resultant of a distributed load assumed to be concentrated?

3-5. Define the locations of the centroids (*a*) of a rectangle and (*b*) of a triangle.

CHAPTER 4

Coplanar, Concurrent Force Systems

4-1 Introduction

In a coplanar, concurrent force system the lines of action of all forces lie in one plane and intersect at a common point, the *point of concurrence.* This type of force system is most commonly encountered in problems involving the determination of stresses in members of trusses and also, as is shown later in this chapter, in certain problems to determine support reactions.

4-2 Resultants of Coplanar, Concurrent Force Systems

Two or more coplanar, concurrent forces may be combined into a single resultant force by three methods: (1) algebraically by summations of rectangular components, as in Prob. 2-5, and graphically (2) by the parallelogram method or (3) by the triangle method, both of which are discussed in Art. 2-8 and are illustrated in Fig. 2-11 *a* and *b*, respectively. *The resultant of a concurrent force system is fully defined when its magnitude, sense, and the direction of its line of action are known.* It always acts through the point of concurrence.

ILLUSTRATIVE PROBLEM

4-1. Determine the resultant force R of the forces P, Q, and S, shown in Fig. 4-1*a*, when $P = 6$ lb, $Q = 7$ lb, and $S = 10$ lb. Find also the angle θ_H it makes with the horizontal. (The algebraic solution by summations of rectangular components is similar to that shown in Prob. 2-5. Hence, only the graphical solutions are shown here.)

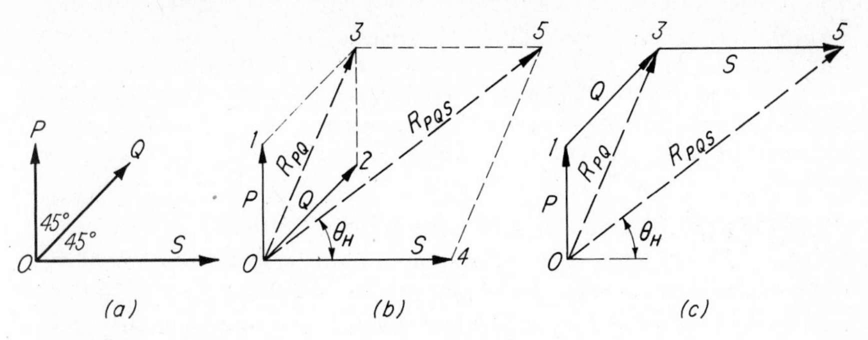

FIG. 4-1 Graphical determination of resultant force. (*a*) Space diagram. (*b*) Parallelogram method. (*c*) Triangle method.

a. Solution by the Parallelogram Method: The three forces are drawn to scale, each in its proper direction and sense, as shown in Fig. 4-1*b*. We obtain the resultant R_{PQ} of *P* and *Q* by completing the parallelogram 0–1–3–2–0. This resultant is then combined with force S by completing the second parallelogram 0–3–5–4–0. The diagonal 0–5 is then the resultant R_{PQS} of forces *P*, *Q*, and S. The angle θ_H may be scaled with a protractor, or its tangent may be scaled by the method explained in Art. 2-8.

b. Solution by the Triangle Method: If the three forces *P*, *Q*, and S are laid out to scale *end to end*, as in Fig. 4-1*c*, the diagonal 0–3 is the resultant of forces *P* and *Q*, and 0–5 is the resultant R_{PQS} of R_{PQ} and S, or of *P*, *Q*, and S. Intermediate resultants such as R_{PQ} are generally omitted in this method. Note that in both methods, the polygons 0–1–3–5–0 are identical. *Ans.* $R = 18.5$ lb ↗; $\theta_H = 36°13'$

PROBLEMS

4-2. By the parallelogram method determine the resultant *R* of the forces shown in Fig. 4-2 and its direction angle θ_H with the horizontal. Check by the triangle method. (Scale on 8½ by 11: 1 in. = 5 kips.)

4-3. Solve for the resultant *R* of the three forces shown in Fig. 4-3 by the algebraic summations of components. (See Prob. 2-5.) Find also its direction angle θ_H with

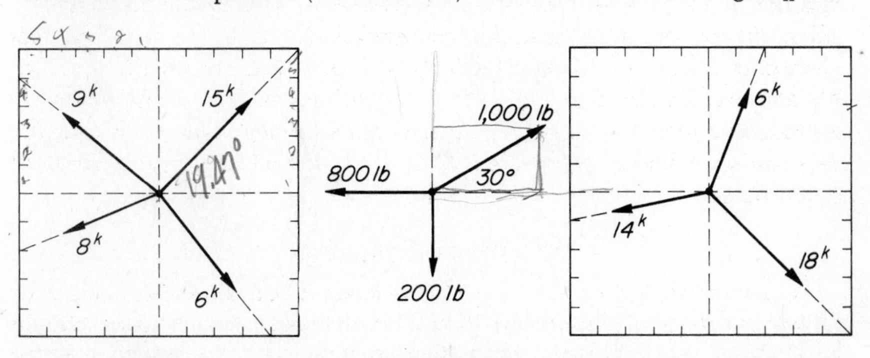

FIG. 4-2 Prob. 4-2. FIG. 4-3 Prob. 4-3. FIG. 4-4 Prob. 4-4.

the horizontal. Check graphically by the triangle method. (Scale on $8\frac{1}{2}$ by 11: 1 in. = 200 lb.) *Ans.* $R = 307$ lb ↗; $\theta_H = 77°36'$

4-4. Determine the resultant R of the forces shown in Fig. 4-4 by the parallelogram method. Find also its direction angle θ_H with the horizontal. Check by the triangle method. (Scale on $8\frac{1}{2}$ by 11: 1 in. = 5 kips.)

4-3 Equilibrium of Coplanar, Concurrent Force Systems

When a system of forces is in equilibrium, the algebraic sums of their vertical and horizontal components must, respectively, equal zero. Since all forces of a concurrent system, acting upon a body at rest, act through the same point, they cannot cause rotation of the body. Therefore, the unknown elements of a concurrent force system may be determined entirely by the following equations of force equilibrium: $\Sigma V = 0$ and $\Sigma H = 0$.

In Fig. 4-5 is shown the graphical equivalent of the algebraic summations obtained by $\Sigma V = 0$, and $\Sigma H = 0$. The four forces P, Q, S, and T are in equilibrium. Their resultant, therefore, is zero. Consequently, when the forces are laid out to scale, end to end, *the resulting force polygon must close.* Further proof of this fact is indicated by the vertical and horizontal components, shown above and to the right of the force polygon. A glance indicates that their algebraic sums are zero. Thus, **for a system of forces in equilibrium, the force polygon must be closed.** This principle is known as the **polygon law.** In such a polygon *the arrowheads will always follow each other,* regardless of the order in which the forces are laid out.

A closed force polygon, however, denotes only free equilibrium and not necessarily moment equilibrium. If, for example, force T is displaced, as shown in Fig. 4-5*c*, force equilibrium still exists, and the force polygon remains unchanged, but the forces are no longer in moment equilibrium.

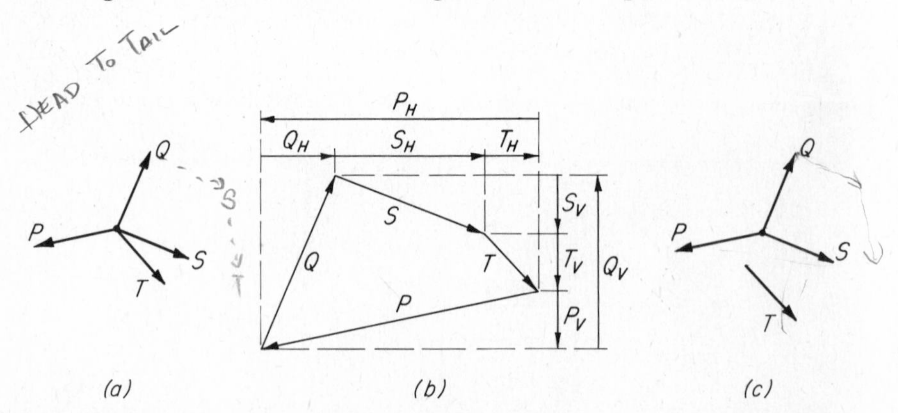

FIG. 4-5 The graphical force polygon. (*a*) System of forces in complete equilibrium. (*b*) Graphical force polygon showing equilibrium of forces. (*c*) System in force equilibrium only.

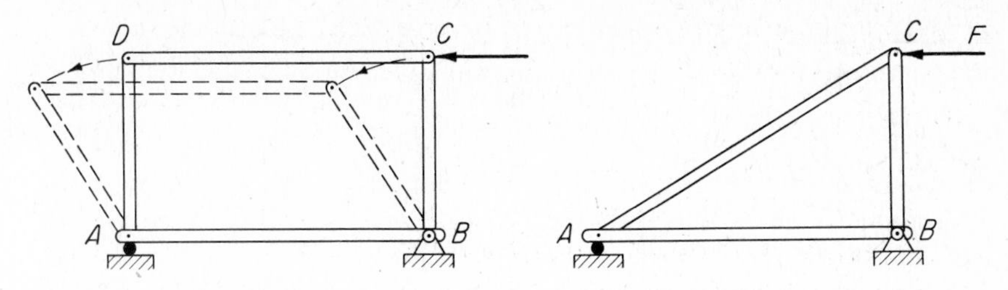

FIG. 4-6 Unstable frame.

FIG. 4-7 Stable frame.

4-4 Trusses

The primary function of a structure is to support loads and to transmit these loads from their points of application through the various parts or members of the structure to its supports. To do this job efficiently requires careful arrangement of the members.

In Fig. 4-6, four members are fastened together at their ends by bolts. This frame, however, is *unstable,* because tightening the bolts sufficiently to prevent the indicated collapse would be difficult. A diagonal member from A to C would obviously overcome the instability. Should the members be so arranged as to form a *triangle,* as is shown in Fig. 4-7, a *stable* structure is obtained, called a truss. When several loads are to be carried, several triangular units are connected together, as in Fig. 4-8.

A truss is a structural unit in which the bars are arranged to form one or more connected triangles. **A joint** is the connection between two or more bars. **A member** is a bar extending from joint to joint.

4-5 Stresses in Members of Trusses

The forces (loads and reactions) acting on a truss are generally applied at the joints and cause **internal forces** or **stresses** to be produced in the various

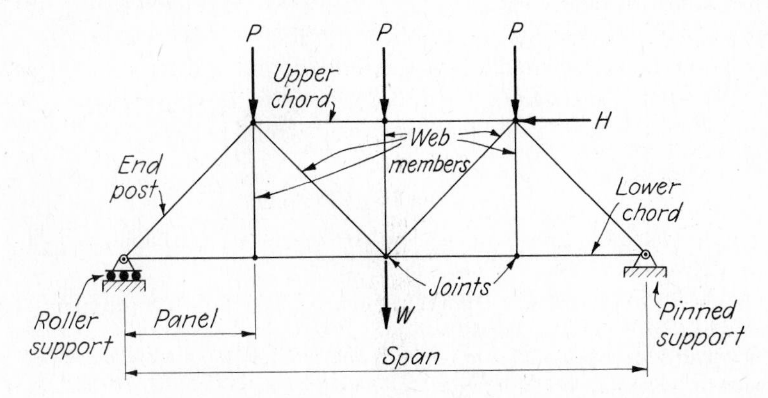

FIG. 4-8 Typical pin-connected truss.

members. Unless the members are rather rigidly connected at the joints, for the purpose of stress analysis they are considered to be connected with smooth, frictionless pins. Such trusses are said to be **pin-connected.** Both ends of each member are presumed to be free to rotate about the connecting pins, and bending in any one member therefore cannot be transmitted to other members. For simplicity of analysis and with no appreciable error, common types of roof and bridge trusses made of timbers and steel rods, and relatively light roof trusses and transmission towers made of steel angles riveted together at the joints, are considered as being pin-connected.

A two-force member is one on which *forces act at two points only,* usually at its ends. When isolated as a free body, a two-force member is then in equilibrium under the action of two forces, one at each end; these forces therefore must be equal, opposite, and collinear. Their common line of action clearly passes through the centers of the two pins and, in a straight member, coincides with the axis of the member. Hence, *when a two-force member is cut, the force (stress) within it is known to act along its axis.* When all forces on a truss are applied at its joints, all members are two-force members.

Type of Stress. If the stress *pulls* on the member, or stretches it, it is called a **tensile stress** and the member is said to be in **tension.** Conversely, if the stress *pushes* on the member, or compresses it, it is called a **compressive stress** and the member is in **compression.** Tension and compression are referred to as *types of stresses* and are usually denoted by T and C. Stresses in two-force members are also called **axial** or **direct stresses,** as distinguished from *bending stresses.*

Axial stresses may be determined by various methods, such as (1) the *geometric,* based on the geometric similarity between the force and space triangles in certain problems, as in Prob. 4-5, and (2) the *trigonometric,* using the sine law, as in Prob. 4-9. The practical applications of these two methods are limited to special cases involving only three forces at a joint. The more general and practical methods, such as the *analytical method of joints* and the *graphical method of joints,* are fully outlined in later articles.

Notation. In analytical solutions, supports and joints of trusses are generally designated by capital letters. A member, as well as the stress within it, is then identified by the two letters appearing at its ends.

ILLUSTRATIVE PROBLEM

4-5. Determine the stresses in members AC and BC of the simple truss shown in Fig. 4-9*a*, caused by a load W of 60 lb.

Solution by Geometry: The two members are cut near joint C, and that joint is isolated as shown in the free-body diagram, Fig. 4-9*b*. The known force W and the unknown stresses AC and BC form a concurrent system in equilibrium whose force polygon (or diagram), shown in Fig. 4-9*c*, then must close. The triangles of the space

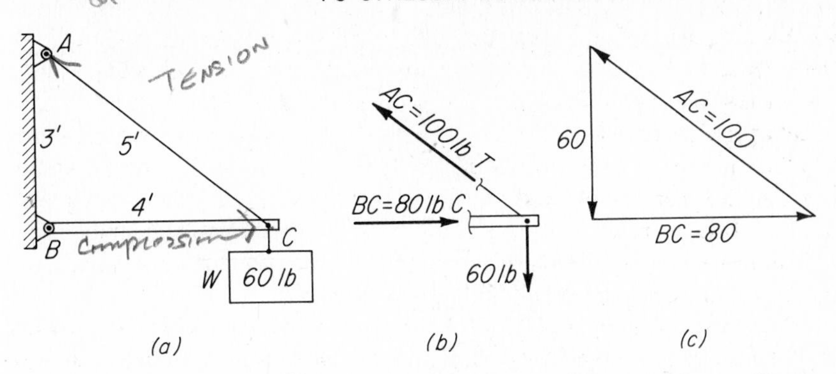

FIG. 4-9 Stresses in truss members by the geometric method. (*a*) Space diagram. (*b*) Free-body diagram. (*c*) Force diagram.

and force diagrams are clearly similar, since corresponding sides are parallel. Therefore, by proportion,

$$\frac{60}{3} = \frac{BC}{4} = \frac{AC}{5}$$

or $\quad AC = \frac{5}{3}(60) = 100 \text{ lb T} \quad$ and $\quad BC = \frac{4}{3}(60) = 80 \text{ lb C} \qquad$ *Ans.*

In the force diagram, the sense of W is known and, because *the arrows must follow each other around the diagram*, BC is seen to be compressive and AC to be tensile. (NOTE: If the space and force diagrams are drawn to scale, a complete graphical solution is obtained, thus providing a convenient check.)

PROBLEMS

4-6. The truss of Fig. 4-7 has dimensions $AB = 15$ ft, $BC = 8$ ft, and $AC = 17$ ft. The force F is 300 lb. Solve by geometry for the stresses AC and AB.

4-7. Determine by geometry the stresses in members AC and BC of the truss shown in Fig. 4-10. *Ans.* $AC = 250$ lb T; $BC = 150$ lb C

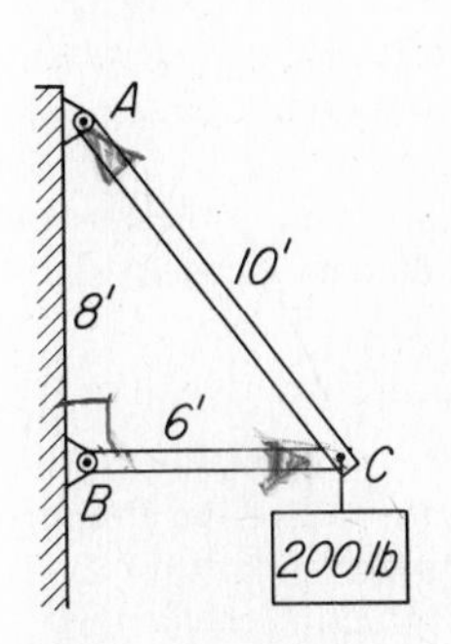

FIG. 4-10 Prob. 4-7.

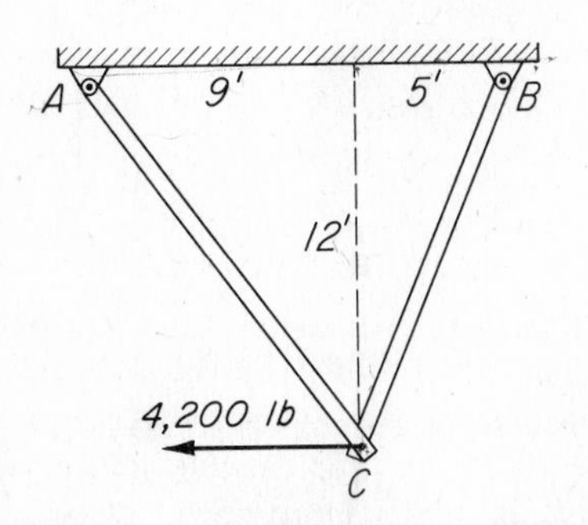

FIG. 4-11 Prob. 4-8.

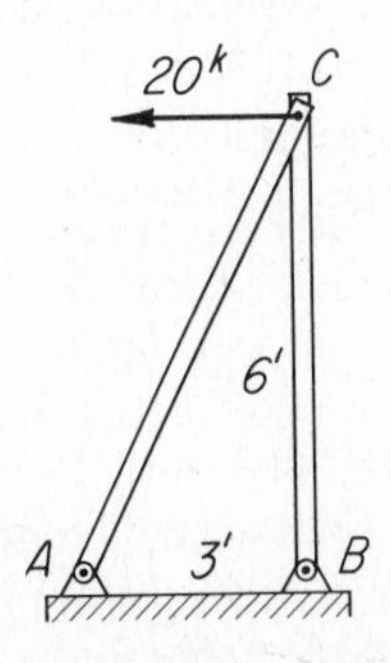

FIG. 4-12 Prob. 4-9.

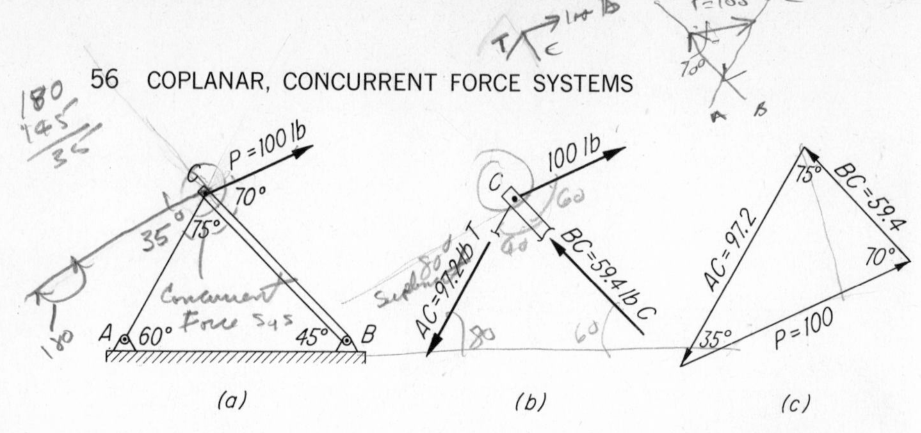

FIG. 4-13 Stresses in truss member by trigonometry. (*a*) Space diagram. (*b*) Free-body diagram. (*c*) Force diagram.

4-8. Solve by geometry for the stresses *AC* and *BC* in the truss shown in Fig. 4-11.

4-9. Find the stresses in *AC* and *BC* in Fig. 4-12, by geometry.
Ans. $AC = 44.7$ kips C; $BC = 40$ kips T

ILLUSTRATIVE PROBLEM

4-10. In members *AC* and *BC* of the truss in Fig. 4-13*a*, solve by trigonometry for the stresses caused by the applied force *P*.

Solution by Trigonometry: When the two members are cut, the joint may be isolated as a free body, as in Fig. 4-13*b*. A force diagram is then constructed (Fig. 4-13*c*) in which the forces *P*, *AC*, and *BC* are drawn parallel, respectively, to force *P* and members *AC* and *BC*, as shown in the space diagram. From the given data, we determine the angles of the force triangle (with great care). Now, with one side and all angles known, by the law of sines, we have

$$\frac{AC}{\sin 70^\circ} = \frac{BC}{\sin 35^\circ} = \frac{100}{\sin 75^\circ} \quad \text{or} \quad \frac{AC}{0.9397} = \frac{BC}{0.5736} = \frac{100}{0.9659}$$

Hence
$$AC = \frac{(100)(0.9397)}{0.9659} = 97.2 \text{ lb T} \qquad \textit{Ans.}$$

and
$$BC = \frac{(100)(0.5736)}{0.9659} = 59.4 \text{ lb C} \qquad \textit{Ans.}$$

The sense of force *P* is known, and the senses of *AC* and *BC*, as found in the force diagram where the arrowheads always follow each other, indicate, respectively, tension and compression.

PROBLEMS

4-11. Solve by trigonometry for the stresses in members *AC* and *BC* of the frame shown in Fig. 4-14.
Ans. $AC = 89.7$ lb T; $BC = 73.2$ lb C

4-12. Calculate by trigonometry the force in the spring and the link in Fig. 4-15.

4-13. Find the stresses *AC* and *BC* in Fig. 4-16, by trigonometry.
Ans. $AC = 153.2$ lb C; $BC = 134.6$ lb T

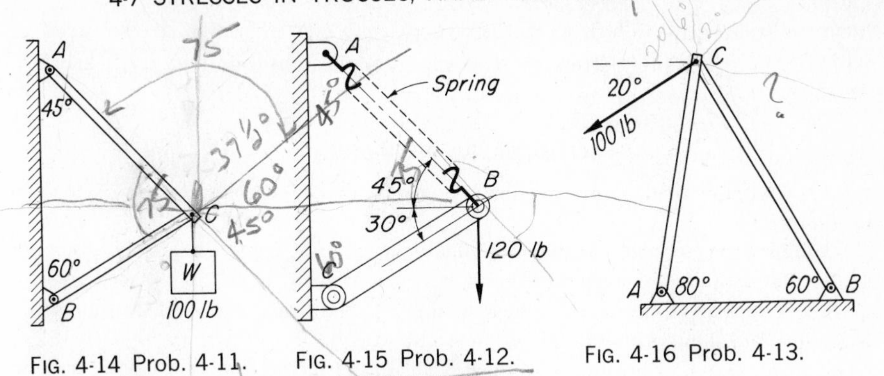

FIG. 4-14 Prob. 4-11. FIG. 4-15 Prob. 4-12. FIG. 4-16 Prob. 4-13.

4-6 Ropes over Sheaves and Pulleys

Sheaves are used to change the direction of a force applied at one end of a rope or flexible cable. They are widely used in hoisting equipment. When a load is applied to a joint of a frame or truss through a sheave, as in Fig. 4-17*a*, **the sheave may be removed and the forces may be applied directly at the joint without change in total effect,** as is indicated in Fig. 4-17*b*. The proof is shown in Fig. 4-17*c*.

The relatively small pin friction is usually neglected. Then T and W are equal, and their resultant R passes through the pin O, where it may again be resolved into the forces T and W. For convenience, T may be shown as pushing on the pin, without change in its point of application.

4-7 Stresses in Trusses; Analytical Method of Joints

Although applicable to all trusses, regardless of size, the analytical method of joints is most conveniently applied to trusses in which most members are at right angles to each other, as in trusses having parallel chords. The joints

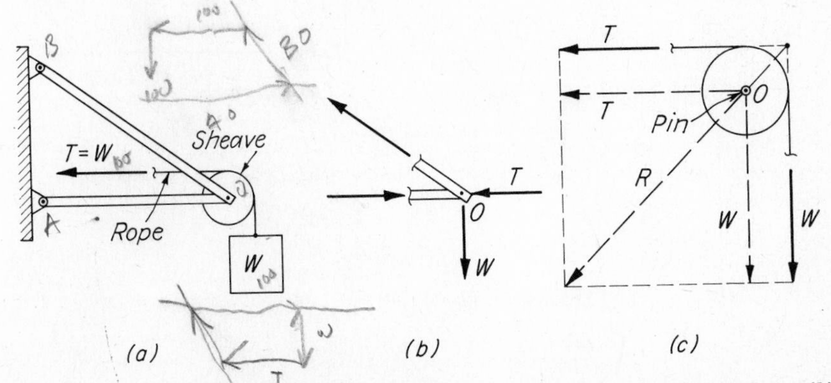

FIG. 4-17 Forces T and W may be applied at pin. (*a*) Rope over sheave. (*b*) Free-body diagram. (*c*) Forces applied at pin.

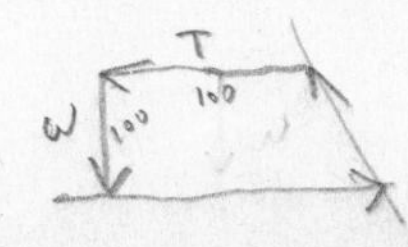

must be isolated in such order that no more than two unknown stresses need be found at any joint. A stress, or its component, is then found by application of the two equations of force equilibrium, $\Sigma V = 0$ and $\Sigma H = 0$.

ILLUSTRATIVE PROBLEMS

4-14. Compute the stresses in members AC and BC of the pin-connected truss shown in Fig. 4-18*a*.

Solution: First, we cut members AC and BC near joint C and isolate that joint. To facilitate summations of vertical and of horizontal forces, BC is replaced with its V and H components, as is shown in the free-body diagram. A *slope triangle* indicating the slope of its line of action is also drawn.

We begin the solution with a summation of forces in the direction having only one unknown force, here the vertical direction. Then, assuming BC_V to be upward-acting, as clearly it is, we have

$[\Sigma V = 0]$ $$BC_V = 600 \text{ lb}$$

If now we imagine BC_H to be placed in the position of the bottom side of the parallelogram of forces, we find, by similarity of the force and slope triangles, that

$$\frac{BC_V}{3} = \frac{BC_H}{4} = \frac{BC}{5} = \frac{600}{3} = 200$$

Hence $$BC_H = (4)(200) = 800 \text{ lb}$$

and $$BC = (5)(200) = 1{,}000 \text{ lb C} \qquad \textit{Ans.}$$

Then, if AC is assumed to be a tensile stress,

$[\Sigma H = 0]$ $$AC = 800 \text{ lb T} \qquad \textit{Ans.}$$

4-15. Solve for the stresses in the four members of the pin-connected derrick shown in Fig. 4-19*a* by the analytical method of joints.

Solution: A free-body diagram of the derrick is shown in Fig. 4-19*b*. The sheave is

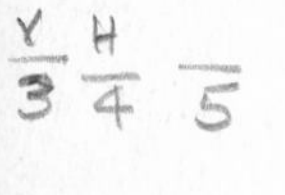

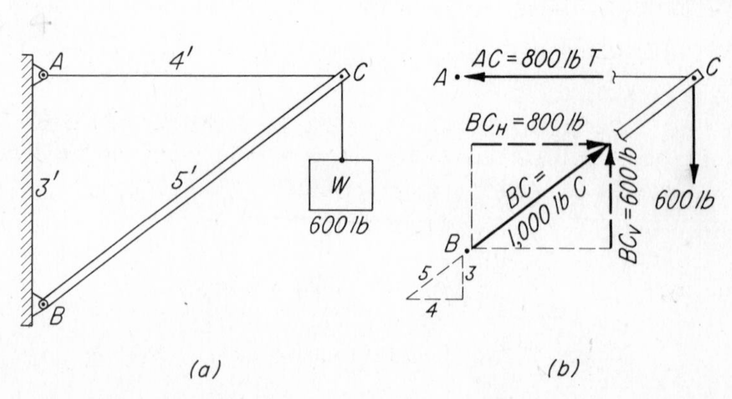

FIG. 4-18 Stresses by the analytical method of joints. (*a*) Space diagram. (*b*) Free-body diagram.

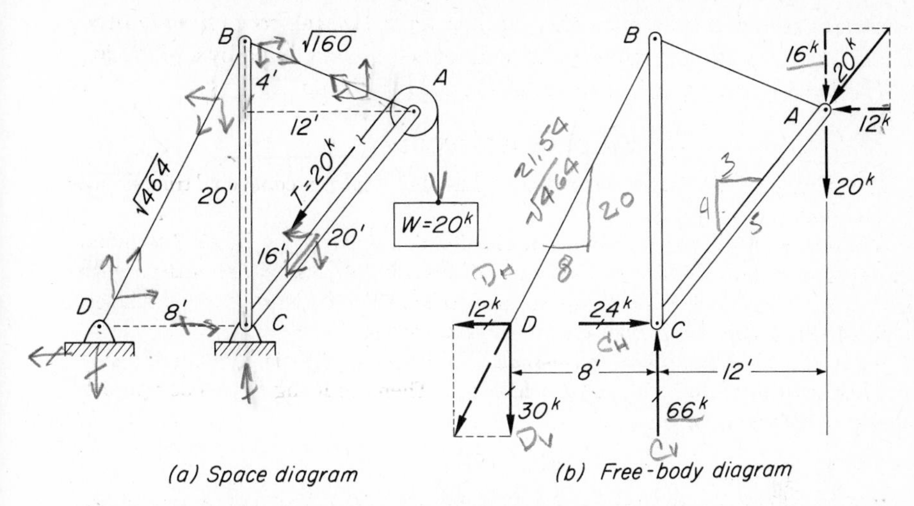

FIG. 4-19 Stresses in derrick by analytical method of joints. (*a*) Space diagram. (*b*) Free-body diagram.

removed, and the two 20-kip forces are applied directly at the pin (see Art. 4-6). We may take moments about C to find the vertical component of the reaction at D.

$[\Sigma M_C] = 0 \qquad 8(D_V) = 12(20) \qquad \text{or} \qquad D_V = 30 \text{ kips}$

By geometry, $\dfrac{D_H}{8} = \dfrac{30}{20} \qquad \text{or} \qquad D_H = 12 \text{ kips}$

The reaction components at C can now be found from equilibrium.

$[\Sigma V = 0] \qquad C_V = 30 + 20 + 16 = 66 \text{ kips}$

$[\Sigma H = 0] \qquad C_H = 12 + 12 = 24 \text{ kips}$

The stress BD can be found from geometry,

$$\frac{BD}{\sqrt{464}} = \frac{12}{8} \qquad \text{or} \qquad BD = 32.4 \text{ kips (T)} \qquad \textit{Ans.}$$

Joint B may be isolated as shown in Fig. 4-20*a*. The horizontal component of the stress in member AB can be found from equilibrium. (AB is shown as a tensile force, which it obviously is.)

$[\Sigma H = 0] \qquad AB_H = 12 \text{ kips}$

By geometry, $\dfrac{AB_V}{4} = \dfrac{12}{12} \qquad \text{or} \qquad AB_V = 4 \text{ kips}$

and $\dfrac{AB}{\sqrt{160}} = \dfrac{4}{4} \qquad \text{or} \qquad AB = 12.65 \text{ kips (T)} \qquad \textit{Ans.}$

The force in the vertical member can now be found from

$[\Sigma V = 0] \qquad BC = 30 + 4 = 34 \text{ kips (C)} \qquad \textit{Ans.}$

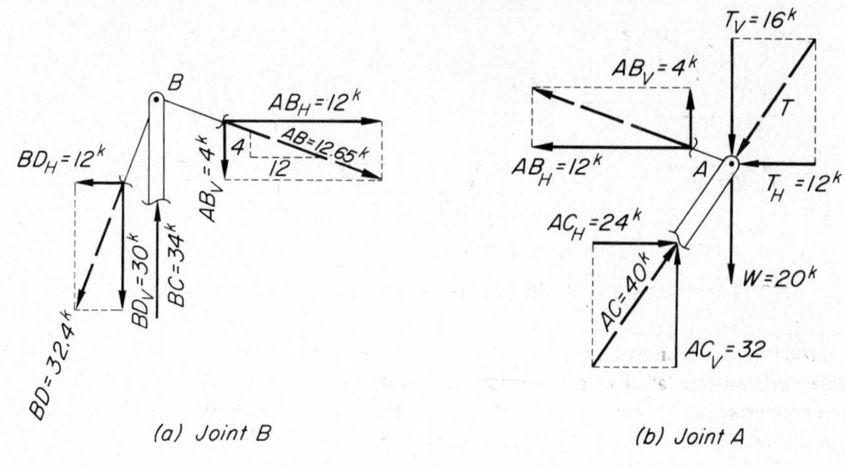

FIG. 4-20 Joint free-body diagrams. (*a*) Joint *B*. (*b*) Joint *A*.

The force components in member *AC* may now be found from the free-body diagram of joint *A* (see Fig. 4-20*b*). *AC* was assumed to be compressive.

$[\Sigma V = 0] \qquad AC_V = 20 + 16 - 4 = 32 \text{ kips}$

$[\Sigma H = 0] \qquad AC_H = 12 + 12 = 24 \text{ kips}$

By geometry, $\frac{AC}{20} = \frac{32}{16}$ or $AC = 40$ kips (C) *Ans.*

When all forces are shown on the free-body diagrams, a glance will show that equilibrium exists. The true senses of the forces were correctly assumed, since all the computed stresses are positive (see Art. 2-18).

4-16. Determine the stresses in all members of the Howe roof truss shown in Fig. 4-21.

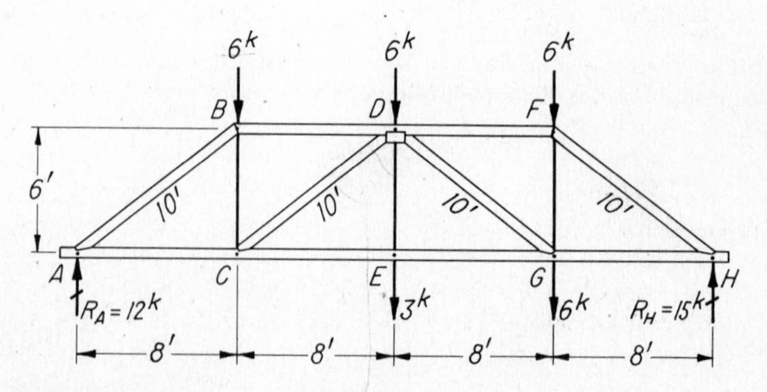

FIG. 4-21 Space diagram. Stresses by the analytical method of joints.

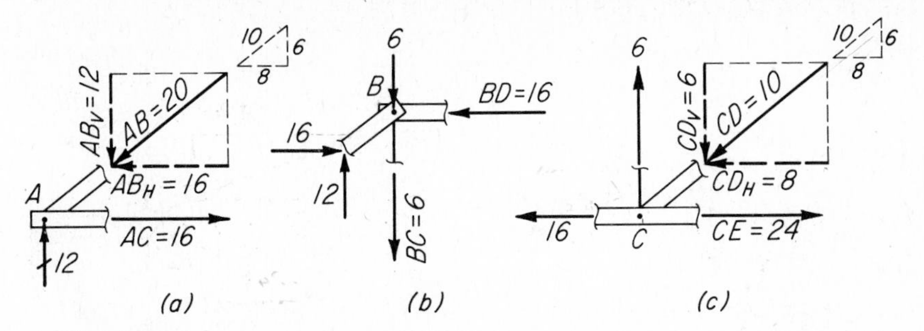

FIG. 4-22 Free-body diagrams. (*a*) Joint *A*. (*b*) Joint *B*. (*c*) Joint *C*.

Solution: The general procedure is similar to that outlined in Probs. 4-14 and 4-15. The joints are isolated in such order that no more than two unknown stresses are to be determined at any joint. Most of the stresses to be determined are dependent upon other stresses previously found, and no positive check on results obtained is available until the last joint is solved. Hence, to minimize cumulative inaccuracies and chances of error, the joints should be solved from each end to the middle, or in the order *A, B, C, E,* and then *H, F, G, D.* Then we obtain a check from a complete free-body diagram of the last joint *D* showing the forces there to be in equilibrium.

The free-body diagram of joint *A* is shown in Fig. 4-22*a*. From $\Sigma V = 0$, AB_V is found. By similarity of slope and force triangles, *AB* and AB_H are obtained, and *AC* is then found by $\Sigma H = 0$. At joint *B*, shown in Fig. 4-22*b*, stresses *BC* and *BD* are unknown. *BC* is found by $\Sigma V = 0$, and *BD* by $\Sigma H = 0$. At joint *C*, CD_V is found by $\Sigma V = 0$. *CD* and CD_H are obtained by similarity of triangles, and *CE* by $\Sigma H = 0$. When more than one or two joints are to be solved, we find it helpful to record the computed stresses on a *stress-summation diagram* as in Fig. 4-23. (Students are urged to complete this problem.)

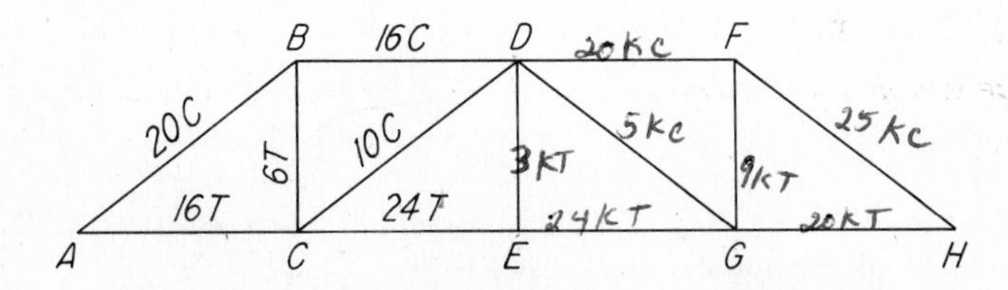

FIG. 4-23 Stress-summation diagram.

PROBLEMS

Stresses in trusses: the analytical method of joints

4-17. Solve for the stresses in members *AC* and *BC* of the truss shown in Fig. 4-24.

Ans. $AC = 340$ lb T; $BC = 160$ lb C

4-18. In Fig. 4-25, determine the stresses in members *AC* and *BC*.

4-19. A flexible hoisting rope passes over a sheave at joint *C* of the truss shown in Fig. 4-26. Find the resulting stresses in members *AC* and *BC*.

Ans. $AC = 30$ kips T; $BC = 30$ kips T

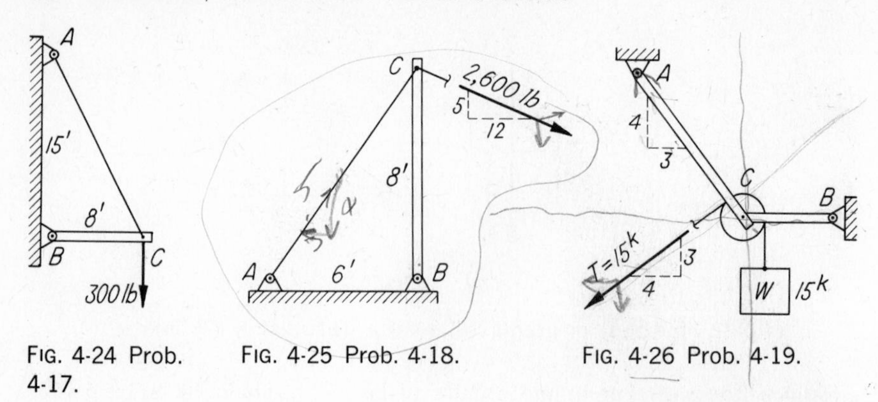

FIG. 4-24 Prob. 4-17.

FIG. 4-25 Prob. 4-18.

FIG. 4-26 Prob. 4-19.

4-20. Determine the stresses in members *AC*, *AD*, and *CD* of the roof truss shown in Fig. 4-27.

4-21. Compute the stresses in all members of the truss shown in Fig. 4-28. Assume pinned joints at *B* and *C*. Draw a stress summation diagram.

Ans. $AB = BC = 7.5$ kips C; $CD = 6$ kips C; $AE = 12.5$ kips T; $EF = 6$ kips T; $FD = 10$ kips T; $BE = 12$ kips C; $CF = 8$ kips C; $CE = 2.5$ kips T

FIG. 4-27 Prob. 4-20.

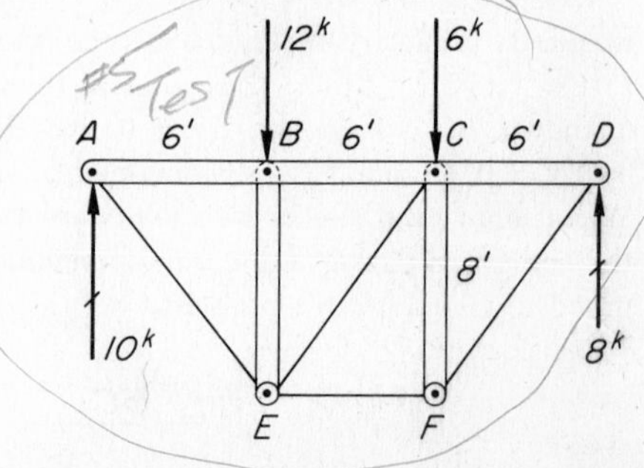

FIG. 4-28 Prob. 4-21.

4-22. Find the value of *W* in Fig. 4-29 required to produce a stress of 3, 250 lb in member *AB*. Determine also the stresses in *BC*, *BE*, and *CD*.

4-23. In Fig. 4-30, a power-hoist cable, passing over sheaves at *C* and *D*, is supported by four coplanar members as shown. Find the stresses in these four members due to the 36-kip load. Neglect all cable and sheave friction.

Ans. $AC = 18$ kips C; $BC = 20$ kips C; $DE = 52$ kips C; $DF = 30$ kips C

4-24. The frame shown in Fig. 4-31 supports two 4-kip loads as shown. Find the vertical and horizontal components of the reactions at *A* and *E*. Then find the stresses in the members by the method of joints. Solve joints in order *A*, *E*, *B*, and *D*. Check answers by use of joint *C*.

Partial Ans. $A_H = 8$ kips←; $A_V = 4$ kips↑; $BD = 8$ kips C; $CD = 20$ kips C

4-25. Find the stresses in the truss shown in Fig. 4-31. Assume that the 4-kip load at *D* acts toward *F* along the line *DF*.

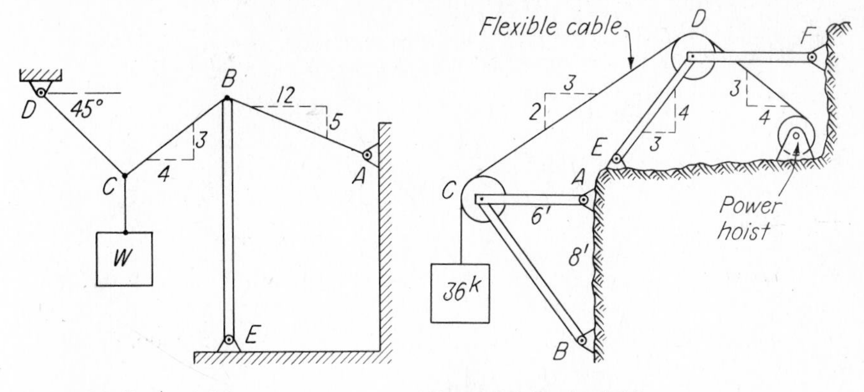

FIG. 4-29 Prob. 4-22.

FIG. 4-30 Prob. 4-23.

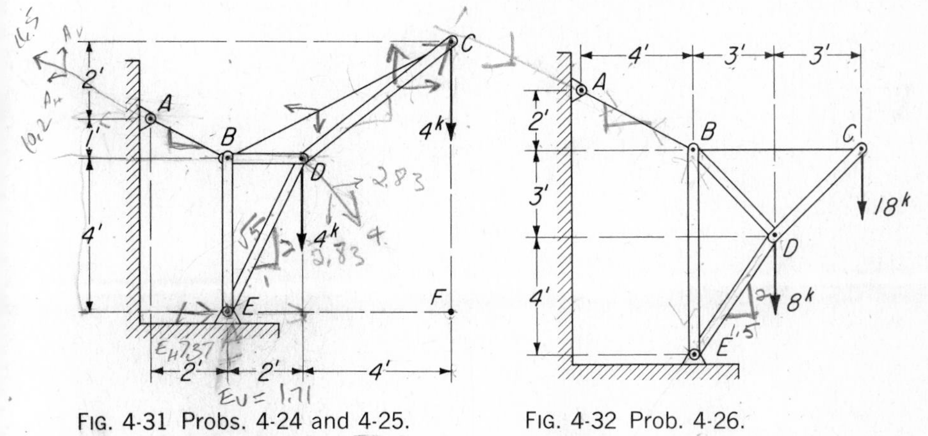

FIG. 4-31 Probs. 4-24 and 4-25.

FIG. 4-32 Prob. 4-26.

Ans. $AB = 11.5$ kips T; $BC = 17.9$ kips T; $CD = 20$ kips C; $BD = 5.76$ kips C; $BE = 13.12$ kips T; $DE = 16.6$ kips C

4-26. The truss shown in Fig. 4-32 supports the two vertical loads shown. Find the vertical and horizontal components of the reactions at A and E. Then find the stresses in all members by the method of joints.

4-27. The truss shown in Fig. 4-33 supports the two vertical loads shown. Find the stresses in all truss members by the analytical method of joints procedure.

Partial Ans. $BC = 30$ kips C; $BD = 24$ kips C; $CE = 40$ kips T

4-28. Find the stresses in all members of the truss shown in Fig. 4-34. Use the analytical method of joints.

4-29. Determine the stresses in all members of the water-tower truss shown in Fig. 4-35. Draw a stress-summation diagram. (The reacting forces at supports E and F are $E_H = 9$ kips←, $E_V = 4$ kips↑, and $F_V = 44$ kips↑.)

Ans. $AB = 6$ kips C; $AC = 24$ kips C; $BC = 10$ kips T; $BD = 32$ kips C; $CE = 16$ kips C; $CD = 9$ kips C; $DE = 15$ kips T; $DF = 44$ kips C

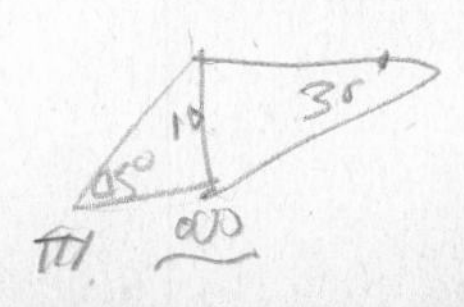

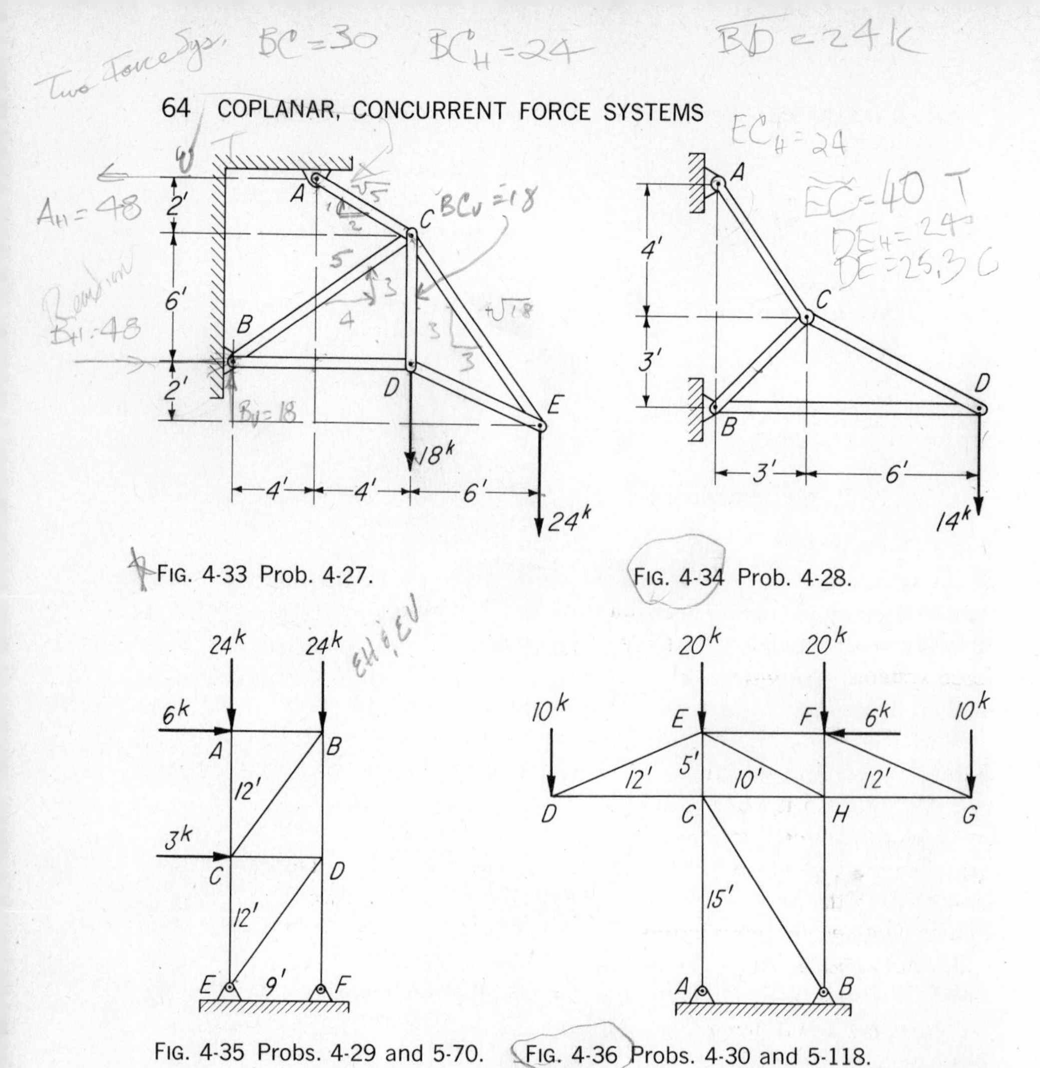

FIG. 4-33 Prob. 4-27.

FIG. 4-34 Prob. 4-28.

FIG. 4-35 Probs. 4-29 and 5-70.

FIG. 4-36 Probs. 4-30 and 5-118.

4-30. The loading-platform truss shown in Fig. 4-36 is subjected to vertical dead loads and to a horizontal wind load, as indicated. Compute the stresses in all members. Draw a stress-summation diagram. (The reacting forces at supports A and B are $A_V = 42$ kips↑, $B_V = 18$ kips↑, and $B_H = 6$ kips→.)

4-8 Stresses in Trusses; the Graphical Method of Joints

This method is the graphical equivalent of the analytical method of joints described in the previous article. In this method we use *the closed graphical force polygon,* which *constitutes a geometric summation of forces in equilibrium,* as was described and illustrated in Art. 4-3 and Fig. 4-5, respectively. In addition to the space and force diagrams, carefully drawn to scale, a free-body diagram of the isolated joint is often helpful, since it shows clearly

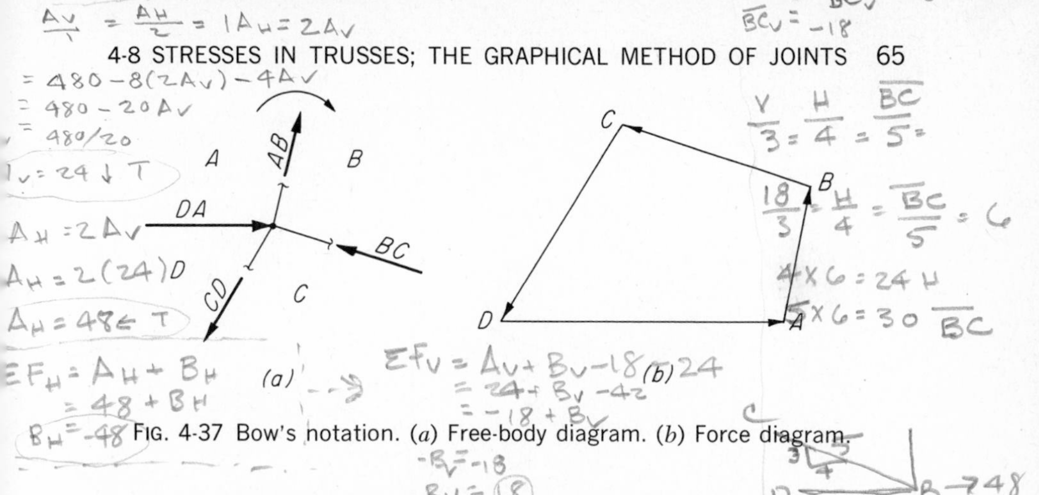

FIG. 4-37 Bow's notation. (*a*) Free-body diagram. (*b*) Force diagram.

all forces acting at the joint. Also, this diagram assists greatly in determining tension and compression. One distinct advantage of this graphical method over the analytical method is that geometric summations of all coplanar, concurrent force systems, *regardless of the direction of the forces*, are made with equal facility.

Bow's Notation. In solutions using the graphical force polygon, a special system of notation, referred to as *Bow's notation* after its originator is universally used and is, indeed, indispensable. In this system, *letters are placed in the spaces between external forces and in each triangular panel of a truss*, generally in a clockwise order around a truss or a joint. Each force and member, and the force within each member, will then lie between two letters and is identified by those letters.

For example, in Fig. 4-37*a* is shown the free-body diagram of an isolated joint. The letters *A*, *B*, *C*, and *D* are placed in the spaces *between* the forces, *usually in clockwise order around the joint*, as is indicated by the curved arrow. The forces are now designated *AB*, *BC*, *CD*, and *DA*. In the force diagram, however, the letters are placed *at the ends* of the lines representing the forces.

ILLUSTRATIVE PROBLEMS

4-31. Determine graphically the stresses in members *CD* and *DA* of the truss shown in Fig. 4-38*a*.

Solution: The space diagram is carefully drawn to scale *with the members shown as single lines*, as in Fig. 4-38*a*. We then cut the two members, isolate the joint as a free body, as in Fig. 4-38*b* and insert the letters *A*, *B*, *C*, and *D* in the spaces between the four forces. The curved arrow indicates that the forces will be considered in *clockwise order* around the joint; that is, first the known forces *AB* and *BC*, then the unknown forces *CD* and *DA*.

In the force diagram, to some convenient scale, *AB* and *BC* are laid off *end to end* in their true directions and magnitudes and with correct senses shown. Here the two letters designating a force now appear at its ends. Because these four forces

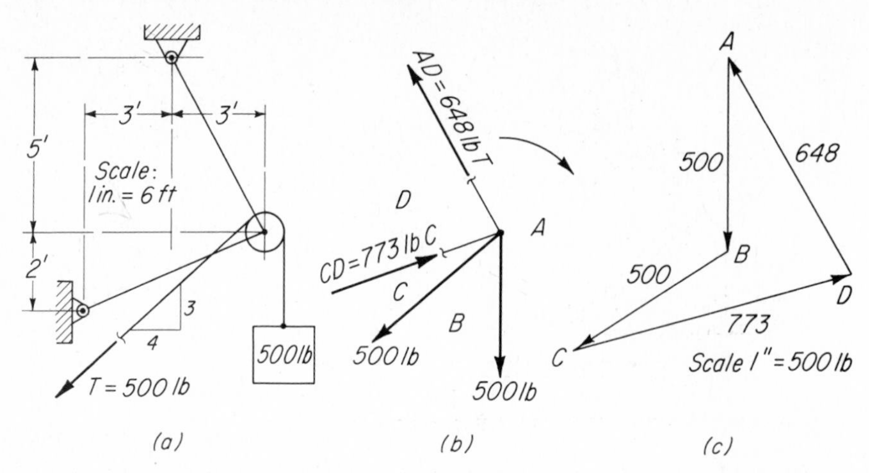

FIG. 4-38 Stresses by the graphical method of joints. (*a*) Space diagram. (*b*) Free-body diagram. (*c*) Force diagram.

are in equilibrium, the force diagram must close. It is closed with a line from *C* *parallel to member CD* and a line from *A* *parallel to member DA*, thus locating the unknown intersection *D*. The magnitude of *CD* is sealed from *C* to *D*, that of *DA* from *D* to *A*. The sense of each force is readily determined, because the *arrows must follow each other around the polygon.* These values and senses are then recorded on the free-body diagram. *CD* scales 773 lb and is seen to be compressive; *DA* scales 648 lb and is tensile.

4-32. Determine graphically the stresses in the boom and the boom cable and in the two structural members of the caterpillar derrick, when all are in the position shown in Fig. 4-39. (The weight of the tackle and of all members has been neglected.)

Solution: The upper right joint is isolated first, since at that joint the only unknown stresses are those in the boom and the boom cable. Because of the arrangement of the tackle at this joint, the tension in the haulback cable is half of the load, or $T = 1.5$ kips. The free-body diagram of that joint (Fig. 4-40*a*) is then drawn. In order to have all known forces in succession, as is preferable, the force *T* is transmitted back along its line of action until it pushes on the joint; this transmission, of course, does not affect the ultimate results. (See Transmissibility of Force, Art. 2-5. When forces are transmitted in this manner, Bow's notation is most conveniently applied to the free-body diagrams only.) The stresses in the boom and the boom cable are readily found in the force diagram Fig. 4-40*b*. *AB* and *BC* are drawn to scale, with proper direction and sense. The diagram is then closed by lines *CD* and *DA*, drawn parallel to those members.

Next the free-body diagram is constructed for the second joint, Fig. 4-40*c*. Here the force *F* equals the stress in the boom cable. *F* may now be transmitted back along its line of action, simply in order not to interfere with the stress *DE*. With the known forces *FA* and *AD* drawn to scale (Fig. 4-40*d*), the stresses *DE* and *EF* are found by closing the diagram. *DE* scales 18.17 kips and is seen to be compressive; *EF* scales 8.80 kips and is tensile.

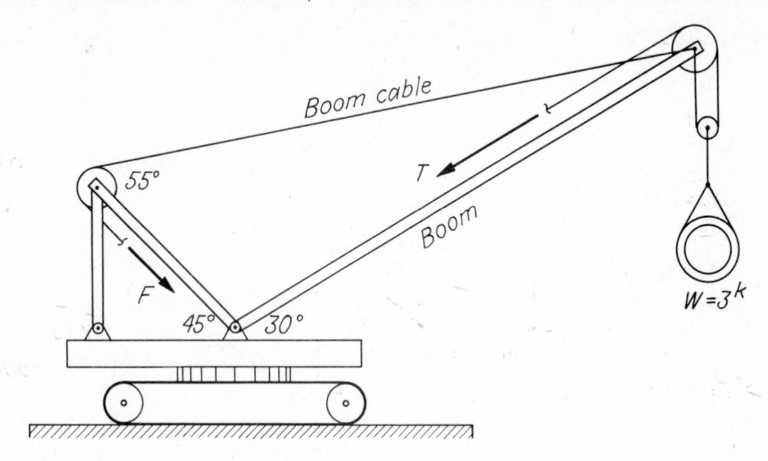

FIG. 4-39 Caterpillar derrick. Stresses by the graphical method of joints.

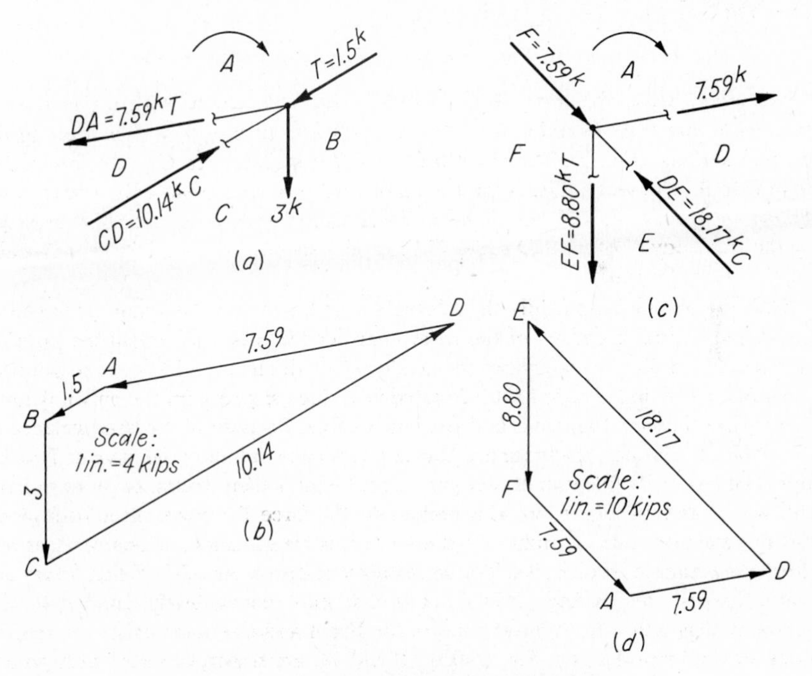

FIG. 4-40 Free-body and force diagrams.

PROBLEMS

Stresses by the graphical method of joints.

4-33. Determine the stresses in members *BC* and *CA* of the truss shown in Fig. 4-41. (Scales on $8\frac{1}{2}$ by 11: 1 in. = 2 ft, 1 in. = 5 kips.)

Ans. BC = 14.4 kips C; *CA* = 20 kips T

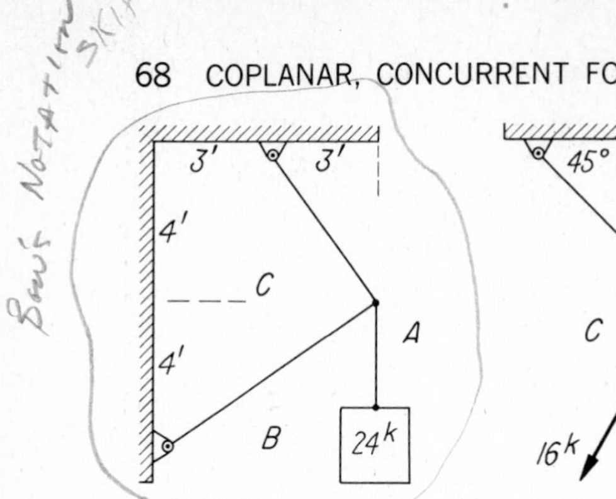

FIG. 4-41 Prob. 4-33.

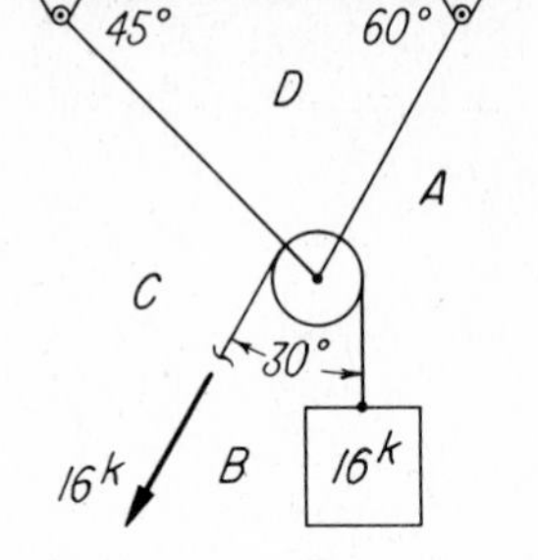

FIG. 4-42 Prob. 4-34.

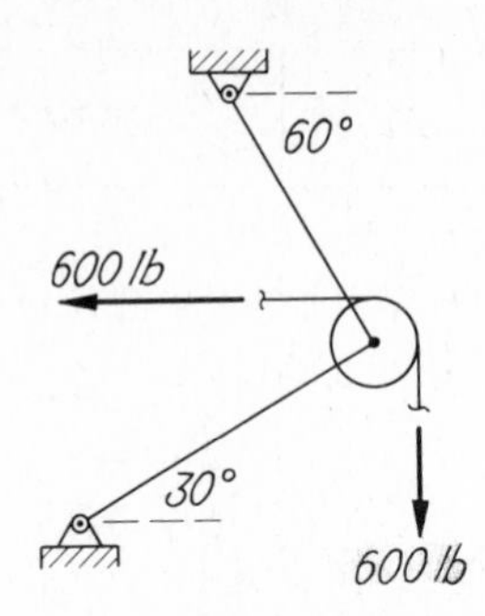

FIG. 4-43 Prob. 4-35.

4-34. Solve for stresses *CD* and *DA* in the truss shown in Fig. 4-42. (Scale on $8\frac{1}{2}$ by 11: 1 in. = 5 kips. Draw no member less than 2 in. long.)

4-35. Find the stresses in the two members of the frame shown in Fig. 4-43. (Scale on $8\frac{1}{2}$ by 11: 1 in. = 200 lb. When forces are transposed, apply Bow's notation to free-body and force diagrams only.) *Ans.* Upper = 220 lb T; lower = 820 lb C

4-36. The cable spool in Fig. 4-44 is supported by a sling as shown. Determine the stresses in the crossbar *AD* and in the sling rope. (Scale on $8\frac{1}{2}$ by 11: 1 in. = 500 lb.)

4-37. Find the stresses in the four members of the truss shown in Fig. 4-45. (Scales on $8\frac{1}{2}$ by 11: 1 in. = 2 ft, 1 in. = 300 lb.)

Ans. *BE* = 721 lb T; *EA* = 400 lb C; *CD* = 541 lb T; *DE* = 180 lb C

4-38. In the construction derrick shown in Fig. 4-46, the stress in the guy wire is maximum when the wire, the boom, and the mast all lie in one plane. Solve for the forces *T*, *F*, and *P*, and for the stresses in the boom and the mast, when *W* is 10 kips. Neglect sheave friction. (Scales on $8\frac{1}{2}$ by 11: 1 in. = 20 ft., 1 in. = 5 kips.)

4-39. Figure 4-47 illustrates a typical yard crane. Find the stresses in all members caused by a load *W* of 4 kips. Neglect sheave friction. (SUGGESTIONS: Apply Bow's notation to each separate joint only. Analyze the three upper joints only. Scales on 11 by 17: 1 in. = 4 ft, 1 in. = 2 kips.)

Ans. Horizontal member, 4 kips T; vertical member, 2.5 kips C

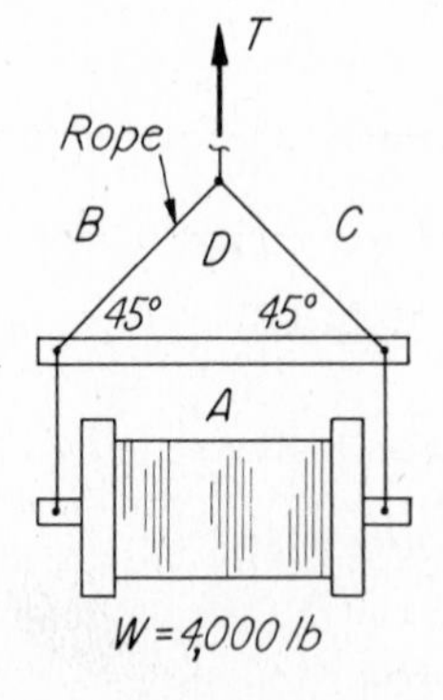

FIG. 4-44 Prob. 4-36.

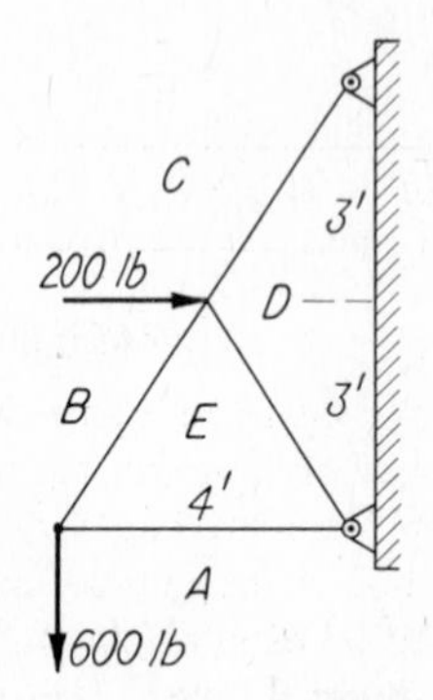

FIG. 4-45 Prob. 4-37.

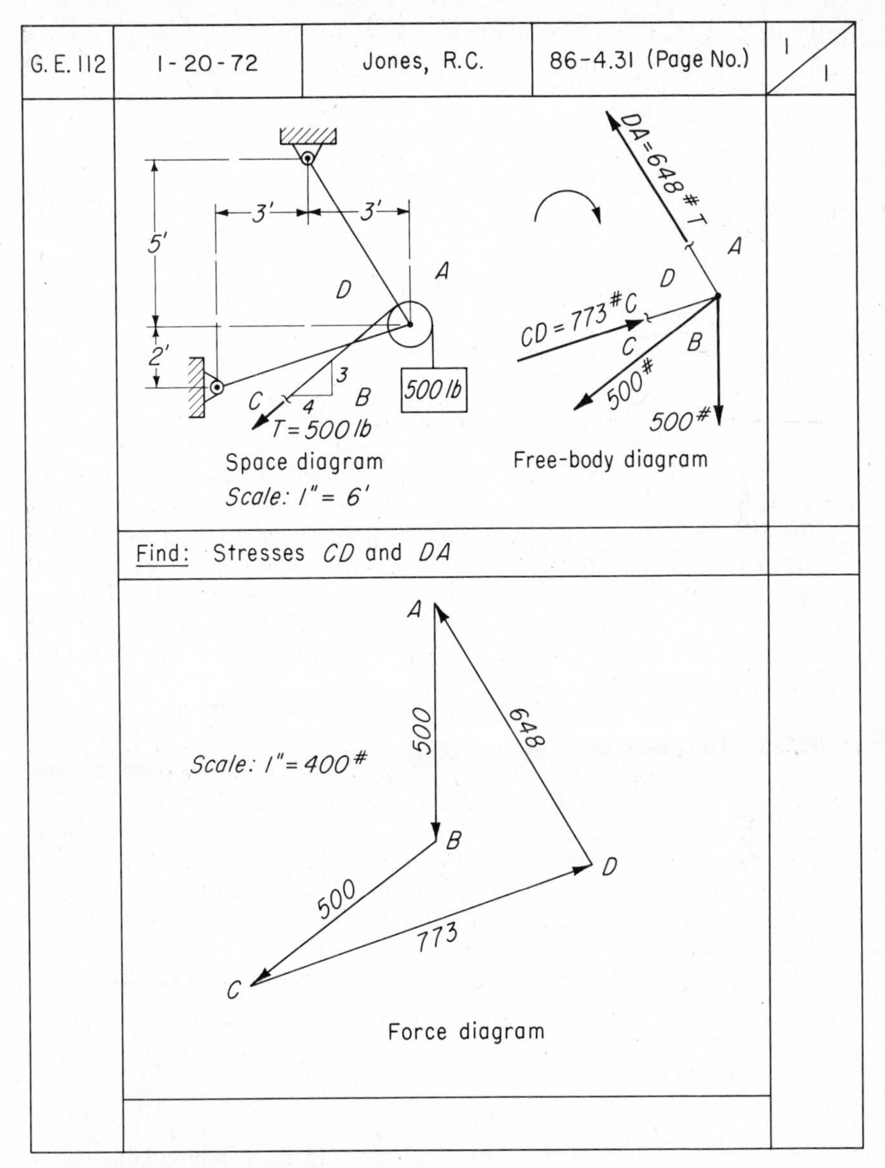

SAMPLE PROBLEM SOLUTION SHEET

Paper: $8\frac{1}{2}$ by 11 in. squared.

Heading: Left to right, course number, date, name, problem number.

Problem Number: The number 86 represents the student's consecutive problem number; 4-31 is the number of the problem in the book; and the third number is the page number in the book.

Upper Right Corner: *Upper,* student's consecutive sheet number; *lower,* total number of sheets in the solution.

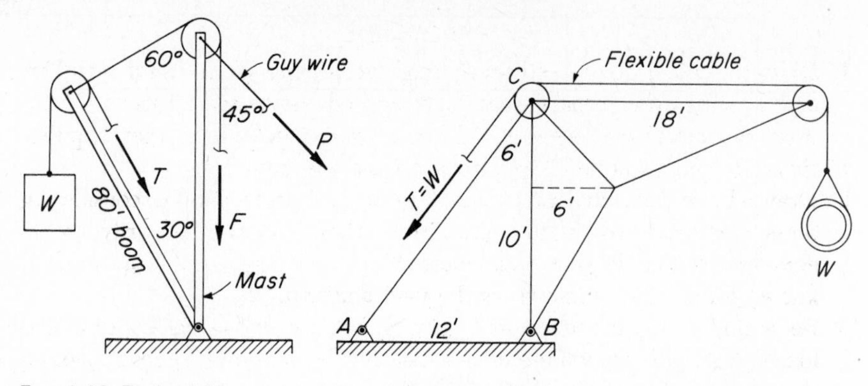

FIG. 4-46 Prob. 4-38.

FIG. 4-47 Prob. 4-39.

4-40. Solve Prob. 4-39 if $W = 12$ kips and if the pull T is vertically downward from the pulley at C.

4-41. Steam exerts a 500-lb force on the piston of Fig. 4-48. What upward vertical force must be applied at joint B to hold the piston stationary?

Ans. $P = 1{,}500$ lb

4-42. In Fig. 4-49 is shown a typical caterpillar construction derrick. Find the stresses in the boom and the boom cable, and in the two structural members when they are in the position shown. (Scale on 11 by 17: 1 in. = 1 kip.)

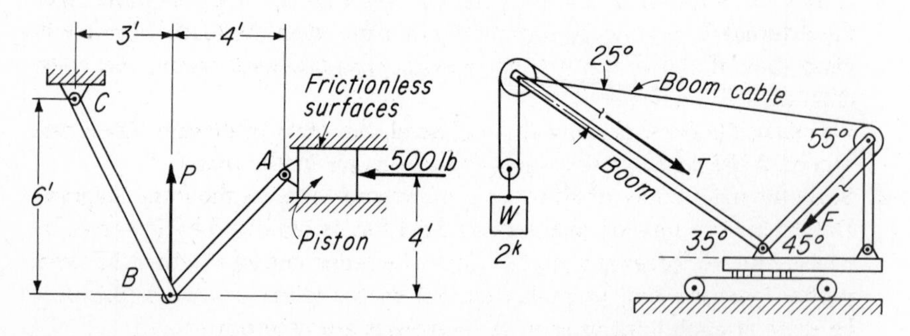

FIG. 4-48 Prob. 4-41.

FIG. 4-49 Prob. 4-42.

4-9 Stresses in Trusses; the Graphical Method of Combined Diagrams

When several joints of a truss need to be solved, as in many commonly used roof and bridge trusses, all the separate force diagrams may be superimposed upon one another until they form a single **combined diagram.** This is possible because the force diagrams of the forces at adjacent joints usually have one or two or three sides in common. Hence, less work is involved than in the joint method, and greater decision is obtainable. This method, therefore, is most commonly used in practice. The general procedure is as follows:

PROCEDURE

1. Draw to scale a good-sized space diagram using *single lines* for members and showing *all* external forces (loads and reactions) and all usual dimensions. Be certain that the external forces are in force and moment equilibrium. Use Bow's notation *clockwise* around the truss.
2. Draw a force diagram of *all* external forces (loads and reactions) considering them *clockwise* around the truss. Since these forces necessarily are in equilibrium, this diagram *must* close. Show magnitude and sense of each known force. This is known as the **load diagram.**
3. Beginning at a joint where only two forces are unknown, locate in the load diagram the known forces and solve for the unknown forces by closing the diagram of the forces at the joint. *Do not show senses of stresses solved for on the force diagram.* Continue this process until all forces are found. All stress lines in the force diagram must be drawn parallel to the corresponding members in the space diagram. The distance by which the *closing line* fails to close on the *final point* is called the **error of closure.** Scale this distance perpendicular from point to line, and note it as a dimension.

   Sources of Error. The error of closure may have two main causes: (1) If it is relatively small, the error may be due to lack of precision in (*a*) the space diagram or (*b*) the load diagram, or to failure in drawing the stress lines *parallel* to the space lines. (2) If the error is relatively large, it may be due to *mistakes* in either the space or the load diagram. Also, the external forces may be in *force* equilibrium; then the load diagram will close; but, if the forces are not also in *moment* equilibrium, the stress diagram will not close.

   Permissible Error of Closure. In general, the error of closure should not exceed *2 per cent of the average of all stresses in the truss.*
4. Scale the magnitudes of all stresses and record them on the force diagram. Determine the *type* of each stress solved for, as indicated by its sense, by considering all forces at a joint in clockwise order, noting whether the stress pushes (compression) or pulls (tension) on the joint. If desired, this may be done at each joint as soon as its stresses are determined.
5. Record magnitude and type of each stress on the space diagram, or on a separate stress-summation diagram.

ILLUSTRATIVE PROBLEMS

4-43. Find the stresses in all members of the Fink roof truss shown in Fig. 4-50*a* by the graphical method of combined diagrams.

Solution: The space diagram is first drawn to scale in accordance with item 1 of the procedure. The load diagram is next constructed (item 2). Since all known forces are parallel, the reactions *DE* and *EA* overlie the loads in the load diagram in Fig. 4-50*b*. Although the load diagram is a straight line, it is nevertheless a closed polygon, since force *EA* closes on the point of beginning *A*. The separate force diagrams are

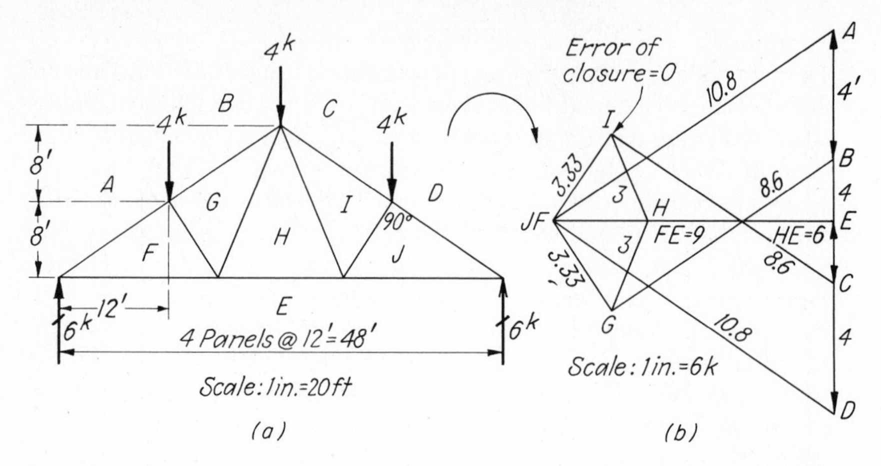

FIG. 4-50 Stresses by the graphical method of combined diagrams. (*a*) Space diagram. (*b*) Force diagram.

then completed (item 3) in the following order, the forces being considered clockwise around each joint: (1) *EAFE*, (2) *FABGF*, (3) *EFGHE*, (4) *DEJD*, (5) *CDJIC*, (6) *IJEHI*. The uppermost joint need not be solved, since all stresses are now known. Failure of the last stress line *HI* to close on point *I* would disclose the *error of closure*, which then should be scaled and noted as a dimension (item 3). Arrowheads are shown on all external forces, but are omitted from all internal stress lines.

The stresses are now scaled (item 4) and their magnitudes are noted on the force diagram. The magnitudes of stresses whose stress lines overlie each other, as do *FE* and *HE*, should be indicated separately as shown. The *kind of stress* is now to be determined. Consider the forces acting at the joint on the left. Clockwisely, they are *EA*, *AF*, and *FE*. Their stress lines are retraced in the force diagram: *EA* is upward acting; *AF* acts downward and to the left, *pushing on the joint, and is therefore compressive;* and *FE* acts from left to right, *pulling on the joint, and is therefore tensile.* Finally (item 5), the magnitude and type of each stress are recorded on the corresponding member on the space diagram, or on a separate stress-summation diagram.

4-44. Find the stresses in all the members of the tower shown in Fig. 4-52 by the graphical method of combined diagrams.

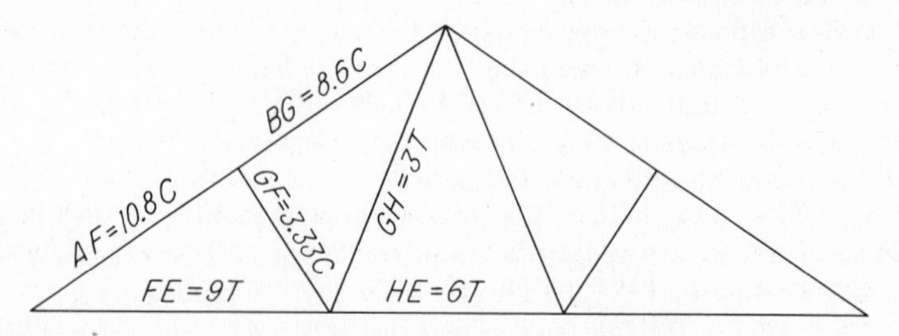

FIG. 4-51 Stress-summation diagram.

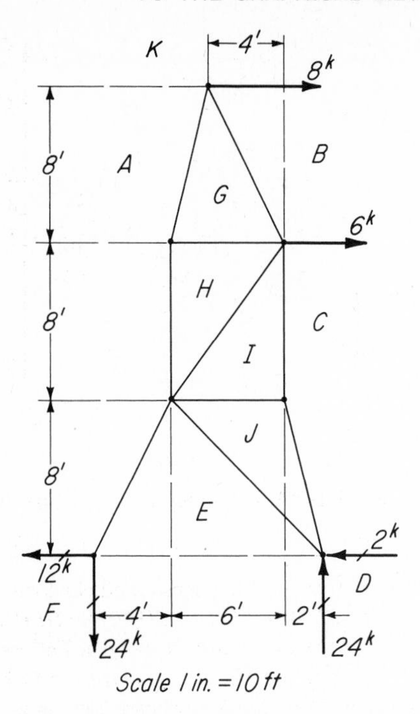

FIG. 4-52 Space diagram.

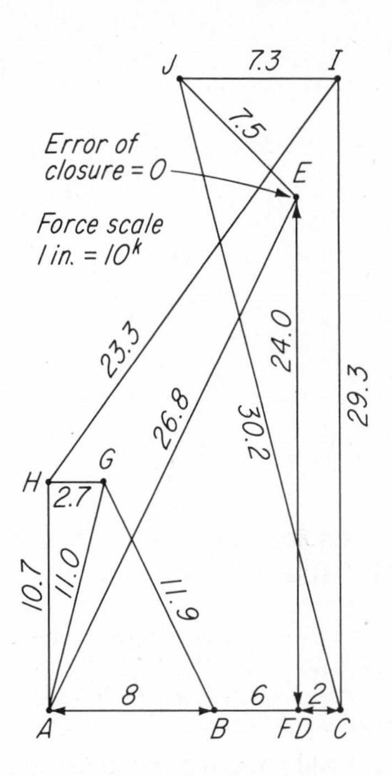

FIG. 4-53 Force diagram.

Solution: The space diagram is drawn to scale first. The load diagram is constructed next. The loads are laid off in the order *AB, BC, CD, DE, EF,* and *FA.* The load diagram is closed. The separate force diagrams are drawn in the following order, the forces being taken consecutively in a clockwise order around each joint: (1) *ABGA,* (2) *AGHA,* (3) *HGBCIH,* (4) *ICJI,* and (5) *AHILEA.* The stresses are scaled and are recorded on the force diagram (Fig. 4-53), and also on the respective members of the space diagram (Fig. 4-52), or on a separate stress-summation diagram similar to those shown in Figs. 4-51 and 4-55.

4-45. Using the graphical method of combined diagrams, find the stresses in all members of the truss shown in Fig. 4-54.

Solution: Having drawn the space diagram to scale, as in Fig. 4-54*a,* we discover that there are a minimum of three unknown stresses to be found at each joint, one more than may be determined by the usual graphical method. This difficulty may be overcome by the temporary use of a **substitute member** as follows:

Temporarily remove members *FG, GH,* and *HF* and substitute for them the single member *XH* indicated by the dash line. We should note here that in all cases the substitute member must be so placed that the truss remains stable and that all stresses can then be solved for by the usual procedure.

If now we cover the left two-thirds of the space diagram, it becomes apparent that the stresses in the members of the right half of the truss do not depend in any

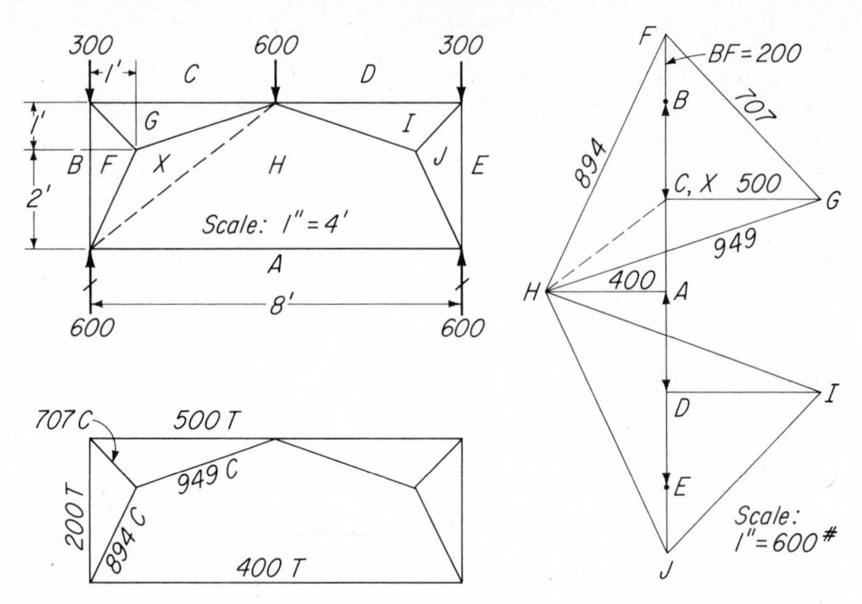

FIG. 4-54 Stresses by the graphical method of combined diagrams using substitute member. (*a*) Space diagram. (*b*) Stress-summation diagram. (*c*) Force diagram.

way upon the particular arrangement of the members in the left half. That is, with a section cut through members *AHID* and the part of the truss on the right then isolated (see Figs. 2-41*b* and 5-62), it becomes apparent that stresses *AH*, *HI*, and *ID* may be found as follows: *AH* by a summation of moments about the center joint of the top chord; the vertical component of *HI* by a summation of vertical forces; and *ID* by a summation of horizontal forces. Consequently, *the stress in member AH,* which member is common to both halves of the truss, *will be the true stress,* whether found by solving the lower right joint with the original arrangement of members, or by solving the lower left joint using the substitute member.

The procedure now is as follows: In the force diagram (Fig. 4-54*c*) lay off to scale all external forces, taking them in order clockwise around the truss. Then remove members *FG*, *GH*, and *HF* and substitute for them the single member *XH*. Next solve the upper left joint for stresses *CX* and *XB*, and then the lower left joint for stresses *XH* and *HA*. Of these stresses only *HA* will be a true stress. Now remove the substitute member *XH* and replace members *FG*, *GH*, and *HF*. A second solution of the lower left joint, superimposed upon the first, and using the true stress *HA*, will give also the true stresses in *BF* and *FH*. From here the force diagram is completed as usual.

It is interesting to note here that this problem may quite easily be solved by the analytical method of joints, using the substitute member as in the graphical solution. The steps are as follows: (1) Remove members, *FG*, *GH*, and *HF* and substitute for them member *XH*. (2) Solve the upper left joint. (3) Solve the lower left joint (*HA* is true stress). (4) Remove *XH*, replace *FG*, *GH*, and *HF*, and resolve the lower left joint using the true stress *HA*. (5) Proceed with the balance of the solution in the normal manner.

PROBLEMS

Stresses by the graphical method of combined diagrams

4-46. Determine the stresses in all members of the truss shown in Fig. 4-55. (Scales on $8\frac{1}{2}$ by 11: 1 in. = 2 ft, 1 in. = 200 lb.)

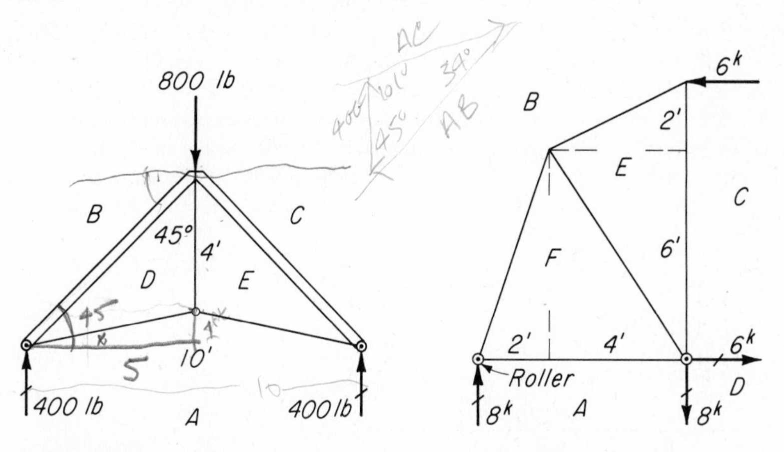

FIG. 4-55 Prob. 4-46.

FIG. 4-56 Prob. 4-47.

4-47. Find the stresses in all members of the truss shown in Fig. 4-56. (Scales on $8\frac{1}{2}$ by 11; 1 in. = 2 ft, 1 in. = 2 kips.) *Ans.* BF = 8.44 kips C; FA = 2.67 kips T; FE = 6.01 kips T; EC = 3 kips T; BE = 6.71 kips C

4-48. Solve Prob. 4-44 if a vertical downward load AK of 12 kips is added at the top of the tower.

4-49. The signboard truss in Fig. 4-57 is subjected to horizontal windloads as shown. Determine the stresses in all members caused by these loads. (Scales on $8\frac{1}{2}$ by 11:

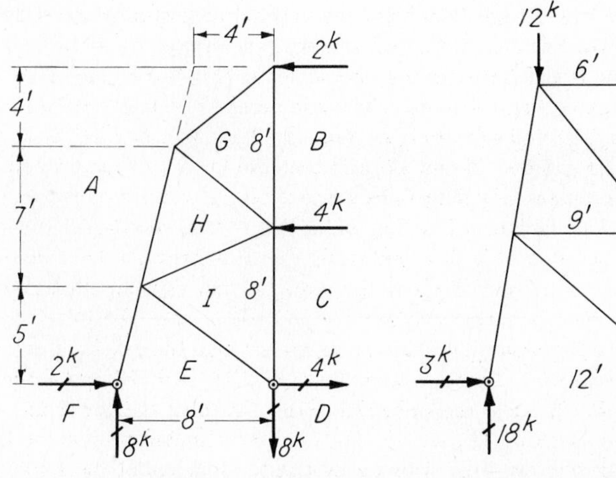

FIG. 4-57 Probs. 4-49 and 4-92.

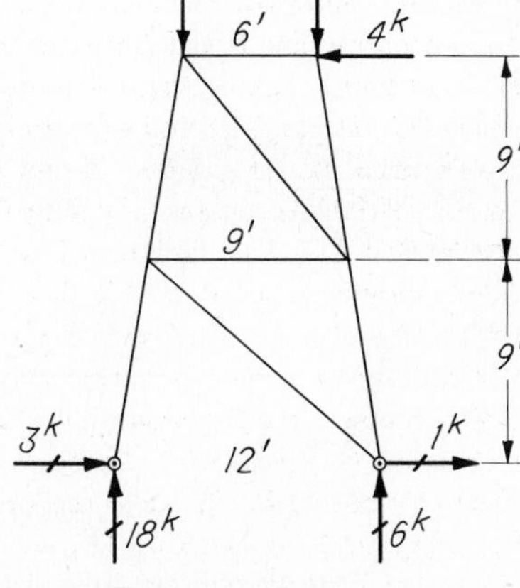

FIG. 4-58 Prob. 4-50.

1 in. = 4 ft, 1 in. = 2 kips.) *Ans.* BG = 1.60 kips T; GA = 2.56 kips C; GH = 1.71 kips T; HA = 2.74 kips C; CI = 5.05 kips T; IH = 5.85 kips C; EI = 4.98 kips T; AE = 8.25 kips C

4-50. The truss shown in Fig. 4-58 is one of two similar trusses supporting a water tank. It is also subjected to a horizontal wind load. Determine the stresses in all members due to these loads. (Scales on 8½ by 11: 1 in. = 4 ft, 1 in. = 4 kips.)

4-51. The loads on the Howe roof truss shown in Fig. 4-59 are caused by the weight of the roof which it supports. Determine the stresses in all members due to these loads (NOTE: Because of symmetry of truss and loads, stresses in members on the right are equal to those in corresponding members on the left. (Scales on 11 by 17: 1 in. = 6 ft, 1 in. = 4 kips.) *Ans.* CH = 26.8 kips C; HA = IA = 24 kips T; DJ = 17.9 kips C; JI = 8.96 kips C; IH = 0; KJ = 8 kips T

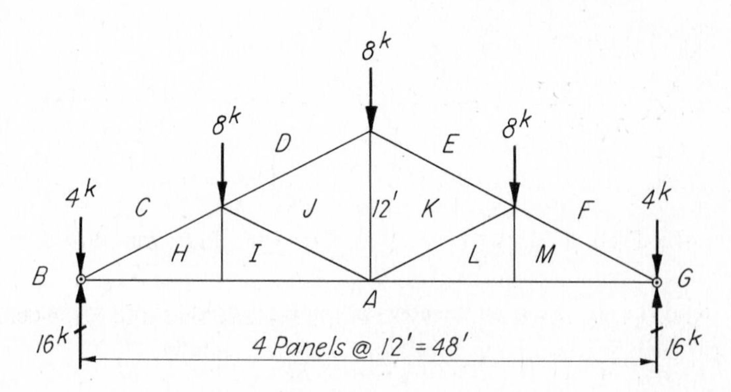

FIG. 4-59 Prob. 4-51. Howe roof truss, dead loads.

4-52. In Fig. 4-60 the wind loads act perpendicular to the sloping chord of the Howe roof truss shown. Assume the reactions at the left and right supports to be parallel to the wind loads. Find the reactions by the inverse-proportion method of Art. 2-13. Then solve for the stresses in all members. (Scales on 8½ by 11: 1 in. = 10 ft, 1 in. = 2 kips.) *Partial Ans.* R_L = 4.13 kips; R_R = 1.87 kips

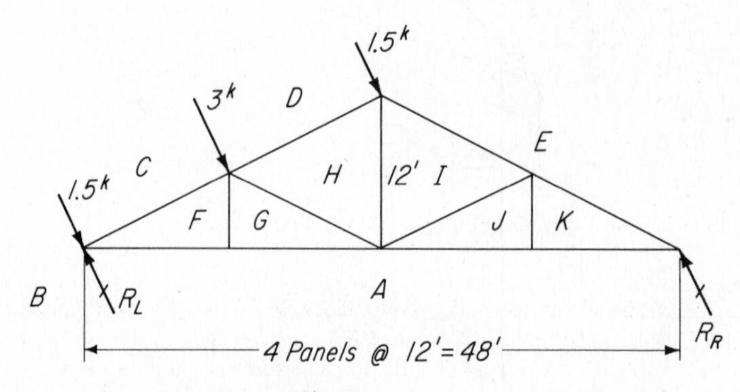

FIG. 4-60 Prob. 4-52. Howe roof truss, wind loads.

4-53. In Fig. 4-61 the wind loads act perpendicular to the sloping chord of the Fink roof truss shown. Determine the stresses in all members. (Scales on 11 by 17: 1 in. = 4 ft, 1 in. = 1 kip.)

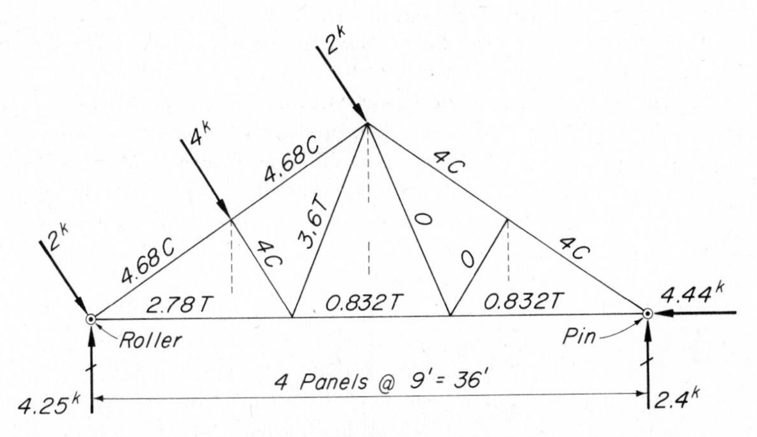

FIG. 4-61 Probs. 4-53 and 4-93. Fink roof truss, wind loads.

4-54. Solve for the stresses in all members of the truss shown in Fig. 5-66. (Scales on $8\frac{1}{2}$ by 11: 1 in. = 5 ft, 1 in. = 5 kips. Use Bow's notation.)

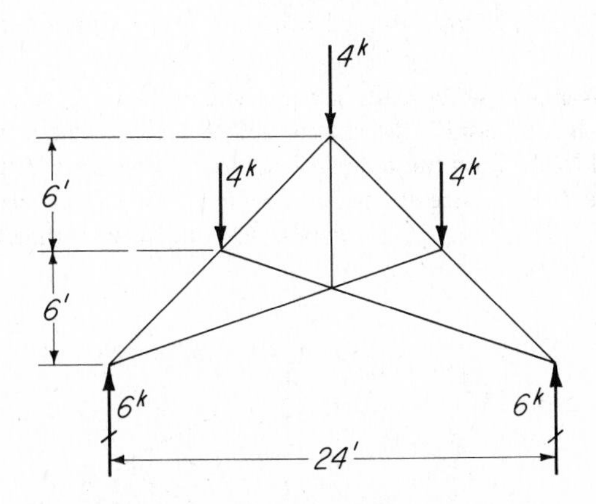

FIG. 4-62 Prob. 4-55. Scissors truss.

4-55. Determine the stresses in all members of the scissors roof truss shown in Fig. 4-62. (Scales on $8\frac{1}{2}$ by 11: 1 in. = 4 ft, 1 in. = 4 kips.)

Partial Ans. Stress in vertical member is 8.25 kips T

4-56. Solve for the stresses in all members of the truss shown in Fig. 5-74. (Scales on $8\frac{1}{2}$ by 11: 1 in. = 10 ft, 1 in. = 5 kips. Use Bow's notation.)

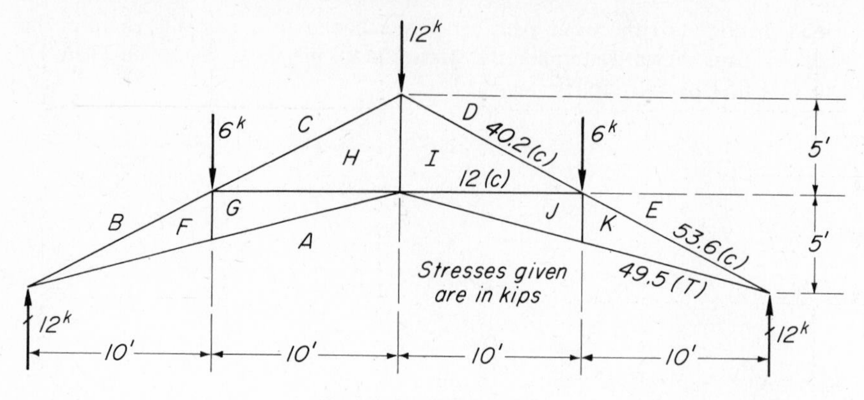

FIG. 4-63 Prob. 4-57.

4-57. Find the stresses in all members of the roof truss shown in Fig. 4-63. (Scales on 11 by 17: 1 in. = 3 ft, 1 in. = 3 kips. Answers are shown on right half of truss.)

4-58. Solve for the stresses in all members of the truss shown in Fig. 4-36. (Scales on 8½ by 11: 1 in. = 6 ft, 1 in. = 10 kips. Use Bow's notation.)

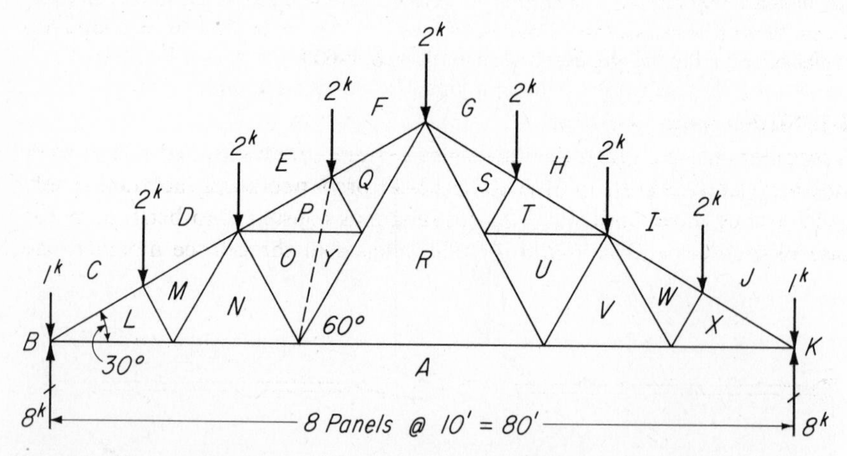

FIG. 4-64 Prob. 4-59. Fink roof truss to be solved by the substitute-member method.

4-59. Using the substitute-member method, solve for the stresses in all members of the compound Fink roof truss shown in Fig. 4-66. (HINT: Remove members *OP* and *PQ* and substitute for them the member *OY* indicated by the dash line. The stress in member *RA* is true stress. Scales on 11 by 17: 1 in. = 10 ft, 1 in. = 2 kips.)

4-60. Using the substitute-member method, do the *alternate graphical solution* suggested in Prob. 5-65. (Space scale: 1 in. = 10 in.; force scale, *AD* by four-force principle, 1 in. = 200 lb; force scale for combined diagram, 1 in. = 100 lb. Use Bow's notation for combined diagram.)

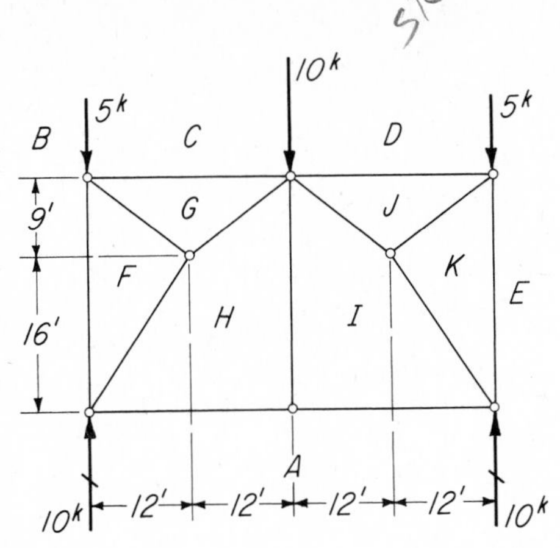

FIG. 4-65 Prob. 4-61.

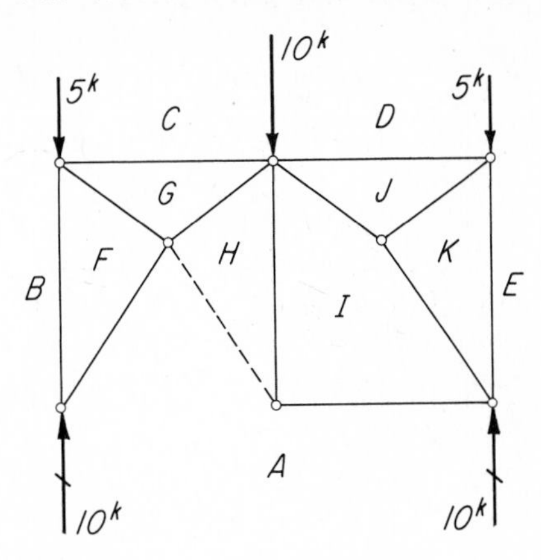

FIG. 4-66 Substitute member, Prob. 4-61.

4-61. Using the substitute-member method, solve for the stresses in all members of the truss shown in Fig. 4-65. (NOTE: In this problem, try substituting the member *AH* indicated by the dash line in Fig. 4-66. Begin the solution at the lower left joint. Stress *AI* will be true stress. Scales on 8½ by 11: 1 in. = 10 ft, 1 in. = 4 kips. See explanation of the *all-graphical solution* of Prob. 5-65.)

4-10 Three-force Members

A member on which forces act at three (or more) points is called a *three-force member* (or a multiple-force member). These forces necessarily act transversely to its axis or have transverse components, thus causing it to bend, as in the case of member *AD* in Fig. 4-67. The stresses in three-force members are

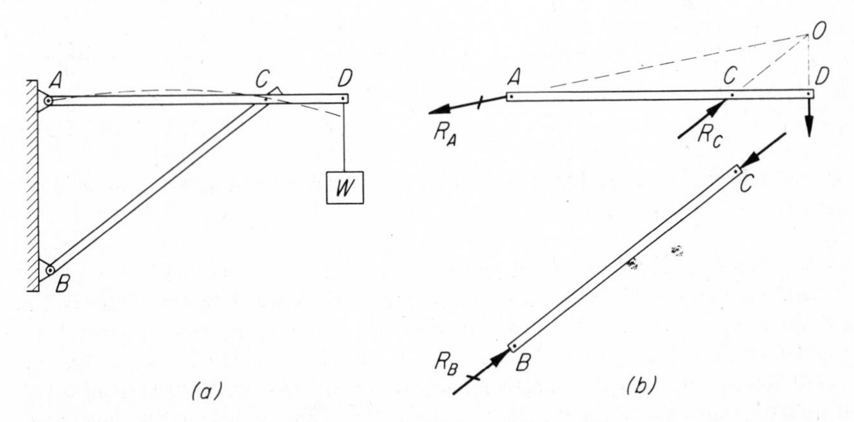

FIG. 4-67 Two- and three-force members. (*a*) Space diagram. (*b*) Free-body diagrams.

complex, being usually a combination of axial, bending, and shearing stresses. Such stresses are considered under the subject of Strength of Materials. *In statics, therefore, a three-force member should never be cut.*

In Fig. 4-67, *BC is clearly a two-force member, since forces act on it at two points only, and the reacting force R_B at B therefore lies parallel to BC.* Because the force W acts transversely to member AD, the reacting forces R_A and R_C also act transversely to AD, or must have transverse components. This is true of all three-force members. Hence, **the reacting force at any support of a three-force member never lies parallel to the axis of that member.** These facts are of importance in the determination of support reactions, outlined in the following article.

4-11 Graphical Determination of Reactions Using the Three-force Principle

Supports and support reactions were discussed in Art. 2-16. As is stated therein, bodies subjected to coplanar force systems are generally supported at two points, at one of which the direction of the reacting force will often be known, while at both the magnitudes will be unknown. The usual graphical problem then is to determine the unknown direction of one support reaction and magnitudes of both. When only three nonparallel forces act on the body, such as one load and two reactions, this unknown direction may be established by the three-force principle, whereafter the two unknown magnitudes are found by a closed graphical force diagram, as shown in Prob. 4-55. When two or more *known* forces act on the body, they must be combined into their single resultant force (Art. 2-18) before the three-force principle can be applied, as shown in Prob. 4-63.

ILLUSTRATIVE PROBLEMS

4-62. Determine graphically the reactions at supports A and B of the frame shown in Fig. 4-68*a*.

Solution: From the space diagram (Fig. 4-68*a*) we recognize that BC is a two-force member (Art. 4-5). The reaction R_B then lies parallel to BC. The extended lines of action of W and R_B then establish O, their point of intersection, through which the action line of reaction R_A also must pass, according to the three-force principle. From the force diagram, we find R_A and R_B to be 118 lb and 165 lb, respectively. Students should record these values on the space diagram.

4-63. Determine graphically the support reactions at A and B of the truss shown in Fig. 4-69*a*.

Solution: Four forces act on this truss, the known forces H and P, and the unknown reactions, R_A and R_B. We must therefore combine H and P into their resultant R. This is done by first extending their action lines backward to their intersection C. From C along the respective action lines we lay off 600 and 500 lb. The diagonal of the completed parallelogram is the action line of R. Now only three forces remain whose action lines must intersect at a common point.

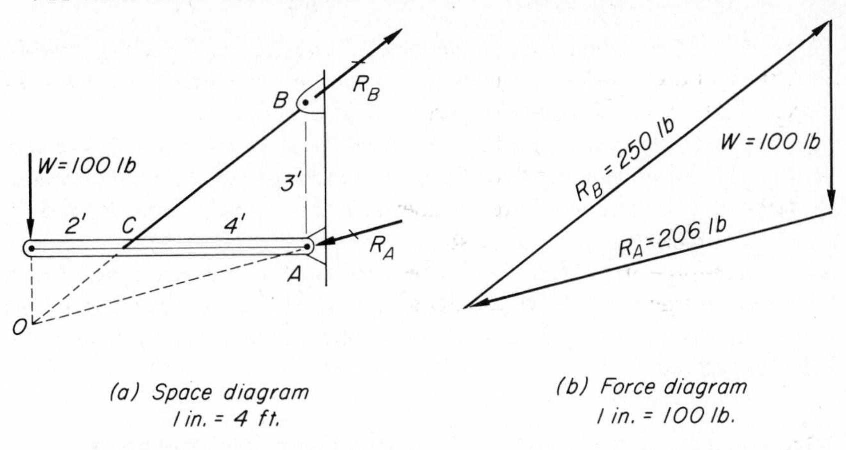

FIG. 4-68 Graphical determination of reactions using the three-force principle. (*a*) Space diagram. 1 in. = 4 ft. (*b*) Force diagram. 1 in. = 100 lb.

Because the rollers at *A* rest on a horizontal surface, the action line of R_A is vertical and intersects that of *R* at *D*, the intersection through which the action line of R_B must then also pass. A force diagram of the four forces, here taken clockwise around the truss, then gives R_A and R_B. Greater accuracy will be obtained by using the original forces *H* and *P* rather than their resultant. As a check against an analytical solution, B_V and B_H may also be scaled.

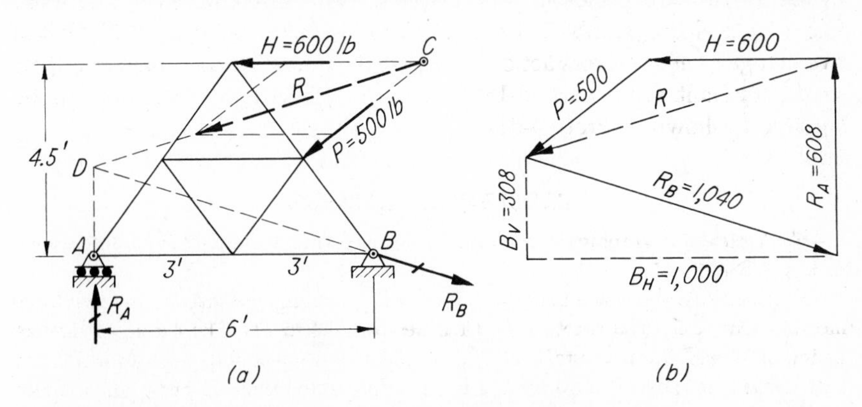

FIG. 4-69 Graphical determination of reactions using the three-force principle. (*a*) Space diagram. 1 in. = 4 ft. (*b*) Force diagram. 1 in. = 600 lb.

PROBLEMS

Graphical determination of reactions

4-64. The cylinder shown in Fig. 4-70 is 4 ft in diameter and weighs 300 lb. Determine the horizontal push *P* required to hold it in the position shown, when block *B* is removed. Neglect friction. (Scale on 8½ by 11: 1 in. = 100 lb.)

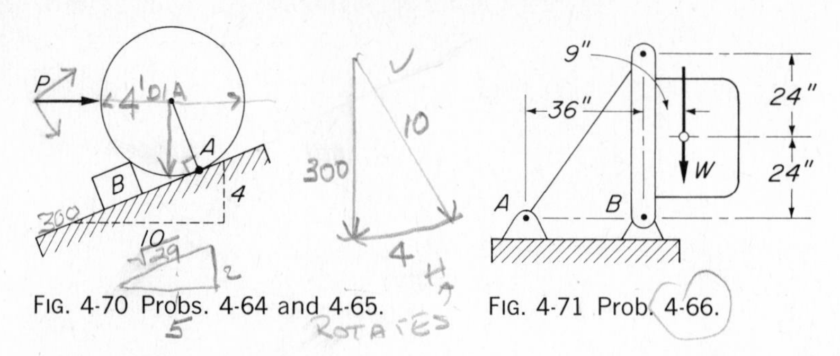

FIG. 4-70 Probs. 4-64 and 4-65.

FIG. 4-71 Prob. 4-66.

4-65. Solve Prob. 4-64 if the push P is parallel to incline. *Ans.* $P = 111.5$ lb

4-66. The weight W in Fig. 4-71 is 480 lb. Determine the support reactions at A and B. (Scale on $8\frac{1}{2} \times 11$: 1 in. = 1 ft; 1 in. = 100 lb.)

4-67. Determine the force F required to start to tip the block shown in Fig. 4-72 about corner C, and the reacting force R_C at C. (Scales on $8\frac{1}{2}$ by 11: 1 in. = 1 ft, 1 in. = 100 lb.)

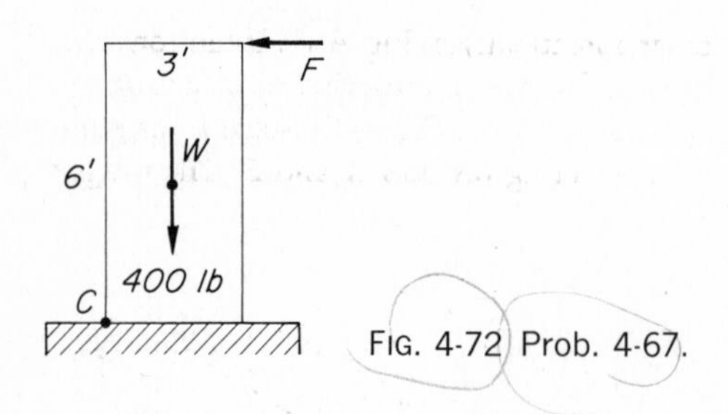

FIG. 4-72 Prob. 4-67.

4-68. The Fink roof truss shown in Fig. 4-73 is subjected to a resultant wind load at the left mid-joint. Determine the support reactions R_A and R_B. (Scales on $8\frac{1}{2}$ by 11: 1 in. = 10 ft, 1 in. = 600 lb.)

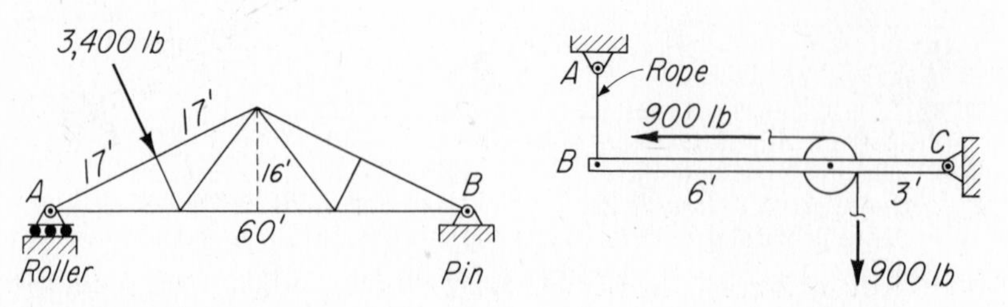

FIG. 4-73 Prob. 4-68.

FIG. 4-74 Prob. 4-69.

4-69. In Fig. 4-74 the 900-lb weight is suspended from a flexible cable passing over a sheave. Find the tension in the rope AB and the reaction at C. (Scales on $8\frac{1}{2}$ by 11: 1 in. = 2 ft, 1 in. = 200 lb.) *Ans.* $T = 300$ lb↑; $R_C = 1{,}082$ lb↗

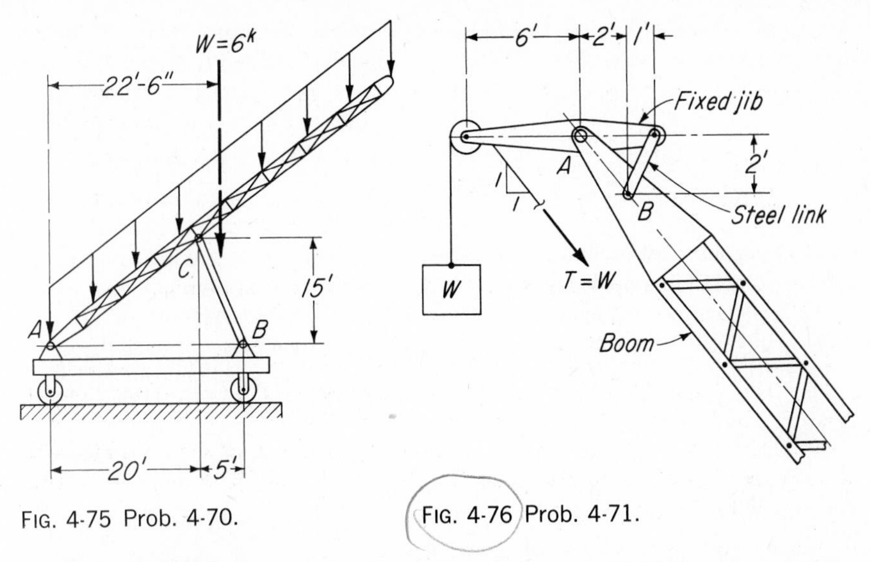

FIG. 4-75 Prob. 4-70.

FIG. 4-76 Prob. 4-71.

4-70. A portable conveyor is shown diagrammatically in Fig. 4-75. If the conveyor and its contents weigh 6 kips, determine graphically the reactions at A and B.

4-71. A crane is equipped with a fixed jib as shown in Fig. 4-76. Make a complete graphical solution and determine the forces acting on pins A and B. The weight W is 3,000 lb.

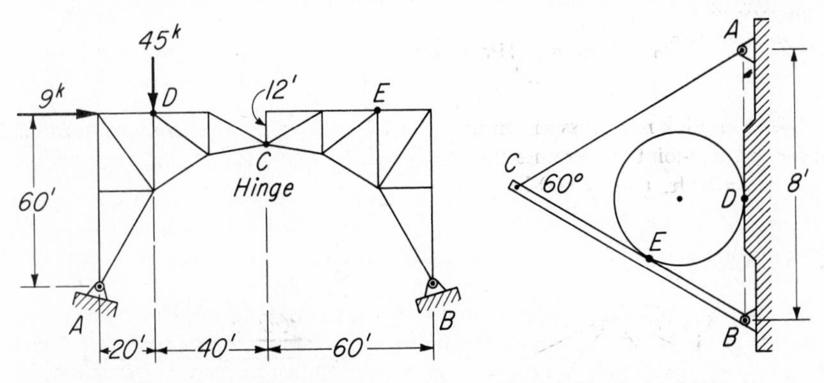

FIG. 4-77 Probs. 4-72 and 4-99.

FIG. 4-78 Prob. 4-73.

4-72. The three-hinged arch shown in Fig. 4-77 is acted upon by two loads which are concurrent through point D. Determine the support reactions R_A and R_B. (Scales on $8\frac{1}{2}$ by 11: 1 in. = 20 ft, 1 in. = 10 kips.)

Ans. $R_A = 33.5$ kips ↗; $R_B = 19.2$ kips ↖

4-73. In Fig. 4-78 is shown an end view of an air-compressor tank 4 ft in diameter and weighing 866 lb, half of which (433 lb) is supported by each of two identical frames ACB. Determine (a) the pressure P exerted by the tank at E and (b) the support reactions at A and B. Neglect friction between tank and supports. [HINT: (*a*) Isolate the tank as a free body. (*b*) Only force P produces reactions at A and B. Scales on $8\frac{1}{2}$ by 11: 1 in. = 2 ft, 1 in. = 100 lb.]

Ans. $P = 500$ lb↙; $R_A = 250$ lb↗; $R_B = 310$ lb↗

SUMMARY

(*By article number*)

4-1. A coplanar, concurrent force system is one in which all forces lie in one plane, and their lines of action all intersect at a common point. The internal forces acting in the members meeting at the joints of trusses illustrate this type of system.

4-2. The resultant of a coplanar, concurrent force system passes through the point of concurrency and may be determined analytically (*a*) by algebraic summation of forces or components parallel to two rectangular axes or (*b*) graphically by the parallelogram or the triangle methods.

4-3. In a system of coplanar, concurrent forces in equilibrium, $\Sigma V = 0$ and $\Sigma H = 0$. Hence, the resultant of such a system is zero. When laid off graphically to scale, *the forces of a system in equilibrium will always form a closed polygon, in which the arrows always follow each other.*

4-4. A truss is a structural unit in which the bars are arranged to form one or more connected triangles. **A joint** is the connection between two or more bars. **A member** is a bar extending from joint to joint.

4-5. Stresses in members of trusses which are pin-connected and are loaded only at the joints always lie parallel to the principal axes of the members. Such members are **two-force members.** When the stress pushes on the member, it is compressive; when it pulls on the member, it is tensile.

4-6. When a load is applied at a joint by means of **a rope passing over a sheave or a pulley,** the sheave or pulley may be removed and the forces may be applied directly at the joint without changing their effect.

4-7. To determine *stresses in trusses* by the **analytical method of joints,** the joints are successively isolated, beginning with one at which only two stresses are unknown, and proceeding in such order that not more than two unknown stresses need be determined at any joint. Sloping forces are replaced by vertical and horizontal components. The unknown stresses are then found by solving the two equations of force equilibrium, $\Sigma V = 0$ and $\Sigma H = 0$.

4-8. In the **graphical method of joints,** stresses are found by drawing to scale a separate closed force diagram of the forces at each joint. The arrowheads always follow each other around the diagram, thus determining whether stresses are tensile or compressive.

4-9. In the **graphical method of combined diagrams,** all external forces are first laid off to scale forming a closed diagram. Thereafter the stresses are found as in the graphical method of joints. The separate force diagrams are joined into a single *combined diagram.*

4-10. A three-force member is one on which forces act at three or more points. Internal stresses in a three-force member are complex, and the total resultant stress does not lie parallel to the axis of the member. Consequently, in statics, a three-force member should never be cut.

4-11. The three-force principle states that, when three nonparallel forces are in equilibrium, they must be coplanar and their lines of action must intersect at a common point. This principle is very useful in *graphical determinations of support reactions,* when only three forces are involved or when they can conveniently be reduced to three. The reaction at the end of a two-force member lies parallel to the member; the reaction at the end of a three-force member does not.

REVIEW PROBLEMS

4-74. Using algebraic summations of rectangular components, solve for the *resultant force R* of the three forces shown in Fig. 4-79 and the angle θ_H it makes with the horizontal. Check graphically by the triangle method. (Scale: 1 in. = 50 lb.)

Ans. $R = 170$ lb↙; $\theta_H = 43°25'$

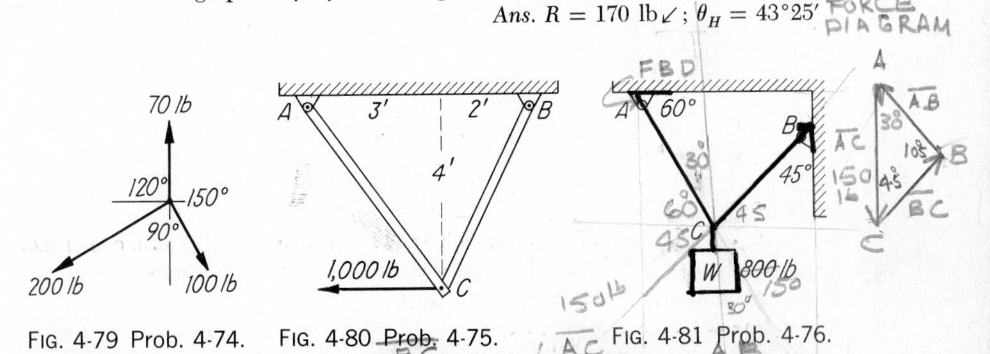

FIG. 4-79 Prob. 4-74. FIG. 4-80 Prob. 4-75. FIG. 4-81 Prob. 4-76.

4-75. By *geometry,* solve for the *stresses* in members *AC* and *BC* of the truss shown in Fig. 4-80. If space and force diagrams are drawn to scale (1 in. = 2 ft, 1 in. = 400 lb), a graphical check is obtained. *Ans. AC* = 1,000 lb C; *BC* = 894 lb T

4-76. Solve by *trigonometry* for the *stresses* in members *AC* and *BC* of the truss in Fig. 4-81. Check graphically (1 in. = 200 lb).

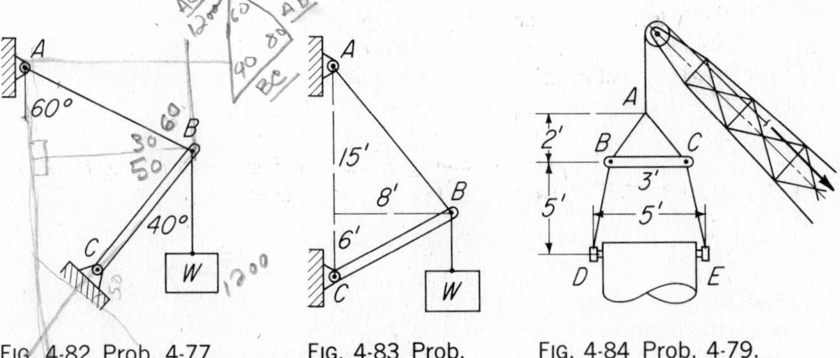

FIG. 4-82 Prob. 4-77. FIG. 4-83 Prob. 4-78. FIG. 4-84 Prob. 4-79.

4-77. The weight *W* shown in Fig. 4-82 is 1,200 lb. Find the *stresses* in members *AB* and *BC* by *trigonometry.* *Ans. AB* = 782 lb T; *BC* = 1,055 lb C

4-78. The truss shown in Fig. 4-83 supports a weight W of 6,300 lb. Using the *geometric method* (similar triangles), solve for the *stresses* in members AB and BC.

4-79. The crane sling shown in Fig. 4-84 supports a concrete bucket weighing 16 kips. Determine the *stress* in the horizontal bar BC by the *analytical method of joints*. The bar BC is 3 ft long. *Ans.* BC = 4.4 kips C

4-80. Using the *analytical method of joints*, solve for *stresses* AB and AC in Fig. 4-85.

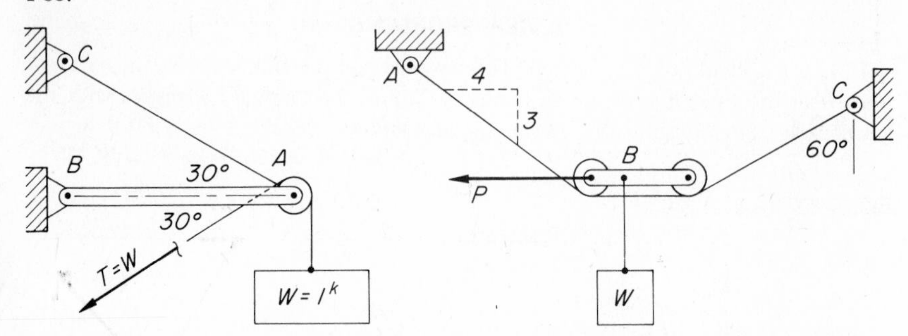

FIG. 4-85 Prob. 4-80. FIG. 4-86 Prob. 4-81.

4-81. Calculate the load W and pull P if the cable tension must not exceed 2,000 lb when the carriage is in the position shown in Fig. 4-86. *Ans.* W = 2,200 lb; P = 132 lb.

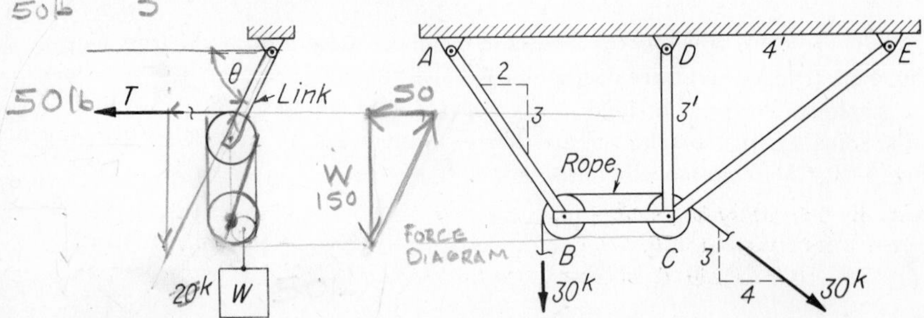

FIG. 4-87 Prob. 4-82. FIG. 4-88 Prob. 4-83.

4-82. In Fig. 4-87 solve analytically for the *stress* in the link, and for the angle θ_H it makes with the horizontal. Assume static conditions and neglect sheave friction. *Ans.* T = 22.36 kips; $\theta_H = 63°27'$

4-83. Using the *analytical method of joints*, solve for the *stresses* in all members of the frame shown in Fig. 4-88. Neglect friction.

4-84. By the *analytical method of joints*, solve for the *stresses* in all members of the truss shown in Fig. 4-89.

4-85. Determine the *stresses* in all members of the truss shown in Fig. 4-90 by the *analytical method of joints*. *Ans.* AB = 15 kips T; AC = 12.4 kips C; BC = 11 kips T; BD = 29.1 kips T; BE = 15.7 kips C; CE = 12 kips C

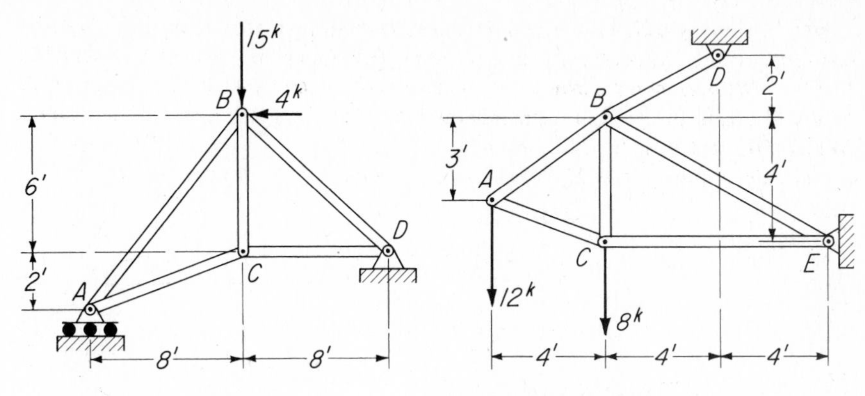

FIG. 4-89 Prob. 4-84.

FIG. 4-90 Prob. 4-85.

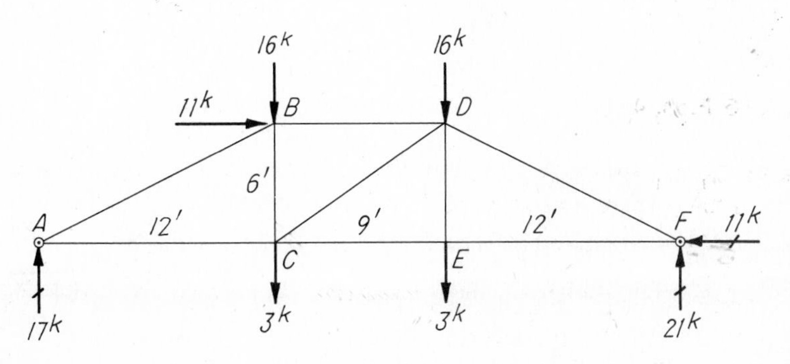

FIG. 4-91 Probs. 4-86, 4-91, 5-69, and 5-70.

4-86. By the *analytical method of joints,* solve for the *stresses* in all members of the roof truss shown in Fig. 4-91. *Ans.* $AB = 38$ kips C; $AC = 34$ kips T; $BC = 1$ kip T; $BD = 45$ kips C; $CD = 3.6$ kips T; $CE = 31$ kips T; $DF = 47$ kips C; $EF = 31$ kips T; $DE = 3$ kips T

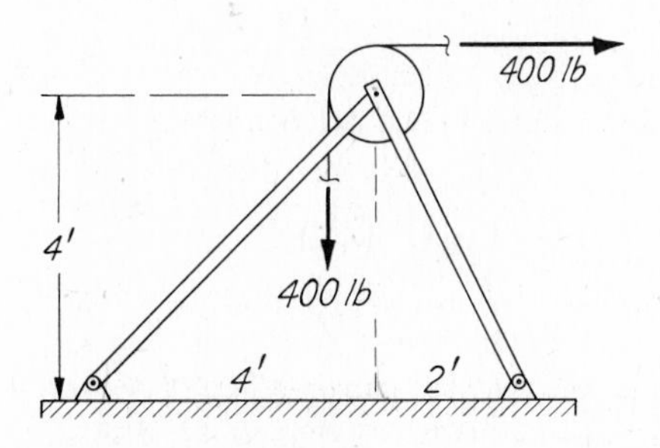

FIG. 4-92 Prob. 4-87.

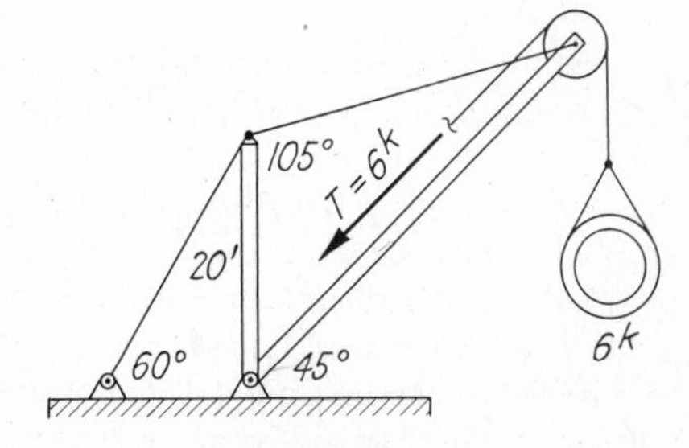

FIG. 4-93 Prob. 4-88.

4-87. Using the *graphical method of joints,* determine the *stresses* in the two members of the frame shown in Fig. 4-92. (Scales on $8\frac{1}{2}$ by 11: 1 in. = 2 ft, 1 in. = 100 lb.) *Ans.* Left = 198 lb T; right = 596 lb C

4-88. Find the *stresses* in all members of the derrick shown in Fig. 4-93, by the *graphical method of joints.* (Scales on 11 by 17: 1 in. = 6 ft, 1 in. = 3 kips.)

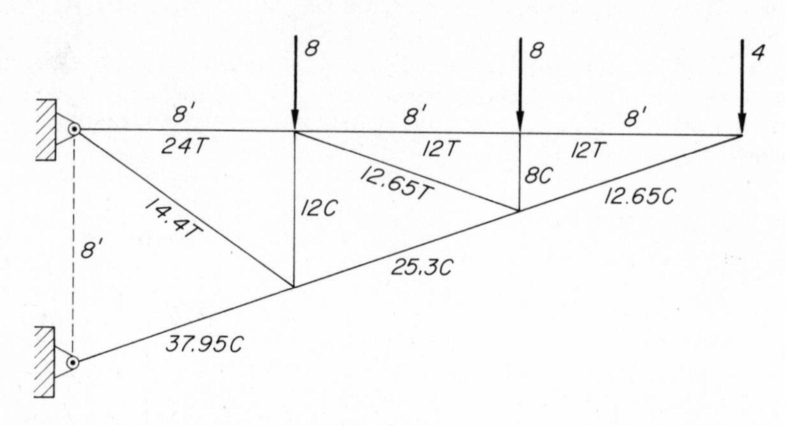

FIG. 4-94 Prob. 4-89.

4-89. Using the combined-diagram method, determine graphically the stresses in all members of the truss shown in Fig. 4-94. Record these stresses on a space diagram and indicate whether *T* or *C*. Force scale: 1 in. = 5 kips.

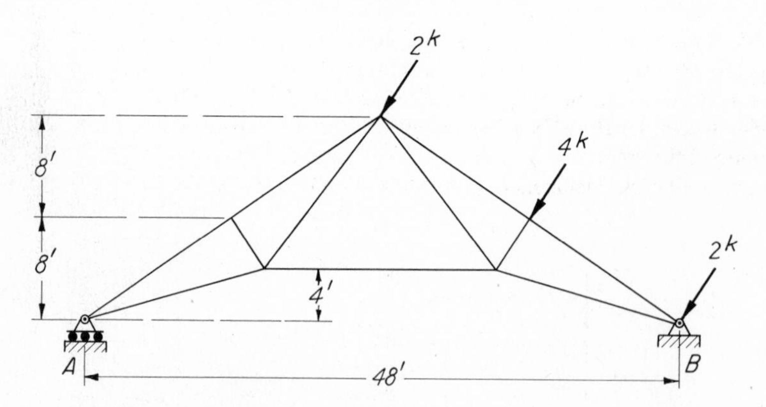

FIG. 4-95 Prob. 4-90. Cambered Fink roof truss, wind loads.

4-90. The cambered Fink roof truss in Fig. 4-95 is subjected to wind loads as shown. By the *three-force principle,* determine the *support reactions.* Then find all *stresses* by the *graphical combined-diagram method.* (Scales on 11 by 17: 1 in. = 6 ft, 1 in. = 2 kips.) *Ans.* R_A = 2.41 kips↑; R_B = 6.14 kips ↗

4-91. By the *graphical combined-diagram method,* solve for all stresses in the roof truss shown in Fig. 4-91. (Scales on 11 by 17: 1 in. = 4 ft, 1 in. = 4 kips.)

Ans. See Prob. 4-86.

4-92. Using the *three-force principle,* check the *support reactions* given in Fig. 4-59. (Scales on 8½ by 11: 1 in. = 4 ft, 1 in. = 2 kips.)

4-93. Check the *support reactions* in Fig. 4-63 by the *three-force principle.* (Scales on 8½ by 11: 1 in. = 10 ft, 1 in. = 2 kips.)

4-94. Using the *three-force principle,* determine the *reactions AD* and *DC* for the truss of Fig. 4-96. Then find the *stresses* in all members by the graphical *combined-diagram method.* Superimpose the combined-diagram solution on the three-force diagram. (Scales on 8½ by 11: 1 in. = 1 ft, 1 in. = 5 kips.)

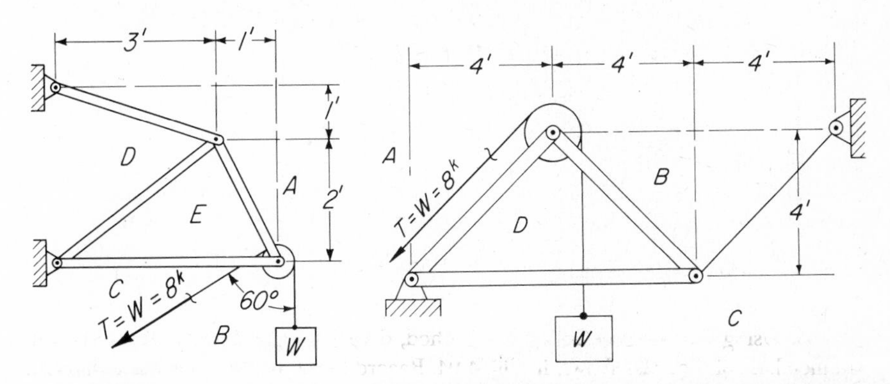

FIG. 4-96 Prob. 4-94. FIG. 4-97 Prob. 4-95.

4-95. Using the *three-force principle,* determine graphically the *reactions* at the supports of the truss shown in Fig. 4-97. Continue the graphical solution to determine the *stresses* in all members. (Scales on 8½ by 11: 1 in. = 4 ft, 1 in. = 4 kips.)

4-96. Using the *three-force principle,* determine graphically the *reactions* at the supports of the truss shown in Fig. 4-98. Continue the graphical solution to find the *stresses* in all truss members. (Scales on 11 by 17: 1 in. = 4 ft, 1 in. = 3 kips.)

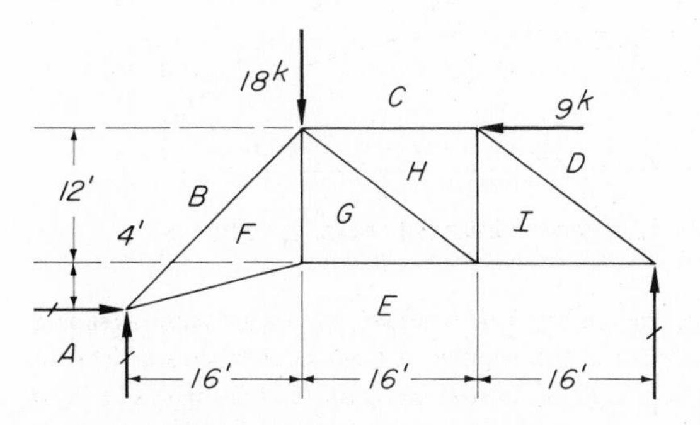

FIG. 4-98 Prob. 4-96.

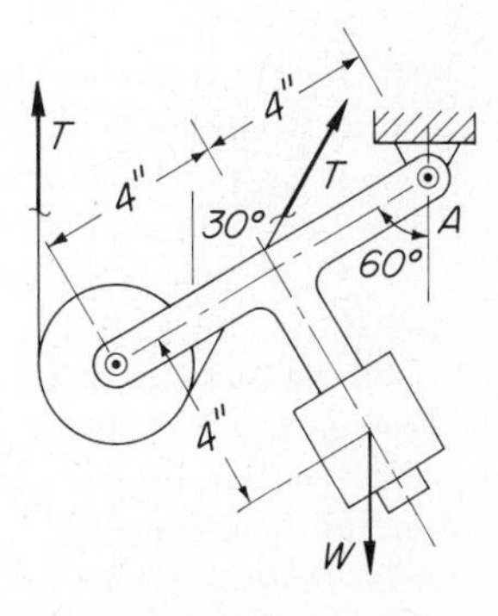

FIG. 4-99 Prob. 4-97.

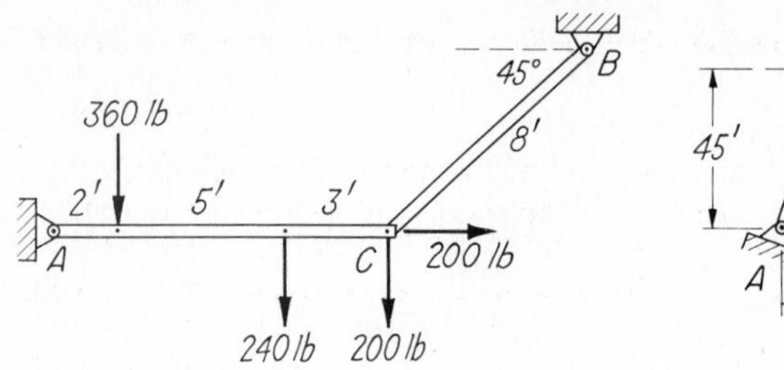

FIG. 4-100 Prob. 4-98.

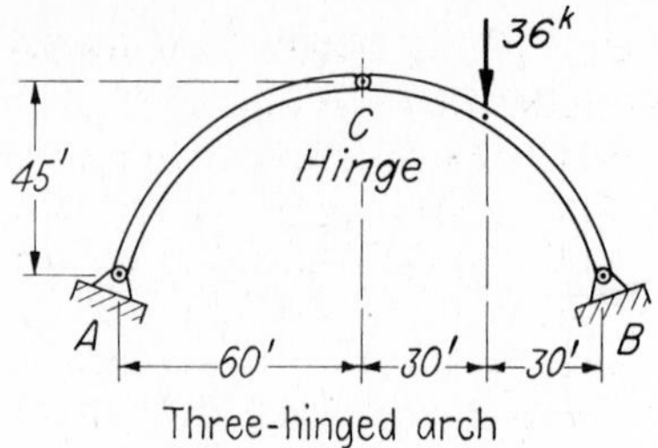

FIG. 4-101 Prob. 4-99.

4-97. Using the *three-force principle*, determine graphically the belt tension T and the reaction at A of the assembly shown in Fig. 4-99 if $W = 100$ lb. (Scales on $8\frac{1}{2}$ by 11: 1 in. = 2 in., 1 in. = 20 lb.) *Ans.* $T = 13.4$ lb, $A_V = 75$ lb, $A_H = 6.7$ lb

4-98. Using the *three-force principle*, determine graphically the *reactions* at supports A and B of the frame shown in Fig. 4-100. (Scales on $8\frac{1}{2}$ by 11: 1 in. = 3 ft, 1 in. = 200 lb.) *Ans.* $R_A = 734$ lb↖; $R_B = 622$ lb↗

4-99. Determine graphically the *reactions* at A and B of the three-hinged arch shown in Fig. 4-101. Use the *three-force principle.* (Scales on $8\frac{1}{2}$ by 11: 1 in. = 30 ft, 1 in. = 10 kips.) *Ans.* $R_A = 15$ kips↗; $R_B = 29.5$ kips↖

4-100. In Fig. 4-77, determine the *reactions* at A and B if an additional vertical load of 30 kips acts downward at E. Use the *three-force principle.* (HINT: Determine separately the reactions due to loads on each half of the arch; then find the resultant reactions. Scales on 11 by 17: 1 in. = 20 ft, 1 in. = 10 kips.)

REVIEW QUESTIONS

4-1. Define a coplanar, concurrent force system.

4-2. What must be known in order to define fully the resultant of a coplanar, concurrent force system?

4-3. By what three methods may the resultant of a coplanar, concurrent force system be determined?

4-4. Why must the graphical force polygon of a system of forces in equilibrium always close?

4-5. In the language of structures, what is a truss, a joint, and a member?

4-6. State the principal feature of a pin-connected truss.

4-7. What is a two-force member?

4-8. What is true about the forces acting on a two-force member?

4-9. What two types of stresses are commonly encountered in pin-connected trusses?

4-10. Name some of the methods by which stresses in members of trusses are determined.

4-11. Outline briefly how stresses in trusses are determined by the analytical method of joints.

4-12. State briefly how stresses in trusses are determined by the graphical method of joints.

4-13. Explain briefly how stresses in trusses are obtained by the graphical combined-diagram method.

4-14. What is a three-force member?

4-15. Explain briefly the types of stresses found in three-force members.

4-16. Can the direction of the reaction at the support of a three-force member be parallel to the member?

4-17. By means of what simple principle of statics may support reactions be determined graphically? State the principle itself.

4-18. In problems to determine support reactions, just what does the three-force principle establish?

CHAPTER 5

Coplanar, Nonconcurrent Force Systems

5-1 Introduction

In a nonconcurrent force system, the action lines of the forces do not meet at a common point. Each force in such a system will then tend to rotate the body, upon which it acts, about some axis. Therefore, *the principle of moments must be used in solving these systems.*

5-2 Resultant of a Coplanar, Nonconcurrent Force System

The resultant of a coplanar, nonconcurrent force system is fully known only when the following items are determined: (1) its magnitude and sense; (2) the angle of inclination θ of its action line with some reference line (usually horizontal); and (3) the location of some point on its line of action. The resultant may be determined analytically, or graphically as in Prob. 5-1. A graphical solution by the string-polygon method is outlined in Prob. 5-12.

ILLUSTRATIVE PROBLEM

5-1. Determine fully the resultant R of the three forces, A, B, and C, acting on bar DE shown in Fig. 5-1*a*, and the angle θ_H which the action line of R makes with the horizontal. All forces are in pounds.

Algebraic Solution: The method followed here is similar to that shown in Prob. 2-5. Sloping forces are replaced by their vertical and horizontal components. Summations of vertical forces and components, giving ΣV, and of horizontal forces and

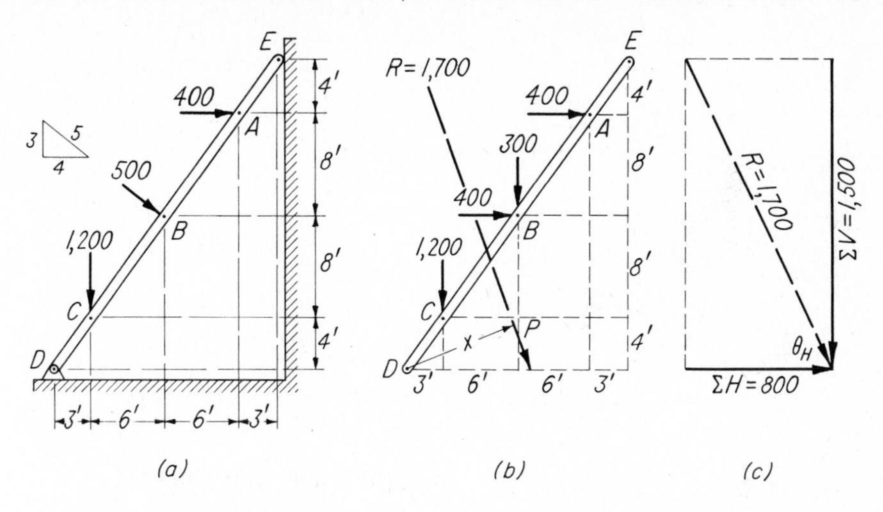

FIG. 5-1 Resultant of coplanar, nonconcurrent forces. Analytical solution. (*a*) Loaded member. (*b*) *V* and *H* components. (*c*) Components of *R*.

components, giving ΣH, are then made. From these we determine (*a*) the magnitude of *R* and (*b*) the inclination θ_H of its line of action with the horizontal.

The *V* and *H* components of force *B* are found to be 300 and 400 lb, respectively, as shown in Fig. 5-1*b*. Then ΣV is 300 + 1,200 or 1,500 lb and ΣH is 400 + 400 or 800 lb (Fig. 5-1*c*). Hence,

$$[R = \sqrt{(\Sigma V)^2 + (\Sigma H)^2}] \qquad R = \sqrt{(1{,}500)^2 + (800)^2} = 1{,}700 \text{ lb} \searrow \qquad \textit{Ans.}$$

Also

$$\tan \theta_H = \frac{\Sigma V}{\Sigma H} = \frac{1{,}500}{800} \qquad \text{or} \qquad \theta_H = 61°56' \qquad \textit{Ans.}$$

A point *P* on the action line of *R* may now be determined as follows: By the principle of moments, the moment of *R* about any point equals the algebraic sum of the moments of the separate forces about the same point. Let *D* be the moment center and *x* the moment arm of *R*, perpendicular to its action line. Then

$$1{,}700x = (1{,}200)(3) + (400)(12) + (300)(9) + (400)(20) = 19{,}100$$

and

$$x = 11.2 \text{ ft} \qquad \textit{Ans.}$$

Graphical Solution: When the bar has been drawn to scale, the three forces are combined into their single resultant by the parallelogram method, as shown in Fig. 5-2. The action lines of forces *A* and *B* are extended backward to their intersection *F* from which *A* and *B* are laid off. The diagonal R_{AB} of the completed parallelogram is the resultant of *A* and *B*. Next we combine R_{AB} and *C* by laying off their magnitudes from *G*, the intersection of their action lines. The diagonal R_{ABC} of the second parallelogram is then the resultant of *A*, *B*, and *C*. It scales 1,700 lb. θ_H scales 62° and *x* scales 11.2 ft.

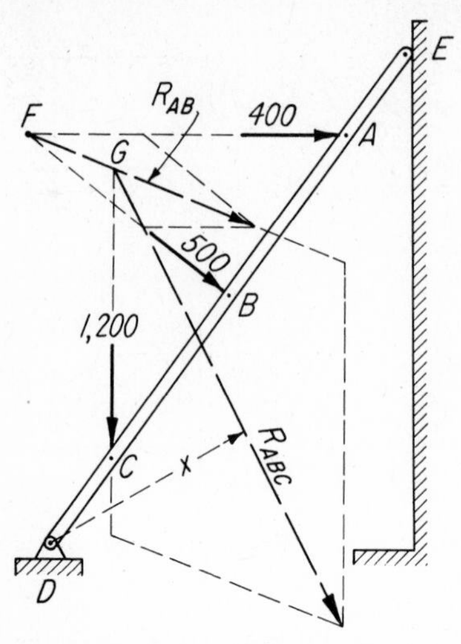

FIG. 5-2 Resultant force; parallelogram method.

PROBLEMS

In the following problems let R be the resultant force, θ_H the angle which the action line of R makes with the horizontal, and x the moment arm from O (or other designated moment center) to the action line of R. All forces are in pounds.

5-2. Determine algebraically R, θ_H, and x (from O) of the four forces shown in Fig. 5-3.

5-3. Find algebraically R, θ_H, and x (from O) of the four forces shown in Fig. 5-4. Check graphically using the parallelogram method. (SUGGESTION: Combine A and B, then C and D, then R_{AB} and R_{CD}. Scales on $8\frac{1}{2}$ by 11: 1 in. = 1 ft, 1 in. = 10 lb.) *Ans.* $R = 12.2$ lb, $\theta_H = 17°$, $x = 2.72$ ft

5-4. Using the graphical parallelogram method, determine R, θ_H, and x (from O) of the four forces shown in Fig. 5-5. (SUGGESTION: Combine A and B, then C and D, then R_{AB} and R_{CD}. Scale on $8\frac{1}{2}$ by 11: 1 in. = 1 ft, 1 in. = 30 lb.)

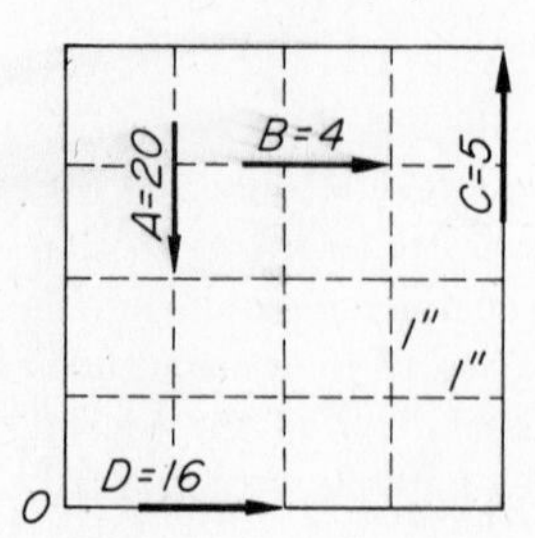

FIG. 5-3 Prob. 5-2.

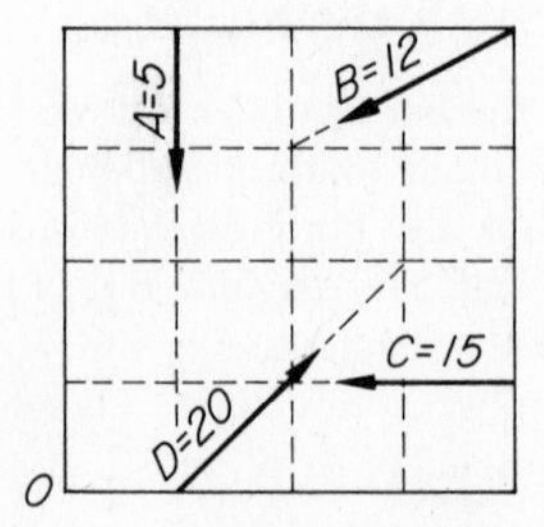

FIG. 5-4 Prob. 5-3.

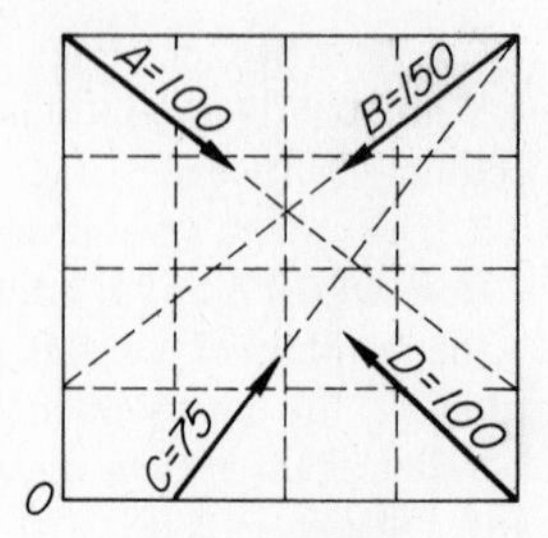

FIG. 5-5 Prob. 5-4.

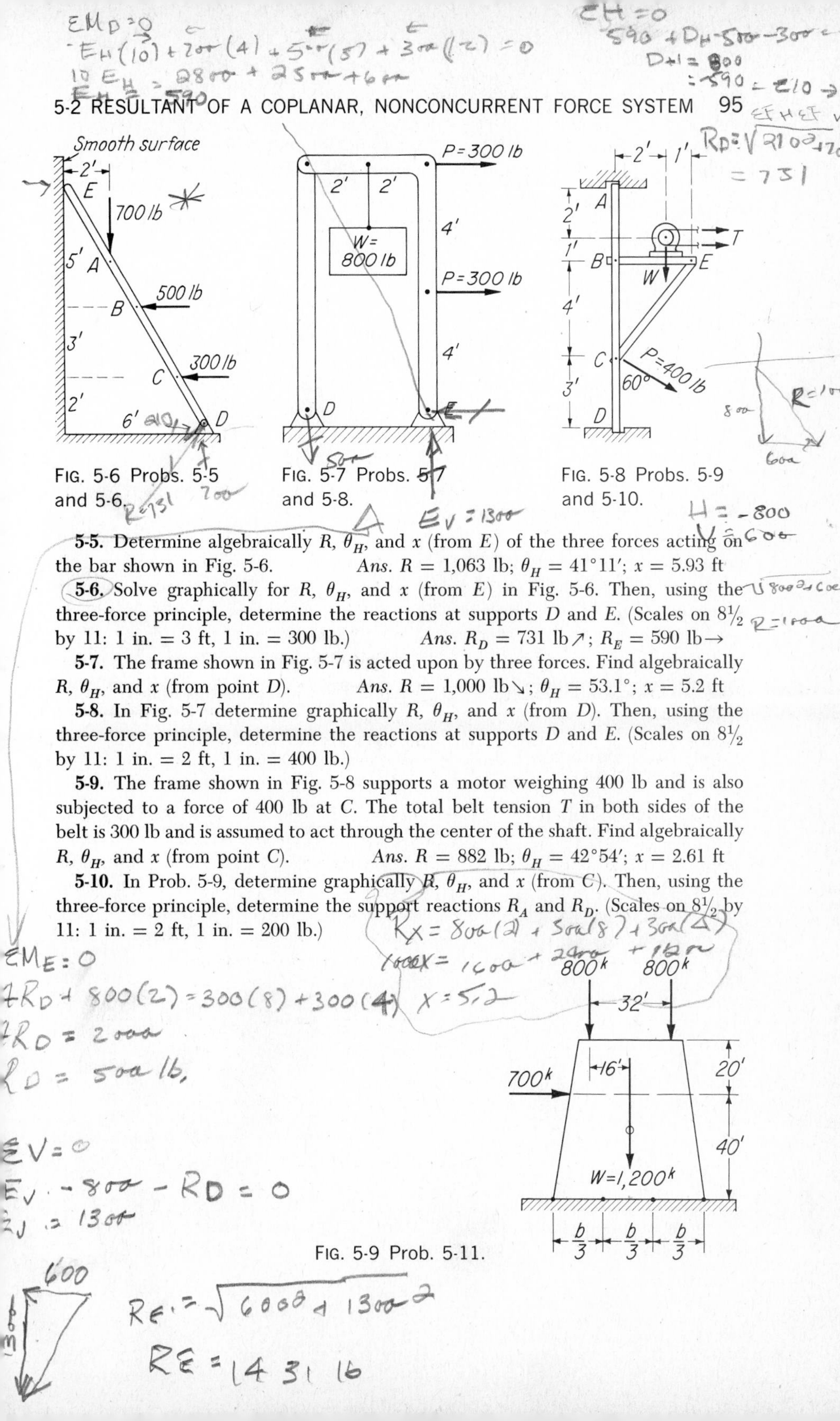

FIG. 5-6 Probs. 5-5 and 5-6.

FIG. 5-7 Probs. 5-7 and 5-8.

FIG. 5-8 Probs. 5-9 and 5-10.

5-5. Determine algebraically R, θ_H, and x (from E) of the three forces acting on the bar shown in Fig. 5-6. *Ans.* $R = 1{,}063$ lb; $\theta_H = 41°11'$; $x = 5.93$ ft

5-6. Solve graphically for R, θ_H, and x (from E) in Fig. 5-6. Then, using the three-force principle, determine the reactions at supports D and E. (Scales on $8\frac{1}{2}$ by 11: 1 in. = 3 ft, 1 in. = 300 lb.) *Ans.* $R_D = 731$ lb ↗; $R_E = 590$ lb →

5-7. The frame shown in Fig. 5-7 is acted upon by three forces. Find algebraically R, θ_H, and x (from point D). *Ans.* $R = 1{,}000$ lb ↘; $\theta_H = 53.1°$; $x = 5.2$ ft

5-8. In Fig. 5-7 determine graphically R, θ_H, and x (from D). Then, using the three-force principle, determine the reactions at supports D and E. (Scales on $8\frac{1}{2}$ by 11: 1 in. = 2 ft, 1 in. = 400 lb.)

5-9. The frame shown in Fig. 5-8 supports a motor weighing 400 lb and is also subjected to a force of 400 lb at C. The total belt tension T in both sides of the belt is 300 lb and is assumed to act through the center of the shaft. Find algebraically R, θ_H, and x (from point C). *Ans.* $R = 882$ lb; $\theta_H = 42°54'$; $x = 2.61$ ft

5-10. In Prob. 5-9, determine graphically R, θ_H, and x (from C). Then, using the three-force principle, determine the support reactions R_A and R_D. (Scales on $8\frac{1}{2}$ by 11: 1 in. = 2 ft, 1 in. = 200 lb.)

FIG. 5-9 Prob. 5-11.

5-11. The concrete bridge pier shown in Fig. 5-9 must support the two vertical loads from the bridge trusses and a horizontal ice load at the water surface as indicated. Satisfactory stability of the pier requires that the resultant R of the four forces pass within the middle third of its base. Determine graphically the minimum permissible width b of the pier. (Scales on $8\frac{1}{2}$ by 11: 1 in. = 10 ft, 1 in. = 500 kips.)

Ans. $b = 60$ ft

Resultant Force; Graphical String-polygon Method. The use of the graphical parallelogram method, described in Prob. 5-1, often becomes impractical, especially so when the action lines of a group of forces whose resultant is desired become nearly parallel. The so-called *string-polygon method,* explained in Prob. 5-12, is readily applied to all problems.

ILLUSTRATIVE PROBLEMS

5-12. Using the *graphical string-polygon method,* determine the resultant force R of the system shown in Fig. 5-10a.

Graphical Solution; String-Polygon Method: To determine R fully, we must find its magnitude and direction, and a point on its line of action must be located. If the forces A, B, C, and D are laid down to scale to form the *load line,* and in the order shown in the *force diagram* (Fig. 5-10b), R will be the magnitude of their resultant shown in its true direction.

We may now locate a point on the action line of R as follows: The point O, called the **pole,** may be located at random. The **rays** AB, BC, CD, DR, and RA are then drawn from O to the respective *intersections* on the load line of forces A and B, B and C, C and D, D and R, and R and A. *These rays are in reality the components of the various forces:* that is, rays AB and BC are the components of force B; rays BC and CD are the components of force C, etc. In like manner, rays DR and RA are the components of the resultant force R. When we study this completed force diagram, we recognize that the AB component of force A is canceled by the AB

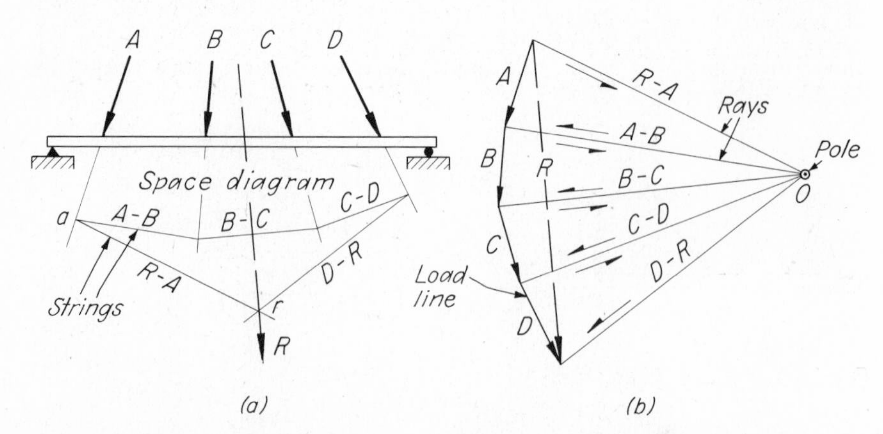

FIG. 5-10 Resultant force; graphical string-polygon method. (a) Action-line or string polygon. (b) Force polygon.

component of force B, because they are identical except for opposite senses. Likewise, the BC and CD components are canceled, leaving only the uncanceled components DR and RA of the resultant force R.

But the opposing components AB, BC, and CD can cancel only if they have common lines of action. To establish these, an *action-line*, or *string, polygon* is drawn, in which *the strings are the action lines of the opposing components*, as shown in Fig. 5-10*a*. We first select any suitable point, such as *a*, on the action line of force A as the starting point of the string polygon. From this point, on the action line of force A, to action line B, draw the **string** AB parallel to ray AB. From action line B to action line C, draw string BC parallel to ray BC. From action line C to action line D draw string CD parallel to ray CD. Point r on the action line of R is then established by drawing *strings* RA and DR parallel, respectively, to *rays* RA and DR. The resultant R in the space diagram is drawn parallel to R in the force polygon.

PROBLEMS

Resultant force; string-polygon method

5-13. Determine the resultant R of forces A, B, and C in Fig. 5-11. (Scales on $8\frac{1}{2}$ by 11: 1 in. = 5 ft, 1 in. = 400 lb.) *Ans.* R = 1,350 lb ↘

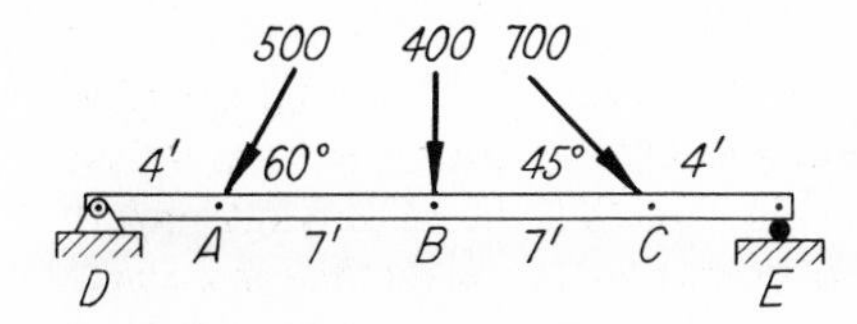

FIG. 5-11 Prob. 5-13.

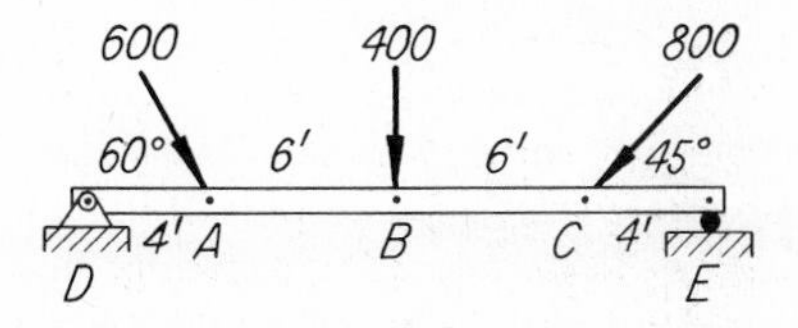

FIG. 5-12 Prob. 5-14.

5-14. In Fig. 5-12 find the resultant R of forces A, B, and C. (Scales on $8\frac{1}{2}$ by 11: 1 in. = 5 ft, 1 in. = 400 lb.)

5-15. Find the resultant R of forces A, B, and C in Fig. 5-13. (Scales on $8\frac{1}{2}$ by 11: 1 in. = 4 ft, 1 in. = 300 lb.)

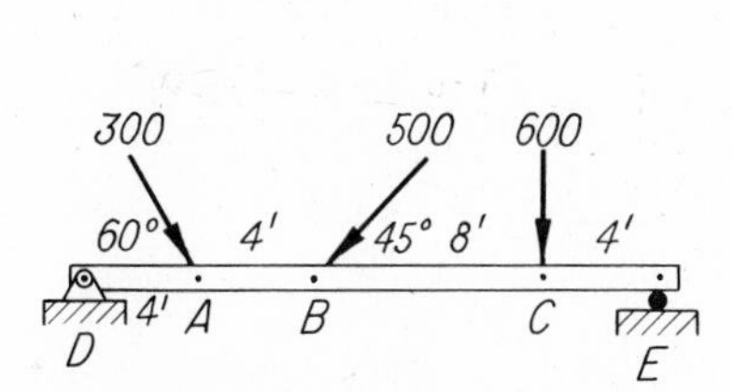

FIG. 5-13 Probs. 5-15 and 5-96.

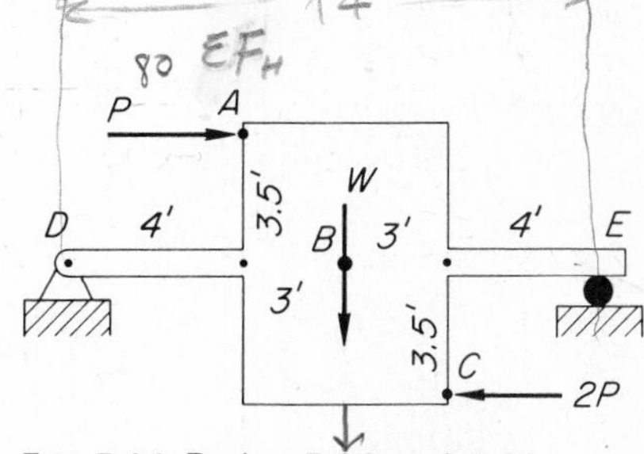

FIG. 5-14 Probs. 5-16 and 5-31.

5-16. Determine the resultant of the forces P and W in Fig. 5-14 if P = 80 lb and W = 200 lb. Force scale on $8\frac{1}{2}$ by 11: 1 in. = 100 lb.

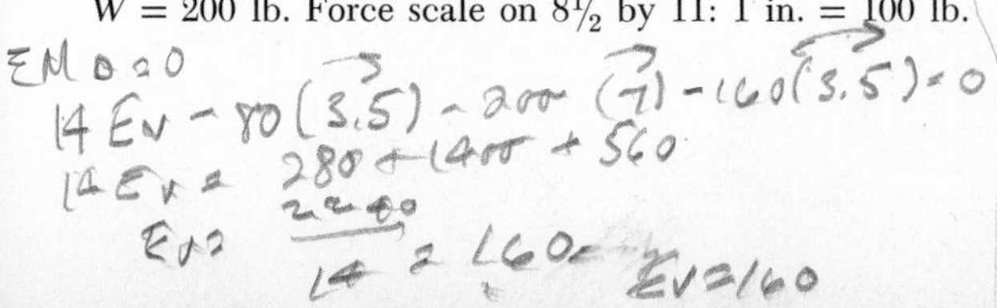

5-3 Equilibrium of Coplanar, Nonconcurrent Force Systems

Nonconcurrent force systems are most frequently encountered in problems involving the determination of unknown externally reacting forces at the supports of structures, structural members, machines, or machine parts. These force systems are also encountered in problems involving determination of stresses in members of trusses, as shown in Art. 5-7. Analytical solutions to determine reactions are based on the three equations of equilibrium, $\Sigma V = 0$, $\Sigma H = 0$, and $\Sigma M = 0$, as in Art 5-5. Graphical solutions employ the three-force principle as in Art 4-11 or the principles of the string and force polygons as is illustrated in Art. 5-4.

5-4 Determination of Reactions; the Graphical String-polygon Method

The basis of this method is similar to that explained in Art. 5-2 for determination of resultant force. When a system of forces is in equilibrium, however, all components of the forces in the force polygon necessarily cancel out; hence, *the string polygon must close with a closing string from the last action line to the point of beginning on the first action line.* The application of the method is most easily understood by noting carefully the various steps outlined in the following example.

ILLUSTRATIVE PROBLEM

5-17. Determine graphically by the string-polygon method the reactions at supports A and D of the beam shown in Fig. 5-15.

Solution: The space diagram is first drawn to scale. Next the loads B and C are laid off to scale to form the beginning of the force polygon, which later will be closed by the unknown reacting forces D and A. The direction of force D is seen to be vertical; that of A is unknown. Now we choose a pole and draw rays AB, BC, and

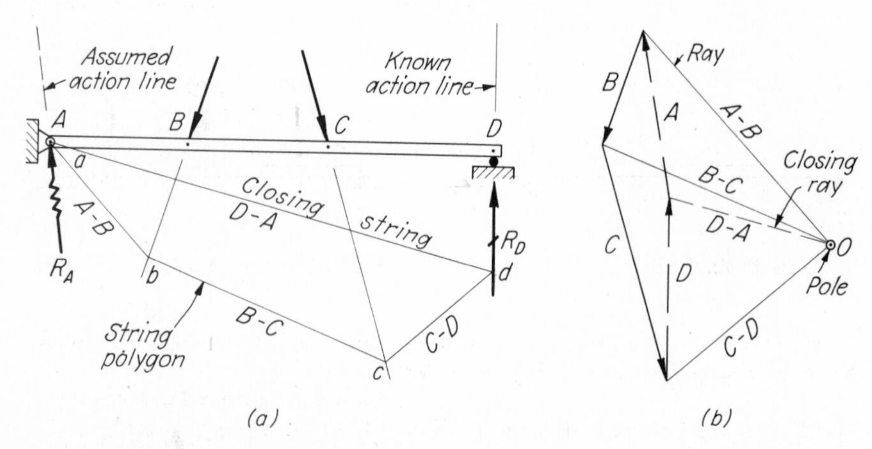

FIG. 5-15 Determination of reactions. String-polygon method. (*a*) Space diagram and string polygon. (*b*) Force polygon.

CD. The *closing ray* DA, the direction of which determines the magnitudes of reactions D and A, can be located only *after* the completion of the string polygon. The unknown direction of reaction A is indicated by the "buckled" arrow.

The string polygon must always start at the point of application of the reaction whose direction is unknown, since that is the only *known* point on the action line of reaction A. Then, from this point on action line A to action line B draw string AB parallel to ray AB; from action line B to action line C draw string BC parallel to ray BC; and from action line C to action line D draw string CD parallel to ray CD.

Only one ray, DA, remains to be located. *This closing ray must be parallel to the closing string* which extends from action line D to the point of beginning on line A. The string polygon may now be closed by string DA, after which ray DA locates the intersection of reactions D and A, thus determining the magnitudes of both and the direction of A.

If the directions of all action lines are known, as in parallel force systems, the string polygon may be started at any point on any action line. *The string polygon is so called because it takes the exact form of a (weightless) string, which is acted upon by forces A, B, C, and D at "joints" a, b, c, and d, respectively.* If now each "joint" is solved, we find that the "stress" in each string is given by the correspondingly lettered force component (ray) in the force polygon. Hence, we recognize that the force polygon (Fig. 5-15*b*) is in reality the combined stress diagram giving the stresses in strings AB, BC, CD, and DA, produced by forces A, B, C, and D.

PROBLEMS

Reactions; string-polygon method

5-18. Determine the reactions at supports A and E of the beam shown in Fig. 5-16. Check by moments. (Scales on $8\frac{1}{2}$ by 11: 1 in. = 5 ft, 1 in. = 500 lb.)

Ans. $R_A = 1{,}075$ lb↑; $R_E = 1{,}325$ lb↑

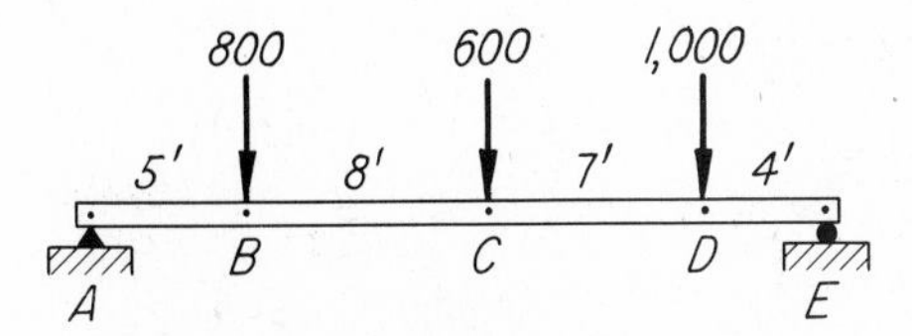

FIG. 5-16 Prob. 5-18.

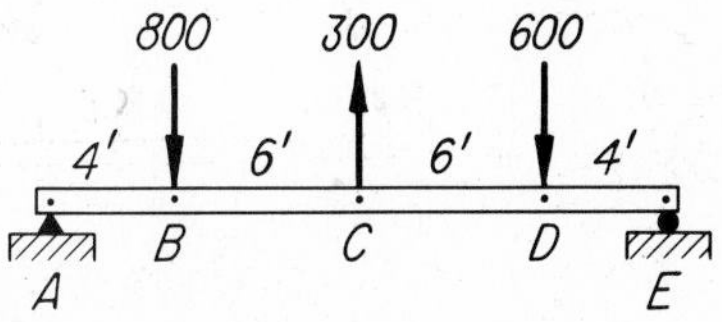

FIG. 5-17 Prob. 5-19.

5-19. In Fig. 5-17, find the reactions at supports A and E. Check by moments. (Scales on $8\frac{1}{2}$ by 11: 1 in. = 4 ft, 1 in. = 300 lb.)

Ans. $R_A = 610$ lb↑; $R_B = 490$ lb↑

5-20. Find the reactions produced by loads B and C at supports A and D of the beam shown in Fig. 5-18. Check R_D by moments about A. (Scales on $8\frac{1}{2}$ by 11: 1 in. = 3 ft, 1 in. = 6 kips.)

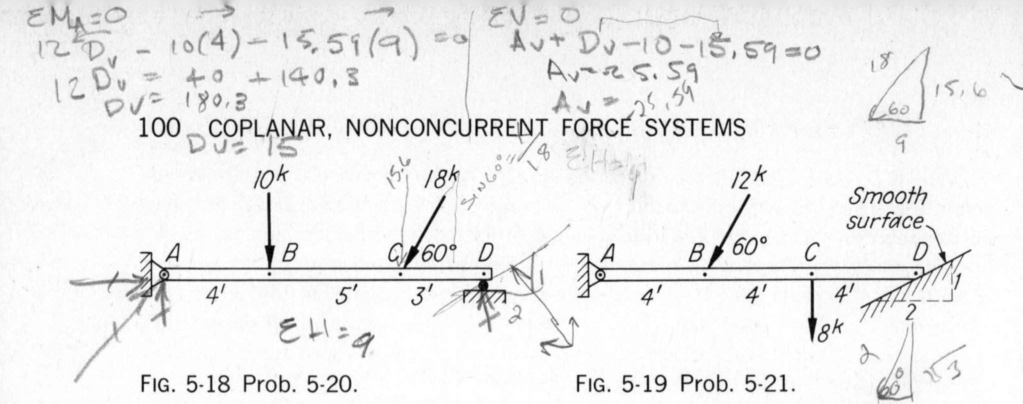

FIG. 5-18 Prob. 5-20.

FIG. 5-19 Prob. 5-21.

5-21. Determine the reactions produced at supports A and D of the beam shown in Fig. 5-19 by loads B and C. Check R_D by moments about A. (Scales on $8\frac{1}{2}$ by 11: 1 in. = 3 ft, 1 in. = 5 kips.) *Ans.* $R_A = 14.2$ kips ↗; $R_D = 9.84$ kips ↖

5-22. The Fink roof truss shown in Fig. 5-20 is acted upon by the resultant dead load B and the resultant wind load C. Find R_A and R_D. (Scales on $8\frac{1}{2}$ by 11: 1 in. = 10 ft, 1 in. = 5 kips.)

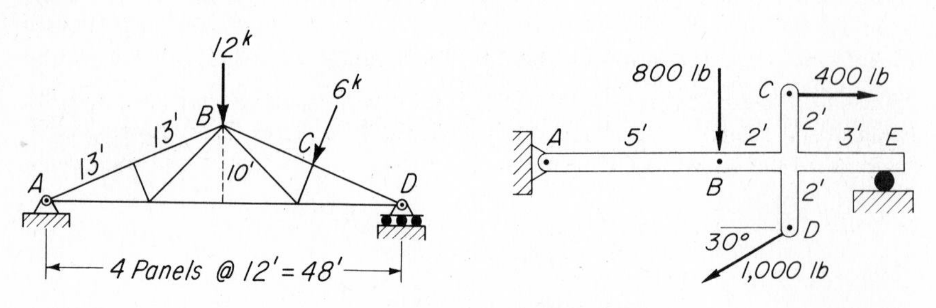

FIG. 5-20 Prob. 5-22.

FIG. 5-21 Probs. 5-23 and 5-28.

5-23. Determine the reactions at supports A and E of the beam shown in Fig. 5-21. (Scales on $8\frac{1}{2}$ by 11: 1 in. = 2 ft, 1 in. = 300 lb.)

Ans. $R_A = 548$ lb, $R_E = 1{,}003$ lb

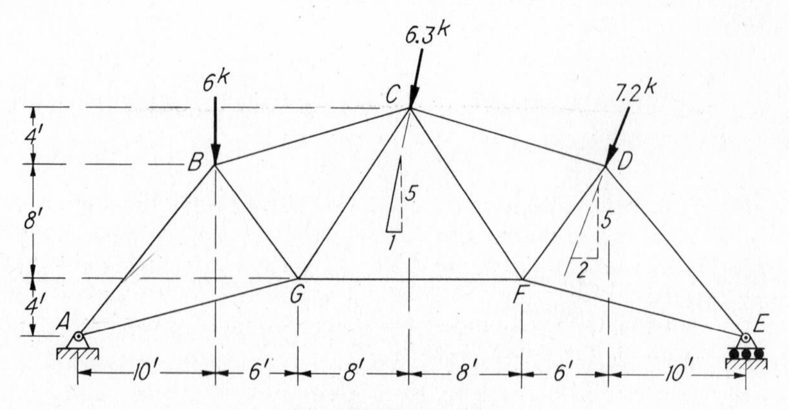

FIG. 5-22 Prob. 5-24.

5-24. In Fig. 5-22 is shown a roof truss acted upon by combined dead and wind loads. Determine first the reactions at supports A and E. Then determine the stresses in all members, using the graphical combined-diagram method. (An all-graphical solution is especially desirable for this type of problem. Scales on 11 by 17: 1 in. = 6 ft, 1 in. = 3 kips.) *Ans.* $R_A = 11.04$ kips ↗; $R_E = 8.55$ kips↑

5-5 Determination of Reactions; the Analytical Method

At every support of a structure subjected to loads there is a single reacting force, the direction of which may or may not be known. If the direction as well as the magnitude of a reaction is unknown, it is generally replaced with two components of known directions, such as vertical and horizontal. Two unknowns, a magnitude and a direction, have then been replaced by two unknown magnitudes. The equations of solution are $\Sigma V = 0$, $\Sigma H = 0$, and $\Sigma M = 0$. In the following problems, suitable graphical solutions have been indicated.

ILLUSTRATIVE PROBLEMS

5-25. The roof truss shown in Fig. 5-23*a* is subjected to wind loads acting perpendicularly to the sloping chord. Solve analytically for the reaction R_A at support A, and the vertical and horizontal components E_V and E_H of the reaction at support E.

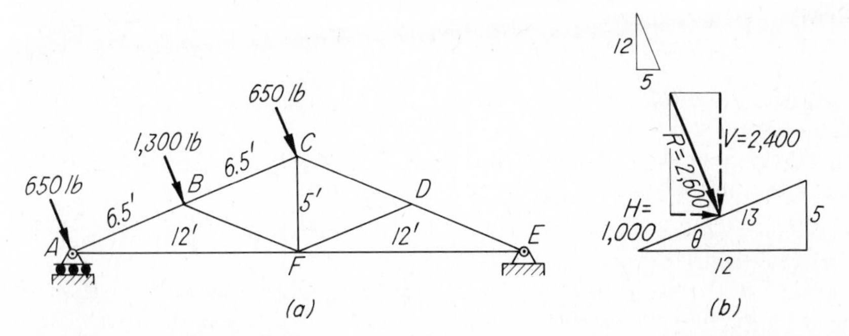

FIG. 5-23 Analytical determination of support reactions. (*a*) Space diagram. (*b*) Components of resultant.

Solution: Since at A a roller is supported on a horizontal surface, the reaction R_A is vertical. The direction of the reaction at E is unknown. Hence, we replace it with, and solve for, its V and H components. This load system is symmetrical about joint B. Therefore, the resultant force ($R = 2{,}600$ lb), applied at B, may replace the three loads. The solution is further simplified by replacing R with its V and H components. From the diagram in Fig. 5-23*b*, we have

$$\frac{V}{12} = \frac{H}{5} = \frac{2{,}600}{13}$$

or $\quad V = \dfrac{(12)(2{,}600)}{13} = 2{,}400 \text{ lb} \quad$ and $\quad H = \dfrac{(5)(2{,}600)}{13} = 1{,}000 \text{ lb}$

If now a separate diagram of the truss is drawn, as in Fig. 5-24, with these components of R acting at B, R_A may be found by moments about E, and E_V by moments about A. A check on these is found by $\Sigma V = 0$. Finally, E_H is found by $\Sigma H = 0$. The equations are

$[\Sigma M_E = 0] \quad 24R_A + (1{,}000)(2.5) = (2{,}400)(18) \quad$ and $\quad R_A = 1{,}696 \text{ lb}\uparrow \quad$ *Ans.*
$[\Sigma M_A = 0] \quad 24E_V = (1{,}000)(2.5) + (2{,}400)(6) \quad$ and $\quad E_V = 704 \text{ lb}\uparrow \quad$ *Ans.*
$[\Sigma V = 0] \quad 1{,}696 + 704 = 2{,}400 \quad$ *Check.*
$[\Sigma H = 0] \quad E_H = 1{,}000 \text{ lb} \leftarrow \quad$ *Ans.*

These answers should then be recorded on the free-body diagram to show *visually* that equilibrium exists.

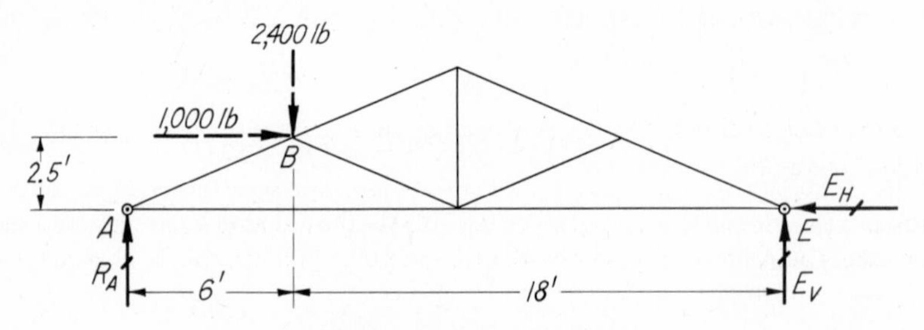

FIG. 5-24 Free-body diagram.

A *graphical solution* is readily obtained by means of the three-force principle. When the resultant R is applied at B, only three forces, R, R_A, and R_E, act on the truss, and their lines of action therefore meet at a common point. Having thus established the action line of R_E, a force polygon will give R_A and R_E.

5-26. Determine the vertical and horizontal reaction components at supports A and B of the frame shown in Fig. 5-25*a*.

Solution: The two supports are removed, and are replaced by the V and H components of the support reaction, as in Fig. 5-25*b*. We shall assume their senses to be as indicated. Because AC is a *three-force member* (Art. 4-10), no known relationship exists between the components A_V and A_H. Member BC, however, is a *two-force member* with a slope of 3 vertical to 2 horizontal. Therefore, $B_V/3 = B_H/2$, from which $B_V = \frac{3}{2}B_H$. We now solve for B_H by taking moments about A. If we assume that R_B acts *against* point B, both of its components will also act *against* B. (Unless this rule is strictly followed, the results obtained will be incorrect, and the mistake cannot be detected until a final check is made by a summation of moments.) Then,

$$[\Sigma M_A = 0] \qquad 10B_H + (4)\left(\frac{3}{2}B_H\right) = (8)(2) + (18)(8)$$

or $\quad 10B_H + 6B_H = 16 + 144 = 160 \quad$ and $\quad B_H = 10 \text{ kips}\rightarrow \quad$ *Ans.*

Hence, $\quad B_V = \dfrac{3}{2}B_H = 15 \text{ kips}\uparrow \quad$ *Ans.*

ΣM_A = 0
4B_V + 10(B_H) − 18(8) − 8(2) = 0

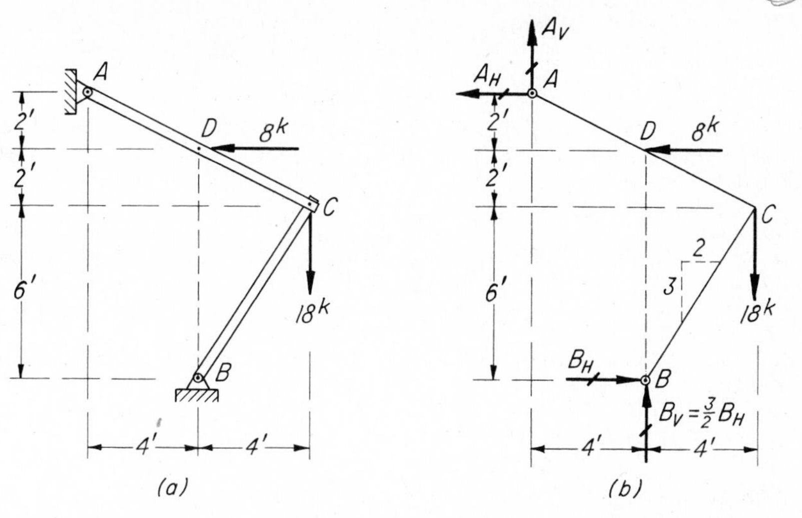

FIG. 5-25 Analytical determination of support reactions. (*a*) Space diagram. (*b*) Free-body diagram.

The positive sign of B_H indicates a correct assumption as to its sense. The reaction components at A are now found by the force-equilibrium equations. That is,

$[\Sigma V = 0]$ $\quad A_V + 15 = 18 \quad$ and $\quad A_V = 3$ kips↑ $\quad$ *Ans.*

$[\Sigma H = 0]$ $\quad A_H + 8 = 10 \quad$ and $\quad A_H = 2$ kips← $\quad$ *Ans.*

The values thus obtained should now be checked by a moment equation in which is included all or most of the computed forces. Choosing D as a moment center, we have

$[\Sigma M_D = 0]$ $\quad (18)(4) - (10)(8) - (2)(2) + (3)(4) = 0 \quad$ or $\quad 84 = 84 \quad$ *Check.*

This problem is easily checked graphically if the two known forces are combined into their resultant. The action line of R_A may then be found by the three-force principle, and R_A and R_B by a simple force diagram.

5-27. Find the V and H reaction components at supports A and B of the two-member frame shown in Fig. 5-26.

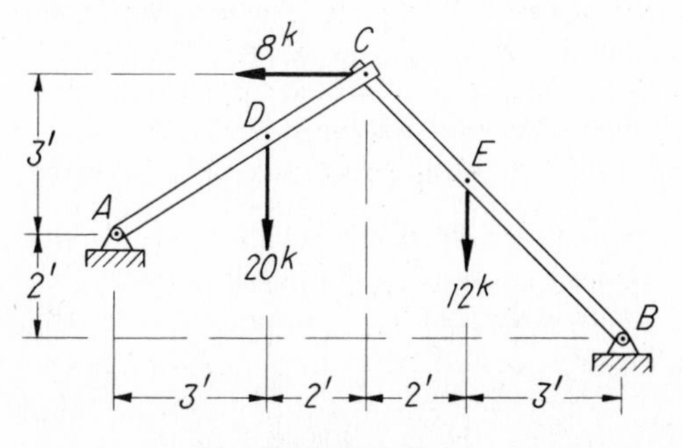

FIG. 5-26 Reactions.

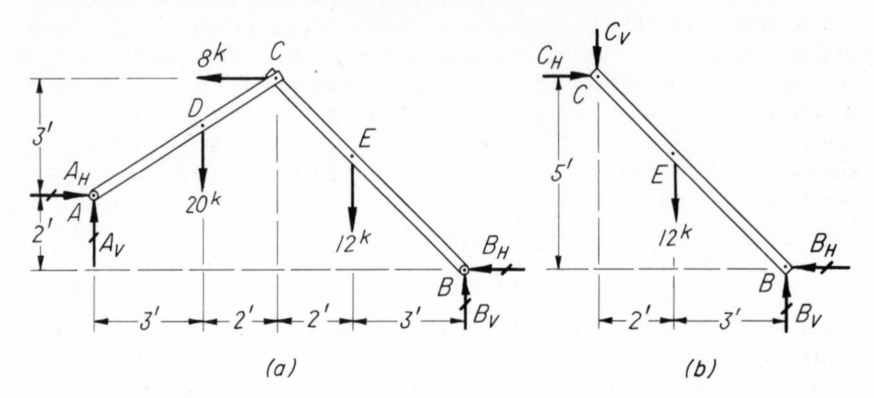

FIG. 5-27 Analytical determination of support reactions. (*a*) Entire frame as free body. (*b*) Member *BC* as free body.

Solution: In this problem, *AC* and *BC* are both three-force members. Therefore, no known relationship exists between the *V* and *H* components at either support, such as simplified the solution of Prob. 5-26. A summation of moments about either *A* or *B* will give an equation involving two unknown forces, and a second and independent equation must be written involving the same two unknowns. A simultaneous solution of these two equations will then give the unknown forces.

Using the entire frame as a free body, we select point *A* as the center of moments, and assume that B_V and B_H both act *against* joint *B*. Then, from Fig. 5-27*a*,

$$[\Sigma M_A = 0] \qquad 10B_V + (8)(3) = 2B_H + (20)(3) + (12)(7)$$

or

$$10B_V - 2B_H = 120 \tag{a}$$

If now member *BC* is isolated, as in Fig. 5-27*b*, we may write a second independent equation by a summation of moments about *C*. The reacting force at *C* is replaced by its *V* and *H* components, as yet unknown. Their moments are zero when *C* is the center of moments. Then,

$$[\Sigma M_C = 0] \qquad 5B_V = 5B_H + (12)(2) \qquad \text{or} \qquad 5B_V - 5B_H = 24 \tag{b}$$

Multiplying (*b*) by 2 and subtracting it from (*a*), we have

$$\begin{aligned} 10B_V - 2B_H &= 120 \qquad (a)\\ 10B_V - 10B_H &= 48 \qquad (c)\\ \hline 8B_H &= 72 \end{aligned} \qquad \text{from which} \qquad B_H = 9 \text{ kips}\leftarrow \qquad \textit{Ans.}$$

Substituting this value of B_H in (*a*), we have

$$10B_V - (2)(9) = 120 \qquad \text{or} \qquad 10B_V = 138 \qquad \text{and} \qquad B_V = 13.8 \text{ kips}\uparrow \qquad \textit{Ans.}$$

By use of the values of B_H and B_V just obtained, the reaction components at *A* are now readily solved for by applying the equations $\Sigma V = 0$ and $\Sigma H = 0$. If we assume A_V to be upward-acting and A_H to act from left to right, we obtain

$$[\Sigma V = 0] \qquad A_V + 13.8 = 20 + 12 \qquad \text{or} \qquad A_V = 18.2 \text{ kips}\uparrow \qquad \textit{Ans.}$$
$$[\Sigma H = 0] \qquad A_H = 9 + 8 \qquad \text{or} \qquad A_H = 17 \text{ kips}\rightarrow \qquad \textit{Ans.}$$

The positive signs of all computed components indicate a correct assumption as to their senses. (If an unknown sense is incorrectly assumed, the correct numerical value may be obtained but will be preceded by a negative sign.) The components A_V and A_H are dependent upon B_V and B_H. Therefore, *a check by summation of moments of all forces acting on the frame, preferably about a point that does not eliminate any computed force, is absolutely essential.* If we select point C as the center of moments, we have

$$[\Sigma M_C = 0] \qquad (18.2)(5) - (17)(3) - (20)(2) + (12)(2) + (9)(5) - (13.8)(5) = 0$$

or

$$160 = 160 \qquad \textit{Check.}$$

This problem is of a type generally referred to as the three-hinged arch. A graphical solution is most easily obtained by removing loads C and E, and solving for the reactions at A and B due to load D only. Next, remove load D and solve for the reactions at A and B due to loads C and E only. The two reactions at each support are then readily combined into a resultant reaction by constructing parallelograms, from which the true reactions or their components may be scaled.

PROBLEMS

Analytical determination of reactions. Graphical check solutions are suggested.

5-28. Solve for the components of the reactions at A and E in Fig. 5-21 by the analytical method.

5-29. Compute the reaction at support G of the Fink roof truss shown in Fig. 5-28, and the V and H reaction components at support A (see Prob. 5-25).

Ans. $R_G = 3{,}300$ lb↑; $A_V = 1{,}500$ lb↑; $A_H = 2{,}400$ lb→

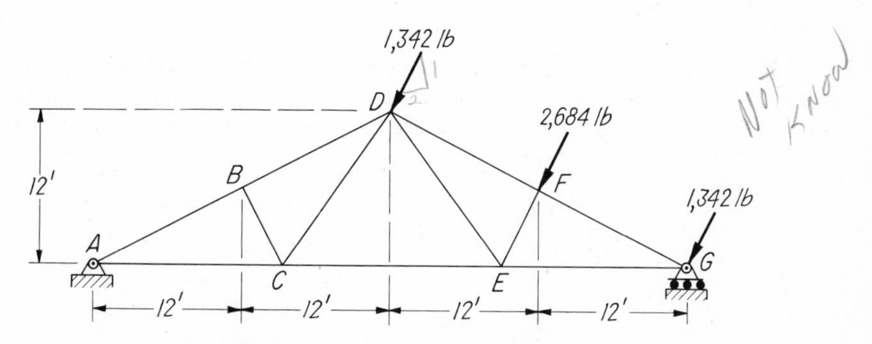

FIG. 5-28 Prob. 5-29. Fink truss, wind loads.

5-30. Find the reaction at support A of the sawtooth truss shown in Fig. 5-29, and the V and H components of the reaction at support G (see Prob. 5-27).

5-31. Solve for the reaction components at D and E in Fig. 5-14. $P = 80$ lb and $W = 200$ lb.

5-32. Determine the V and H reaction components at supports A and D of the truss shown in Fig. 5-30.

Ans. $A_V = 5$ kips↑; $A_H = 20$ kips←; $D_V = 10$ kips↑; $D_H = 20$ kips→

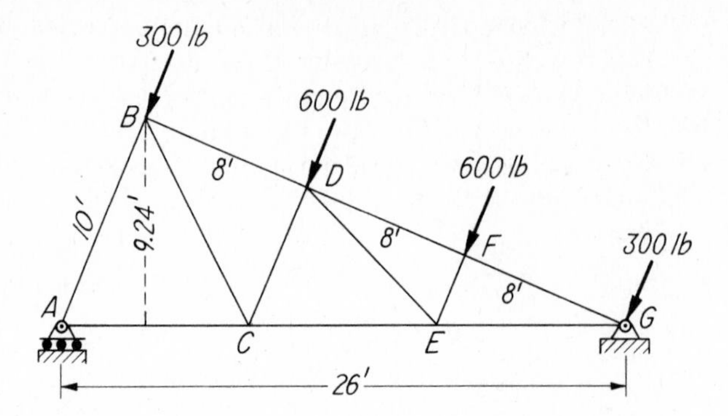

FIG. 5-29 Prob. 5-30. Sawtooth roof truss, wind loads.

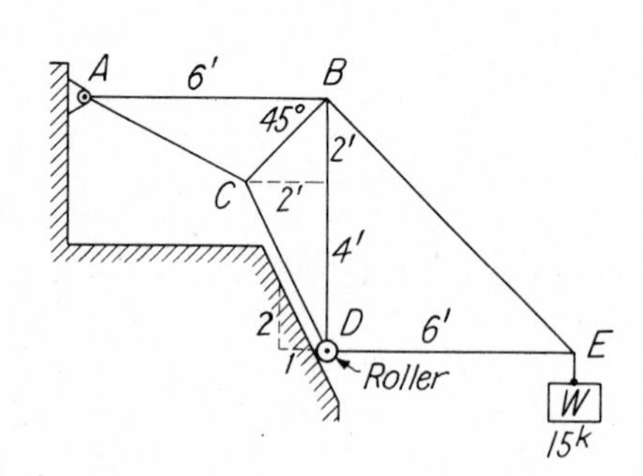

FIG. 5-30 Probs. 5-32 and 5-125.

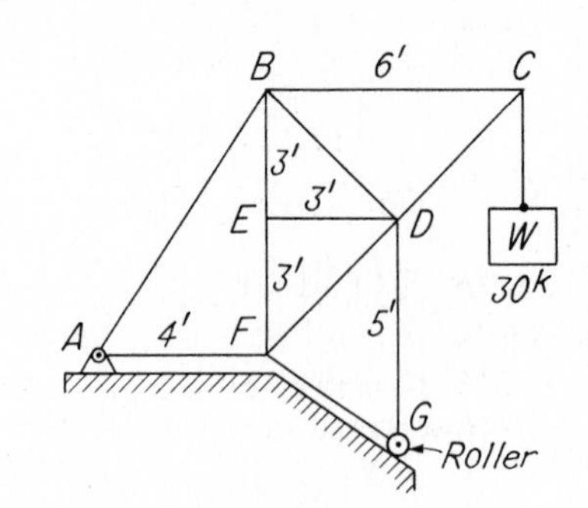

FIG. 5-31 Prob. 5-33.

5-33. Find the V and H reaction components at supports A and G of the truss shown in Fig. 5-31.

5-34. Determine the vertical and horizontal components of the reactions at A and B of the crane shown in Fig. 5-32. $W = 12$ kips.

Ans. $A_V = 20$ kips↑; $A_H = 8$ kips←; $B_V = 8$ kips↓; $B_H = 8$ kips→

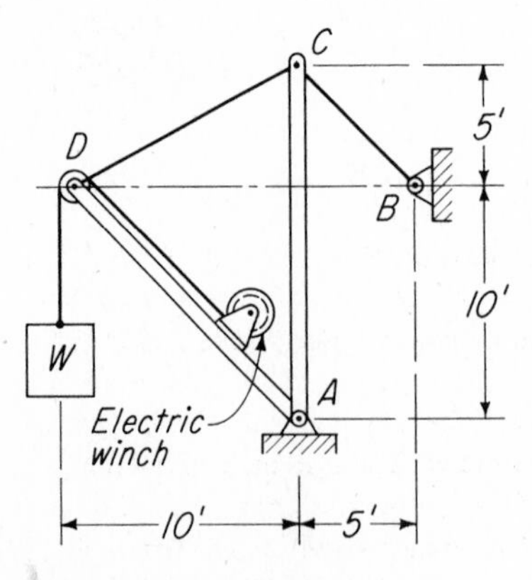

FIG. 5-32 Prob. 5-34.

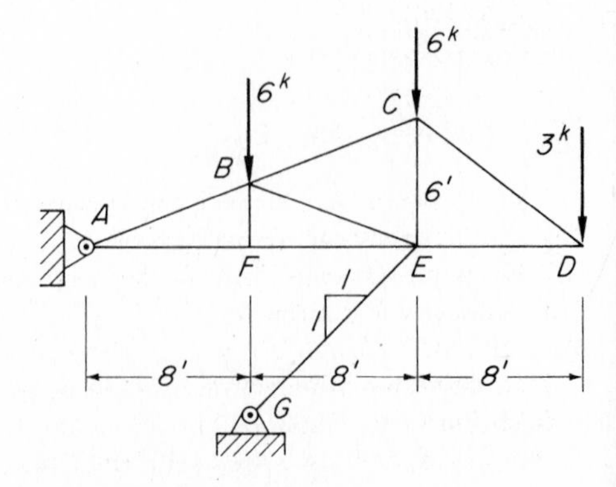

FIG. 5-33 Prob. 5-35.

5-35. Determine the vertical and horizontal components of the reactions at A and G of the truss shown in Fig. 5-33.

5-36. Compute the V and H reaction components at supports A and B of the truss shown in Fig. 5-34.

Ans. $A_V = 2$ kips↑; $A_H = 3$ kips←; $B_V = 10$ kips↑; $B_H = 15$ kips→

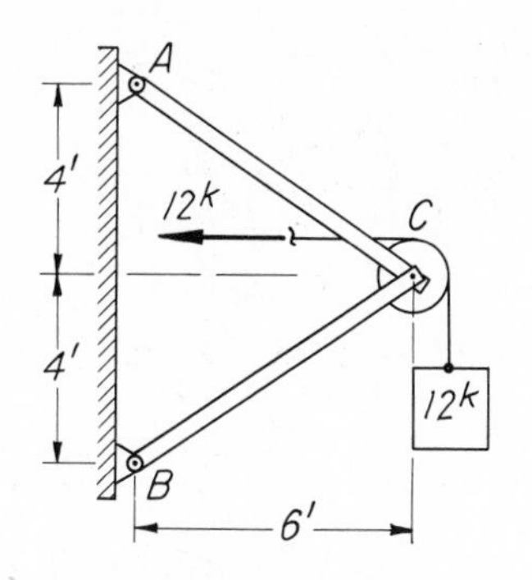

FIG. 5-34 Prob. 5-36.

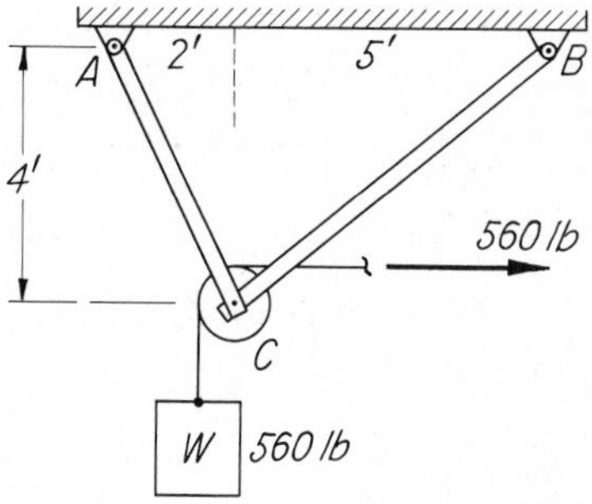

FIG. 5-35 Prob. 5-37.

5-37. Find the reactions R_A and R_B at the two supports of the truss shown in Fig. 5-35.

5-38. Compute the vertical and horizontal reaction components at A and B for the truss shown in Fig. 5-36.

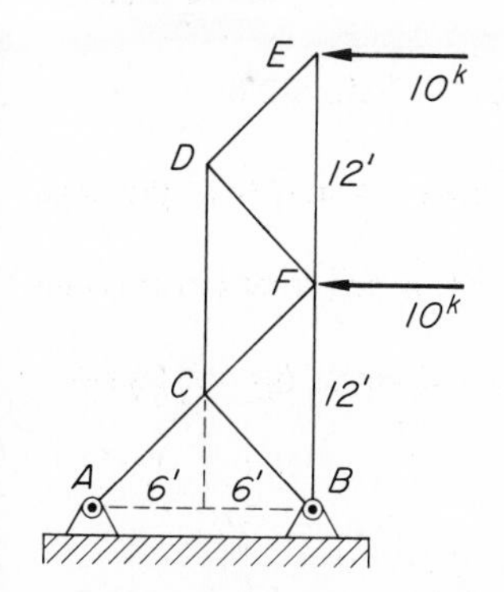

FIG. 5-36 Prob. 5-38.

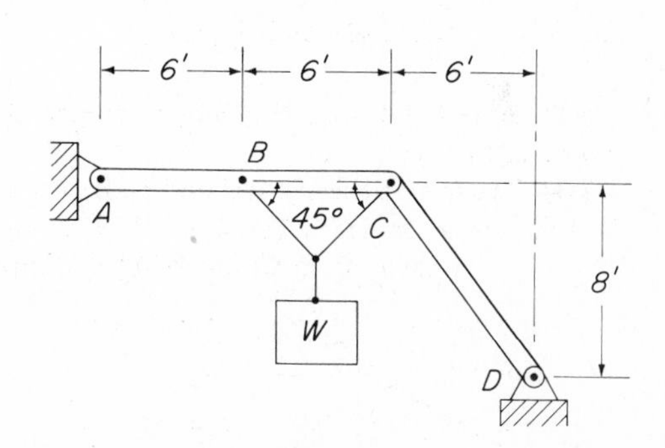

FIG. 5-37 Prob. 5-39.

5-39. Compute the vertical and horizontal reaction components at A and D in Fig. 5-37. The weight $W = 1{,}200$ lb.

5-40. Determine the V and H reaction components at supports at A and B of the truss shown in Fig. 5-38.

Ans. $A_V = 3.78$ kips↑; $A_H = 3.03$ kips←; $B_V = 10.22$ kips↑; $B_H = 13.03$ kips→

5-41. Solve for the reactions R_A and R_B at supports A and B in Fig. 5-39.

5-42. Find the V and H reaction components at supports A and F of the truss shown in Fig. 5-40. (SUGGESTION: Detach member EF and solve for the forces at E.)

Ans. $A_V = 12$ kips↓; $A_H = 4$ kips→; $F_V = 36$ kips↑; $F_H = 24$ kips←

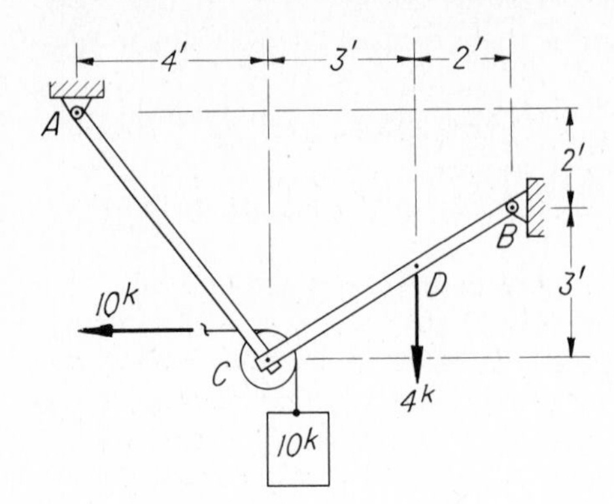

FIG. 5-38 Prob. 5-40.

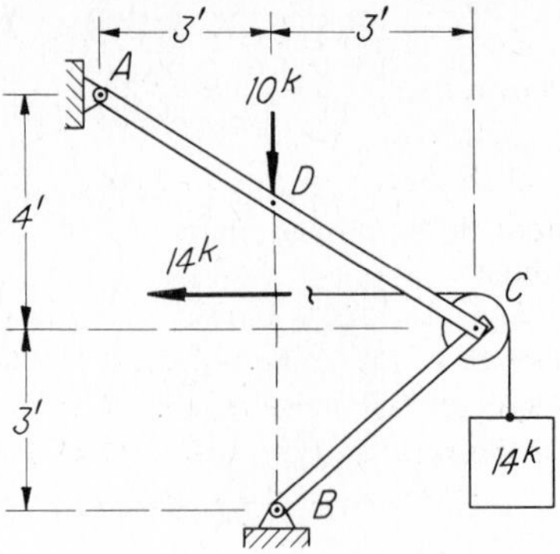

FIG. 5-39 Prob. 5-41.

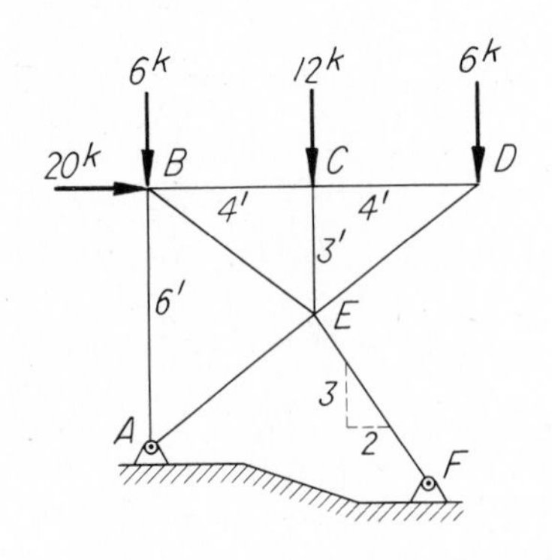

FIG. 5-40 Probs. 5-42, 5-92, and 5-124.

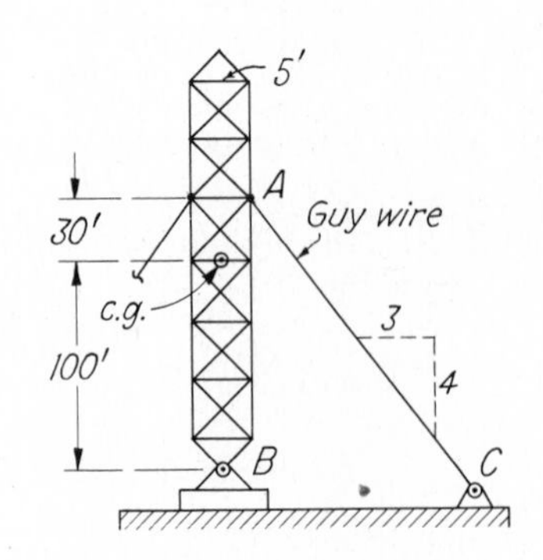

FIG. 5-41 Prob. 5-43.

5-43. The symmetrical steel radio tower shown in Fig. 5-41 is 5 ft square and 200 ft high and it weighs 60 kips. It is supported against horizontal wind loads by four guy wires attached 130 ft above the central base B. Its vertical projected area of 1,000 sq ft is subjected to horizontal wind pressure of 15 psf. When the wind blows from right to left, only guy wire AC is active. Compute the V and H reaction components at A and B and the tension T in AC.

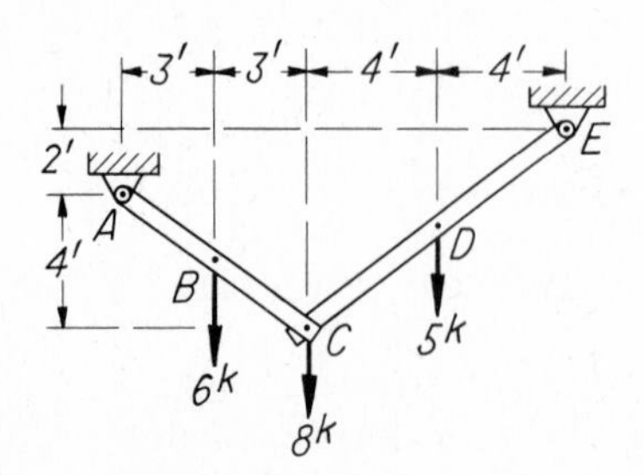

FIG. 5-42 Prob. 5-44.

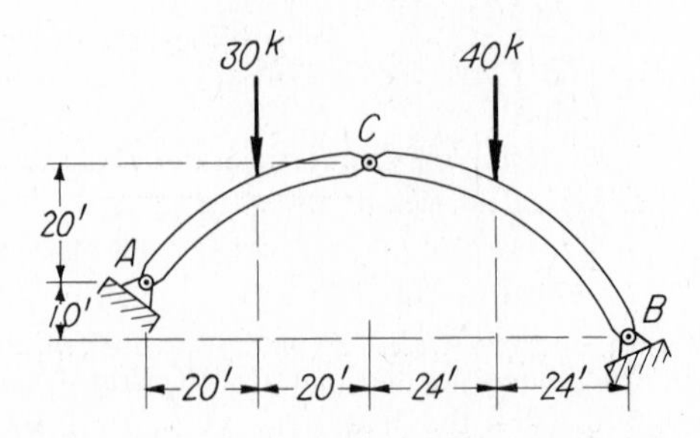

FIG. 5-43 Prob. 5-45.

5-44. Compute the V and H reaction components at supports A and E of the frame shown in Fig. 5-42 (see Prob. 5-27).

Ans. $A_V = 9.35$ kips↑; $A_H = 9.53$ kips←; $E_V = 9.65$ kips↑; $E_H = 9.53$ kips→

5-45. Solve for the reactions R_A and R_B at supports A and B of the three-hinged arch shown in Fig. 5-43, and for the V and H reaction components at the crown hinge C (see Prob. 5-27).

5-46. In Fig. 5-44 is shown a loading crane with its supporting framework. The weight of the horizontal truss is 5 kips and is concentrated at its center of gravity. Compute the V and H reaction components at supports A, B, and C due to the weight of the truss and the load at F as shown. Neglect weight of other members.

Ans. $A_V = 0.4$ kips↑; $A_H = 16.8$ kips→; $B_V = 25.2$ kips↑; $B_H = 8.4$ kips→
$C_V = 12.6$ kips↓; $C_H = 25.2$ kips←

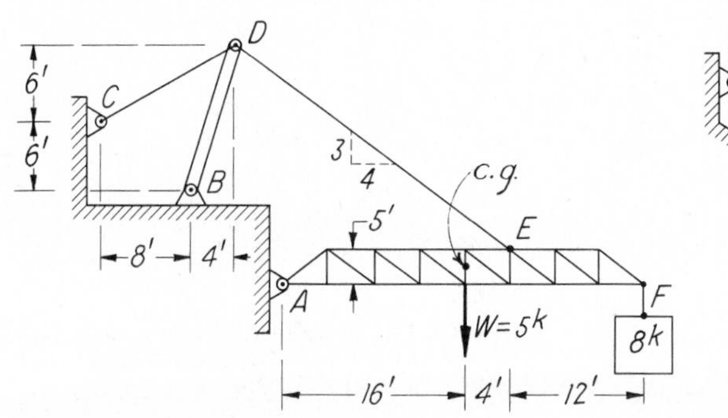

FIG. 5-44 Prob. 5-46.

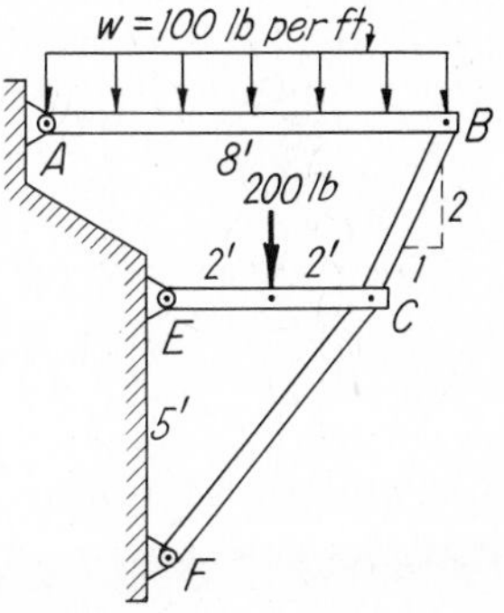

FIG. 5-45 Prob. 5-47.

5-47. Find the V and H reaction components at supports A, E, and F of the frame shown in Fig. 5-45. (SUGGESTION: Isolate member AB and consider first the unknown forces at A and B. Next isolate EC and CF as a unit.) *Ans.* $A_V = 400$ lb↑

5-48. In Fig. 5-46 is shown the boom of a concrete-road paver with its supporting framework. At E the flexible wire rope passes over a sheave, the small diameter of which may be disregarded. The weight of the boom is 600 lb which may be concen-

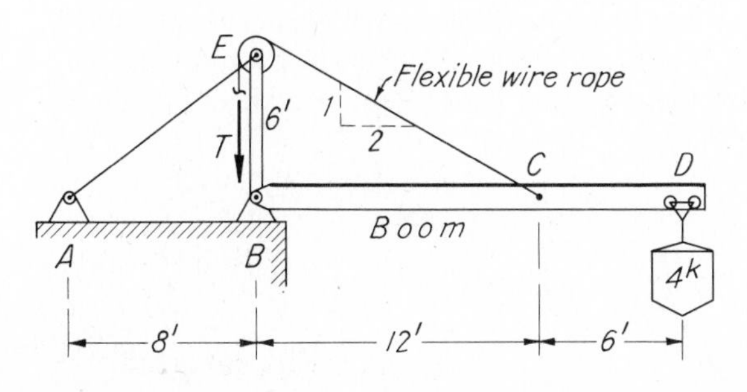

FIG. 5-46 Prob. 5-48.

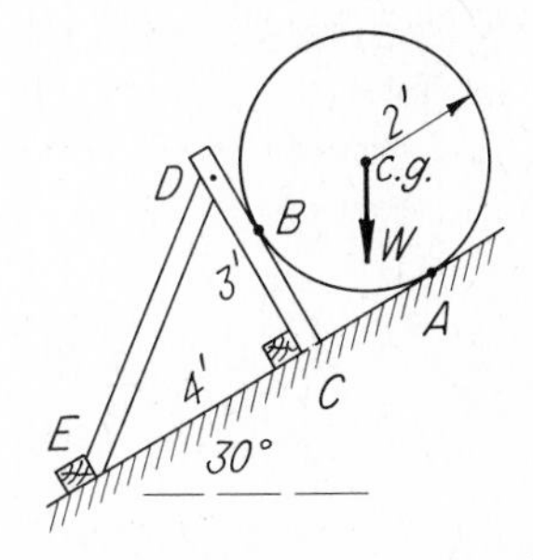

FIG. 5-47 Probs. 5-49 and 5-50.

trated at its center. Compute the tension T in the wire rope, and the V and H reaction component at supports A and B due to the weight of the boom and the load as shown.

Ans. $T = 14.42$ kips; $A_V = 9.68$ kips↓; $A_H = 12.9$ kips←; $B_V = 28.7$ kips↑; $B_H = 12.9$ kips→

5-49. A cylindrical tank, 4 ft in diameter and weighing 150 lb, is supported on an incline by a bracket, as shown in Fig. 5-47. Compute the reacting forces R_A and R_B acting perpendicularly to the supporting surfaces at A and B. Disregard possible friction.

5-50. Compute the stress in brace DE of the bracket shown in Fig. 5-47 (see Prob. 5-49).

Ans. 62.5 lb C

5-6 Pin Reactions; the Method of Members

Trusses are generally so constructed, loaded, and supported that all forces act at the joints, thereby producing simple axial or direct stresses (tensile or compressive) in the members which, then, are two-force members. In the design of structures and equipment, and in machines, three-force members (Art. 4-10) are avoided whenever possible, because the material in such members is less efficiently used than in two-force members. Frames in which three-force members *must* be used are often connected by pins. Forces produced by loads on such frames are transmitted through members to the pins at the joints, and by the pins to other connected members, and are referred to as **pin reactions.** In computing these pin reactions, any possible pin friction and friction between members connected at a joint are disregarded.

The usual procedure to determine these pin reactions is as follows: *All determinable support reactions, or their components, are first computed, using the frame as a whole.* Each three-force member is then isolated as a free body, beginning with one on which *known* forces act. When *all* forces acting on any single member are considered, they constitute a system in equilibrium. Hence, the unknown forces, or their components, may be found by application of the equations of static equilibrium, $\Sigma V = 0$, $\Sigma H = 0$, and $\Sigma M = 0$. From these components the actual pin reactions are readily computed. Possible difficulty concerning the senses of these forces may be overcome by remembering that *a force acting on any given member represents the action, pushing or pulling, of another member on the given member.* Since each member is thus considered individually, the method is logically referred to as the **method of members.**

ILLUSTRATIVE PROBLEM

5-51. The frame shown in Fig. 5-48 is held in place by a pin connection at A and rests on a smooth surface at F. Determine the reactions at A and F, and the pin reactions at B and C.

Solution: The support reactions at A and F are first determined, R_F by moments about A, A_V by moments about F, and A_H by $\Sigma H = 0$. Then $R_A = \sqrt{A_V^2 + A_H^2}$. The equations are

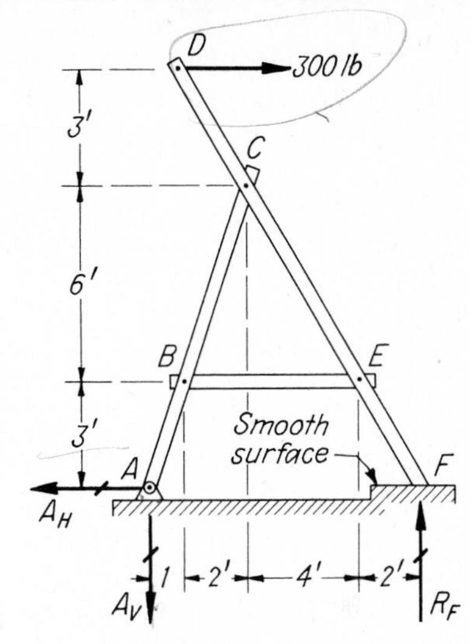

FIG. 5-48 Pin reactions.

$$[\Sigma M_A = 0] \quad 9R_F = (300)(12) = 3{,}600 \quad \text{and} \quad R_F = 400 \text{ lb}\uparrow \quad \textit{Ans.}$$
$$[\Sigma M_F = 0] \quad 9A_V = (300)(12) = 3{,}600 \quad \text{and} \quad A_V = 400 \text{ lb}\downarrow$$
$$[\Sigma H = 0] \quad A_H = 300 \text{ lb}\leftarrow$$

Then $\quad R_A = \sqrt{(400)^2 + (300)^2} = \sqrt{250{,}000} \quad$ or $\quad R_A = 500 \text{ lb}\swarrow \quad$ *Ans.*

Next we draw free-body diagrams of the three members as in Fig. 5-49. The unknown forces to be determined are the *V* and *H* components at *B* and *C*. *BE* is clearly a two-force member, since forces act on it only at *B* and *E*, and the pin reactions at *B* and *E* are therefore horizontal. Also, *BE* is apparently in tension, which we shall assume. Using member *AC*, we may determine force B_H by moments about

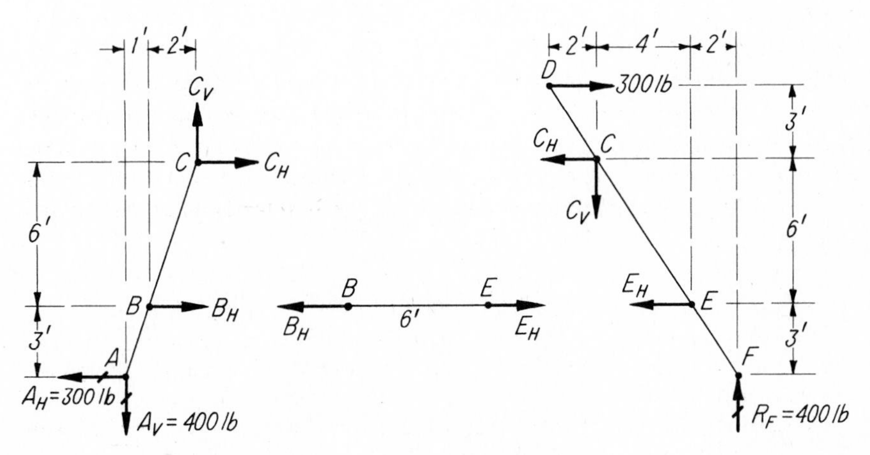

FIG. 5-49 Free-body diagrams.

C. Thereafter the components at C are found by ΣV and ΣH. Using member AC, we have

$$[\Sigma M_C = 0] \quad 6B_H + (400)(3) = (300)(9) \quad \text{and} \quad B_H = 250 \text{ lb} \quad \textit{Ans.}$$
$$[\Sigma V = 0] \quad C_V = 400 \text{ lb} \quad C_V = 400 \text{ lb}$$
$$[\Sigma H = 0] \quad C_H + 250 = 300 \quad \text{and} \quad C_H = 50 \text{ lb}$$

The pin reaction R_C at C then is

$$R_C = \sqrt{(400)^2 + (50)^2} = \sqrt{162{,}500} \quad \text{or} \quad R_C = 403 \text{ lb} \quad \textit{Ans.}$$

These forces acting at B and C are now transferred to member DF but *with opposite senses,* since, if they are considered to be *actions* on member AC, they must be *reactions* when applied to member DF, and *action and reaction are always equal, opposite, and collinear.*

We now obtain the final check by a moment equation about D. That is,

$$[\Sigma M_D = 0] \quad (50)(3) + (400)(2) + (250)(9) = (400)(8) \quad \text{or} \quad 3{,}200 = 3{,}200 \quad \textit{Check.}$$

Graphical Solution: The action lines of forces D and R_F are extended to their intersection through which the action line of R_A then must pass, according to the three-force principle. R_A and F are then found by a closed diagram of forces D, R_A, and R_F. Similarly, using a scale-drawn free-body diagram of member AC, the action line of R_C passes through the intersection of the known action lines of forces R_A and B. Forces B and R_C are then found from a closed diagram of forces R_A, B, and R_C.

PROBLEMS

5-52. Determine the reactions at supports A and F of the frame shown in Fig. 5-50, and the pin reactions at B and C.

Ans. $R_A = 200$ lb↑; $R_F = 40$ lb↑; $R_B = 150$ lb; $R_C = 250$ lb

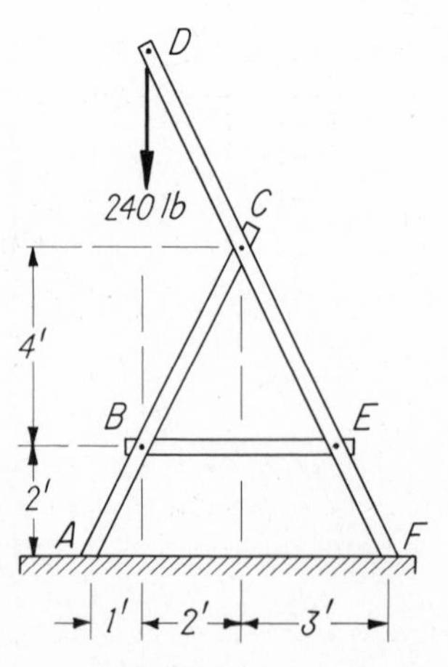

FIG. 5-50 Prob. 5-52.

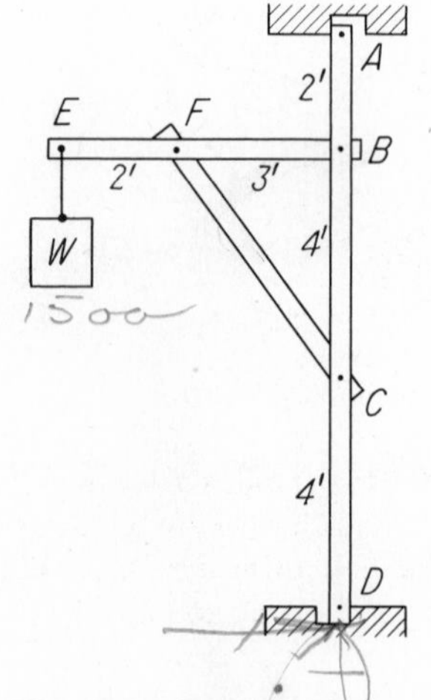

FIG. 5-51 Prob. 5-53.

5-53. The crane shown in Fig. 5-51 is supported at the top and bottom in loose-fitting sockets. Compute the horizontal reactions at A, the V and H reaction components at D, the pin reaction at B, and the stress in member CF if $W = 1{,}500$ lb.

5-54. The A-frame shown in Fig. 5-52 rests on a smooth surface at A and E. Find the reactions at A and E, and the V and H components of the pin reactions at B, C, and D. *Ans.* $R_A = 320$ lb↑; $R_E = 280$ lb↑; $B_V = 360$ lb; $B_H = 180$ lb; $C_V = 40$ lb; $C_H = 180$ lb; $D_V = 240$ lb; $D_H = 180$ lb

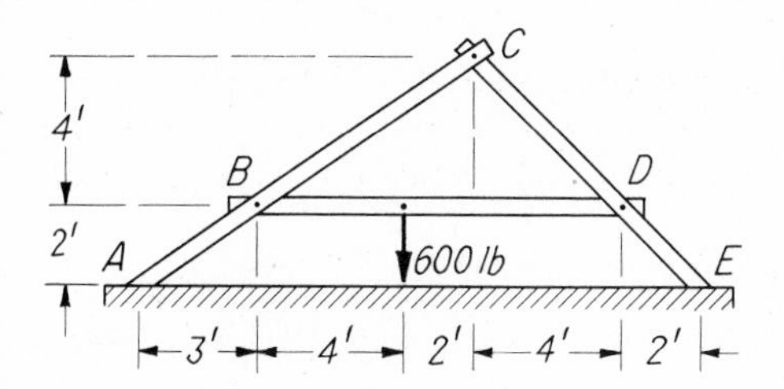

FIG. 5-52 Prob. 5-54.

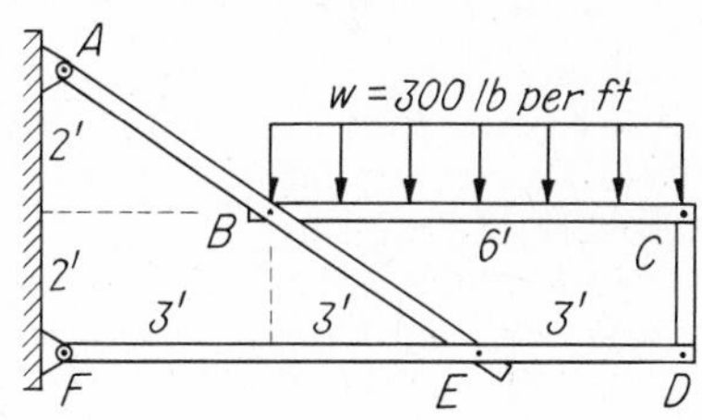

FIG. 5-53 Prob. 5-55.

5-55. The frame shown in Fig. 5-53 supports a uniformly distributed load of 300 lb per ft on member BC. Determine the V and H components of the pin reactions at all pins.

5-56. The horizontal member of the frame of Fig. 5-54 supports a uniformly distributed load of 2 kips per ft. Determine the vertical and horizontal components of the reactions at A and B.

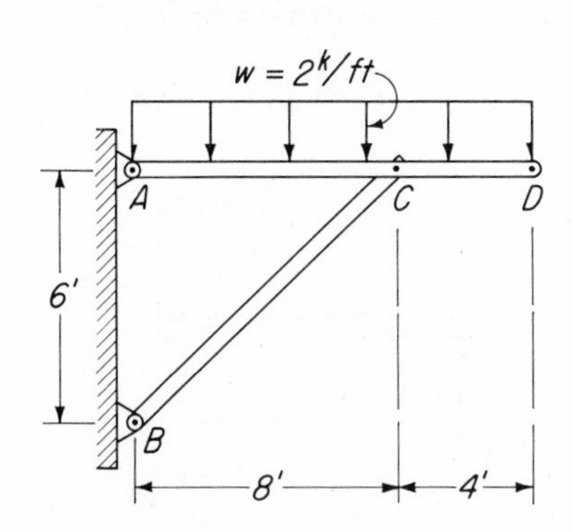

FIG. 5-54 Prob. 5-56.

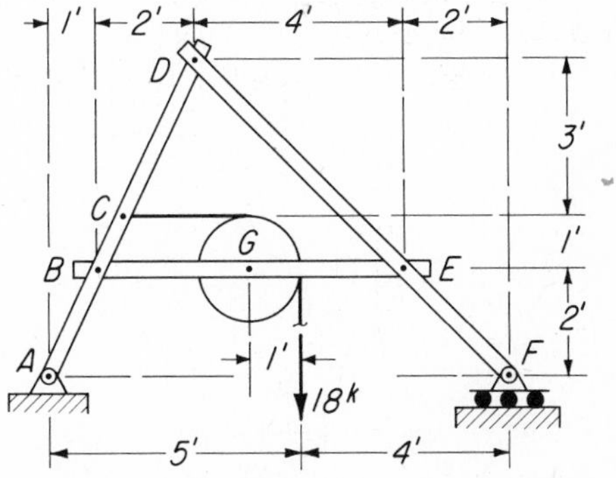

FIG. 5-55 Prob. 5-57.

5-57. The cable shown in Fig. 5-55 passes over the drum G and is fastened to the frame at C. Determine the vertical and horizontal components of the reactions at A and F. Also determine the vertical and horizontal components of the pin reactions at B, D, and E.

Ans. $A_V = 8$ kips; $F_V = 10$ kips; $BV = 9$ kips; $B_H = 12$ kips; $D_V = 1$ kip; $D_H = 6$ kips; $E_V = 9$ kips; $E_H = 6$ kips

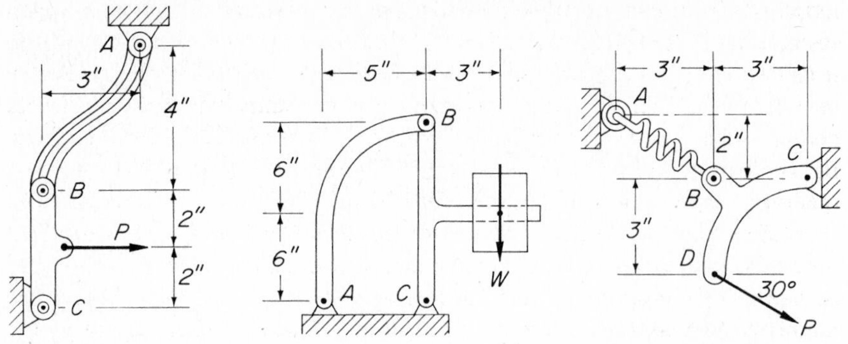

FIG. 5-56 Prob. 5-58. FIG. 5-57 Prob. 5-59. FIG. 5-58 Prob. 5-60.

5-58. The pull P in Fig. 5-56 is 60 lb. Calculate the reactions at A and C.

5-59. The weight W in Fig. 5-57 is 200 lb. Calculate the reactions at A and C.

Ans. $R_A = 130$ lb; $R_C = 357$ lb

5-60. The load P in Fig. 5-58 is 100 lb. Calculate the tensile force in the spring and the reaction at C.

5-61. Determine the V and H components of the pin reactions at all pins of the frame shown in Fig. 5-59.

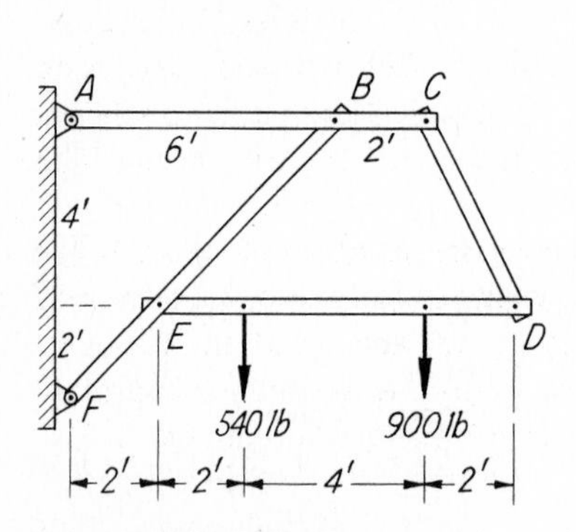

FIG. 5-59 Prob. 5-61.

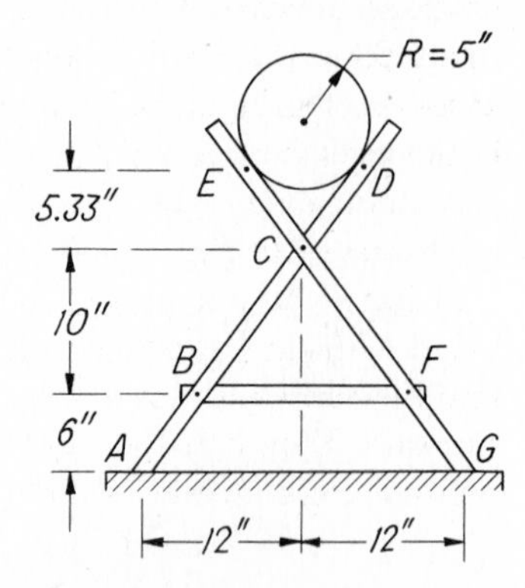

FIG. 5-60 Prob. 5-62.

5-62. A log is supported by two identical frames of a sawbuck, one of which is shown in Fig. 5-60. Each frame supports a weight of 90 lb and rests on a smooth surface. Determine the reactions at A and G, the stress in member BF, and the pin reaction at C. Neglect all friction.

Ans. $R_A = R_G = 45$ lb↑; $BF = 104$ lb T; $R_C = 164$ lb

5-7 Stresses in Trusses; the Method of Sections

In Art. 4-7 was described an analytical method of determination of stresses in trusses—the analytical method of joints—in which the joints were successively isolated as free bodies. The forces acting at each joint then constituted a concurrent system of forces in equilibrium, the unknown quantities of which were found by the equations $\Sigma V = 0$ and $\Sigma H = 0$. In the **method of sections** *a section is cut through the entire truss and one of the two parts is isolated as a free body.* The forces acting on each part, namely, the external forces acting on the part and the internal forces, or stresses, in the cut members, then constitute a *nonconcurrent system of forces in equilibrium,* the unknown quantities of which are found by the equations $\Sigma V = 0$, $\Sigma H = 0$, and $\Sigma M = 0$.

In the method of sections, each stress may be found independently of all others. A truss may therefore be cut at any section, and *any single stress may be found independently of all other stresses in the truss.* Not more than three members with *unknown* stresses may be cut at any section, except in the special case where all but one of the cut members intersect at a common point. In general, *a section should cut the least number of members, and the smaller of the two parts should be isolated as a free body.*

Trusses in which the two main chords are parallel, as in Prob. 5-63 below, are solved by a special adaptation called the **shear method of sections.** Other trusses are most easily solved by the **component method of sections,** as in Prob. 5-64.

In some instances, the members of a truss are so arranged that there will be three or more unknown stresses at some or all joints, as, for example, in Fig. 5-68, and the truss cannot, therefore, be solved by the analytical method of joints or by either of the graphical methods. In such problems, one stress in the truss is readily determined by the method of sections, and the solution may then be completed graphically, or analytically by the joint method. This is illustrated in Prob. 5-65.

A graphical solution to obtain three unknown stresses at a section is possible by use of the four-force principle. All known external forces are first combined into their resultant force. The resultant of this force and one of the unknown stresses is then equal, opposite, and collinear with the resultant of the other two unknown stresses. Problem 5-65 may be solved in this manner.

ILLUSTRATIVE PROBLEMS

5-63. Using the *shear method of sections,* determine the stresses in members *BD, CD,* and *CE* of the Howe roof truss shown in Fig. 5-61*a*. Determine also stresses *AB, AC, BC, DE,* and *EG*. Draw a stress-summation diagram for recording the stresses, letting *T* and *C*, respectively, denote tension and compression. (Students should solve for the remainder of the stresses in this truss.)

Solution: Vertical section 3 is cut through the second panel of the truss containing members *BD, CD,* and *CE,* whose stresses we must find. The part on the left is then isolated as a free body, as shown in Fig. 5-62. Six forces act on this part, the reaction at *A*, the loads *B* and *C*, and the unknown stresses in the three cut members.

The stress in the diagonal *CD* should first be determined. Its vertical component CD_V is easily found and is seen to be downward-acting. That is,

$$[\Sigma V = 0] \qquad CD_V + 6 + 6 = 18 \qquad \text{and} \qquad CD_V = 6 \text{ kips}$$

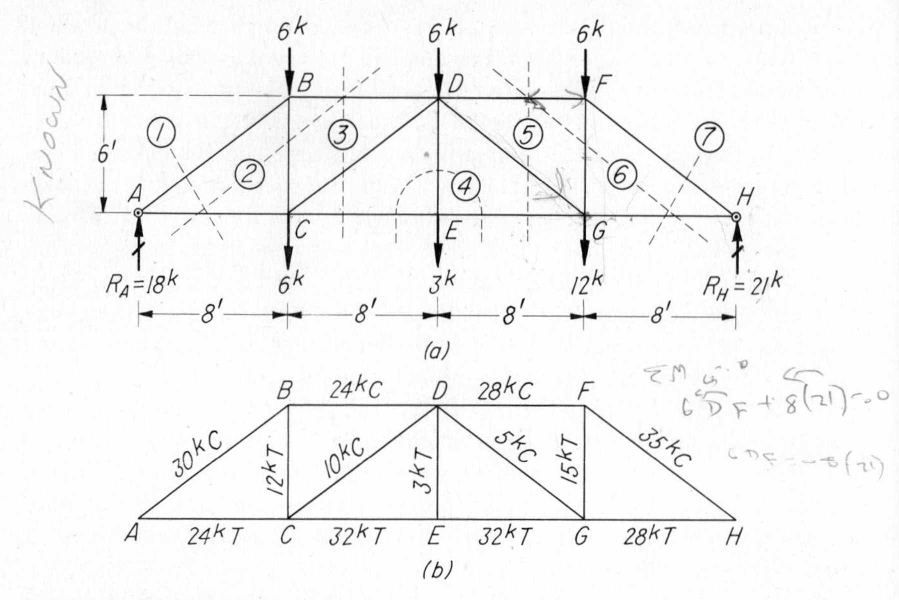

FIG. 5-61 Determination of stresses in trusses. The shear method of sections. (*a*) Space diagram. (*b*) Stress-summation diagram.

Then, by similarity of slope and force triangles,

$$\frac{CD_V}{6} = \frac{CD_H}{8} = \frac{CD}{10} = \frac{6}{6} = 1$$

or $\quad CD_H = (8)(1) = 8 \text{ kips} \quad$ and $\quad CD = (10)(1) = 10 \text{ kips C} \quad$ *Ans.*

Because the vertical component of *CD* acts *against* member *CD*, the horizontal component and the stress *CD* also act *against* it, indicating compression.

To find *BD independently* of *CD* and *CE*, we must take moments about *C*, thus eliminating the moments of *CD* and *CE*. Likewise, we find *CE independently* by moments about *D*, thus eliminating the moments of *BD* and *CD*. If we assume *BD* to be compressive and *CE* to be tensile (as they apparently are), we have

$[\Sigma M_C = 0] \quad 6BD = (18)(8) = 144 \quad$ and $\quad BD = 24 \text{ kips C} \quad$ *Ans.*

$[\Sigma M_D = 0] \quad 6CE + (6)(8) + (6)(8) = (18)(16) \quad$ and $\quad CE = 32 \text{ kips T} \quad$ *Ans.*

The positive signs of *BD* and *CE* indicate that the *type of stress* in each was correctly assumed. When these forces are shown on the free-body diagram, as was done in Fig. 5-62, a summation of horizontal forces clearly indicates force equilibrium, which is a sufficient check. (Students should *always* record computed values on the free-body diagrams to obtain a visual check.) That is,

$[\Sigma H = 0] \quad 32\rightarrow = 24\leftarrow + 8\leftarrow \quad$ or $\quad 32\rightarrow = 32\leftarrow \quad$ *Check.*

We see clearly now that any stress at any section may be found *independently* of all other stresses in the truss.

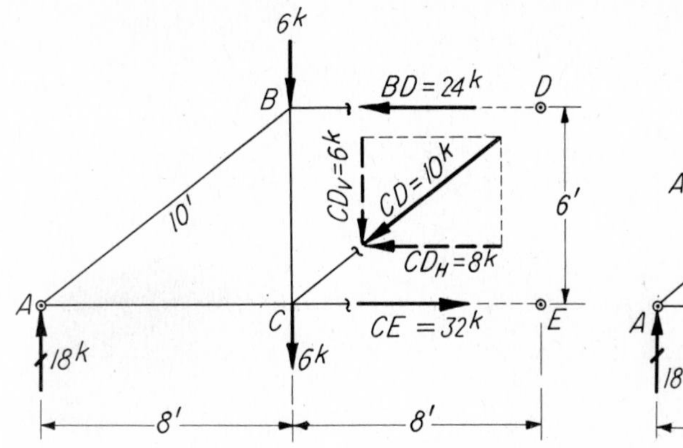

FIG. 5-62 Free-body diagram, section 3.

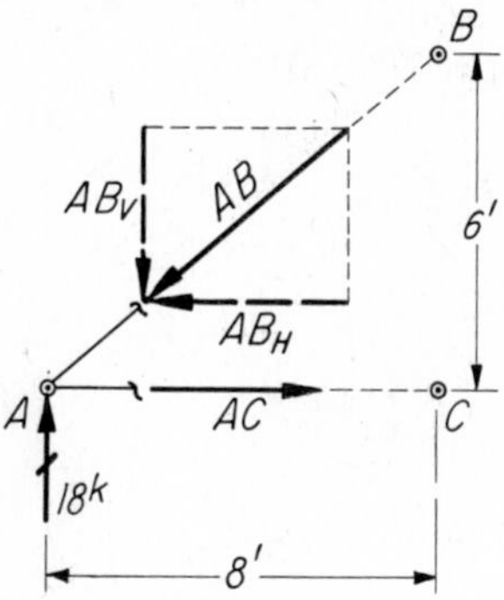

FIG. 5-63 Free-body diagram, section 1.

If all stresses in this truss are desired, a number of sections are cut such as those indicated by the encircled numbers in Fig. 5-61*a*. A free-body diagram is then drawn, usually of the smallest part, and the unknown stresses are determined. To find stresses *AB* and *AC*, for example, section 1 is cut and the portion on the left is isolated as in Fig. 5-63. From $\Sigma V = 0$, AB_V is found to be 18 kips. Then, by similarity of slope and force triangles, AB_H and *AB* are found to be 24 and 30 kips, respectively. *AC* is found independently by moments about *B*. A check of horizontal forces then indicates force equilibrium.

To find *AC*, *BC*, and *BD*, section 2 is cut, isolating the part shown in Fig. 5-64. *BC* is found from $\Sigma V = 0$, *BD* by moments about *C*, and *AC* by moments about *B*. The stress *DE* is most easily found by cutting the circular section 4 (Fig. 5-65) and using $\Sigma V = 0$. *EG* is then found from $\Sigma H = 0$. *In general, joints should not be isolated* except as in sections 1, 4, and 7 in Fig. 5-61*a*. Sections are generally cut from left to right in the order of the numbers 1 to 7.

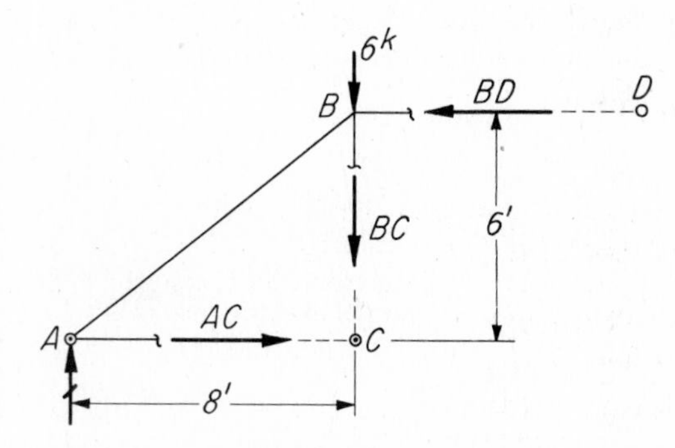

FIG. 5-64 Free-body diagram, section 2.

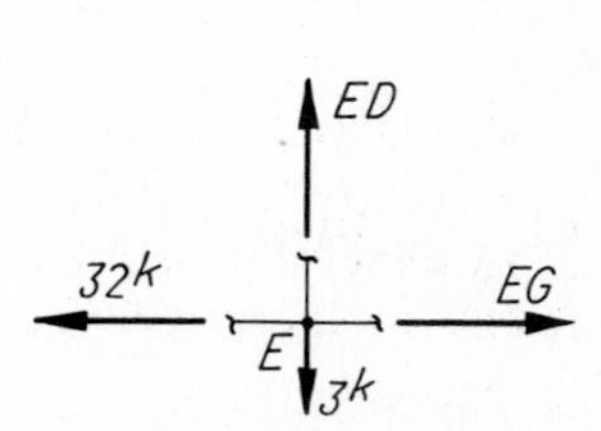

FIG. 5-65 Free-body diagram, section 4.

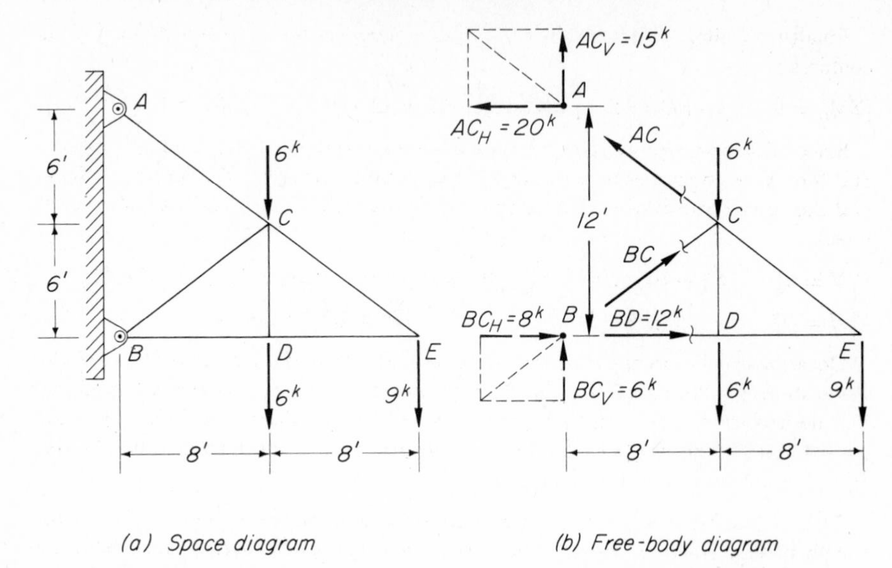

FIG. 5-66 Determination of stresses in trusses; the component method of sections. (*a*) Space diagram. (*b*) Free-body diagram.

5-64. By means of the *component method of sections*, find the stresses in members *AC*, *BC*, and *BD* of the truss shown in Fig. 5-66.

Solution: The most suitable section to cut is one through members *AC*, *BC*, and *BD*. We then use the part on the right as a free body, as shown in Fig. 5-66*b*. *By proper selection of moment centers, we may determine each of the three stresses independently of the other two.* That is, we find *AC* by moments about *B*, *BC* by moments about *E*, and *BD* by moments about *C*. In each of these three moment equations, the moments of the other two stresses are zero about the moment center selected and are thus eliminated. Assuming *AC* to be a tensile stress, as apparently it is, and with *B* as a moment center, thus eliminating *BC* and *BD*, we have

$$[\Sigma M_B = 0] \qquad 21AC_H = 6(8) + 6(8) + 9(16) \qquad \text{and} \qquad AC_H = 20 \text{ kips}$$

By similar triangles

$$\frac{AC_H}{8} = \frac{AC}{10} \qquad \text{and} \qquad AC = \frac{10}{8}(20) = 25 \text{ kips T} \qquad \textit{Ans.}$$

Next, to find *BC*, we select *E* as the moment center and replace *BC* with its *V* and *H* components, applied at either *B* or *C*, but most conveniently at *B*. Assuming *BC* to be a compressive stress, we solve for its *vertical* component BC_V. That is,

$$[\Sigma M_E = 0] \qquad 16BC_V = 6(8) + 6(8) \qquad \text{and} \qquad BC_V = 6 \text{ kips}$$

By similar triangles

$$\frac{BC_V}{6} = \frac{BC}{10} \qquad BC = \frac{10(6)}{6} = 10 \text{ kips C} \qquad \textit{Ans.}$$

Finally, to obtain BD, we select C as the moment center. Then, assuming BD is compressive,

$$[\Sigma M_C = 0] \qquad 6BD = 9(8) \qquad \text{and} \qquad BD = 12 \text{ kips C} \qquad \textit{Ans.}$$

Since all results are positive, the type of stresses were correctly assumed. We can find the other components of AC and BC by similar triangles. This has been done, and the results are shown on the free-body diagram shown in Fig. 5-66*b*. As a final check,

$$[\Sigma V = 0] \qquad 15\uparrow + 6\downarrow + 9\downarrow + 6\downarrow + 6\uparrow = 0 \qquad \text{or} \qquad 21\uparrow = 21\downarrow \qquad \textit{Check.}$$

$$[\Sigma H = 0] \qquad \overleftarrow{20} + \overrightarrow{8} + \overrightarrow{12} = 0 \qquad \text{or} \qquad \overleftarrow{20} = \overrightarrow{20} \qquad \textit{Check.}$$

Occasionally, the moment center which will eliminate the moments of two of the three stresses to be found lies outside the truss and may be inconvenient to locate. The desired stress may then be found by ΣV and ΣH, provided a final check is made by ΣM about a center so chosen that no moment of a computed stress is eliminated. In any event, *a final check is essential.*

5-65. A graphical solution to determine the stresses in the members of the truss shown in Fig. 5-67*a* is desired, but a start on it cannot be made, since there are three or more unknown stresses at every joint. Using the *method of sections,* determine analytically the stress in one member in order that the solution may then be completed graphically. All forces are in pounds.

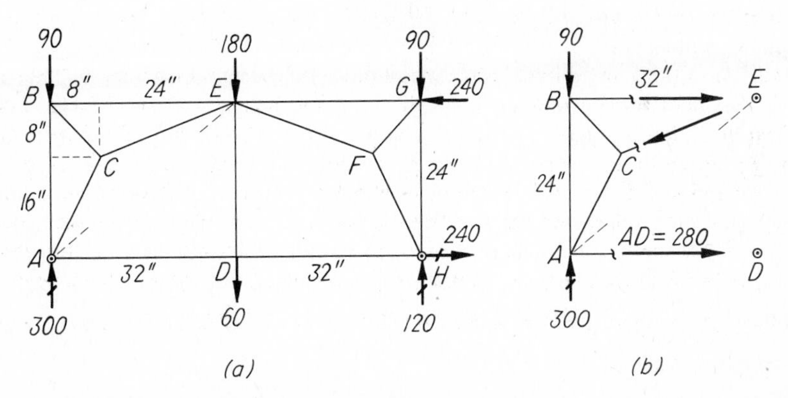

FIG. 5-67 Determination of a single stress by the method of sections. (*a*) Loaded truss. (*b*) Free-body diagram.

Solution: The simplest section to cut is clearly through members BE, CE, and AD. When the portion on the left is isolated as a free body, as shown in Fig. 5-67*b*, the stress in AD is readily found by a summation of moments about point E. That is,

$$[\Sigma M_E = 0] \qquad 24AD + (90)(32) = (300)(32) \qquad \text{and} \qquad AD = 280 \text{ lb T} \qquad \textit{Ans.}$$

An all-graphical solution of this problem is made possible by the **substitute-member method** (see Prob. 4-45) in which the three members, AC, BC, and CE are removed and are replaced with a single straight member *AE*, partly indicated by the dash lines. From study of Fig. 5-67*b*, we see that *the stress AD is unaffected by this substitution,* since, by moments about *E*, *AD* is still 280 lb T. Because only two unknown stresses now exist at joint *B*, joints *B* and *A* may be solved graphically. Of the stresses so found, *only AD will be a true stress,* and the others are therefore disregarded. The original members, *AC*, *BC*, and *CE*, are now replaced, and the graphical solution may then begin at joint *A*—since stress *AD* is now known—and may proceed through joints *B*, *C*, *D*, *E*, *F*, *G*, and *H*.

An alternate graphical solution is to obtain stress *AD* by use of the four-force principle (Art. 2-15), in which the resultant of stresses *BE* and *CE* is applied at *E* in Fig. 5-67*b* and its equal and opposite resultant is applied at *A*, thus: Remove members *AC*, *BC*, and *CE* and substitute for them a member *AE*. When joint *B* is now solved, the stress in *AB* is found to be 90 lb compression. The resultant of this stress *AB* and the 300-lb reaction at *A* is 210 lb upward. When this 210-lb force is laid off upward from *A* as the vertical side of a parallelogram of forces, the horizontal side will be the stress *AD* and the diagonal will be the stress in the substitute member *AE*, which stress, then, is equal and opposite to the resultant of stresses *BE* and *CE*, applied at *E*. When *AD* has thus been determined, the stresses in other members may then be determined by the usual combined-diagram method. (See Prob. 4-60 for scales to use.)

PROBLEMS

Solve Probs. 5-66 through 5-74 by the *shear method of sections.*

5-66. Determine the stresses in members *BD*, *CD*, and *CE* of the roof truss shown in Fig. 5-68. *Ans.* *BD* = 24 kips C; *CD* = 15 kips T; *CE* = 12 kips T

5-67. Find the stresses in members *AC*, *BC*, and *BD* of the truss shown in Fig. 5-68.

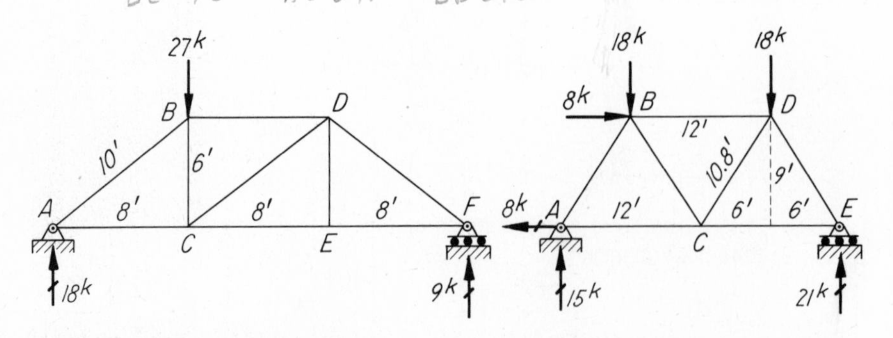

FIG. 5-68 Probs. 5-64 and 5-65.

FIG. 5-69 Probs. 5-66 and 5-117.

5-68. Solve for the stresses in members *BD*, *CD*, and *CE* of the Warren roof truss shown in Fig. 5-71. *Ans.* *BD* = 16 kips C; *CD* = 3.6 kips T; *CE* = 14 kips T

5-69. In Fig. 4-90, solve for the stresses in members *AC*, *BC*, and *BD*.
Ans. *AC* = 34 kips T; *BC* = 1 kip T; *BD* = 45 kips C

5-70. In Fig. 4-90 solve for the stresses in members *BD*, *CD*, and *CE*.

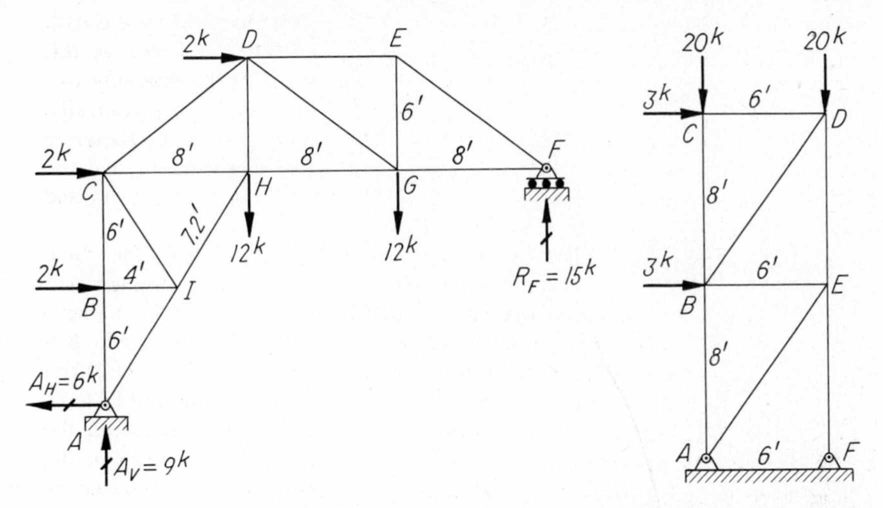

FIG. 5-70 Probs. 5-71, 5-80, 5-81, and 5-116.

FIG. 5-71 Prob. 5-73.

5-71. In Fig. 5-70 is shown one truss of a traveling crane. Determine the stresses in members *DE, DG,* and *GH.*

Ans. *DE* = 20 kips C; *DG* = 5 kips C; *GH* = 24 kips T

5-72. Find the stresses in members *BD, CD,* and *CE* of the tower truss shown in Fig. 4-35.

5-73. The vertical tower truss shown in Fig. 5-71 supports a water tank and is also subjected to horizontal wind loads. Compute the stresses in all members.

Ans. *AB* = 16 kips C; *BD* = 5 kips T; *EF* = 32 kips C

5-74. Find the stresses *BD, CD,* and *CE* of the truss shown in Fig. 5-72. Loads are P_1 = 36 kips and P_2 = 16 kips.

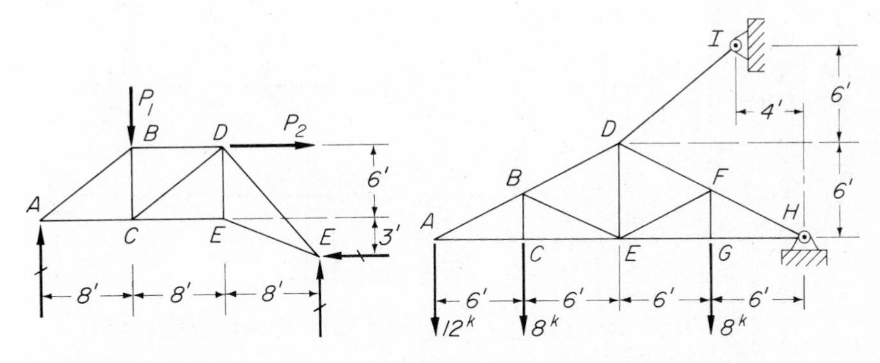

FIG. 5-72 Prob. 5-74.

FIG. 5-73 Probs. 5-75 and 5-101.

Solve Probs. 5-75 through 5-86 by the *component method of sections.*

5-75. Compute the stresses *BD, BE,* and *CE* of the truss shown in Fig. 5-73.

Ans. *BD* = 35.7 kips T; *BE* = 9.0 kips C; *CE* = 24.0 kips C

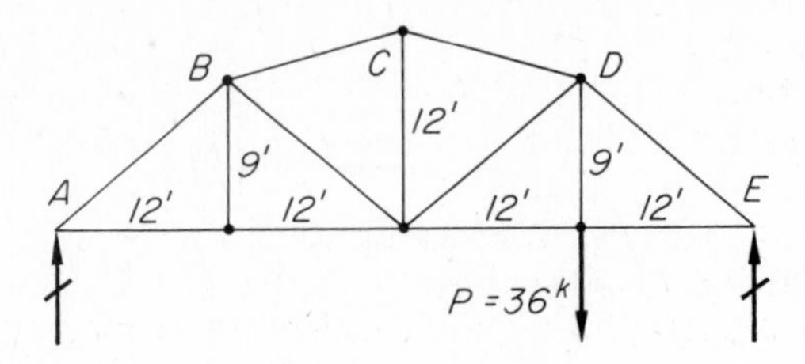

FIG. 5-74 Probs. 5-76 and 5-77.

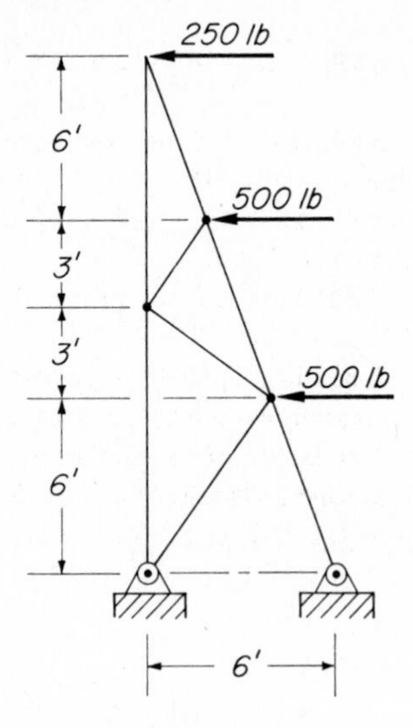

FIG. 5-76 Prob. 5-79.

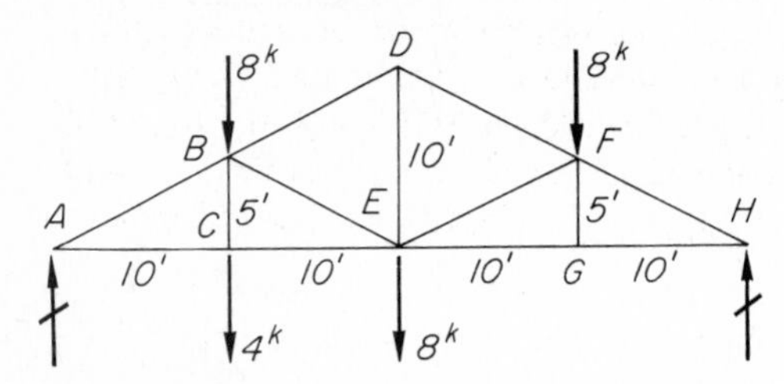

FIG. 5-75 Prob. 5-78.

5-76. The bowstring truss in Fig. 5-74 supports a live load of 36 kips at joint *F*. Compute the support reactions and the stresses in members *AB*, *BH*, and *GH* produced by this load.

5-77. Calculate the support reactions at *A* and *E*, and determine the stresses in members *BC*, *BG*, and *GH* of the bowstring truss shown in Fig. 5-74 produced by the 36 kip load at joint *F*.

Ans. *BC* = 18.6 kips C; *BG* = 7.5 kips T; *GH* = 12 kips T

5-78. Find the stresses *DF* and *EF* in Fig. 5-75.

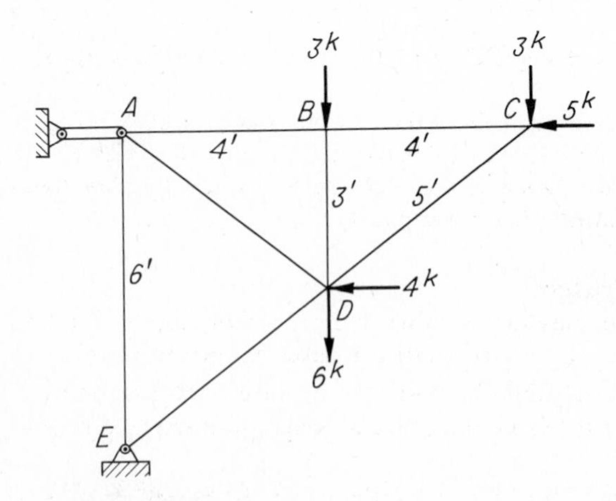

FIG. 5-77 Prob. 5-82.

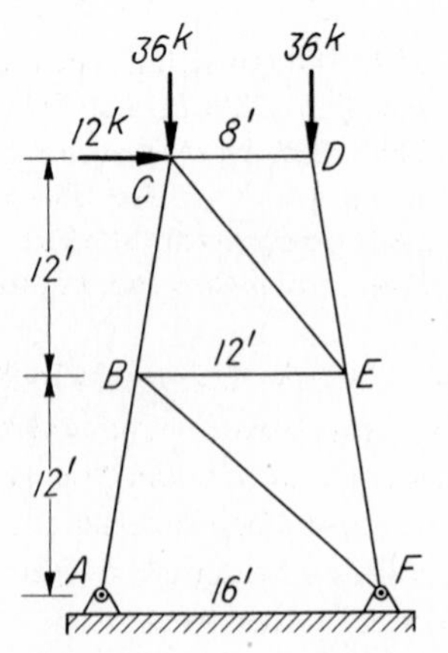

FIG. 5-78 Probs. 5-83, 5-86, and 5-98.

5-79. Find the stresses *DF*, *DE*, and *CE* in Fig. 5-76.

Ans. *DF* = 2,370 lb T; *DE* = 902 lb C; *CE* = 1,500 lb C

5-80. Solve for the stresses in members *BC*, *CI*, and *HI* in the traveling-crane truss shown in Fig. 5-70.

5-81. Find the stresses in members *CD*, *CH*, and *HI* of the traveling-crane truss shown in Fig. 5-70. *Ans.* *CD* = 27.5 kips C; *CH* = 19 kips T; *HI* = 9 kips T

5-82. Find the stresses in members *AB*, *AD*, and *DE* of the truss shown in Fig. 5-77.

5-83. In Fig. 5-78 is shown one vertical truss of a tower supporting a heavy construction crane. The horizontal load at *C* represents the total wind pressure against the crane and truss. Determine the stresses in members *BC*, *CE*, and *DE*. (HINT: Note that the action lines of *BC* and *DE* intersect at a point 24 ft above the center of member *CD*. Use upper part. Apply the components of *BC* at *B*, of *CE* at *E*, and of *DE* at *D*.) *Ans.* *BC* = 24.3 kips C; *CE* = 15.6 kips C; *DE* = 36.5 kips C

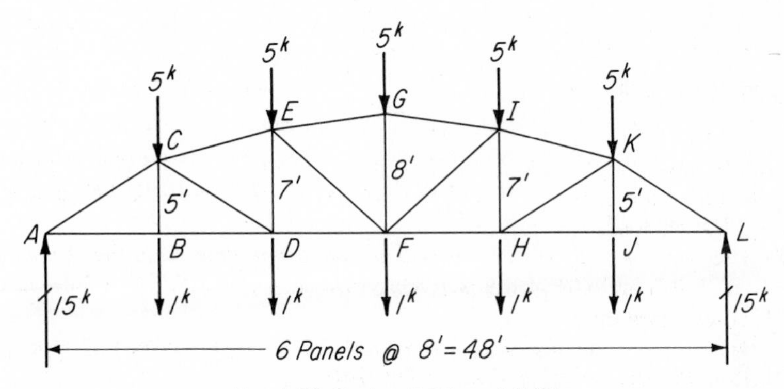

FIG. 5-79 Probs. 5-84 and 5-85.

5-84. The bowstring roof truss of Fig. 5-79 is loaded as shown. Find the *stresses* in members *CE*, *CD*, and *BD*.

5-85. Find the stresses in members *EG*, *EF*, and *DF* of the bowstring truss of Fig. 5-79. *Ans.* *EG* = 27.2 kips C; *EF* = 0.6 kips C; *DF* = 27.4 kips T

5-86. Compute the support reactions at *A* and *F* in Fig. 5-78 and determine the stresses in members *BC*, *BE*, and *EF*. (Use lower part.)

5-8 Counter Diagonals in Trusses

In many trusses subjected to moving or other varying loads, some of the diagonals may undergo a reversal of stress from tension to compression, or vice versa. Such reversal of stress might be produced by a train moving across a railroad bridge or by the change in direction of wind pressure against a vertical tower truss.

In trusses such as the one shown in Fig. 5-80, the inside diagonals *BE*, *DG*, *GH*, and *IJ* are usually designed to carry tensile stress only and are apt to be so slender that they could easily buckle under a compressive stress. To

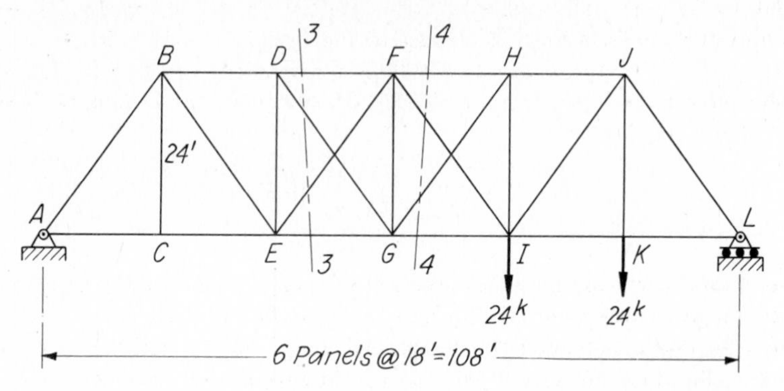

FIG. 5-80 Probs. 5-87, 5-88, and 5-119. Counter diagonals in trusses.

prevent possible failure of this type, *counter diagonals EF* and *FI* are inserted in all panels where a reversal of stress might occur, due either to wind loads or to moving loads. The assumption then is that *only the tension diagonal is active and carries stress.*

Let the truss in Fig. 5-80 be one of a pair supporting a railroad bridge, and consider the two loads at panel points I and K to be caused by a load moving across the bridge. The resulting reaction at A is 12 kips upward. If now we cut section 4-4 through panel 4, and isolate the left half as a free body, the vertical component of the *active diagonal stress* is 12 kips downward, indicating that member FI is the active tension diagonal. GH is in compression and is therefore assumed to be inactive.

When the two loads are applied at G and I, R_A will be 20 kips upward and the vertical component in the active diagonal at section 4 will be 4 kips upward, indicating that GH is now the active tension diagonal.

PROBLEMS

5-87. Determine the stress in the active tension diagonal at section 4-4 of the truss shown in Fig. 5-80, and the stresses in the chords FH and GI.

Ans. $FI = 15$ kips T; $FH = 36$ kips C; $GI = 27$ kips T

5-88. Solve Prob. 5-87, with the two 24-kip loads applied at panel points G and I.

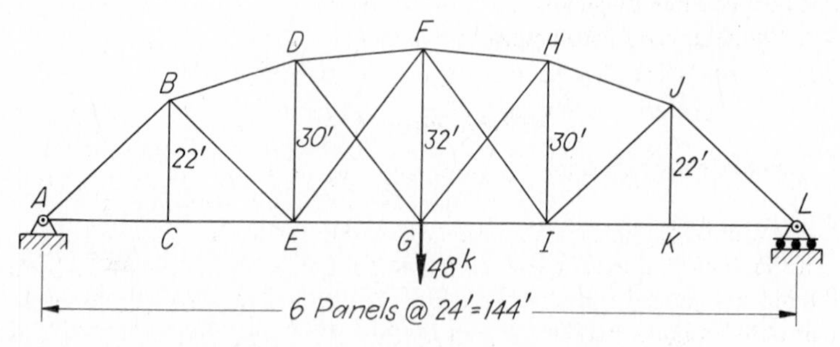

FIG. 5-81 Probs. 5-89 and 5-90.

5-89. In Fig. 5-81, find the stress in the active tension diagonal in the third panel from the left, and the stresses in chords DF and EG. (Isolate the left half as a free body.) *Ans.* $DG = 25$ kips T; $DF = 54.2$ kips C; $EG = 38.4$ kips T

5-90. Solve Prob. 5-89 when the 48-kip load is moved from G to I. (Isolate the left half as a free body.)

SUMMARY

(By article number)

5-1. A coplanar, nonconcurrent force system is one in which all forces lie in one plane but whose action lines do not meet at a common point.

5-2. The resultant force may be determined analytically by algebraic summations of forces and of moments, and graphically by the parallelogram method or by the string-polygon method.

5-3. The unknown quantities of **coplanar, nonconcurrent force systems in equilibrium** are generally solved analytically by algebraic summations of forces and of moments. Graphical solutions are also available. Typical problems are (1) determination of support reactions; (2) determination of the forces acting at the pins of certain pin-connected frames by the **method of members;** and (3) determination of stresses in members of trusses by the **method of sections.**

5-4. In determinations of **reactions by the graphical string-polygon method,** the string polygon must always start at the point of application of the reaction whose direction is unknown. Problems in which most of the forces are neither vertical nor horizontal are most suitably solved by this method.

5-5. Analytical determination of reactions is most suitable for problems in which most of the forces are either vertical or horizontal. The equations of solution are $\Sigma V = 0$, $\Sigma H = 0$, and $\Sigma M = 0$.

5-6. Pin reactions at connections of frames containing three-force members are obtained by the **method of members.** First, all determinable support reactions are found. Thereafter, each three-force member is isolated as a free body. Unknown forces are then readily determined, because the forces acting on each member so isolated are in equilibrium.

5-7. The method of sections enables us to determine the stress in any single member of a truss *independently* of all other stresses. The truss is cut in two through the least number of members, and either part—usually the smaller—is then isolated as a free body in equilibrium. Each unknown stress is then determined by one of the equations of equilibrium, $\Sigma V = 0$, $\Sigma H = 0$, or $\Sigma M = 0$.

5-8. When **counter diagonals** are used in trusses, *only the tension diagonal is considered to be active.* The compressive diagonal is assumed to be inactive.

REVIEW PROBLEMS

5-91. Calculate the reactions at A and E in Fig. 5-84.
Ans. $A_V = 240$ lb↑; $A_H = 360$ lb→; $E = 300$ lb

5-92. Determine analytically the *resultant* R of the four forces acting on the truss shown in Fig. 5-40. Find also the direction angle θ_H that the action line of R makes with the horizontal, and the distance x (from C) to its line of action. Check by the *graphical parallelogram method.* *Ans.* $R = 31.2$ kips↘; $\theta_H = 50°12'$; $x = 0$

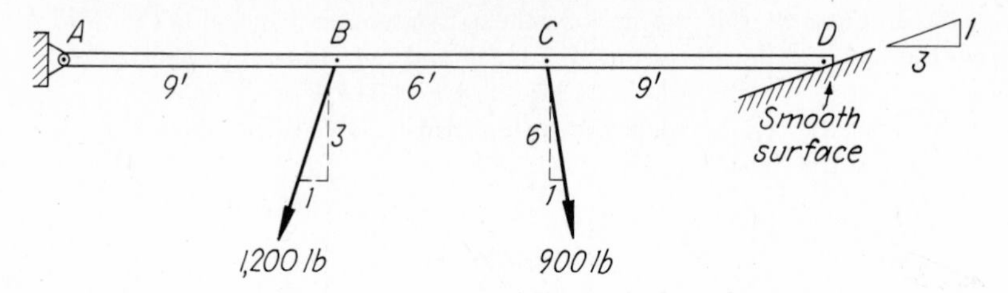

FIG. 5-82 Probs. 5-93 and 5-94.

5-93. Using the graphical *string-polygon method,* determine the *resultant* R of forces B and C in Fig. 5-82. (Scales on $8\frac{1}{2}$ by 11: 1 in. = 2 ft, 1 in. = 500 lb.)

5-94. Determine the *reactions* A and D at supports A and D in Fig. 5-82, using the graphical *string-polygon method.* (Scales on $8\frac{1}{2}$ by 11: 1 in. = 5 ft, 1 in. = 500 lb.) *Ans.* A = 1,185 lb ↗; D = 1,035 lb ↖

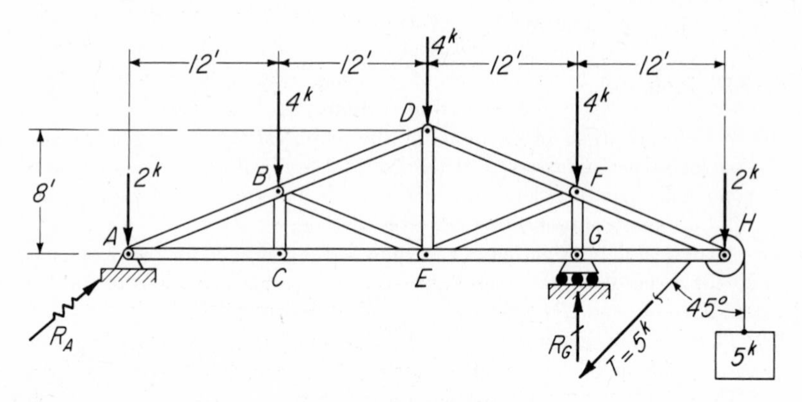

FIG. 5-83 Prob. 5-95.

5-95. The truss shown in Fig. 5-83 is subjected to the roof loads and the hoist load shown. Determine the *reactions* R_A and R_G, using the graphical *string-polygon method.* (Scales on 11 by 17: 1 in. = 5 ft, 1 in. = 5 kips.)

5-96. Find graphically the *reactions* at supports D and E of the beam shown in Fig. 5-13. Use the *string-polygon method.* (Scales on $8\frac{1}{2}$ by 11: 1 in. = 4 ft, 1 in. = 300 lb.) *Ans.* D = 577 lb ↗; E = 673 lb↑

5-97. The cart shown in Fig. 5-85 has four wheels, and the loads $W_1 = W_2$ = 1,200 lb are midway between the near and far set of wheels. Calculate the reaction forces under the near wheels C and D. *Ans.* C = 150 lb; D = 1,050 lb

5-98. Solve analytically for the V and H *reaction components* at supports A and F of the truss shown in Fig. 5-78.

Ans. A_V = 18 kips↑; A_H = 3 kips→; F_V = 54 kips↑; F_H = 15 kips←

5-99. The truss of Fig. 5-86 supports the hoist sheave shown. Solve analytically for the V and H *reaction components* at supports A and B. As a check on A_V, select a moment center which will eliminate B_V, B_H, and A_H, and recalculate A_V.

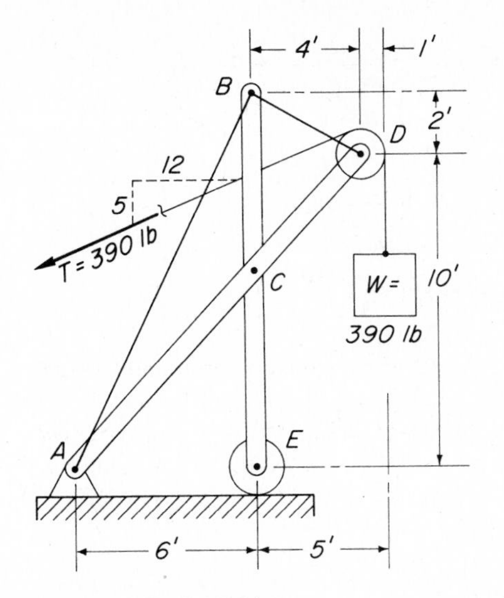

FIG. 5-84 Prob. 5-91.

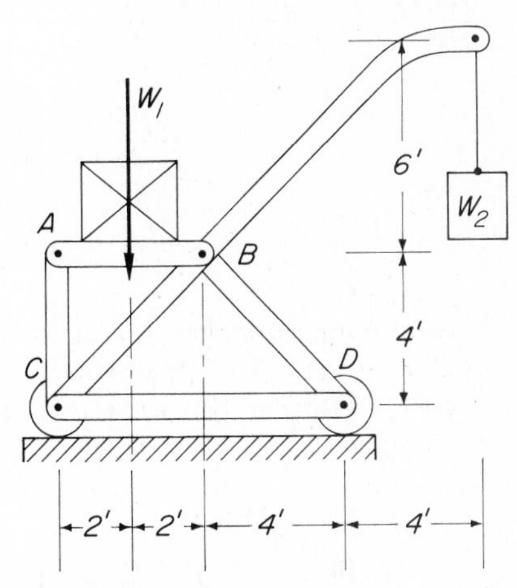

FIG. 5-85 Prob. 5-97.

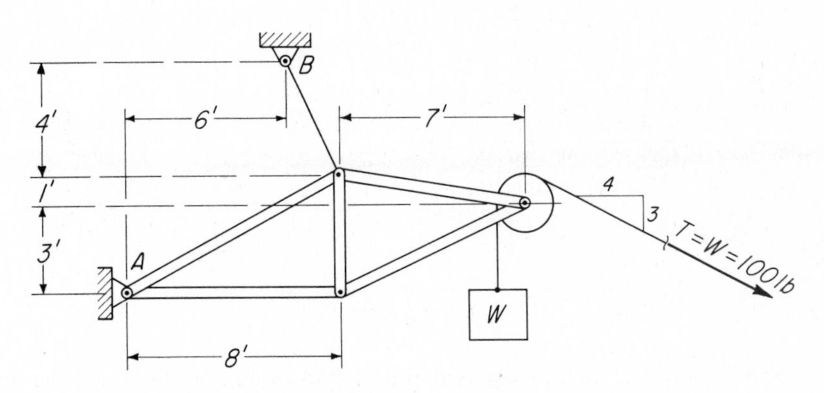

FIG. 5-86 Prob. 5-99.

5-100. In Fig. 5-87 solve analytically for the V and H *reaction components* at supports A and E.

Ans. $A_V = 19$ kips↑; $A_H = 8$ kips→; $E_V = 3$ kips↑; E_H 12 kips←

5-101. Calculate the stresses DF, EF, and EG in the truss of Fig. 5-73.

Ans. $DF = 0$; $EF = 9$ kips C; $EG = 24$ kips C

5-102. The frame shown in Fig. 5-88 is of the three-hinged-arch type. Determine analytically the V and H *reaction components* at supports A and E. Check by moments about C.

Ans. $A_V = 3.94$ kips↑; $A_H = 1.44$ kips→; $E_V = 7.06$ kips↑; $E_H = 8.44$ kips←

5-103. The roof hoist shown in Fig. 5-89 is resting on blocks at A and B and is prevented from sliding and overturning by a small block at B and a counterweight

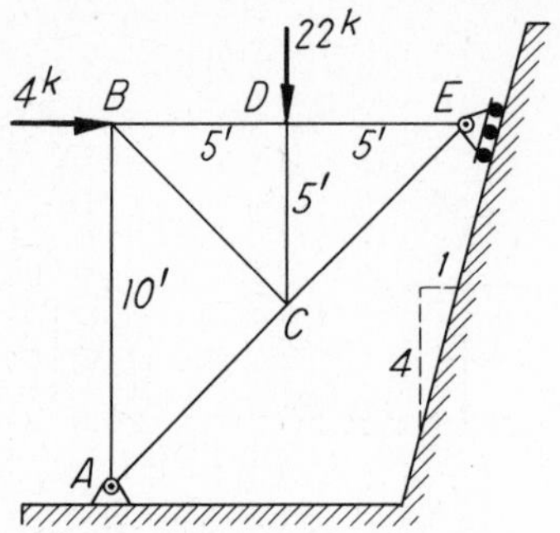

FIG. 5-87 Probs. 5-100 and 5-123.

FIG. 5-88 Prob. 5-102.

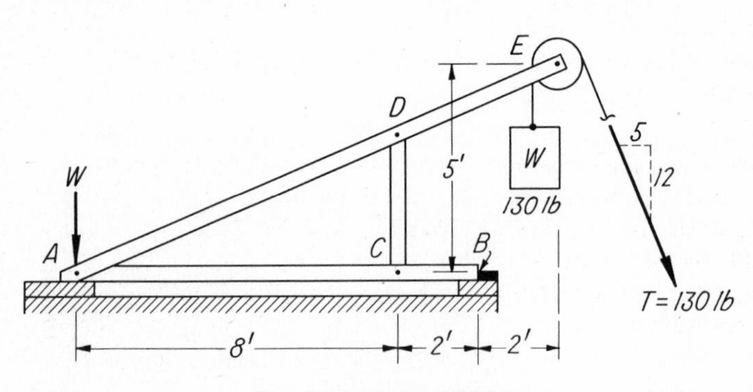

FIG. 5-89 Prob. 5-103.

W at A. Compute the minimum value of the counterweight W required to prevent overturning, and the V and H *reaction components* at B. Neglect possible friction.

5-104. In Fig. 5-90 is shown one of two identical frames supporting an air-compressor tank having a diameter of 4 ft and weighing 1,200 lb. The load W on each frame then is 600 lb. Compute the V and H components of the force exerted by the tank on bar AB, and the V and H *reaction components* at supports A and at pin B. Neglect friction. *Ans.* $F_V = 600$ lb↓; $F_H = 347$ lb→; $A_V = 213$ lb↑; $A_H = 218$ lb←; $B_V = 387$ lb↑; $B_H = 129$ lb←

5-105. Determine the *reactions* at supports D and E of the frame shown in Fig. 5-90. (See Prob. 5-104 for values of B_V and B_H.)

5-106. In Fig. 5-93 the rigid bar AB is resting against smooth surfaces at A and B and is held at C by the rope CD. Compute the *reactions* at A and B, and the stress in CD. *Ans.* $R_A = 620$ lb↑; $R_B = 447$ lb↖; $CD = 500$ lb T

5-107. Solve Prob. 5-106 graphically, using the *four-force principle.* (HINT: Find first the resultant of E and B, then the resultant of A and CD. Scales on $8\frac{1}{2}$ by 11: 1 in. = 5 ft, 1 in. = 300 lb.)

5-108. Solve Prob. 5-106 when an additional horizontal force of 260 lb, acting from left to right, is applied at C. *Ans.* $R_A = 395$ lb↑; $R_B = 313$ lb↖; $CD = 25$ lb T

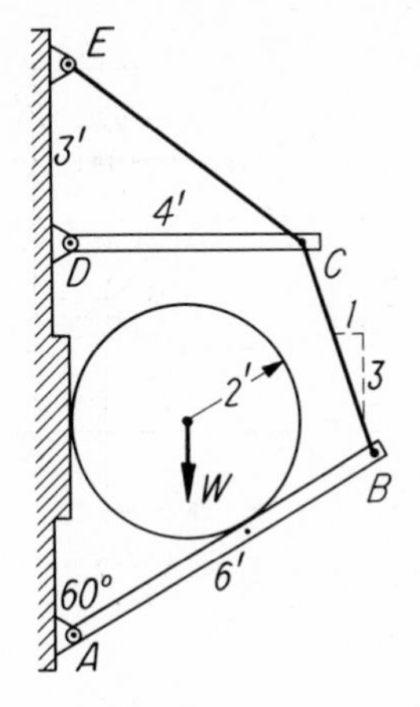

FIG. 5-90 Probs. 5-104 and 5-105.

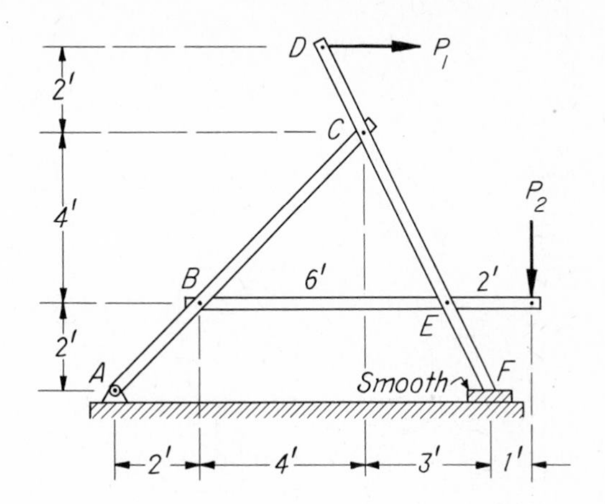

FIG. 5-91 Prob. 5-109.

5-109. Using the *method of members* compute the *pin reactions* at A, B, and E in Fig. 5-91. The frame rests on a smooth block at F and the loads are $P_1 = 90$ lb and $P_2 = 270$ lb. *Ans.* $R_A = 120.3$ lb; $R_B = 108$ lb; $R_C = 36$ lb; $R_E = 365$ lb

5-110. By the *method of members*, compute the V and H components of the forces acting at B, C, D, and E of the frame shown in Fig. 5-92. Find also the stresses in members AB and DF.

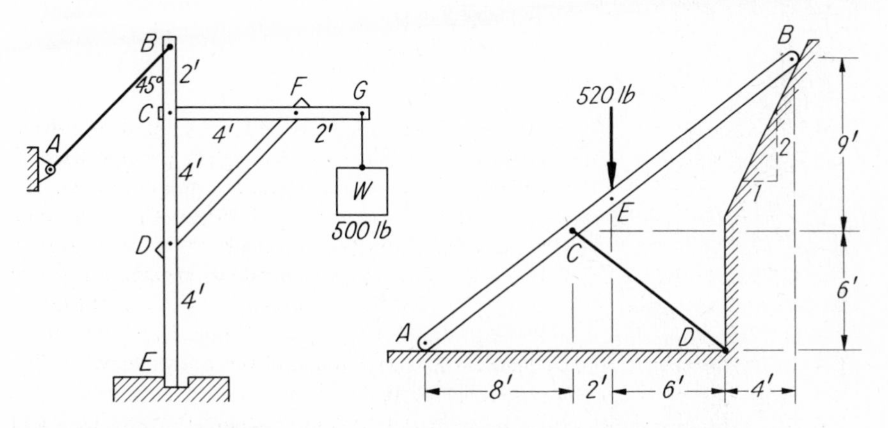

FIG. 5-92 Prob. 5-110.

FIG. 5-93 Probs. 5-106 and 5-108.

5-111. The forces P and W in Fig. 5-94 are 210 lb and 420 lb, respectively. Using the *method of members*, compute the vertical and horizontal components of the pin *reactions* at B, C, and D. *Ans.* $B_V = 280$ lb; $B_H = 245$ lb; $C_V = 406$ lb; $C_H = 245$ lb; $D_V = 70$ lb; $D_H = 245$ lb

5-112. Solve Prob. 5-111 if $P = 300$ lb and $W = 150$ lb.

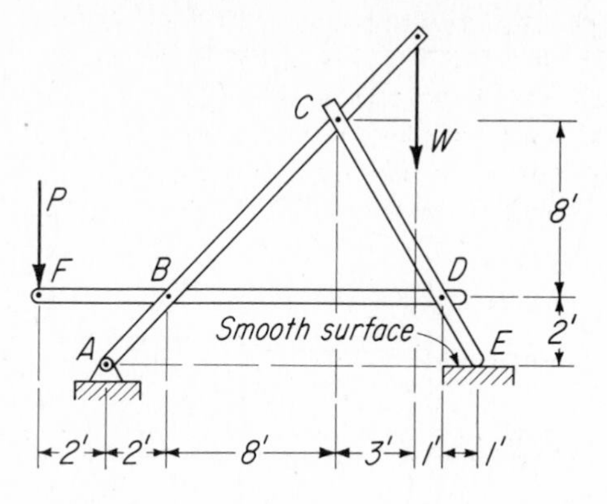

FIG. 5-94 Probs. 5-111 and 5-112.

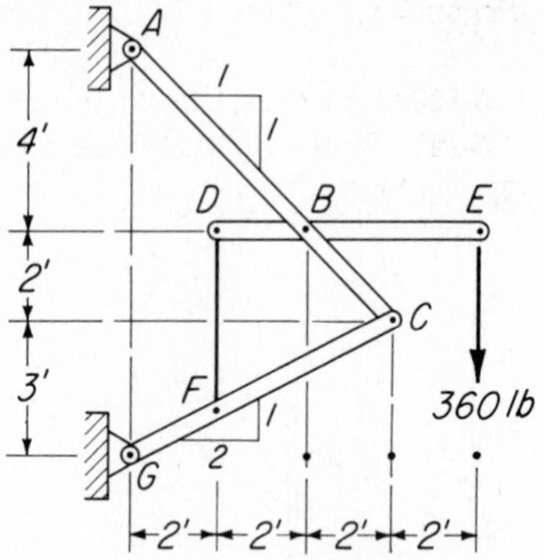

FIG. 5-95 Prob. 5-113.

5-113. Using the *method of members* compute the vertical and horizontal components of the pin *reactions* at *B* and *C* of Fig. 5-95.

5-114. The sliding joint at *D* in Fig. 5-96 is frictionless. Calculate the vertical and horizontal components of the pin reactions at *E*, *C*, and *D*.

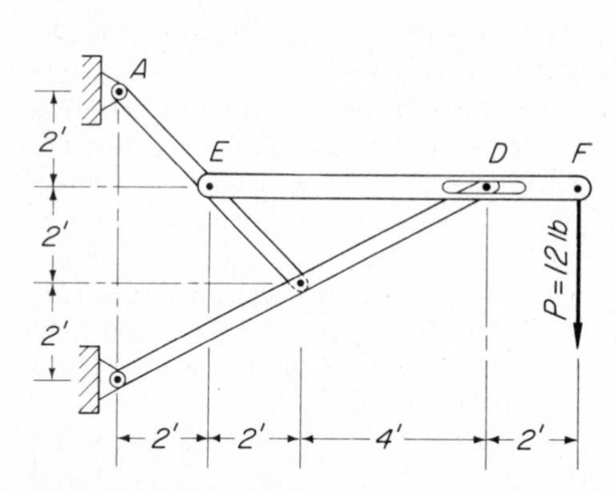

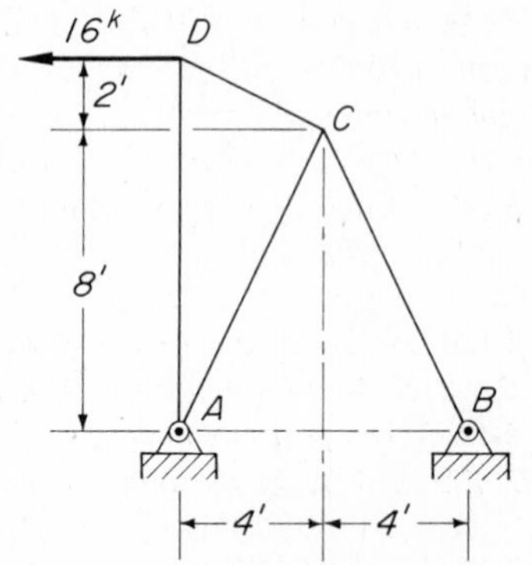

FIG. 5-96 Prob. 5-114.

FIG. 5-97 Probs. 5-115 and 5-120.

5-115. The truss shown in Fig. 5-97 supports the horizontal load of 16 kips. By the *shear method of sections* calculate the force in members *AC* and *BC*.

Ans. $AC = 13.4$ kips C; $BC = 22.4$ kips T

5-116. Using the *shear method of sections,* compute the *stresses* in members *DE*, *EG*, and *GF* of the truss shown in Fig. 5-70.

5-117. By the *shear method of sections,* determine the *stresses* in members *AC*, *BC*, and *BD* of the Warren truss shown in Fig. 5-69.

Ans. $AC = 18$ kips T; $BC = 3.6$ kips C; $BD = 16$ kips C

5-118. Using the *shear method of sections,* find the *stresses* in members *AC*, *BC*, and *BH* of the loading-platform truss shown in Fig. 4-36.

5-119. Using the *shear method of sections,* determine the stress in the *active tension*

diagonal at section 3-3 in Fig. 5-80 and the stresses in the chords *DF* and *EG*.
Ans. $DG = 15$ kips T; $DF = 27$ kips C; $EG = 18$ kips T

5-120. Calculate the stresses in members *CD* and *AD* of Fig. 5-97.

5-121. By the *component method of sections,* solve for the *stresses* in members *BD, CD,* and *CE* of the truss shown in Fig. 5-98. ($E_V = 2$ kips↓; $E_H = 4$ kips→.)
Ans. $BD = 13.4$ kips C; $CD = 4$ kips C; $CE = 5.67$ kips T

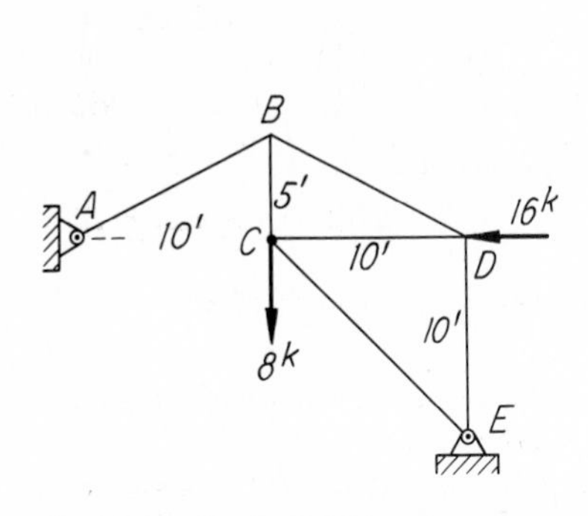

FIG. 5-98 Prob. 5-121.

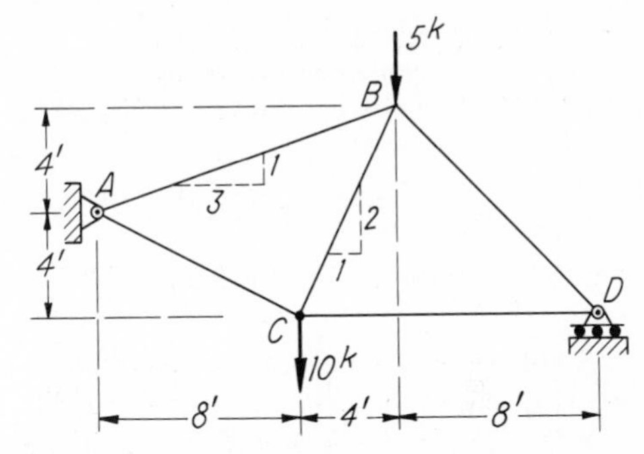

FIG. 5-99 Prob. 5-122.

5-122. Using the *component method of sections,* determine the *stresses* in members *AB, BC,* and *CD* of the truss shown in Fig. 5-99. (SUGGESTION: Use part to the right of cut section and apply the *V* and *H* components of *AB* at *A*, and those of *BC* at *C*.) *Ans.* $AB = 10.12$ kips C; $BC = 5.81$ kips T; $CD = 7$ kips T

5-123. Using the *component method of sections,* solve for the *stresses* in members *AC, BC,* and *BD* of the truss shown in Fig. 5-87. (See Prob. 5-100. HINT: Isolate right half.) *Ans.* $AC = 11.3$ kips C; $BC = 15.6$ kips T; $BD = 15$ kips C

5-124. By the *component method of sections,* find the *stresses* in members *BC, BE,* and *AE* of the truss shown in Fig. 5-40. (See Prob. 5-42. HINT: Isolate right half.)

5-125. Using the *component method of sections,* compute the *stresses* in members *CD, BD,* and *BE* of the truss shown in Fig. 5-30 (see Prob. 5-32).
Ans. $CD = 11.2$ kips T; $BD = 20$ kips C; $BE = 21.2$ kips T

5-126. In Fig. 5-80, which is the *active tension diagonal* in the third panel from the left, and what is its stress? *Ans.* $DG = 15$ kips T

5-127. If, in Fig. 5-80, the two 24-kip loads are applied at panel points *C* and *E*, which will be the *active tension diagonal* in the third panel from the left, and what will be its stress? *Ans.* $EF = 15$ kips T

REVIEW QUESTIONS

5-1. Define a coplanar, nonconcurrent force system.

5-2. What items must be determined before the resultant of a coplanar, nonconcurrent force system is fully known?

5-3. In the string-polygon method, what do the rays represent and what do the strings represent?

5-4. When support reactions are determined by the string-polygon method, at which point must the string polygon be started? What is the only exception to this rule?

5-5. What is the direction of the reaction at the end of a two-force member?

5-6. What is known about the direction of the reaction at the end of a three-force member?

5-7. What is the direction of the reaction at a roller support and at a smooth-surface support?

5-8. What is meant by a pin reaction?

5-9. In problems to determine pin reactions, which forces must always be computed before the frame is dismembered?

5-10. What are the purpose and function of the method of sections?

5-11. To what type of truss is the shear method of sections especially applicable? Explain briefly the usual procedure.

5-12. When is the component method of sections generally used? Explain briefly the usual procedure.

5-13. Is it always possible, by the method of sections, to determine any single stress in a truss independently of any other stress, provided not more than three members are cut at a section?

5-14. What is a counter diagonal, and what purpose does it serve?

5-15. In any panel of a truss containing two diagonals, which of the two is considered to be active and to carry stress?

CHAPTER 6

Noncoplanar, Parallel Force Systems

6-1 Introduction

The most common system of noncoplanar, parallel forces is that of the gravity forces (weights) of the several parts of a structure or machine such as a building, an airplane, or an automobile. A common type of problem is that of determining the *resultant force* of the weights of all parts of an airplane, and the location of its line of action, which, for proper balance of the plane, must pass through a given point in the plane. Other problems concern the *equilibrium of parallel forces in space;* such arise in the determination of loads on the columns of a building produced by the dead weight of the floors they support and the live loads thereon. These problems are discussed in the next two articles.

6-2 Resultant of a Noncoplanar, Parallel Force System

The magnitude of the resultant of a parallel force system in space is equal to the algebraic sum of the component forces. The **center of a system of parallel forces** is that point through which the resultant passes, or about which the algebraic sum of the moments of the forces is zero. The magnitude of a resultant force is readily determined by simple algebraic summation of the forces comprising the system, and the location of the center is readily found by moments, because *about any axis* **the moment of the resultant force equals the algebraic sum of the moments of the separate forces.** Summations of moments must, however, be made in two planes, usually at right angles to each other.

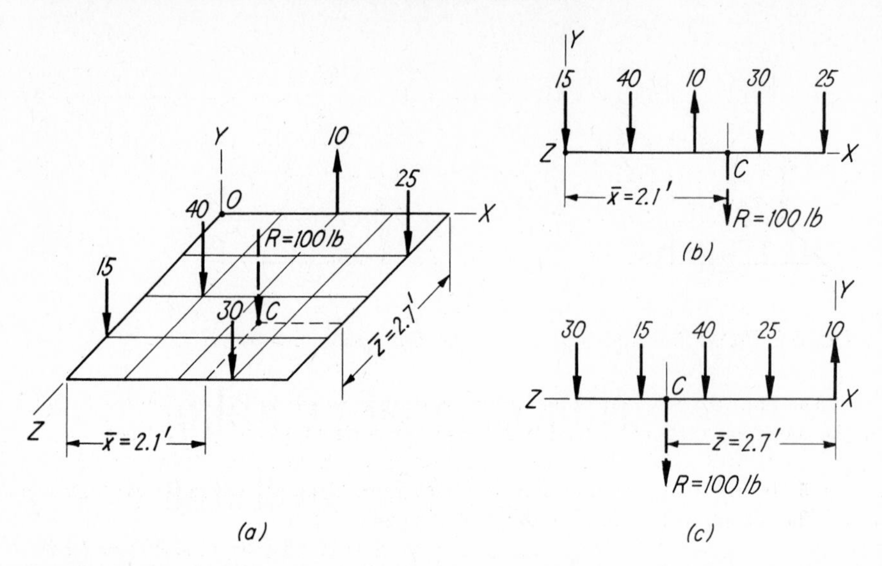

FIG. 6-1 Resultant of a system of parallel forces in space. (*a*) *XZ* plane is horizontal and is 4 ft square. (*b*) Forces projected into *XY* plane. (*c*) Forces projected into *YZ* plane.

ILLUSTRATIVE PROBLEM

6-1. The magnitude and the location of the center *C* of the system of parallel forces shown in Fig. 6-1*a* are to be determined. The side of each square is 1 ft in length. All forces are in pounds.

Solution: The center of these forces is fully located if we determine its position in each of two planes, usually perpendicular. First let all forces be projected back into the vertical *XY* plane as shown in Fig. 6-1*b*. In this view the *Z* axis appears as a point. Let *Z* be the moment center and let $\bar{x}$ (bar *x*) be the distance from *Z* to the center of forces. The resultant *R* is the algebraic sum of the forces. Hence

$[R = \Sigma F]$ $\quad R = 15 + 40 - 10 + 30 + 25 \quad$ or $\quad R = 100$ lb $\quad$ *Ans.*
$[R\bar{x} = \Sigma M_Z]$ $\quad 100\bar{x} = (40)(1) - (10)(2) + (30)(3) + (25)(4)$
and $\quad \bar{x} = 2.1$ ft $\quad$ *Ans.*

Next let all forces now be projected into the vertical *YZ* plane, as in Fig. 6-1*c*. Here, axis *X* appears as a point. Let *X* now be the moment center and let $\bar{z}$ be the distance from *X* to the center of forces. Then

$[R\bar{z} = \Sigma M_X]$ $\quad 100\bar{z} = (25)(1) + (40)(2) + (15)(3) + (30)(4)$
and $\quad \bar{z} = 2.7$ ft $\quad$ *Ans.*

PROBLEMS

6-2. Solve Prob. 6-1 if the 30-lb force is upward-acting.

6-3. Solve Prob. 6-1 if a downward force of 30 lb is added at $x = 3$ and $z = 3$.
Ans. $R = 130$ lb; $\bar{x} = 2.31$ ft; $\bar{z} = 2.77$ ft

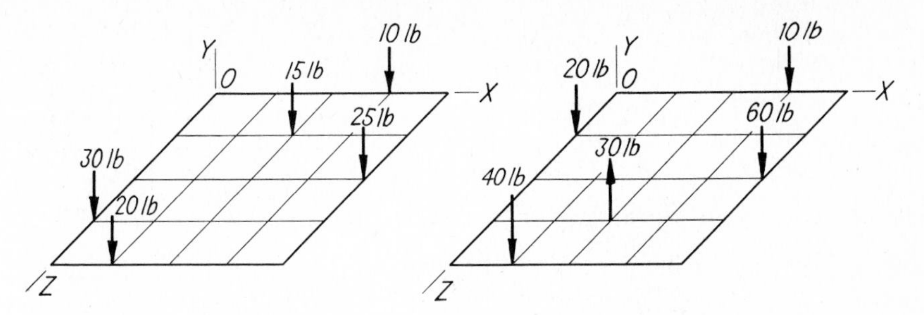

FIG. 6-2 Probs. 6-4 and 6-6.

FIG. 6-3 Probs. 6-5 and 6-7.

6-4. Find the resultant R of the force system shown in Fig. 6-2 and determine the location of its center C. The XZ plane is horizontal and is 4 ft square.

Ans. $R = 100$ lb$\downarrow$; $\bar{x} = 1.8$ ft; $\bar{z} = 2.35$ ft

6-5. In Fig. 6-3, determine the resultant force R and the location of its center C. The XZ plane is horizontal and is 4 ft square.

Ans. $R = 100$ lb$\downarrow$; $\bar{x} = 2.5$ ft; $\bar{z} = 2.1$ ft

6-6. In Fig. 6-2, let the 20-lb force be upward-acting. Compute R, $\bar{x}$, and $\bar{z}$.

6-7. In Fig. 6-3, let the 10-lb force be upward-acting. Compute R, $\bar{x}$, and $\bar{z}$.

Ans. $R = 80$ lb$\downarrow$; $\bar{x} = 2.375$ ft; $\bar{z} = 2.625$ ft

6-3 Equilibrium of Noncoplanar, Parallel Force Systems

Since their directions are known, only the magnitudes of the unknown forces of a parallel system in equilibrium need to be determined. They are usually determined by summations of moments about two axes at right angles to each other. In statics, not more than three unknown forces may be so determined.

ILLUSTRATIVE PROBLEM

6-8. The square, horizontal plate shown in Fig. 6-4 supports a vertical load of 300 lb at D and is in turn supported by vertical reactions at A, B, and C. Compute these vertical reacting forces at supports A, B, and C. The side of each square is 1 ft in length. Neglect weight of plate.

Solution: In Figs. 6-4*b* and *c*, the forces have been projected, respectively, into the XY and YZ planes. In each view three forces A, B, and C are unknown. With point A as the moment center in both views, reaction A is eliminated from consideration. Two independent equations may now be written which, then, are solved simultaneously, yielding reactions B and C. Reaction A may then be found by moments about C. Finally, a summation of vertical forces provides a sufficient check. Hence, from Fig. 6-4*b*, we have

$$[\Sigma M_A = 0] \qquad 2B + 6C = (300)(3) = 900 \qquad (a)$$

and from Fig. 6-4*c* we obtain

$$[\Sigma M_A = 0] \qquad 6B + 3C = (300)(2) = 600 \qquad (b)$$

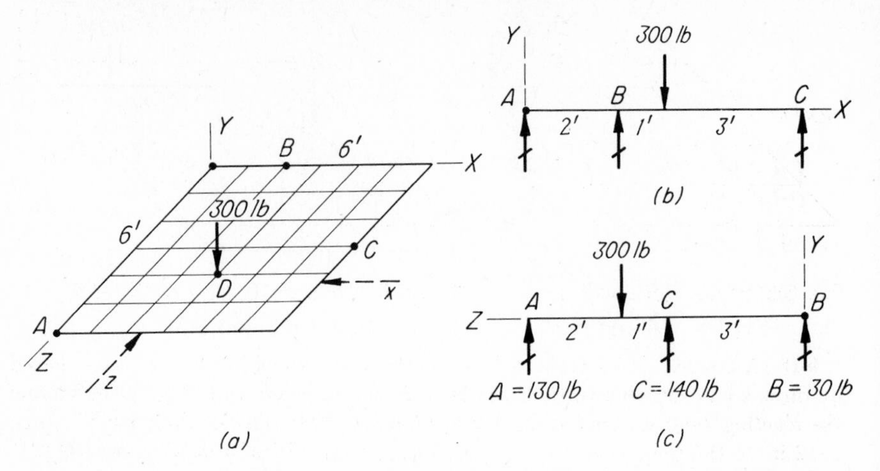

FIG. 6-4 Determination of reactions. Parallel forces in space. (*a*) XZ plane is horizontal and is 6 ft square. (*b*) Forces projected into XY plane. (*c*) Forces projected into YZ plane.

Multiplying (*a*) by 3 and subtracting (*b*),

$$6B + 18C = 2700 \qquad (c)$$
$$6B + 3C = 600 \qquad (b)$$
$$15C = 2100 \qquad \text{and} \qquad C = 140 \text{ lb} \qquad \textit{Ans.}$$

Substituting this value of C in Eq. (*a*) to find B, we have

$$2B + (6)(140) = 900 \qquad \text{and} \qquad B = 30 \text{ lb} \qquad \textit{Ans.}$$

Taking moment about C, in Fig. 6-4*c*, to find A, we obtain

$$[\Sigma M_C = 0] \qquad 3A = (300)(1) + (30)(3) = 390 \qquad \text{and} \qquad A = 130 \text{ lb} \qquad \textit{Ans.}$$
$$[\Sigma V = 0] \qquad 300 = 140 + 30 + 130 \qquad \text{or} \qquad 300 = 300 \qquad \textit{Check.}$$

(NOTE: When two of three unknown forces overlie each other in one view, as do A and B in Fig. 6-5, the third force may be solved for directly.)

PROBLEMS

6-9. The square, level plate shown in Fig. 6-5 weighs 120 lb, which may be concentrated at its center. Compute the vertical reactions at supports A, B, and C, due to this weight and to the two vertical loads at D and E.

Ans. $A = 115$ lb; $B = 165$ lb; $C = 140$ lb

6-10. The horizontal plate shown in Fig. 6-6 is subjected to two vertical loads at D and E. Neglecting the weight of the plate, compute the vertical reactions at supports A, B, and C. The side of each square is 1 ft in length.

6-11. Solve Prob. 6-10 if the support B is moved 1 ft to the left.

Ans. $A = 20$ lb; $B = 80$ lb; $C = 90$ lb

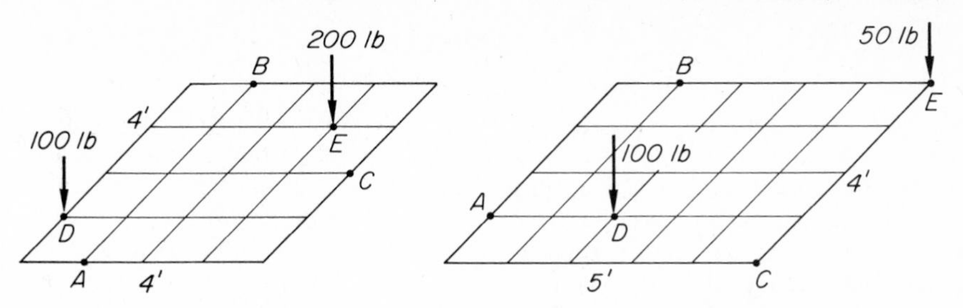

FIG. 6-5 Probs. 6-9 and 6-17.

FIG. 6-6 Probs. 6-10, 6-11, and 6-18.

6-12. A circular, level table, 4 ft in diameter and weighing 150 lb, is supported on three equally spaced vertical legs at A, B, and C, as shown in Fig. 6-7. Compute the reacting force in each leg, if a vertical load of 180 lb is concentrated at D, which is 12 in. to the right of A. *Ans.* $A = 170$ lb; $B = C = 80$ lb

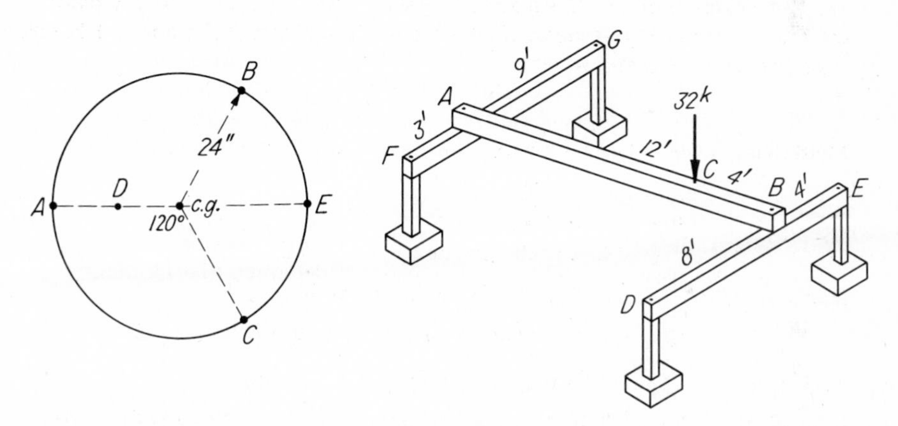

FIG. 6-7 Probs. 6-12 and 6-13.

FIG. 6-8 Probs. 6-14, 6-15, and 6-19.

6-13. If the circular table shown in Fig. 6-7 weighs 120 lb, what maximum downward-acting vertical force P may be applied at E without tipping the table?

6-14. Three horizontal beams, arranged as shown in Fig. 6-8 and resting on vertical posts, support a 32-kip vertical load at C. Neglecting the weights of the beams, compute the load on each of posts D, E, F, and G. Draw a separate free-body diagram of each of the three beams.

Ans. $D = 8$ kips; $E = 16$ kips; $F = 6$ kips; $G = 2$ kips

6-15. In Fig. 6-8, the beams are of uniform cross section and density. Beam AB weighs 1,200 lb, and each of beams DE and FG weighs 600 lb. Compute the loads on each of posts D, E, F, and G due to these weights and the 32-kip load.

6-16. Solve Prob. 6-8 if the support C is moved 2 ft further away from the X axis (parallel to the z axis) and if the load is moved to $x = 2$, $z = 3$.

SUMMARY

(By article number)

6-2. The magnitude of the resultant of a system of parallel forces in space is their algebraic sum. Its location may be determined by summations of moments about two axes, usually perpendicular, since, about any axis, the moment of the resultant equals the algebraic sum of the moments of the separate forces.

6-3. The unknown forces (usually three reactions) of a system of noncoplanar, parallel forces in equilibrium may be determined by summations of moments about two axes, usually perpendicular.

REVIEW PROBLEMS

6-17. In Prob. 6-9, disregard the weight of the plate and compute the *reactions* A, B, and C required for equilibrium of the remaining system of noncoplanar, parallel forces. *Ans.* $A = 75$ lb; $B = 125$ lb; $C = 100$ lb

6-18. In Prob. 6-10, let E be an upward-acting force. Compute the vertical *reactions* A, B, and C required to maintain *equilibrium*.

6-19. In Prob. 6-14, let an additional downward-acting vertical force of 24 kips be acting on beam AB at a point 4 ft from A. Compute the *reactions* at beam ends D, E, F, and G. *Ans.* $D = 10$ kips; $E = 20$ kips; $F = 19.5$ kips; $G = 6.5$ kips

6-20. Solve Prob. 6-8 if a 200-lb downward-acting force is added at the origin O (along the Y axis). *Ans.* $A = 170$ lb; $B = 270$ lb; $C = 60$ lb

REVIEW QUESTIONS

6-1. What is meant by the center of a system of parallel forces?

6-2. By what basic principle of statics may we determine the location of this center?

6-3. Describe briefly the method used to determine the location of the center of a noncoplanar, parallel system of forces.

6-4. Explain briefly the method used to determine the unknown forces of a noncoplanar, parallel force system in equilibrium.

CHAPTER 7

Noncoplanar, Concurrent Force Systems

7-1 Introduction

A simple example of a noncoplanar, concurrent force system is an ordinary tripod supporting a weight at the intersection of the three legs. Another example is the simple three-member frame shown in Fig. 7-4. The weight W and the stresses in members A, B, and C constitute such a system: concurrent because their action lines all meet at O, and noncoplanar because they do not all lie in one plane.

The resultant of such a system is seldom required in ordinary engineering practice. Problems in determining the unknown forces, reactions, or stresses required to maintain equilibrium are, however, frequently encountered.

7-2 Components of a Force in Space

In Fig. 7-1, F is a force in space represented as a vector of length F. For the purpose of reference, let O arbitrarily be called its *near end* and A its *far end*. Through O are passed three axes, X, Y, and Z, all mutually perpendicular, of which Y is vertical, and X and Z are horizontal.

The distances x, y, and z, which locate the far end A of the force with respect to its near end O, are called the **space coordinates** of point A. These x, y, and z coordinates are measured, respectively, parallel to the X, Y, and Z axes from the origin O. They are positive when measured to the right, upward, and toward the observer.

We may now imagine Fig. 7-1 to represent a transparent box with visible edges whose dimensions are the coordinates x, y, and z. Force F is the diagonal

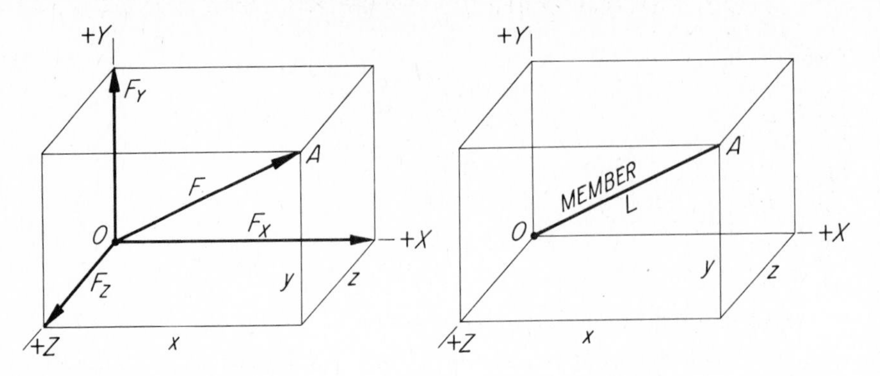

FIG. 7-1 A force in space.

FIG. 7-2 A two-force member in space.

of this box, and F_X, F_Y, and F_Z are, respectively, the X, Y, and Z components of F, as a little study will show. Obviously, then, **the X, Y, and Z components of a force in space are proportional to the x, y, and z coordinates of its far end with respect to its near end.** The component always bears the sign, positive or negative, of its corresponding coordinate.

Since F is the diagonal of a parallelepipedon,

$$F^2 = F_X^2 + F_Y^2 + F_Z^2 \qquad \text{or} \qquad F = \sqrt{F_X^2 + F_Y^2 + F_Z^2} \tag{7-1}$$

Figures 7-1 and 7-2 are identical. In both, the coordinates of the far end of the diagonal with respect to its near end are x, y, and z. In Fig. 7-2, however, let this diagonal represent a two-force structural member of length L. Let F. whose action line passes through points O and A, be the *axial stress* in this member. Clearly, then, **when the near end of a two-force member in space is the origin O of the rectangular X, Y, and Z axes, the components F_X, F_Y, and F_Z of the axial stress within it are proportional to the x, y, and z space coordinates of its far end.** This proportionality is an important aid in the determination of axial stresses in space frames, discussed in the following article. In equation form, the proportions can be written

$$\frac{F_X}{x} = \frac{F}{L} \quad \frac{F_Y}{y} = \frac{F}{L} \quad \frac{F_Z}{z} = \frac{F}{L} \qquad \text{or} \qquad \frac{F_X}{x} = \frac{F_Y}{y} = \frac{F_Y}{z} \tag{7-2}$$

ILLUSTRATIVE PROBLEM

7-1. In Fig. 7-1, let $F = 100$ lb, and let $x = 6$, $y = 4$, and $z = 3$. Then determine the components F_X, F_Y, and F_Z.

Solution: The components can be determined by Eq. (7-2). The length L is

$$[L = \sqrt{x^2 + y^2 + z^2}] \qquad L = \sqrt{(6)^2 + (4)^2 + (3)^2} = \sqrt{61} = 7.81$$

Then

$$\left[\frac{F_X}{x} = \frac{F}{L}\right] \qquad F_X = x\frac{F}{L} = 6\left(\frac{100}{7.81}\right) = 76.8 \text{ lb} \qquad \textit{Ans.}$$

$$\left[\frac{F_Y}{y} = \frac{F}{L}\right] \qquad F_Y = y\frac{F}{L} = 4\left(\frac{100}{7.81}\right) = 51.2 \text{ lb} \qquad \textit{Ans.}$$

$$\left[\frac{F_Z}{z} = \frac{F}{L}\right] \qquad F_Z = z\frac{F}{L} = 3\left(\frac{100}{7.81}\right) = 38.4 \text{ lb} \qquad \textit{Ans.}$$

A check on these computations may be obtained from Eq. (7-1) as follows:

$$[F^2 = F_X{}^2 + F_Y{}^2 + F_Z{}^2] \quad 10{,}000 = 5{,}900 + 2{,}623 + 1{,}477 = 10{,}000 \qquad \textit{Check.}$$

PROBLEMS

7-2. In Fig. 7-1, let $F = 340$ lb, and let $x = 12$, $y = 9$, and $z = 8$. Determine the force components F_X, F_Y, and F_Z.

7-3. The member OA shown in Fig. 7-2 has x, y, and z components of length of 6, 8, and 24 ft respectively. Calculate the force in the member if the X-component of the force is 1,200 lb. *Ans.* $F = 5{,}400$ lb

7-4. Solve Prob. 7-3 if the Z-component of the force is 600 lb.

7-3 Equilibrium of Noncoplanar, Concurrent Force Systems

The usual problem is that of determining the unknown stresses in trusses similar to those shown in Probs. 7-5 and 7-6. A number of methods of solution are available, both analytical and graphical. Only analytical methods are presented here.

The force method is based on three algebraic summations of forces parallel to three rectangular X, Y, and Z axes, thus requiring the simultaneous solution of three equations.

The moment method requires only two algebraic summations of moments to be made in two perpendicular planes, thus requiring the simultaneous solution of only two equations. This method, therefore, is generally preferred.

We may project the space frame and all forces acting on it onto one of the three planes associated with the coordinate axes. The planes used are the vertical XY and YZ planes and the horizontal XZ plane, containing the three intersecting and mutually perpendicular X, Y, and Z axes as illustrated in Fig. 7-3. An important point to note is that a force will not appear on a plane perpendicular to its line of action. In other words, on the XY plane only forces parallel to the X and Y axes will be considered, only Y and Z forces are considered on the YZ plane, and only X and Z forces are considered on the XZ plane.

Instead of projecting all members and forces onto the mutually perpendicular planes, we may adopt an alternate viewpoint. The space frame, its loads, and the reaction components may be visualized as a three-dimensional

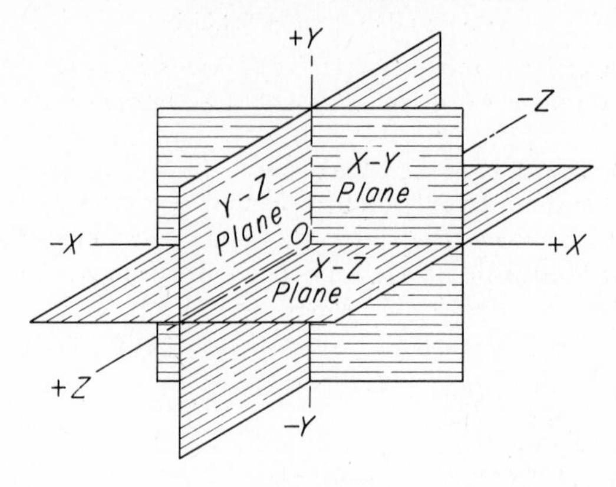

FIG. 7-3 Planes formed by three intersecting and mutually perpendicular axes.

concept in which forces may be summed in any direction or moments can be taken about any axis. When taking moments, all forces whose lines of action *pass through* the moment axis and all forces *parallel* to the moment axis will not contribute to the summation.

Both the "projected on coordinate planes" and the "three-dimensional" concepts will be used in the following illustrative problems.

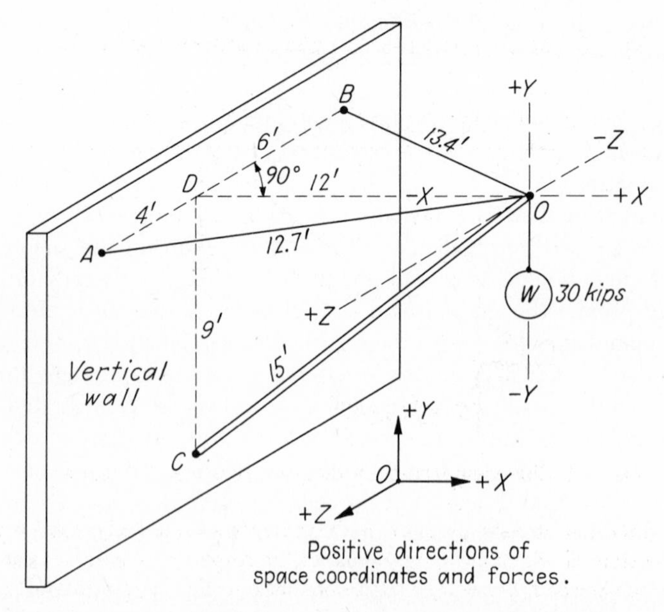

FIG. 7-4 Determination of stresses in a space frame.

ILLUSTRATIVE PROBLEMS

7-5. The three-member frame shown in Fig. 7-4 is supported at *A*, *B*, and *C* on a vertical wall. Determine the stresses in the three members caused by the vertical load *W*.

Solution by Moment Method: Let each member, and the stress within it, be designated by the letter which also designates the support as its far end. In Figs. 7-5 and 7-6, are shown, respectively, a top view and a side view of the frame. Since *A*, *B*, and *C* are *axial stresses*, they equal, respectively, the reactions at supports *A*, *B*, and *C*. The latter may therefore be solved for.

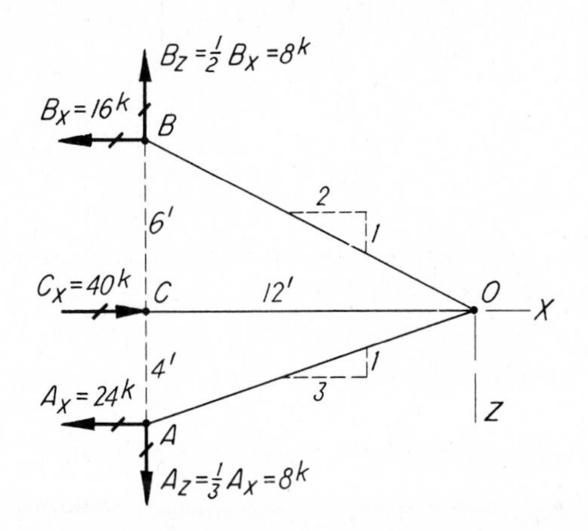

FIG. 7-5 Top-view free-body diagram, horizontal *XZ* plane.

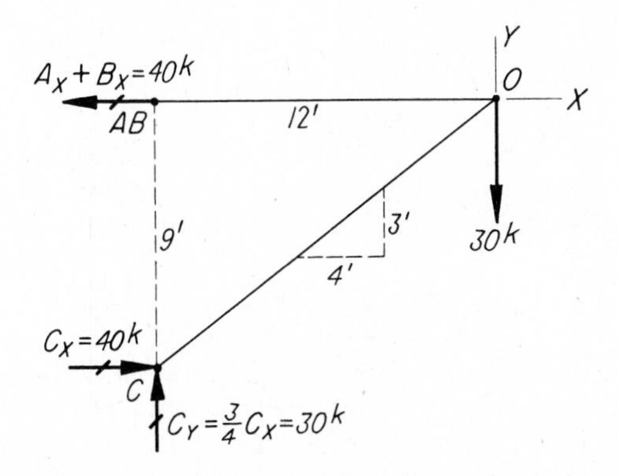

FIG. 7-6 Side-view free-body diagram, vertical *XY* plane.

Let the direction of each member in these two views be indicated by a small triangle, the sides of which are proportional to the respective coordinates of the far end of the member and, therefore, proportional also to the components of the stress in the member. We then express the *Y* and *Z* components in terms of the *X* component. In Fig. 7-6 the Z components are perpendicular to the vertical XY plane and hence produce no reactions in that plane. The components A_X and B_X overlie each other and C_Y is seen to equal $\frac{3}{4}C_X$. In Fig. 7-5 the Y components act perpendicularly to the horizontal XZ plane and hence produce no reactions in that plane. From the direction triangles, we see that A_Z is $\frac{1}{3}A_X$ and B_Z is $\frac{1}{2}B_X$. A and B are clearly tensile stresses, and C is compressive. In this problem, one component of each of the three stresses A, B, and C may now be solved for directly by moments. As usual, no sign convention is necessary in moment solutions. Solving for C_X by moments about point AB, in Fig. 7-6, we have

$$[\Sigma M_{AB} = 0] \qquad 9C_X = (30)(12) = 360 \qquad \text{and} \qquad C_X = 40 \text{ kips}$$

Then

$$C_Y = \frac{3}{4}C_X = \left(\frac{3}{4}\right)(40) = 30 \text{ kips and } C = \sqrt{(40)^2 + (30)^2} = 50 \text{ kips C} \qquad \textit{Ans.}$$

Solving for B_X by moments about point A, in Fig. 7-5, we obtain

$$[\Sigma M_A = 0] \qquad 10B_X = (40)(4) = 160 \qquad \text{and} \qquad B_X = 16 \text{ kips}$$

Then

$$B_Z = \frac{1}{2}B_X = \left(\frac{1}{2}\right)(16) = 8 \text{ kips and } B = \sqrt{(16)^2 + (8)^2} = 17.9 \text{ kips T} \qquad \textit{Ans.}$$

Solving for A_X by moments about point B, in Fig. 7-5, we get

$$[\Sigma M_B = 0] \qquad 10A_X = (40)(6) = 240 \qquad \text{and} \qquad A_X = 24 \text{ kips}$$

Then

$$A_Z = \frac{1}{3}A_X = \left(\frac{1}{3}\right)(24) = 8 \text{ kips and } A = \sqrt{(24)^2 + (8)^2} = 25.3 \text{ kips T} \qquad \textit{Ans.}$$

The positive signs of the computed stresses indicate that the type of each stress was correctly assumed. We must now check the results just obtained. To do so, we record all V and H components on the free-body diagrams, as was done in Figs. 7-5 and 7-6. If $\Sigma V = 0$ and $\Sigma H = 0$, a sufficient check is obtained, since the components were found by moments.

Alternate Solution by Moment Method: The frame, load, and reactions are shown in Fig. 7-7. Members AO and BO have been assumed to be in tension and member CO is assumed to be in compression. We may take moments about the line AB. The reactions or reaction components at A and B will not appear in the resulting equation because their line of actions pass through AB. The reaction component C_Y will likewise not appear because its line of action also passes through AB. We have, then,

$$[\Sigma M_{AB} = 0] \qquad 10(30) = 9C_X \qquad \text{or} \qquad C_X = 40 \text{ kips}$$

The force C can now be found from Eq. (7-2).

$$\left[\frac{F}{L} = \frac{F_X}{x}\right] \qquad \frac{C}{15} = \frac{40}{12} \qquad \text{or} \qquad C = \frac{40(15)}{12} = 50 \text{ kips C} \qquad \textit{Ans.}$$

Taking moments about a vertical line through A gives

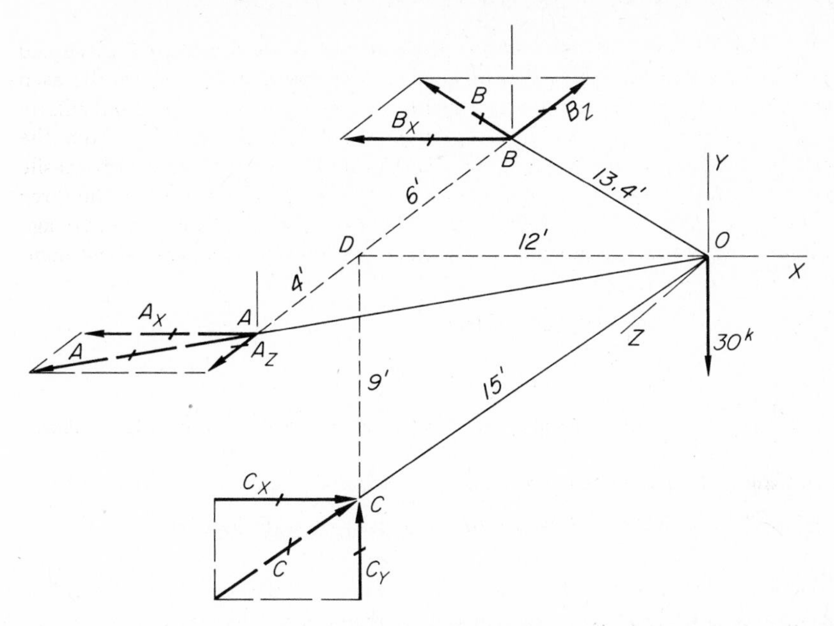

FIG. 7-7 Stresses in space frame by alternate solution.

$$[\Sigma M_{A_Y} = 0] \qquad 4C_X = 10B_X \qquad \text{or} \qquad B_X = \frac{4(40)}{10} = 16 \text{ kips}$$

The force B is then

$$\left[\frac{F}{L} = \frac{F_X}{x}\right] \qquad \frac{B}{13.4} = \frac{16}{12} \qquad \text{or} \qquad B = \frac{16(13.4)}{12} = 17.9 \text{ kips T} \qquad \textit{Ans.}$$

Finally, taking moments about a vertical line through B gives

$$[\Sigma M_{B_Y} = 0] \qquad 6C_X = 10A_X \qquad \text{or} \qquad A_X = \frac{6(40)}{10} = 24 \text{ kips T} \qquad \textit{Ans.}$$

The force A is then

$$\left[\frac{F}{L} = \frac{F_X}{x}\right] \qquad \frac{A}{12.65} = \frac{24}{12} \qquad \text{or} \qquad A = \frac{24(12.65)}{12} = 25.3 \text{ kips T} \qquad \textit{Ans.}$$

7-6. Fig. 7-8 shows a three-member frame supported at A, B, and C. The locations of the three supports are given by x, y, and z space coordinates measured from O, parallel to the rectangular X, Y, and Z axes which pass through O. The frame is subjected to a 1,200-lb load applied at D and parallel to the X-axis. Determine the stresses in all members.

Solution by the Moment Method: Let the stress in each member be designated by A, B, and C, which letter also designates the support end of each member. Let the X, Y, and Z components of all stresses be applied at the respective supports. Assume members AO and CD are tensile and BD compressive. This seems reasonable, and

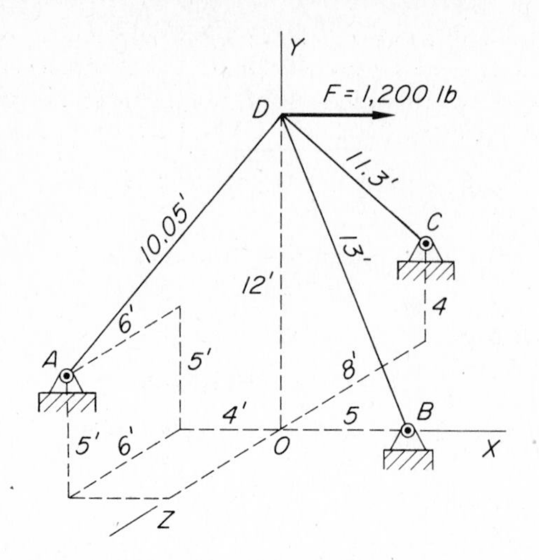

FIG. 7-8 Determination of stresses in space frame.

if the assumption is wrong, the member or members having stresses differing from those assumed will have negative computed components.

Figure 7-9 shows two free-body diagrams of the frame projected onto the vertical XY and YZ planes. The x, y, and z components of the lengths of each member are shown in the appropriate sketch. The X, Y, and Z force components of each reaction are shown in terms of the Y component, using Eq. (7-2). We may now solve for the stresses by either the force method or the moment method. The solution by the moment method follows.

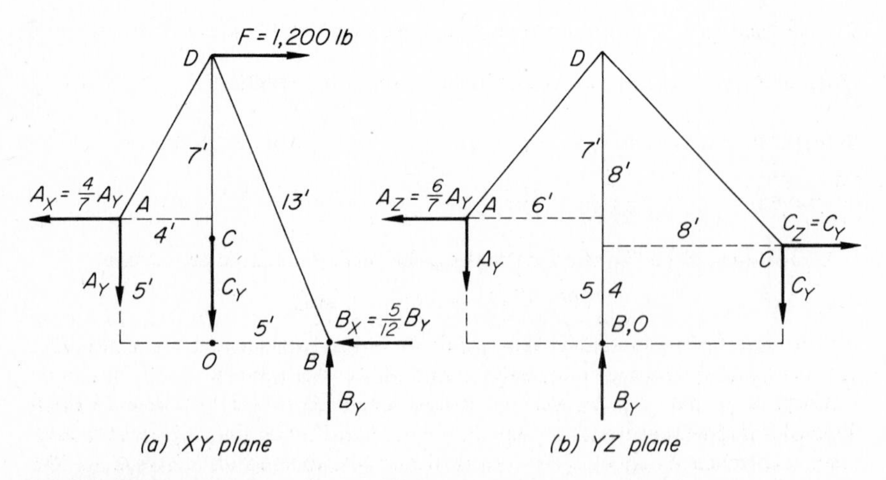

FIG. 7-9 Stresses in space frame. Projected members and forces. (*a*) XY plane. (*b*) YZ plane.

Using the free-body diagrams in Fig. 7-9, two independent moment equations can be written, one in the XY plane and one in the ZY plane, each with point A as the moment center. All components of stress A are thus automatically eliminated from both moment equations. Then, in the XY plane,

$$[\Sigma M_A = 0] \qquad 9B_Y - 5\left(\frac{5}{12}B_Y\right) - 4C_Y - 7(1{,}200) = 0$$

$$9B_Y - 2.08B_Y - 4C_Y = 8{,}400$$

$$6.92B_Y - 4C_Y = 8{,}400 \qquad (a)$$

Similarly in the YZ plane,

$$[\Sigma M_A = 0] \qquad 6B_Y + C_Y - 14(C_Y) = 0$$

$$6B_Y - 13C_Y = 0 \qquad \text{or} \qquad B_Y = \frac{13}{6}C_Y \qquad (b)$$

Substituting the value of B_Y from (b) into (a) gives

$$6.92\left(\frac{13}{6}C_Y\right) - 4C_Y = 8{,}400$$

$$15C_Y - 4C_Y = 8{,}400 \qquad \text{or} \qquad C_Y = 765 \text{ lb}$$

By proportion,

$$\left[\frac{F}{L} = \frac{F_Y}{y}\right] \qquad \frac{C}{11.3} = \frac{765}{8} \qquad \text{or} \qquad C = 1{,}080 \text{ lb T} \qquad \textit{Ans.}$$

Also, from Eq. (b), $B_Y = {}^{13}\!/_6\, C_Y = 1{,}657$ lb

$$\left[\frac{F}{L} = \frac{F_Y}{y}\right] \qquad \frac{B}{13} = \frac{1{,}657}{12} \qquad \text{or} \qquad B = 1{,}795 \text{ lb C} \qquad \textit{Ans.}$$

We may find A by a moment summation about C in the YZ plane.

$$[\Sigma M_C = 0] \qquad 8(1{,}657) = \frac{6}{7}A_Y + 14A_Y \qquad \text{or} \qquad A_Y = 892 \text{ lb}$$

By proportion

$$\left[\frac{F}{L} = \frac{F_Y}{y}\right] \qquad \frac{A}{10.05} = \frac{892}{7} \qquad \text{or} \qquad A = 1{,}280 \text{ lb T} \qquad \textit{Ans.}$$

A partial check on the computations can be made by a force summation.

$$[\Sigma F_Y = 0] \qquad 1{,}657 = 765 + 892 \qquad \textit{Check.}$$

Note that the computations showed all force components to be positive. This indicated that the tensile-compressive assumptions were correct.

Alternate Solution by the Moment Method Using Fig. 7-10: Note that in three dimensional structures we take moments about an axis rather than a point. We may, then, take moments about both a vertical and a horizontal axis through A. The resulting equations can be solved for C_Y and B_Y. A moment summation about a *vertical* line through A in Fig. 7-10 gives

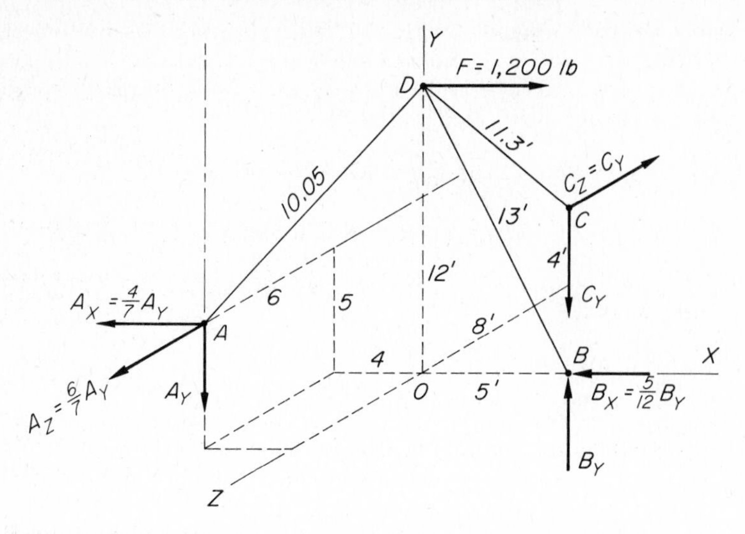

FIG. 7-10 Stresses in space frame by alternate method.

$$[\Sigma M_{A_Y} = 0] \qquad 6(1{,}200) = 4C_Y + 6\left(\frac{5}{12}B_Y\right)$$

$$4C_Y + 2.5B_Y = 7{,}200 \qquad (c)$$

Next, a moment summation about a line *parallel* to the Z axis through A gives,

$[\Sigma M_{A_Z} = 0]$

$$7(1{,}200) + 4C_Y + 5\left(\frac{5}{12}B_Y\right) - 9B_Y = 0 \qquad \text{or} \qquad 4C_Y = 6.92B_Y - 8{,}400 \qquad (d)$$

Substituting Eq. (d) into (c)

$$6.92B_Y - 8{,}400 + 2.5B_Y = 7{,}200$$

This gives

$$B_Y = \frac{15{,}600}{9.42} = 1{,}657 \text{ lb}$$

Substituting this value of B_Y in (d) gives

$$4C_Y = 6.92(1{,}657) - 8{,}400 \qquad \text{or} \qquad C_Y = \frac{3{,}060}{4} = 765 \text{ lb}$$

Summing moments about a vertical line through B gives

$$[\Sigma M_{B_Y} = 0] \qquad 5(765) + 6\left(\frac{4}{7}A_Y\right) - 9\left(\frac{6}{7}A_Y\right) = 0$$

$$3{,}825 + 3.43A_Y - 7.73A_Y = 0 \qquad \text{or} \qquad A_Y = \frac{3{,}825}{4.3} = 892 \text{ lb}$$

These are the same force components found in the first solution and so would yield $A = 1{,}280$ lb T; $B = 1{,}792$ lb C; and $C = 1{,}080$ lb T by proportion.

Solution by the Force Method: Summing forces in the X, Y, and Z directions in Fig. 7-10 yields

$$[\Sigma F_X = 0] \qquad \frac{4}{7}A_Y + \frac{5}{12}B_Y = 1{,}200 \qquad (e)$$

$$[\Sigma F_Y = 0] \qquad A_Y + C_Y - B_Y = 0 \qquad (f)$$

$$[\Sigma F_Z = 0] \qquad C_Y = \frac{6}{7}A_Y \qquad (g)$$

Equations (e) to (g) may be solved simultaneously for A_Y, B_Y, and C_Y. The stresses A, B, and C may then be found by proportion.

PROBLEMS

7-7. Determine the stresses in members A, B, and C of the frame shown in Fig. 7-11 produced by the vertical load of 32 kips. The X and Z axes are horizontal.

Ans. $A = 40$ kips T; $B = C = 13.42$ kips C

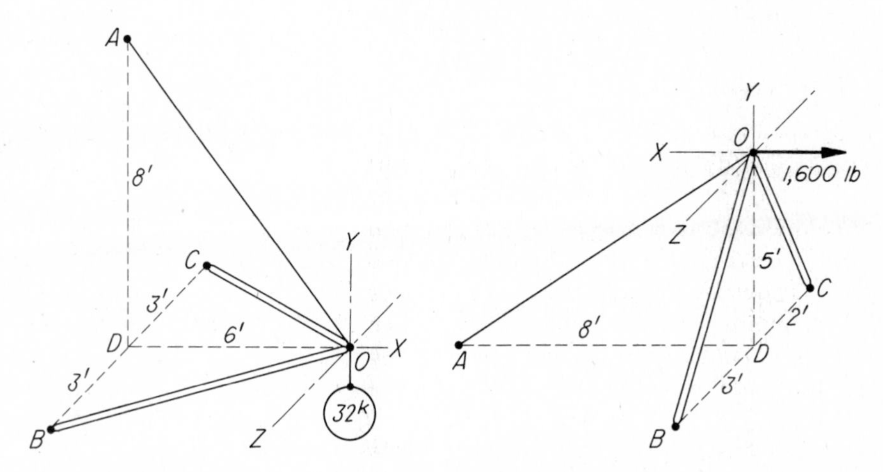

FIG. 7-11 Prob. 7-7.

FIG. 7-12 Prob. 7-8.

7-8. In members A, B, and C of the frame shown in Fig. 7-12, find the stresses produced by the horizontal force of 1,600 lb. The X and Z axes are horizontal.

7-9. Solve for the stresses in members A, B, and C of the frame shown in Fig. 7-13. (HINT: Solve by summations of moments in the XY and XZ planes.) The load is vertical. *Ans.* $A = 26.7$ kips T; $B = 14.3$ kips T; $C = 45$ kips C

7-10. The frame shown in Fig. 7-14 supports a fixed load of 30 kips at F. Find the stresses in legs A and B, when boom DF and cable OF are in the vertical plane $COFDC$. Determine also the stress in the mast OD. (SUGGESTIONS: Consider the entire frame in the XY plane. Solve for the X and Y components at AB and D. Then consider the frame in the XZ plane.) Finally, isolate joint D to find OD.

Ans. $A = B = 24.2$ kips T; $OD = 24$ kips C

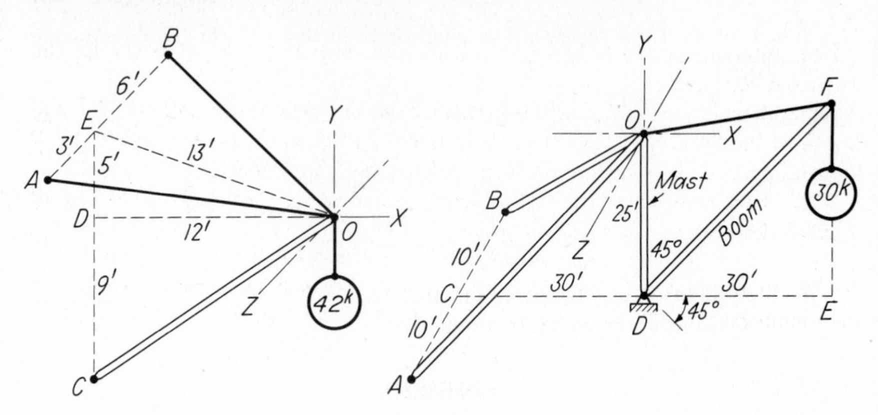

FIG. 7-13 Prob. 7-9.

FIG. 7-14 Probs. 7-10 to 7-12.

7-11. The maximum stress in leg *A* of the frame shown in Fig. 7-14 occurs when the vertical plane containing the boom is perpendicular to the vertical plane containing leg *B*. Solve for this maximum stress in *A*. *Ans.* $A = 76.5$ kips

7-12. In Fig. 7-14 the boom of the frame shown can be swung horizontally 90° to either side of the vertical plane *COFED*. (*a*) Determine the stresses in legs *A* and *B* when the boom has been swung into a position which is 45° toward the observer. Will motion of the boom in a vertical plane affect the stresses (*b*) in the mast and (*c*) in the two legs? Will motion in a horizontal plane affect the stresses (*d*) in the mast and (*e*) in the two legs?

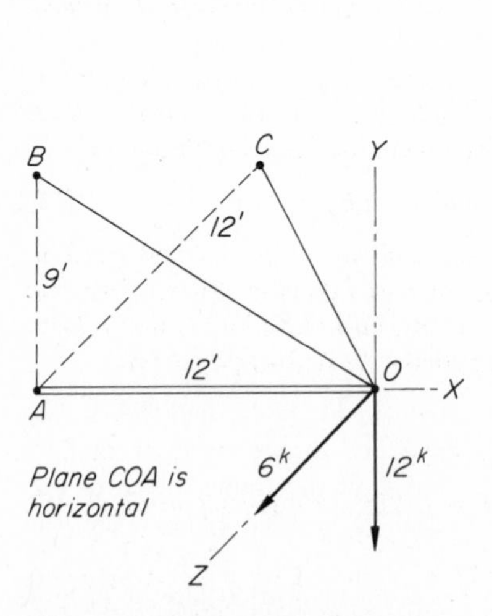

FIG. 7-15 Probs. 7-13 and 7-14.

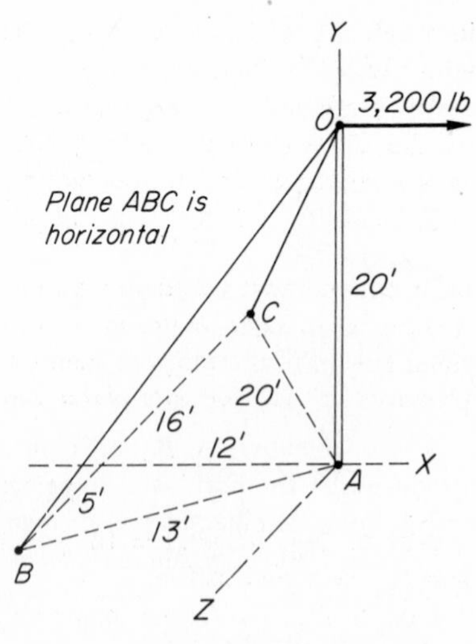

FIG. 7-16 Probs. 7-15 and 7-16.

7-13. The frame shown in Fig. 7-15 is subjected to a force applied at *O* having a vertical component of 12 kips and a horizontal component of 6 kips as shown. Find the stress in each of members *A*, *B*, and *C*.

7-14. Interchange the 6- and 12-kip loads shown in Fig. 7-15 and solve for the stresses *A*, *B*, and *C*.

7-15. Member *OA* in Fig. 7-16 is vertical. Find the stresses in members *A*, *B*, and *C* due to the horizontal load of 3,200 lb applied at *O* as shown, parallel to the *X* axis. Supports *A*, *B*, and *C* all lie in the horizontal *XZ* plane.

7-16. Determine the stresses in the frame shown in Fig. 7-16 if the 3,200-lb load is tilted down at 30° to the *X*-direction.

Ans. $A = 6{,}220$ lb C; $B = 4{,}200$ lb T; $C = 1{,}560$ lb T

7-17. Raise point *B* in Fig. 7-4 vertically 4 ft and then solve for the stresses in the members. *Ans.* $A = 21.6$ kips T; $B = 15.9$ kips T; $C = 42.5$ kips C

SUMMARY

(By article number)

7-2. The axial stress *F* in a member of a three-dimensional space frame equals the square root of the sum of the squares of its three rectangular *X*, *Y*, and *Z* components. That is,

$$F = \sqrt{F_X^2 + F_Y^2 + F_Z^2} \tag{7-1}$$

If one of the three rectangular components of a stress *F* is known, the other two components may readily be computed from the following relationship:

$$\frac{F_X}{x} = \frac{F_Y}{y} = \frac{F_Z}{z} \tag{7-2}$$

in which *x*, *y*, and *z* are the space coordinates of the far end of the stressed member with respect to its near end.

7-3. Three unknown **axial stresses in members of space frames** may be determined by the **force method,** based on *algebraic summations of forces parallel to three rectangular X, Y, and Z axes,* in which the three equations of equilibrium are

$$\Sigma F_X = 0 \qquad \Sigma F_Y = 0 \qquad \text{and} \qquad \Sigma F_Z = 0 \tag{7-5}$$

or by the **moment method** in which *algebraic summations of moments are made in two planes at right angles to each other.* Since only two equations need be solved simultaneously, the moment method is preferred. An important fact is that **a force produces no reaction in a plane that is perpendicular to its line of action.**

REVIEW PROBLEMS

7-18. In Fig. 7-1 let $x = 12$, $y = 4$, and $z = 3$. Compute the *components* F_X, F_Y, and F_Z when $F = 260$ lb. *Ans.* $F_X = 240$ lb; $F_Y = 80$ lb; $F_Z = 60$ lb

7-19. In Fig. 7-1, let $x = 6$, $y = 9$, and $z = 12$. Compute the *x* and *z* components of the force if the *y* component is 180 lb. Also compute the magnitude of the force.

Ans. $F_X = 120$ lb; $F_Z = 240$ lb; $F = 323$ lb

7-20. The line *OA* in Fig. 7-2 represents a structural member of a space frame. The *x*, *y*, and *z* coordinates of its far end A, measured from *O*, are, respectively, 12, 6, and 4. The *Y* component of the stress in the member has been calculated and is 300 lb. Compute the *stress* S in the member. *Ans.* S = 280 lb

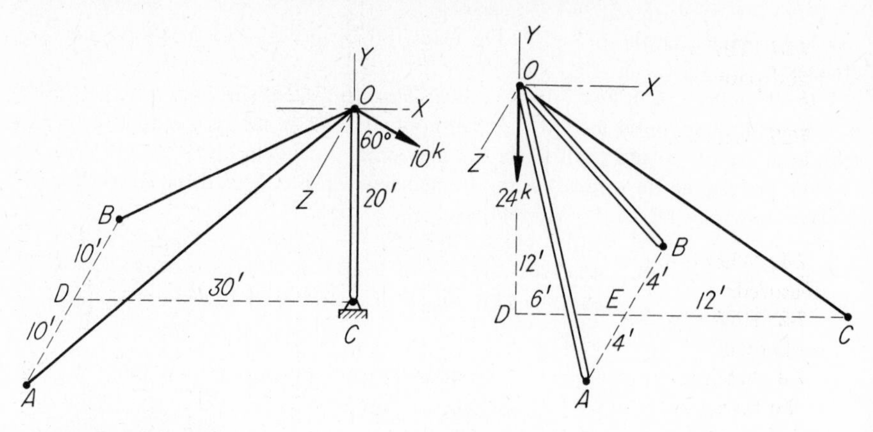

FIG. 7-17 Prob. 7-21.

FIG. 7-18. Prob. 7-22.

7-21. Using the *moment method* determine the stresses in the guy wires *OA* and *OB*, and in the mast *OC*, of the frame shown in Fig. 7-17. The 10-kip force lies in the *XY* plane and is tilted downward at an angle of 30° measured from the *X* axis.

Ans. $OA = OB = 5.4$ kips T; $OC = 10.77$ kips C

7-22. Find the *stresses* in legs *A* and *B* and in the backstay *C*, of the frame shown in Fig. 7-18. *Ans.* $A = B = 21$ kips C; $C = 21.63$ kips T

7-23. The tripod shown in Fig. 7-19 supports a vertical load of 5 kips. Find the stresses in the three legs.

Ans. $A = 2{,}585$ lb C; $B = 1{,}710$ lb C; $C = 1{,}520$ lb C

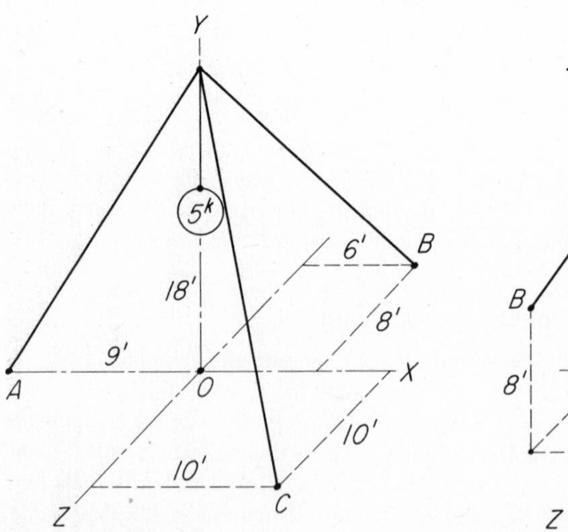

FIG. 7-19 Prob. 7-23.

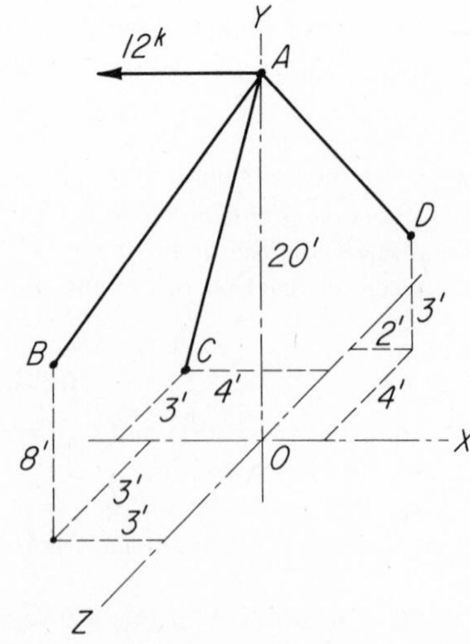

FIG. 7-20 Prob. 7-24.

7-24. The tripod shown in Fig. 7-20 is subjected to a horizontal load applied at A and parallel to the X axis. Find the stresses in the three legs.

Ans. AB = 8.84 kips T; AC = 48.9 kips C; AD = 40.3 kips T

7-25. Let the 1,200-lb load in Fig. 7-8 be in the Z direction. Solve for the stresses.

Ans. A = 534 lb C; B = 553 lb C; C = 1,248 lb T

REVIEW QUESTIONS

7-1. What is the function of space coordinates? From which point are they always measured?

7-2. How can the magnitude of a force be found if its x, y, and z components are known?

7-3. What important relationship exists between the x, y, and z space coordinates of the far end of a two-force member (with respect to its near end) and the X, Y, and Z stress components within that member?

7-4. Outline briefly the essential steps in determining stresses in the three members of a space frame (a) by the force method, and (b) by the moment method.

7-5. In the moment method, what forces and components of forces are considered (a) in the XY plane, (b) in the YZ plane, and (c) in the XZ plane?

CHAPTER 8

Noncoplanar, Nonconcurrent Force Systems

8-1 Introduction

The noncoplanar, nonconcurrent force system is most commonly encountered in the more advanced phases of design of structures and machines and is therefore treated only briefly here. The resultant of such a system may be a force, or a moment, or a force and a moment. The necessity of determining such a resultant is seldom encountered in ordinary engineering practice. The usual problem is that of determining the unknown reacting forces of a system in equilibrium.

8-2 Equilibrium of Noncoplanar, Nonconcurrent Force Systems

Unknown forces are determined by solving for their X, Y, and Z components in three mutually perpendicular planes, usually the vertical XY and YZ planes, and the horizontal XZ plane. When signs of space coordinates and forces are used, they are positive in the directions indicated in Fig. 8-1. Much of the

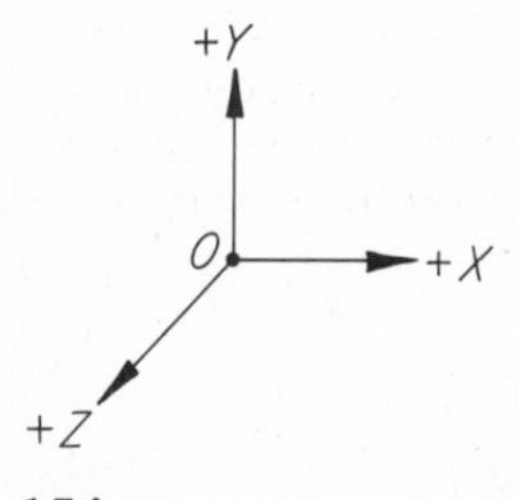

FIG. 8-1 Positive direction of space coordinates and forces.

difficulty usually encountered in treating these force systems is overcome, if it is clearly understood that *a force produces no reaction in a plane that is perpendicular to its line of action.* This point is clearly illustrated in the following example.

ILLUSTRATIVE PROBLEM

8-1. A rectangular shelf, 12 by 20 in., of uniform thickness and weighing 80 lb, is supported in a horizontal position by *loose-fitting hinges* at *A* and *B*, and by the cord *CD*, as is shown in Fig. 8-2. If the weight *W* of the shelf is considered to be concentrated at its center, determine the magnitudes of the reacting forces at supports *A*, *B*, and *C*. Let it be assumed that all thrust parallel to the *Z* axis is resisted by hinge *A*.

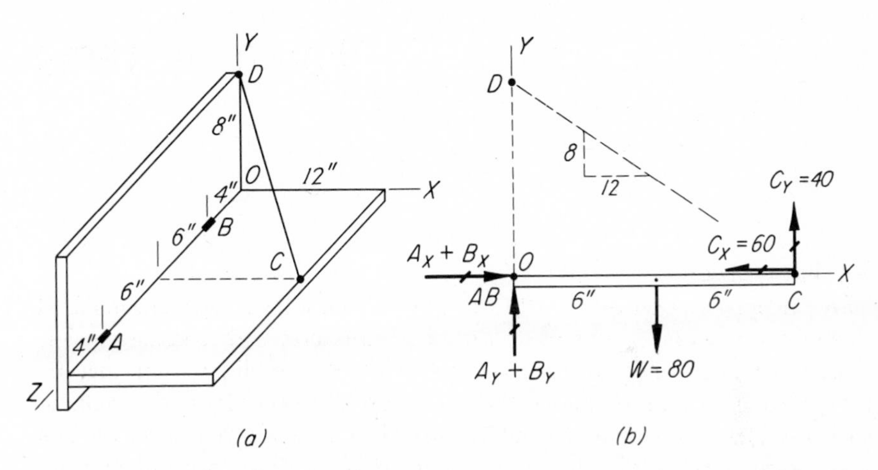

FIG. 8-2 Determination of reactions. Nonconcurrent forces in space. (*a*) Space diagram. (*b*) Free-body diagram, *XY* plane.

Solution: The reaction at *C* is clearly the stress in the cord *CD*. It has three components, C_X, C_Y, and C_Z. The magnitudes of these components are proportional to the *x*, *y*, and *z* space coordinates of point *D* which, with respect to point *C*, are -12, $+8$, and -10. Since only the magnitude of *C* is desired, the signs of these space coordinates have no significance.

In Fig. 8-2*b*, the shelf has been isolated as a free body, and the *X* and *Y* components acting at *A*, *B*, and *C* are shown projected onto the vertical *XY* plane. In this plane, the *Z* components are zero. Since *B* is directly behind *A*, components A_X and B_X overlie each other, as do A_Y and B_Y. *W* is applied at the center of the shelf, and C_X and C_Y at *C*. *CD* is clearly a tensile force. With *AB* as the moment center, we may now find C_Y. That is,

$[\Sigma M_{AB} = 0]$ $$12C_Y = (80)(6) = 480$$

and $$C_Y = 40 \text{ lb T}$$

Then, by proportion, when $x = 12$, $z = 10$, and $y = 8$, we have

$$\frac{C_X}{12} = \frac{C_Z}{10} = \frac{C_Y}{8} = \frac{40}{8} = 5$$

or
$$C_X = (5)(12) = 60 \text{ lb}$$

and
$$C_Z = (5)(10) = 50 \text{ lb}$$

Next, we project the forces into the vertical *YZ* plane, as in Fig. 8-3*a*. In this plane, the *X* components are zero. C_Y and C_Z are known forces. We may now find A_Y by moments about *B*, and B_Y by moments about *A*. Both are clearly upward-acting forces. Since we assumed originally that hinge *A* takes all reaction in the *Z* direction, B_Z is zero and A_Z is clearly opposite to C_Z. Then, by moments we obtain

$[\Sigma M_B = 0]$ $\quad 12A_Y + (40)(6) = (80)(6) \quad$ and $\quad A_Y = 20 \text{ lb}\uparrow$

$[\Sigma M_A = 0]$ $\quad 12B_Y + (40)(6) = (80)(6) \quad$ and $\quad B_Y = 20 \text{ lb}\uparrow$

Then, to find A_Z,

$[\Sigma F_Z = 0]$ $\quad A_Z = 50 \text{ lb}\leftarrow$

Finally, to determine the *X* components, we project the forces into the horizontal *XZ* plane, as is shown in Fig. 8-3*b*. C_X and C_Z are known. By moments about *A*,

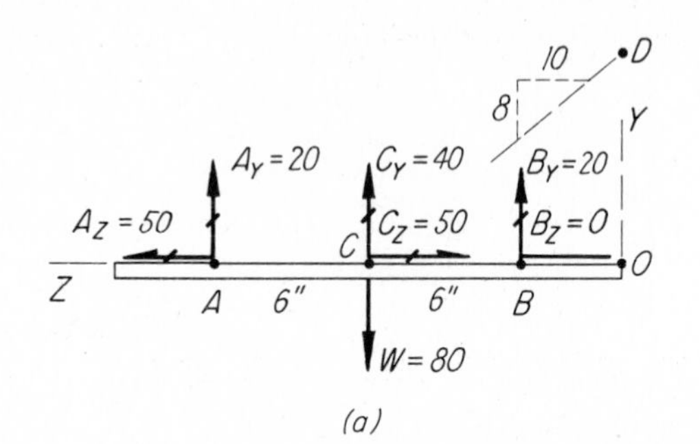

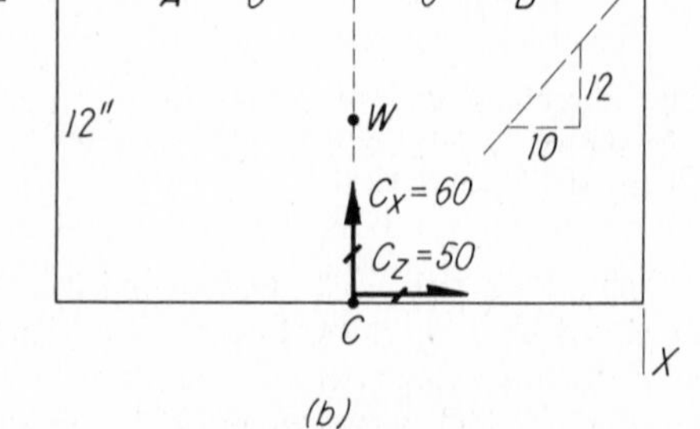

FIG. 8-3 Free-body diagrams. (*a*) *YZ* plane. (*b*) *XZ* plane.

we find B_X, whose sense is clearly as shown. By moments about B, we find A_X, whose sense apparently is as shown. That is,

$$[\Sigma M_A = 0] \qquad 12B_X = (50)(12) + (60)(6) = 960 \qquad \text{and} \qquad B_X = 80 \text{ lb}\downarrow$$
$$[\Sigma M_B = 0] \qquad 12A_X + (60)(6) = (50)(12) \qquad \text{and} \qquad A_X = 20 \text{ lb}\uparrow$$

Since all components of forces A, B, and C are now known, we find these forces as follows:

$$A = \sqrt{A_X^2 + A_Y^2 + A_Z^2} = \sqrt{(20)^2 + (20)^2 + (50)^2} = 57.4 \text{ lb} \qquad Ans.$$
$$B = \sqrt{B_X^2 + B_Y^2 + B_Z^2} = \sqrt{(80)^2 + (20)^2 + 0} = 82.5 \text{ lb} \qquad Ans.$$
$$C = \sqrt{C_X^2 + C_Y^2 + C_Z^2} = \sqrt{(60)^2 + (40)^2 + (50)^2} = 87.8 \text{ lb} \qquad Ans.$$

To emphasize again the importance to students of *visual equilibrium* in their work, all numerical values of components have been recorded in the several free-body diagrams. A careful study of the solution will, of course, indicate which forces in any free-body diagram are originally known and which forces must be solved for.

PROBLEMS

8-2. The horizontal shaft shown in Fig. 8-4 is 8 ft long between the centers of journals A and D and is subjected to horizontal belt tensions from pulley B and to vertical belt tensions from pulley C, as indicated, all acting perpendicularly to the shaft. Neglect all journal friction. Compute the magnitudes of the journal reactions A and D. *Ans.* $A = 101$ lb; $D = 138$ lb

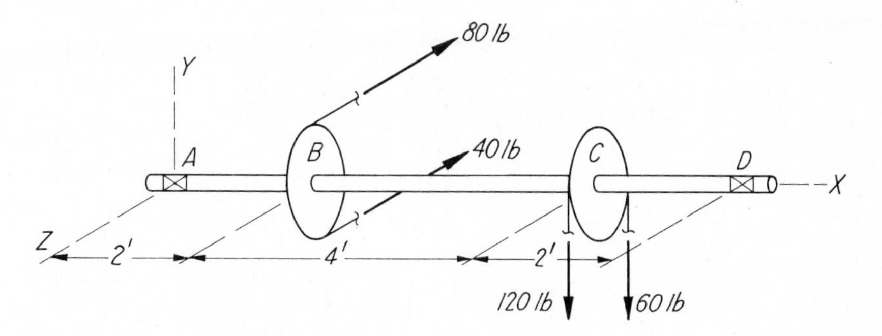

FIG. 8-4 Prob. 8-2.

8-3. A truck tailgate is rectangular, weighs 90 lbs, is supported by loose-fitting hinges at A and B as shown in Fig. 8-5, and is held in the position shown by the chain CD. Compute the tensile force in the chain. Assume the weight acts through F. *Ans.* $T = 38.4$ lb

8-4. In Fig. 8-6 is shown a circular water tank supported by a structure which is symmetrical in all planes about a vertical center line. The tank has an inside diameter of 10 ft and a net inside depth of 12 ft; it weighs 5,100 lb. Compute the total weight W of the tank when filled with water (62.5 pcf), and the vertical forces *acting on the supporting truss* at B, C, F, and G. Disregard horizontal wind load

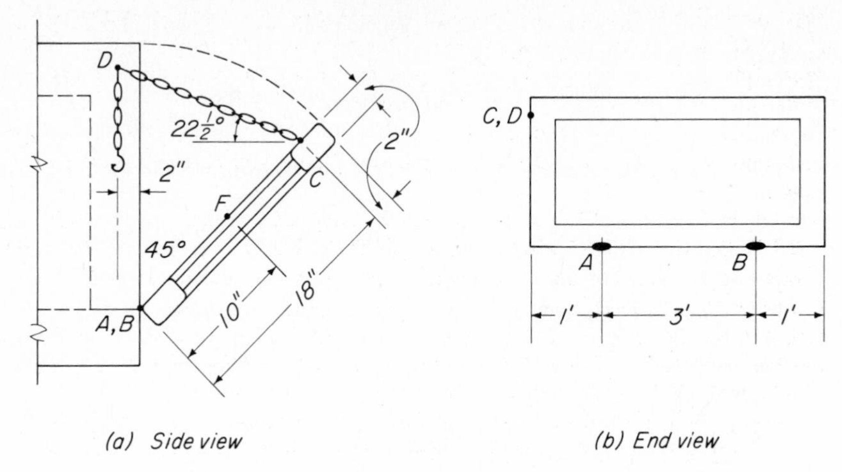

FIG. 8-5 Probs. 8-3, 8-11, and 8-12. (*a*) Side view. (*b*) End view.

H. (NOTE: Although no actual roller would exist at supports *H* and *E*, freedom of lateral motion at these supports must be assumed, for the purpose of stress analysis.)

Ans. $W = 64{,}000$ lb; $R_B = R_C = R_F = R_G = 16{,}000$ lb↓

8-5. Determine analytically the stresses in the sloping corner post *GH*, of the space frame shown in Fig. 8-6 and in the horizontal members, *AH*, *HE*, *BG*, and *GF*, produced by the vertical load at *G* only. (NOTE: See answer to Prob. 8.4.) All diagonal braces are inactive when only vertical loads are considered.

Ans. $GH = 16{,}970$ lb C; $AH = HE = 4{,}000$ lb T; $BG = GF = 4{,}000$ lb C

8-6. Calculate the vertical and horizontal forces produced *on the truss* at *B*, *C*, *F*, and *G* by a maximum horizontal wind load *H* of 2,400 lb against the side of the tank, as shown in Fig. 8-6. In this problem disregard the weight *W*. The usual manner of fastening the floor-system timbers with steel driftpins makes it reasonable to assume

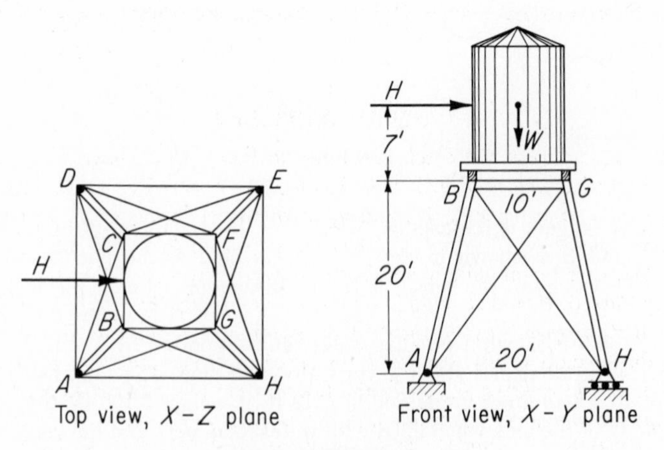

FIG. 8-6 Probs. 8-4 to 8-8.

that one-fourth of *H*, or 600 lb, is resisted horizontally at each of supports *B*, *C*, *F*, and *G*. (SUGGESTIONS: Draw free-body diagrams of both tank and truss *ABGH*, separating tank from truss vertically by 2 in. Then solve for the reaction components *under the tank* and show the magnitude and sense of each force. One-half of the value of *H* may be used to obtain the reaction components at *B* and *G* only. These components may then be transferred to joints *B* and *G* of the truss *with opposite senses*.)

Ans. $B_Y = C_Y = 840$ lb↑; $F_Y = G_Y = 840$ lb↓; $B_X = C_X = F_X = G_X = 600$ lb→

8-7. Using the wind-load forces on truss *ABGH*, Fig. 8-6, as found in Prob. 8-6, compute the *X* and *Y* reaction components at supports *A* and *H*. (Draw a complete front-view free-body diagram of the truss.) Then compute the stresses in members *GH, AH, HE, BG, GF,* and *AG*. Diagonal braces are slender steel rods and will resist tension only. Hence, brace *BH* is now inactive. (Draw front-view and top-view free-body diagrams of each of joints *H* and *G*.)

Ans. $A_Y = 1{,}620$ lb↑; $H_Y = 1{,}620$ lb↑; $A_X = 1{,}200$ lb←; $GH = 1{,}718$ lb C; $AH = HE = 405$ lb T; $BG = 390$ lb C; $GF = 210$ lb C; $AG = 994$ lb T

8-8. Consider the stresses found in Probs. 8-5 and 8-7; what total stresses (dead load plus wind load) should be used in designing one corner post (*GH*), one upper horizontal member (*BG*), one lower horizontal member (*AH*), and one diagonal brace (*AG*)? (NOTE: The *design stress* for any member is *the sum of the stresses of similar type*, tensile *or* compressive, produced by all the load systems.)

Ans. $GH = 18{,}690$ lb C; $BG = 4{,}390$ lb C; $AH = 4{,}405$ lb T; $AG = 994$ lb T

SUMMARY

8-2. The *X*, *Y*, and *Z components* of a force vector in space are proportional to the *x*, *y*, and *z coordinates* of its far end with respect to its near end (its point of application).

An important fact is that *a force produces no reaction in a plane that is perpendicular to its line of action.*

The unknown forces of a noncoplanar, nonconcurrent system of forces in equilibrium are determined by moment and force summations using the components of the forces.

REVIEW PROBLEMS

8-9. In Fig. 8-7 is shown a solid rectangular block 40 in. long, 18 in. wide, and 12½ in. high. A force *F* of 325 lb is applied at *D* and acts in the direction *DE*. Compute the *components* F_X, F_Y, and F_Z of this force. Supports *A*, *B*, and *C* lie in a horizontal plane.

Ans. $F_X = 240$ lb; $F_Y = 125$ lb; $F_Z = 180$ lb

8-10. The solid rectangular block shown in Fig. 8-7 weighs 650 lb and is supported at three points, *A*, *B*, and *C*, lying in one horizontal plane. The ball hinge at *A* allows rotation but prevents displacement. The ball-point supports at *B* and *C* prevent vertical displacement only. Compute the *reaction components* A_X, A_Y, and A_Z, the magnitude of the reaction R_A at *A*, and the vertical reactions at supports *B* and *C*, caused by the weight of the block and the force *F*.

Ans. $A_X = 240$ lb; $A_Y = 325$ lb; $A_Z = 180$ lb; $R_A = 442$ lb; $B = 50$ lb; $C = 400$ lb

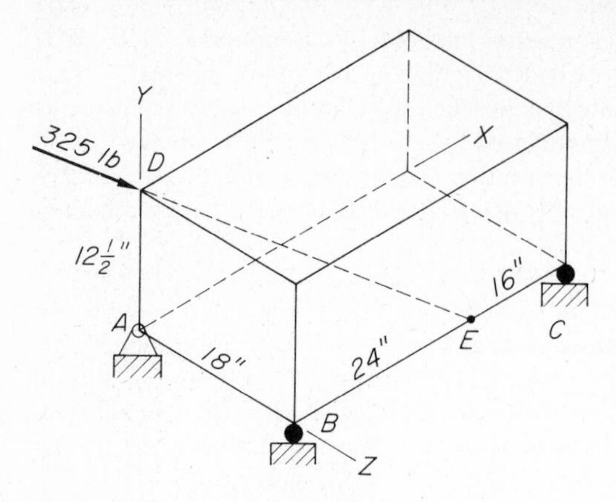

FIG. 8-7 Probs. 8-9 and 8-10.

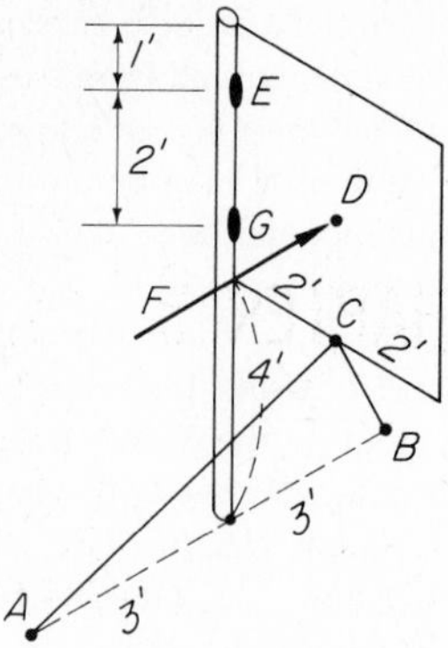

FIG. 8-8 Prob. 8-13.

8-11. Refer to Prob. 8-3. Move point D in Fig. 8-5*a* 6 in. towards the observer, and solve. *Ans.* $T = 51.9$ lb

8-12. Solve Prob. 8-3 if the tailgate is lowered into a horizontal position.

8-13. The sign shown in Fig. 8-8 is 4-ft square and is supported by hinges E and G and the guy wires AC and BC. Calculate the tension in wire AC due to the lateral wind force F of 400 lb. Note that wire BC will be slack when the wind blows from the left.

8-14. Solve Prob. 8-1 if the lower end of the cord is attached to the shelf opposite to hinge A. (Point D will be moved 6 in. in the Z direction.)

Ans. $A = 82.5$ lb; $B = 89.5$ lb; $CD = 107.75$ lb

REVIEW QUESTIONS

8-1. Give the equation for obtaining the magnitude of a force F in terms of its three components F_X, F_Y, and F_Z.

8-2. Explain briefly the moment method of determining the unknown forces of a noncoplanar, nonconcurrent force system in equilibrium.

CHAPTER 9

Friction

9-1 Introduction

When two objects are in contact under pressure, motion or attempted motion of either object with respect to the other in a direction *parallel to the contacting surfaces* will be resisted. This resistance is called *frictional resistance* or simply **friction.** The total frictional resistance between two contacting surfaces is called the **friction force,** since force is required to overcome it. It may be determined experimentally by measuring the force required to overcome it. The friction force always acts parallel to the two surfaces in contact, and its sense is such as to oppose any motion, or attempted motion, in that direction of either body with respect to the other.

Friction is generally thought of as being an adversary of mankind. This impression, of course, is not entirely correct; friction variously acts as friend and foe. But for friction between its wheels and the roadbed, an automobile could not start. On the other hand, friction consumes all the energy put out by the motor, in driving at uniform speed along a level road. Again, the stopping of the automobile depends upon the frictional resistance developed by its brakes and between the wheels and the roadbed. When the surfaces in contact are at rest with respect to each other, this resistance is called **static friction;** when they are in motion with respect to each other, **kinetic friction.**

Let the body shown in Fig. 9-1*a* be resting on a level surface. The total pressure on the supporting surface equals the force N, the reaction of the supporting surface, which is called the **normal pressure.** Let P be the force required to place the body on the verge of sliding (motion impending relative

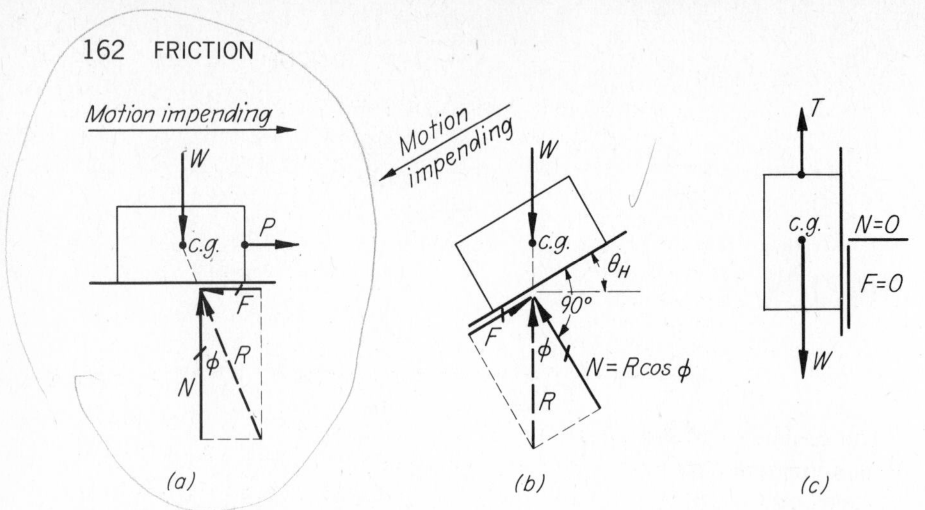

FIG. 9-1 Friction forces and impending motion. (*a*) ϕ is angle of friction. (*b*) θ_H is angle of inclination. (*c*) $\theta_H = 90°$, $N = 0$.

to its supporting surface), and let F be the *friction force* resisting the tendency of P to slide the body. Since frictional resistance is essentially a reaction, it cannot exist except when induced by a force producing or tending to produce motion. Evidently, therefore, F equals P for all values of P from zero to the highest F can reach without motion occurring. This highest value of F for any given condition is called the **limiting friction.** The *reaction* R is the resultant of N and F.

In Fig. 9-1*b*, a body of weight W is resting upon an inclined surface. Evidently the normal pressure N varies with the *angle of inclination* θ_H of the supporting surface, being maximum when θ_H is zero and minimum when θ_H is 90 degrees, as is shown in Fig. 9-1*c*.

9-2 Coefficient of Friction, Angle of Friction, and Angle of Repose

Experiments have shown that, if a body of known weight, as in Fig. 9-1*a*, is acted upon by a force P as shown, the magnitude of the friction force F when motion impends, called the *limiting friction*, depends upon (1) the normal pressure N between the contacting surfaces, (2) the kinds of materials, and (3) the roughness of these surfaces. Since N can be computed and F may be determined experimentally, the ratio of the limiting friction to the normal pressure, F/N, for any two surfaces of like or unlike materials is readily found. This ratio is called the **coefficient of static friction** and is denoted by the symbol f (Table 9-1). In Fig. 9-1*a* and *b* the angle ϕ between the reaction R and the normal pressure N, *when motion is impending*, is called the **angle of friction,** denoted by the Greek letter ϕ (phi). Its tangent is seen to be F/N, which ratio was just defined as the coefficient of static friction. Therefore,

$$\textbf{Coefficient of static friction: } f = \frac{F}{N} = \tan \phi \qquad \textbf{(9-1)}$$

Table 9-1 AVERAGE COEFFICIENTS OF STATIC FRICTION FOR DRY SURFACES

Wood on wood	0.30 to 0.60
Wood on metal	0.20 to 0.60
Metal on metal	0.15 to 0.30
Leather on wood	0.30 to 0.50
Leather on metal	0.30 to 0.60
Stone on concrete	0.50 to 0.70
Rubber on concrete	0.60 to 0.80

The coefficient of static friction f at the contacting surfaces of bodies of different materials may also be determined experimentally as follows: A body is placed on an incline, as in Fig. 9-1*b*. The incline is then tipped until the body is on the verge of sliding. *The tangent of the angle* θ_H, when *sliding impends, is the desired coefficient of static friction.* The angle of inclination θ_H, when motion impends, is called the **angle of repose.**

These coefficients are for *dry surfaces* and are of course rather approximate, because of the varying degrees of roughness of the contacting surfaces. Coefficients of kinetic (sliding) friction at low velocities for dry surfaces are about equal to those for static friction, but tend to decrease at higher velocities. Frictional resistance between lubricated surfaces differs materially from that between dry surfaces, as is pointed out in Art. 9-3.

9-3 Laws of Friction

Experiments carried on by scientists and engineers for more than 150 years have led to the following general conclusions, known as the **laws of friction.**

Dry surfaces

1. The friction force *is dependent* on the kinds of the materials and the degree of roughness of the two surfaces in contact.
2. The maximum friction force (limiting friction) which can be developed *is dependent* upon the normal pressure between the contacting surfaces, and is proportional to it.
3. The friction force *is not dependent* upon the areas of the surfaces in contact.
4. The friction force *is dependent* on the velocity (to a limited extent), and *decreases* as the velocity increases.
5. The friction force *is not dependent* on temperature in any important degree.

Lubricated surfaces

1. The friction force *is not dependent* on the kinds of materials nor (within limits) on the degree of roughness of the contacting surfaces.
2. The friction force *is not dependent* upon the normal pressure between the contacting surfaces.
3. The friction force *is dependent* upon the areas of the surfaces in contact.
4. The friction force *is dependent* upon the velocity, and *increases* as the velocity increases.
5. The friction force *is dependent* on temperature to a considerable degree.

These laws indicate clearly that the frictional effects produced on lubricated surfaces are almost the opposite of those produced on dry surfaces. This reversal of effect is due to the presence of a film of lubricant *between* the two surfaces, which therefore are no longer in contact, so that the frictional resistance is then largely dependent on the properties of the lubricant itself.

9-4 Friction Problems

Numerous problems involving friction, and relating to both static and kinetic friction, arise in engineering practice. This subject, therefore, is deserving of careful study.

Few solid bodies remain in continuous motion. The problem of setting into motion a body that is at rest is therefore a common one. In such problems, we are naturally interested in *the total frictional resistance at the moment the body is on the verge of motion*. These, then, are problems in *static friction when motion is impending*. Graphical solutions are especially adapted to problems involving static friction, as is illustrated in the following examples. These examples, and the problems that follow, indicate the importance of a good understanding of static friction.

ILLUSTRATIVE PROBLEMS

9-1. In Fig. 9-2*a* is shown a block *B* weighing 1,000 lb which is being moved slowly to the right by the action of the vertical force *P* on the wedge *A*. Determine the force *P* required barely to move the block, if the coefficient of friction *f* is 0.3 for

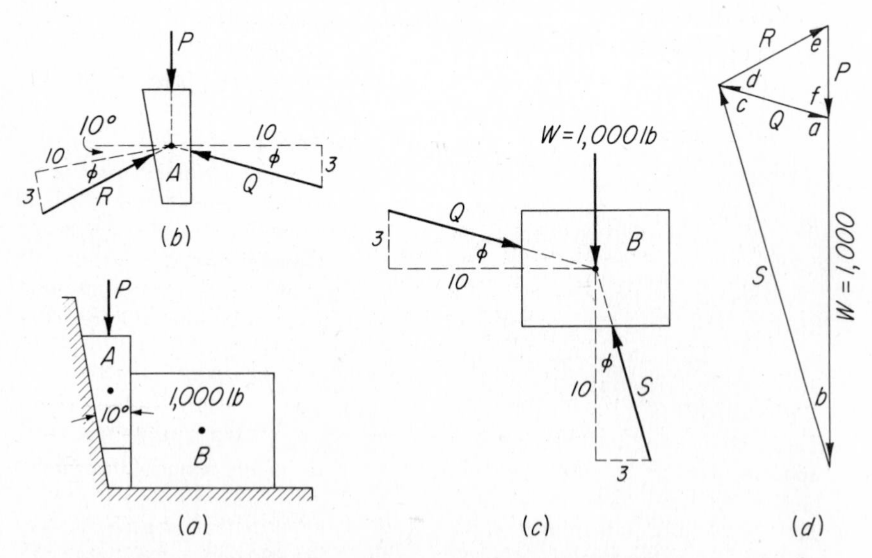

FIG. 9-2 Friction forces. Wedge-and-block problem. (*a*) Wedge and block. (*b*) Wedge *A* as free body. (*c*) Block *B* as free body. (*d*) Force diagram.

all contacting surfaces. (The friction, and hence P, may be considered to be the same for impending or very slow motion.) Weight of wedge may be neglected.

Graphical Solution: When the free-body diagrams of the wedge and the block are drawn, as in Fig. 9-2*b* and *c*, we recognize that the problem must be solved in two parts, beginning with the object on which a *known* force acts, which here is block B. Let the reacting forces at the contacting surfaces be R, Q, and S. Since only three forces act on each free body, they must be concurrent. By careful reasoning we may now determine the *proper directions* of these forces. Apparently, from Fig. 9-2*a*, block B tends to move to the right. Hence reaction S will be inclined to the left, so as to *oppose* the motion. Since the tangent of the angle of friction ϕ equals the coefficient of friction f, which is 0.3, the true direction of S may be shown by laying off to scale 10 units *normal* to the surface of the body and 3 units *parallel* to it. Clearly, wedge A in Fig. 9-2*b* tends to move downward; R and Q are therefore inclined upward. Force Q on block B is of course equal, opposite, and collinear with Q on wedge A. We are ready now to solve for P. Since only W is known, the force diagram (Fig. 9-2*c*) begins with forces W, S, and Q, acting on block B. With respect to block A the force Q is oppositely directed, and P is solved for by completing the polygon of forces Q, R, and P. P scales 265 lb.

Trigonometric Solution: For more precise results, we may obtain a trigonometric solution by determining the angles at a, b, c, d, e, and f in the force diagram in Fig. 9-2*d*, then P is found through two applications of the law of sines. Since $\tan\phi = 0.3$, $\phi = 16°42'$. Then $a = 90° + \phi = 106°42'$; $b = \phi = 16°42'$; $c = 90° - 2\phi = 56°36'$; $d = 10° + 2\phi = 43°24'$; $e = 90° - 10° - \phi = 63°18'$; $f = 90° - \phi = 73°18'$. As a check on these angles, $a + b + c = 180°$ and $d + e + f = 180°$. Solving first for Q and then for P by the law of sines, we have

$$\frac{Q}{\sin b} = \frac{W}{\sin c} \qquad \text{and} \qquad Q = \frac{W \sin b}{\sin c} = \frac{(1{,}000)(0.287)}{0.835} = 344 \text{ lb}$$

Also,

$$\frac{P}{\sin d} = \frac{Q}{\sin e} \qquad \text{and} \qquad P = \frac{Q \sin d}{\sin e} = \frac{(344)(0.687)}{0.893} = 265 \text{ lb} \qquad \textit{Ans.}$$

9-2. The ladder shown in Fig. 9-3*a* is 12 ft long and is supported by a horizontal floor and a vertical wall as shown. The coefficient of friction f at the wall is 0.2; at the floor, 0.4. The weight W of the ladder is 36 lb, considered as concentrated at its center B. The ladder supports a vertical load P of 180 lb, at C. Determine the V and H reaction components at A and D, and compute the lowest value of the angle α at which the ladder may be placed without slipping to the left.

Analytical Solution: The V and H reaction components at supports A and D are shown in the free-body diagram, Fig. 9-3*b*. Let N_A be the *normal pressure*, or V component, at A. The H component at A is the friction force, which, since the ladder is on the verge of slipping, is $f_A \cdot N_A$ or $0.4N_A$, and is directed to the right, since the bottom of the ladder tends to slide to the left. At D, the horizontal normal pressure is N_D, and the friction force is $f_D \cdot N_D$ or $0.2N_D$, directed upward, since at D the ladder tends to move downward. We now have two unknown forces, N_A and N_D, which must be determined through two force equations, since α is also unknown.

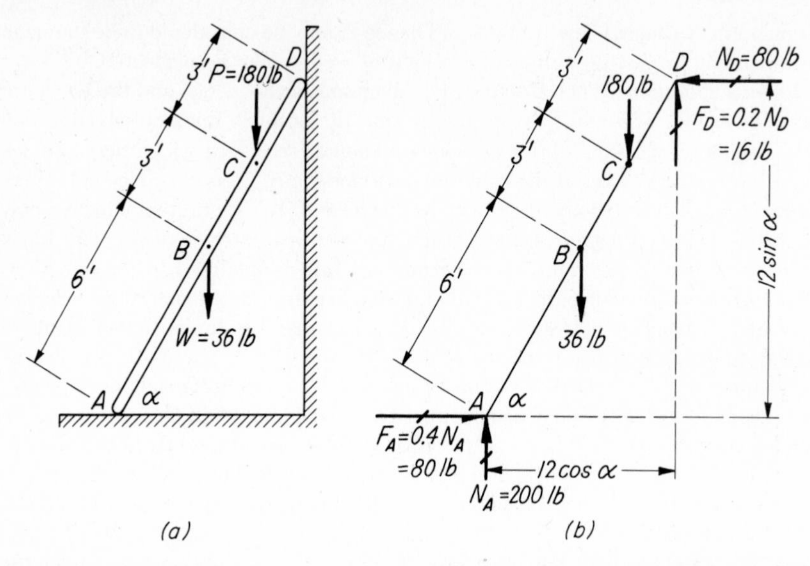

FIG. 9-3 Friction forces. Ladder problem, analytical solution. (*a*) Space diagram. (*b*) Free-body diagram.

Therefore, α is found by a summation of moments. The force equations are

$$[\Sigma H = 0] \qquad N_D = 0.4N_A \tag{a}$$

$$[\Sigma V = 0] \qquad 0.2N_D + N_A = 180 + 36 = 216 \tag{b}$$

Substituting Eq. (*a*) in Eq. (*b*), we have

$$(0.2)(0.4N_A) + N_A = 216$$

or

$$1.08N_A = 216 \qquad \text{and} \qquad N_A = 200 \text{ lb} \qquad \textit{Ans.}$$

Using this value in Eq. (*a*), we obtain

$$N_D = 0.4N_A = (0.4)(200) \qquad \text{and} \qquad N_D = 80 \text{ lb} \qquad \textit{Ans.}$$

The friction forces then are

$$F_A = (0.4)(200) = 80 \text{ lb} \qquad \text{and} \qquad F_D = (0.2)(80) = 16 \text{ lb} \qquad \textit{Ans.}$$

When these values are recorded on the free-body diagram, equilibrium is seen to exist. Finally, we determine α by moments about A. That is,

$$[\Sigma M_A = 0] \qquad (36)(6 \cos \alpha) + (180)(9 \cos \alpha) = (80)(12 \sin \alpha) + (16)(12 \cos \alpha)$$

Hence,

$$216 \cos \alpha + 1{,}620 \cos \alpha = 960 \sin \alpha + 192 \cos \alpha$$

or

$$1{,}644 \cos \alpha = 960 \sin \alpha$$

Then,

$$\frac{\sin \alpha}{\cos \alpha} = \tan \alpha = \frac{1{,}644}{960} = 1.712 \qquad \text{and} \qquad \alpha = 59°43' \qquad \textit{Ans.}$$

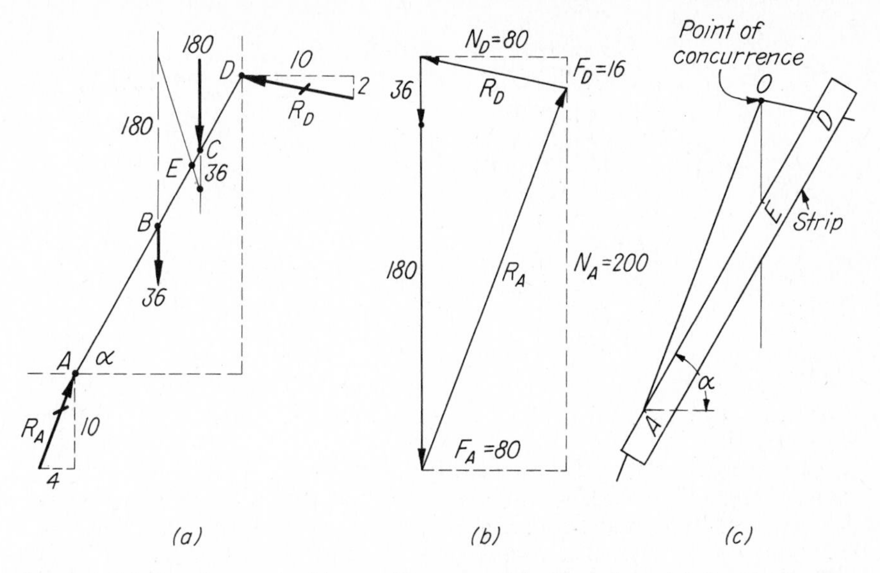

FIG. 9-4 Friction forces. Ladder problem, graphical solution. (*a*) Free-body diagram. (*b*) Force diagram. (*c*) Action-line diagram.

Graphical Solution: In Fig. 9-4*a* the ladder is drawn in its true length, and all forces acting on it are shown. The tangents of the angles of friction at A and D are 0.4 and 0.2, equaling the respective coefficients of friction. These tangents establish the directions of the reacting forces at A and D. Consequently, we may now draw the complete force diagram, thus solving for R_A and R_D. The V and H components are then scaled as indicated in Fig. 9-4*b*.

To determine the angle α, we may resort to a simple trial-and-error type of solution. The reacting forces at A and D, and the resultant of W and P, form a system of three forces in equilibrium which, then, must be concurrent. The point of concurrence O is selected at random. From O the lines of action of the three forces are drawn in their true directions, as shown in Fig. 9-4*c*.

By the inverse-proportion method (Fig. 9-4*a*) we determine the point of application E of the resultant of W and P. Then, from the free-body diagram, we transfer the three points of application, A, E, and D, to the straight edge of a strip of paper. These three points must naturally lie on their respective lines of action. The strip may then be adjusted on the action-line diagram until they do, as shown in Fig. 9-4*c*. If a line is then drawn through the points along the edge of the strip, α is the angle desired. This graphical solution is, of course, a special one applicable only to this and similar problems. However, the reasoning by which the solution is arrived at is so fundamental that its presentation here seems justified.

PROBLEMS

9-3. The right end of the inclined plane surface shown in Fig. 9-5 is slowly raised until the block slips when α is 19°. Determine the coefficient of friction f.

Ans. $f = 0.344$

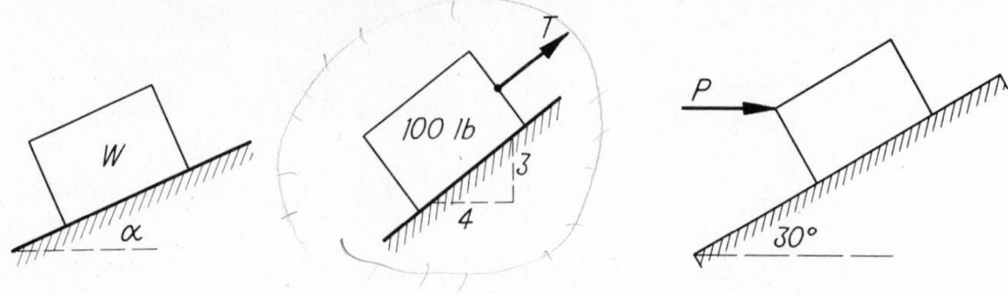

FIG. 9-5 Probs. 9-3 and 9-4.

FIG. 9-6 Prob. 9-5.

FIG. 9-7 Prob. 9-6.

9-4. If the coefficient of friction of cardboard on smooth sheet metal is 0.3, at what angle α, as shown in Fig. 9-5, should a metal chute be placed so that cardboard boxes will slide on it at a slow uniform speed?

9-5. In Fig. 9-6, the coefficient of friction f between the block and incline is 0.4. Determine the force T which will cause impending motion (*a*) up the incline and (*b*) down the incline. Compute also the friction force F (*c*) when $T = 40$ lb, (*d*) when $T = 60$ lb, and (*e*) when $T = 70$ lb. (NOTE: Separate free-body diagrams should be drawn for each of the five parts.)

Ans. (*a*) $T = 92$ lb; (*b*) $T = 28$ lb; (*c*) $F = 20$ lb ↗; (*d*) $F = 0$; (*e*) $F = 10$ lb ↙

9-6. Compute the horizontal force P required to cause motion of the block shown in Fig. 9-7 to impend up the incline, if $W = 100$ lb and $f = 0.25$.

9-7. A jackscrew has a threaded screw which turns in a threaded base, see Fig. 9-8*a*. The load P is applied to an arm at a distance a from the centerline. For analysis, a single thread may be shown as an incline up which the load W is pushed (see Fig. 9-8*b*).

If the pitch p is 1 in. and the pitch diameter D is 3 in., calculate the required force P to raise $W = 4$ tons. Assume $a = 20$ in. and $f = 0.12$.

HINT: Angle $CAB = \phi + \theta$, $BC = F$, and $AB = W$. *Ans.* $P = 137.5$ lb

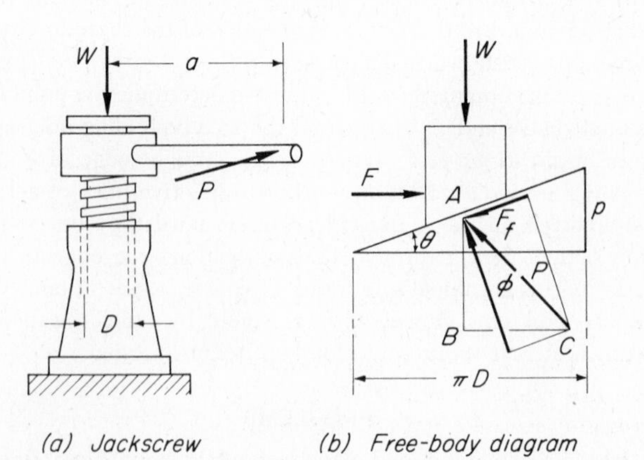

FIG. 9-8 Probs. 9-7, 9-8, and 9-38. (*a*) Jackscrew. (*b*) Free-body diagram.

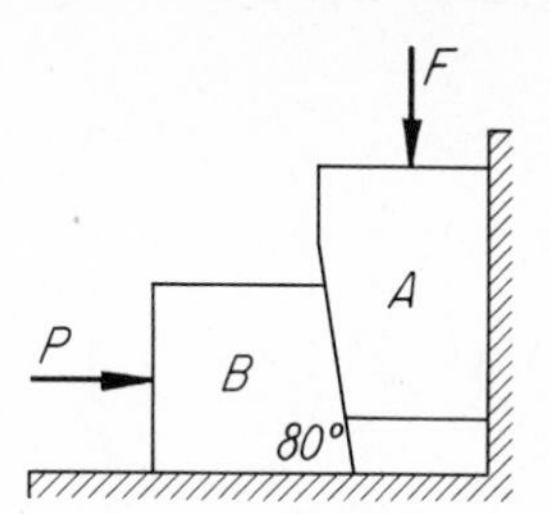

FIG. 9-9 Probs. 9-9 and 9-39.

9-8. Solve Prob. 9-7 if $W = 6{,}000$ lb, $a = 25$ in., $f = 0.15$, $D = 2\frac{1}{2}$ in., and $p = \frac{3}{4}$ in.

9-9. The block A in Fig. 9-9 is acted upon by a force F of 100 lb. Assume f to be 0.2 between blocks A and B, and 0.3 between the other contacting surfaces. Determine the *minimum* horizontal force P required to prevent block B from slipping to the left. Disregard weights of blocks. *Ans.* $P = 128.4$ lb

9-10. Compute the horizontal force P required to cause motion of the block shown in Fig. 9-10 to impend down the incline if $W = 156$ lb and f is assumed to be 0.627. *Ans.* $P = 26$ lb ←

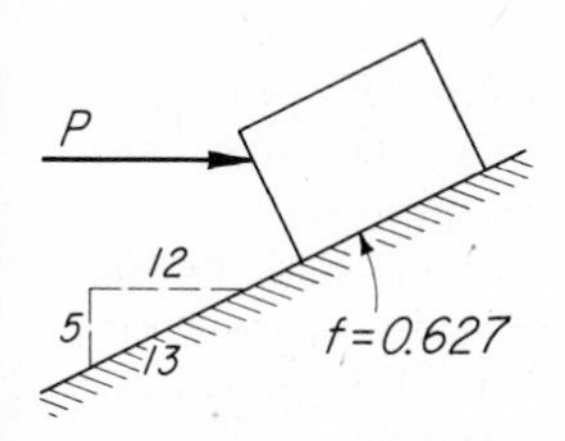

FIG. 9-10 Probs. 9-10 and 9-11.

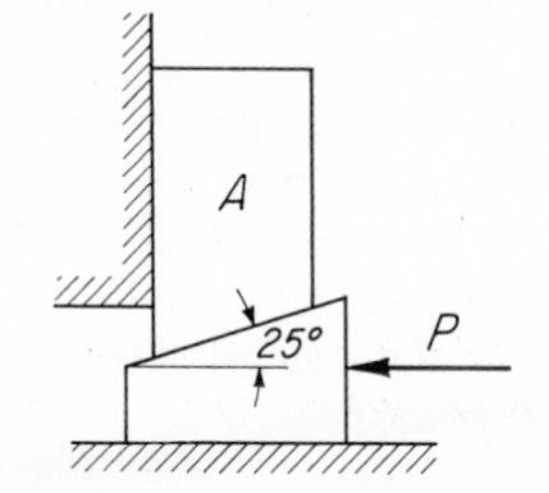

FIG. 9-11 Prob. 9-12.

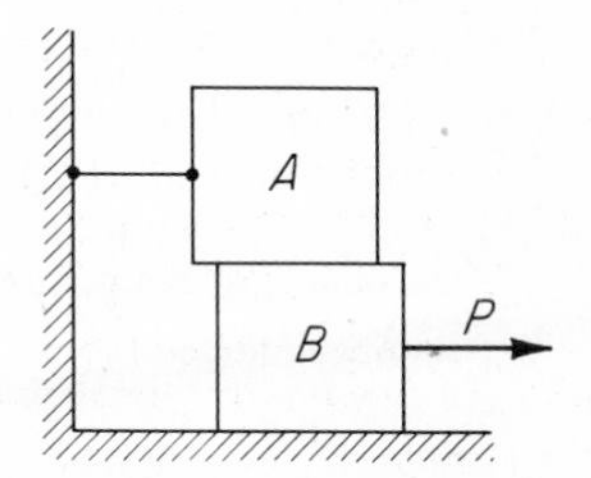

FIG. 9-12 Prob. 9-13.

9-11. Solve Prob. 9-10 if motion is to impend up the incline. *Ans.* $P = 221$ lb→

9-12. Block A in Fig. 9-11 weighs 200 lb. Find the force P to lift block A if $f = 0.3$ for all surfaces of contact. Disregard the weight of the wedge.

9-13. Block A in Fig. 9-12 weighs 200 lb and block B weighs 300 lb. Find the force P required to move block B. Assume the coefficient of friction f for all surfaces is 0.3. *Ans.* $P = 210$ lb

9-14. In Fig. 9-13, the heavy rectangular stone block C weighs 1,000 lb. It is being raised slightly by means of two wooden wedges A and B, and by sledge-hammer blows P on wedge B. The angle between the contacting surfaces of the wedges is 5°. If $f = 0.3$ for all surfaces, compute the value of P required to cause upward motion of the block to impend. Neglect weight of wedges. Force scale: 1 in. = 200 lb.

9-15. Two wooden blocks A and C, as shown in Fig. 9-14, are held in position through the pressures exerted on their sides by two steel plates held together by the tensile force in the bolt B. Assume that the centers of pressure between plates and blocks are as indicated by the dots. (a) Compute the maximum value which

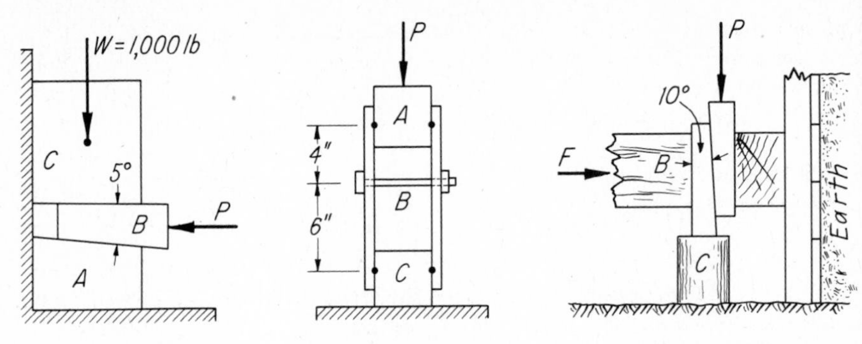

FIG. 9-13 Prob. 9-14.

FIG. 9-14 Probs. 9-15 and 9-40.

FIG. 9-15 Prob. 9-16.

the load P may reach without having either block slip vertically between the plates when the tensile force in the bolt is 500 lb and $f = 0.4$. (*b*) On which block will slipping occur first, if P is increased? *Ans.* (*a*) $P = 160$ lb; (*b*) on block C

9-16. In Fig. 9-15 two opposing timber walls used as shoring in an excavated trench are held tightly against the earth by the horizontal strut B, which is wedged tightly by sledge-hammer blows on one of two like wedges. Block C is used merely to give vertical support to the left wedge while the right wedge receives the blow. The angle between the two faces of each wedge is 10°, and f between all contacting surfaces is 0.4. Determine the horizontal compressive force F in the strut when the force P of a blow of 500 lb drives the wedge slightly. Disregard any possible friction between wedge and block C. Consider reaction at block C to be vertical. Force scale: 1 in. = 200 lb.

9-17. The steel bracket shown in Fig. 9-16 may slide freely on the vertical shaft. Let A and B be the points of application of the resultant pressures on the shaft, caused by the force P. Assume that $f = 0.2$ and that the friction forces at A and

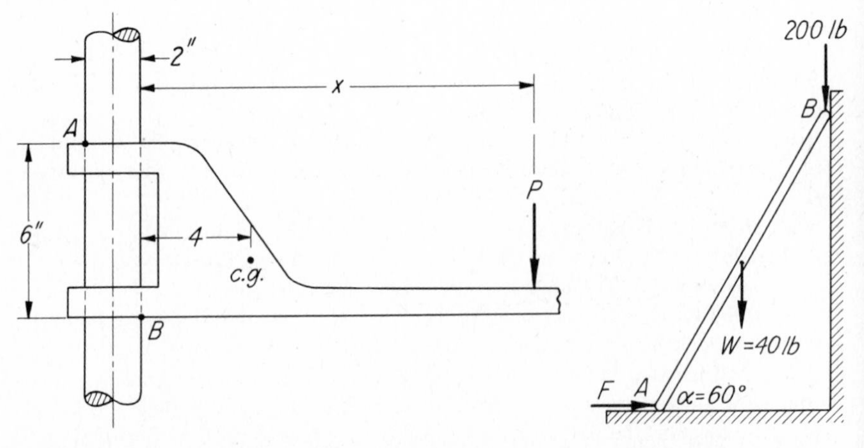

FIG. 9-16 Probs. 9-17, 9-18, and 9-41.

FIG. 9-17 Prob. 9-19.

B are equal. (*a*) Compute the *minimum* distance x from the edge of the shaft at which a force P can be applied without having the bracket slide on the shaft, if its weight is neglected. (*b*) Find x if P is 40 lb and if the weight W of the bracket is 20 lb, concentrated at its center of gravity *c.g.*

Ans. (*a*) $x = 14$ in.; (*b*) $x = 19$ in.

9-18. Solve Prob. 9-17 graphically. (HINT: In the space diagram, consider only the points of application of the forces. Scales on $8\frac{1}{2}$ by 11: 1 in. = 4 in.; 1 in. = 20 lb.)

9-19. A ladder 10 ft long, weighing 40 lb, is placed against a wall as shown in Fig. 9-17 and is prevented from slipping at the bottom by friction and by an additional horizontal force F. The coefficient of friction at the wall is 0.25; at the floor, 0.35. (*a*) Find the minimum force F that will prevent slipping, if the weight of the ladder is concentrated at its mid-point and if it supports at the top a man weighing 200 lb. (*b*) Find the minimum angle α at which the ladder may be placed to prevent slipping, if the force F is removed. *Ans.* (*a*) $F = 36.6$ lb; (*b*) $\alpha = 68°56'$

9-20. A ladder resting on a horizontal floor leans against a vertical wall at an angle of 60° with the floor. If f is 0.25 at all surfaces and the weight of the ladder is neglected, what percentage of its length can a person ascend without causing the ladder to slip? Solve graphically. *Ans.* 46.6 percent

9-5 Belt Friction

The transmission of power by belt drives and the braking effect obtained by band-type brakes depend upon the frictional resistance developed between a flexible band and a cylindrical surface. Now consider power transmission by means of a flat belt (see Fig. 9-18*a*). A driving force is transmitted from the surface of the driving pulley to the belt because of the frictional resistance between the two surfaces. The tension in the belt will vary throughout its length of contact with the pulley (see Fig. 9-18*b*).

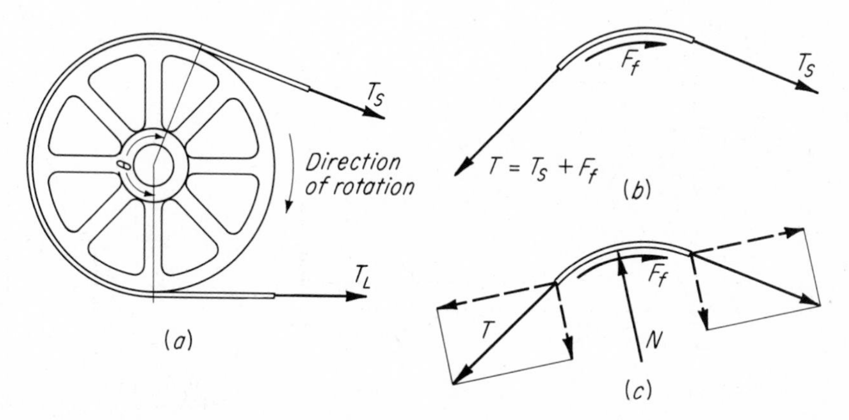

FIG. 9-18 Belt friction. Flat belt. (*a*) Driving pulley. (*b*) Tension varies along the belt. (*c*) Normal force varies with belt tension.

The normal force exerted by the pulley on a small increment of belt clearly depends upon the tension in the belt (see Fig. 9-18*c*). If slipping is impending, the friction force between the belt and pulley will depend upon the normal force and the coefficient of friction f for the two surfaces. The friction force per unit of length of belt will vary along the surface of contact because of the variation in normal force, which increases in magnitude in the direction from T_S to T_L.

It can be shown that, when slipping is impending,

$$\log_{10} \frac{T_L}{T_S} = \frac{f\theta}{132} \tag{9-2}$$

This equation may be written in the alternate form

$$\log_{10} T_L - \log_{10} T_S = \frac{f\theta}{132} \tag{9-3}$$

In the above equations

T_L = larger belt tension, lb
T_S = smaller belt tension, lb
f = coefficient of friction
θ = angle of contact between belt and cylinder, degrees

Small belt drives are often of the V-belt type (see Fig. 9.19). The wedging action of the belt tension increases the normal force on the belt and hence also increases the maximum allowable friction force.

From Fig. 9-19*c*, it can be seen that the normal force N is given by

$$N = \frac{F/2}{\sin \beta}$$

Since the friction force, when motion is impending, is given by $f \cdot N$,

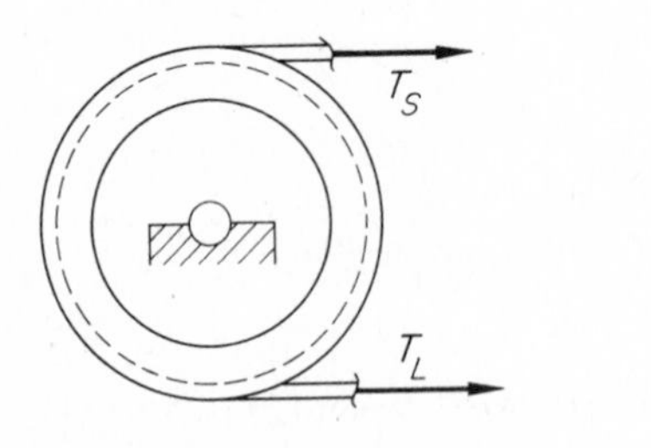

(a) Pulley and belt

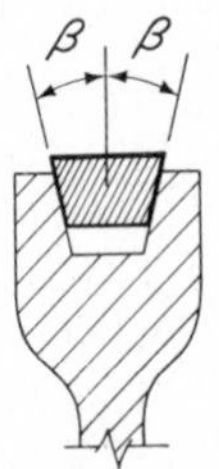

(b) V-belt in groove

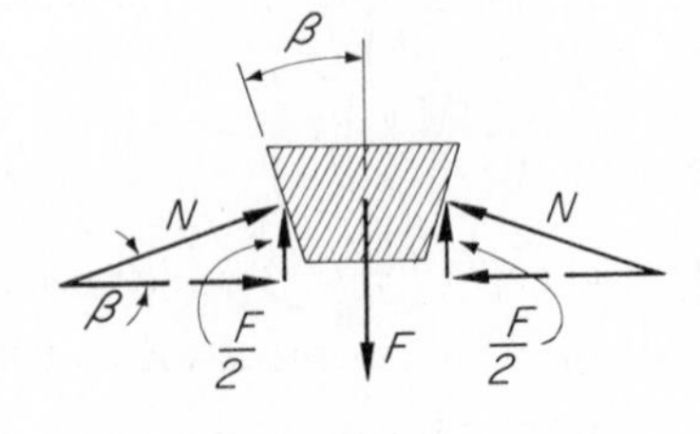

(c) Free-body diagram of V-belt

FIG. 9-19 V-belt pulley and belt. Wedging action. (*a*) Pulley and belt. (*b*) V-belt in groove. (*c*) Free-body diagram of V-belt.

$$F_f/\text{side} = \frac{fF}{2 \sin \beta} \qquad \text{or} \qquad F_f = \frac{fF}{\sin \beta}$$

Since the total friction force F_f is equal to $T_L - T_S$, we have

$$\log_{10} \frac{T_L}{T_S} = \log_{10} T_L - \log_{10} T_S = \frac{f\theta}{132 \sin \beta} \tag{9-4}$$

ILLUSTRATIVE PROBLEMS

9-21. The coefficient of friction between a certain kind of belting and a wooden pulley is to be determined by the apparatus shown in Fig. 9-20*a*. The spring indicates a belt tension of 180 lb in the horizontal portion of the belt when slipping is impending. Determine the coefficient of friction *f*.

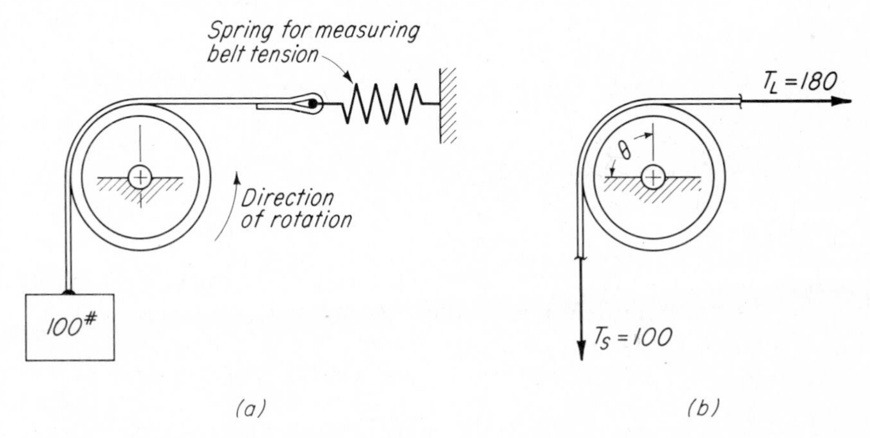

FIG. 9-20 Determination of coefficient of friction between belt and pulley. (*a*) Apparatus. (*b*) Belt tensions.

Solution: The ratio of the two belt tensions is known (see Fig. 9-20*b*). Equation (9-2) may be used to find the unknown coefficient of friction *f*. The angle of contact is 90°.

$$\left[\log_{10} \frac{T_L}{T_S} = \frac{f(\theta)}{132}\right] \qquad \log_{10} \frac{180}{100} = \frac{f(90)}{132}$$

The log of 1.8 is 0.255; therefore

$$0.255(132) = 90(f) \qquad \text{or} \qquad f = 0.374 \qquad \textit{Ans.}$$

9-22. What is the maximum weight that can be slowly lowered by a man who can exert a 75-lb pull on a rope if the rope is wrapped $1\frac{1}{4}$ turns around a horizontal spar? Assume the coefficient of friction between the surfaces is 0.3. See Fig. 9-21.

Solution: The smaller tension in the rope is 75 lb, the angle of contact is $1\frac{1}{4}$ turns or (1.25)(360) = 450°, and the friction coefficient *f* is 0.3. The larger tension T_L may be found from Eq. (9-3).

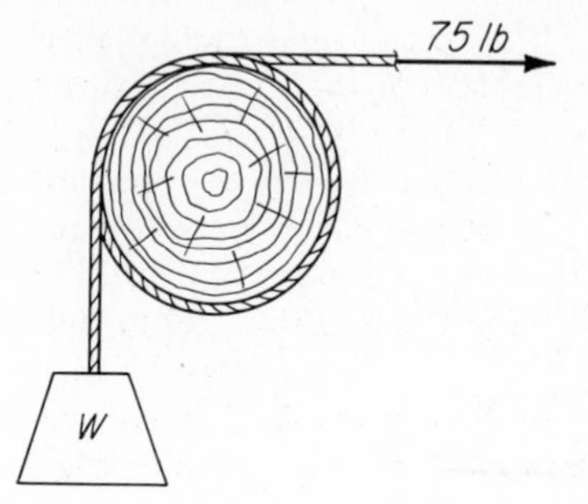

FIG. 9-21 Heavy weight lowered by $1\frac{1}{4}$ turns on spar.

$$\left[\log_{10} T_L - \log_{10} T_S = \frac{f(\theta)}{132}\right] \qquad \log_{10} T_L - \log_{10} 75 = \frac{0.3(450)}{132}$$

or

$$\log_{10} T_L = \frac{0.3(450)}{132} + 1.875 = 2.898$$

The tension T_L may now be found by finding the antilog of 2.898, which is 790. This tension T_L is caused by the weight W. Hence,

$$W = 790 \text{ lb} \qquad \textit{Ans.}$$

PROBLEMS

9-23. Find the tension in the horizontal portion of the belt of Prob. 9-21 if the direction of rotation of the pulley is reversed. Assume the coefficient of friction f is 0.374 as found in Prob. 9-21. *Ans.* $T = 55.6$ lb

9-24. Solve Prob. 9-22 if the rope is given $2\frac{1}{4}$ turns on the spar.

9-25. A rope makes three turns around a capstan on a pier. What is the maximum permissible force on the loaded end of the rope if the pull on the other end of the rope is not to exceed 60 lb? Assume the coefficient of friction f is 0.35.

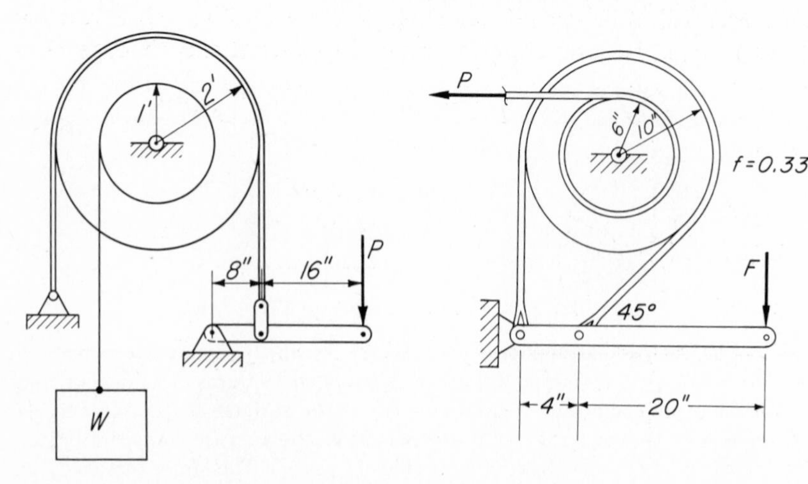

FIG. 9-22 Probs. 9-26 and 9-27.

FIG. 9-23 Probs. 9-28 and 9-44.

9-26. The coefficient of friction between the brake drum and the flexible brake band shown in Fig. 9-22 is 0.44. What is the maximum load W that can be lowered slowly if the force P is 25 lb? *Ans.* $W = 112.3$ lb

9-27. Solve Prob. 9-26 if the line supporting the weight comes off the opposite side of the drum. (The direction of the tendency for rotation will be reversed from that of Prob. 9-26.) *Ans.* $W = 447$ lb

9-28. What force F must be exerted on the brake arm shown in Fig. 9-23 to hold the drum against rotation if $P = 800$ lb? *Ans.* $F = 77.8$ lb

9-29. A belt makes contact with a driving pulley through one half its circumference. If $T_S = 10$ lb and $f = 0.3$, calculate the value of T_L for impending slippage if the belt is (*a*) flat, and (*b*) V-type with $\beta = 20°$.

Ans. (*a*) $T_L = 25.7$ lb; (*b*) $T_L = 158.2$ lb

9-30. A V-belt pulley, $\beta = 15°$ and $f = 0.4$, has belt tensions $T_S = 10$ lb and $T_L = 100$ lb. What is the minimum value of θ that will prevent slipping?

9-6 Rolling Resistance

Our everyday experiences tell us that rolling resistance of objects is less than sliding resistance. This at least partially explains the common use of wheels on moving vehicles, and of ball and roller bearings in machinery. If sliding surfaces are separated by a lubricant, the molecules of the lubricant may act as small "balls" between the surfaces and thus reduce the frictional resistance.

Consider a wheel on a "rigid" surface as shown in Fig. 9-26. We wish to determine the force P required to maintain a small constant velocity to the right. A moment summation about point O will show that force F must equal zero. Therefore, since $\Sigma H = 0$, the force P must equal zero. Evidently, once started, the wheel would roll forever on such a surface. We know, however, that no material is perfectly rigid. Any force, no matter how small, applied to a material will cause some deformation. The weight of the wheel and any load that it might be supporting will cause the material under the wheel to

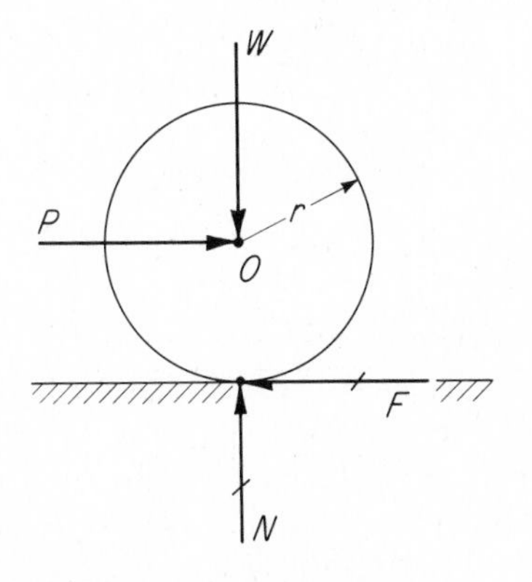

FIG. 9-24 "Rigid" surface.

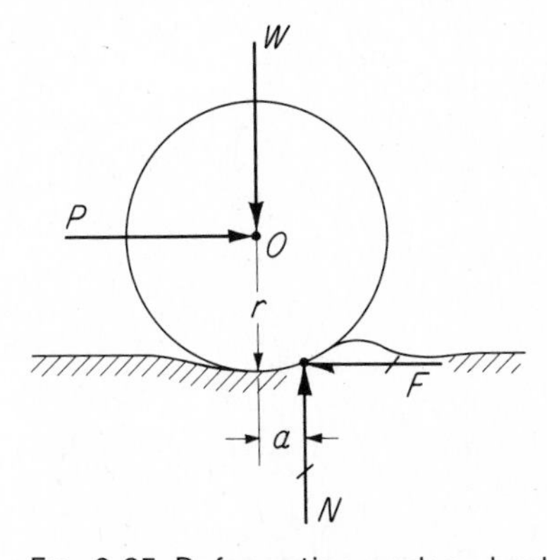

FIG. 9-25 Deformation under wheel.

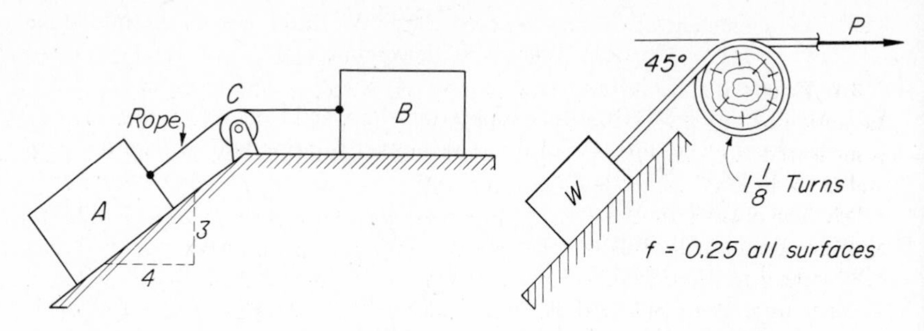

FIG. 9-26 Probs. 9-35 and 9-36.

FIG. 9-27 Prob. 9-37.

yield. The deformation shown in Fig. 9-25 seems realistic, though admittedly greatly exaggerated, for motion of the wheel to the right. A moment summation about point O gives

$$[\Sigma M_O = 0] \qquad Fr = Na \quad \text{or} \quad a = \frac{Fr}{N} \tag{9-5}$$

The dimension a, in inches, is called the **coefficient of rolling resistance.**

ILLUSTRATIVE PROBLEM

9-31. A railway freight car is supported by eight wheels. The coefficient of rolling resistance a for steel wheels on steel rails is about 0.02 in. What force P is required to keep an 80-ton car rolling on a straight level track? The wheels are 32 in. in diameter, and P is applied parallel to the rails.

Solution: We may assume that the 160,000-lb load is distributed equally to all eight wheels. The load per wheel then is 20,000 lb. The radius of each wheel is 16 in. and the coefficient of rolling resistance is 0.02. From Eq. (9-4) we then obtain

$$\left[F = \frac{Na}{r}\right] \qquad F = \frac{20{,}000(0.02)}{16} = 25 \text{ lb}$$

For all eight wheels, the total required force $P = (25)(8) = 200$ lb. *Ans.* Note that the entire weight could have been assumed to be concentrated on one wheel without changing the final answer. Hence, the distribution of weight among several wheels is immaterial.

PROBLEMS

9-32. An automobile has 30-in. wheels (outside diameter) and weighs 3,200 lb. If the coefficient of rolling resistance a is $\frac{3}{8}$ in. for paved surfaces, what horizontal force would be required to keep the car rolling at low speed on a level paved street?

9-33. A horizontal pull of 213 lb is required to tow the car of Prob. 9-32 on a level gravel road. What is the coefficient of rolling resistance a for this case?

Ans. $a = 1.0$ in.

SUMMARY

(By article number)

9-1. Friction is the resistance offered by two surfaces, in contact under pressure, to motion or attempted motion of one surface upon the other. Frictional resistance is measured in terms of force which always acts parallel to the contacting surfaces and is so directed as to oppose the motion.

9-2. The **coefficient of friction** f is the numerical ratio of the *limiting friction force* F to the *normal pressure* N between the contacting surfaces. That is, $f = F/N$, where the limiting friction F is the maximum that can be and is developed when *motion is impending*. The *reaction* R is the resultant of F and N. The angle ϕ between the forces R and N is called the *angle of friction*, and its tangent is F/N, or f. The highest angle with the horizontal to which an inclined plane may be tipped without causing a body resting thereon to slip is called the *angle of repose;* hence, the angle of repose equals the angle of friction.

9-3. According to the **laws of static friction** for dry surfaces, the limiting friction force (1) is dependent on the kinds of materials and on the degree of roughness of the contacting surfaces; (2) is independent of the areas in contact; and (3) is proportional to the normal pressure.

9-5. The maximum friction force that may be developed between a flexible belt and a cylindrical surface depends upon the angle of contact θ and the coefficient of friction f between the contacting surfaces. The difference between the belt tensions at the initial and final points of contact is equal to this friction force. Flat-belt tensions are related by:

$$\log_{10} \frac{T_L}{T_S} = \log_{10} T_L - \log_{10} T_S = \frac{f\theta}{132} \qquad \text{(9-2 and 9-3)}$$

and V-belt tensions by

$$\log_{10} \frac{T_L}{T_S} = \log_{10} T_L - \log_{10} T_S = \frac{f\theta}{132 \sin \beta} \qquad \text{(9-4)}$$

9-6. The **coefficient of rolling resistance** a measures the resistance of a rolling object to continued motion. The coefficient is large for materials that yield considerably under load. Theoretically, rigid surfaces offer no resistance to rolling. Practically, they do offer relatively low resistance.

REVIEW PROBLEMS

9-34. A man is standing on a roof which rises 5 ft vertically in every 10 ft measured horizontally. (*a*) What must be *the minimum coefficient of friction f* between the roof and his shoes to prevent slipping? (*b*) If f is 0.36, what maximum slope could the roof have without slipping? *Ans.* (*a*) $f = 0.5$; (*b*) $\theta_H = 19°48'$

9-35. The blocks A and B shown in Fig. 9-26 are connected with a flexible cable passing over a sheave C, the friction of which may be neglected. Let $f = 0.2$, between the blocks and their supporting surfaces. If block A weighs 100 lb, compute *the minimum weight W_B of block B which will barely prevent sliding.*

Ans. $W_B = 220$ lb

9-36. If block B in Fig. 9-26 weighs 480 lb and if $f = 0.25$, find the weight W_A of block A in order that *motion will be impending*. Neglect rope and sheave friction.
Ans. $W_A = 300$ lb

9-37. Calculate the pull P in Fig. 9-27 that must be exerted to slowly lower the weight $W = 1{,}000$ lb down the incline. *Ans.* $P = 90.6$ lb

9-38. Calculate the load that can be raised by a 50-lb force applied to the arm of the jackscrew shown in Fig. 9-8 if $a = 15$ in., $D = 2$ in., $p = \frac{1}{2}$ in., and $f = 0.1$.

9-39. In Fig. 9-9, let the force $F = 100$ lb, and let the weight of each of blocks A and B be 100 lb. Determine graphically *the least value of the force P* that will prevent block B from sliding to the left, if $f = 0.3$. *Ans.* $P = 182$ lb

9-40. Assume in Prob. 9-15 (Fig. 9-14) that another bolt is inserted through the side plates 3 in. below bolt B, and that the tension in each bolt is now 300 lb. (a) Compute P. (b) On which block will slipping occur?
Ans. (a) $P = 216$ lb; (b) on block A

9-41. Solve part (b) of Prob. 9-17 if P is 40 lb *upward-acting* instead of downward-acting, as illustrated in Fig. 9-16. (HINT: Points A and B now move to opposite sides of the shaft.) *Ans.* $x = 9$ in.

9-42. A rope is thrown over a horizontal overhead pipe. (a) How heavy a weight can a man lift with a 100-lb pull? (b) What weight can be lowered slowly if the man can maintain a 100-lb pull on his end of the rope while the weight is being lowered? Assume the friction coefficient f is 0.4.
Ans. (a) $W = 28.5$ lb; (b) $W = 351$ lb.

9-43. A power-driven capstan can maintain a 10,000-lb pull on a hawser. What is the minimum number of turns that the hawser must make on the capstan drum if the tension in the slack end of the hawser is to be 10 lb and the coefficient of friction is 0.25? *Ans.* $\theta = 4.4$ turns

9-44. The cable drum of Fig. 9-23 is to be held against rotation by the brake shown. What is the maximum permissible cable tension P if the brake force F is limited to 50 lb? *Ans.* $P = 515$ lb

9-45. The cylinders shown in Fig. 9-28 have a total weight of 100 lb. Assume $f = 0.2$ for all surfaces of contact. Calculate the pull P to slowly rotate the cylinders.
Ans. $P = 31.7$ lb

9-46. A strap wrench is shown in Fig. 9-29. Calculate the minimum arm a that will prevent slipping if $f = 0.1$. Include the additional friction at the left end of the wrench due to the radial force at that point.

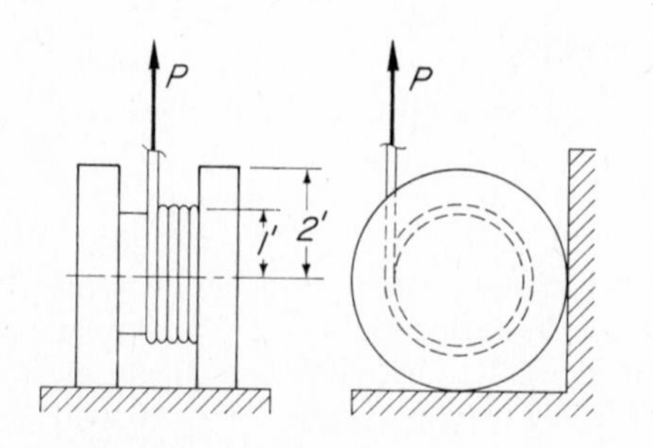

FIG. 9-28 Prob. 9-45.

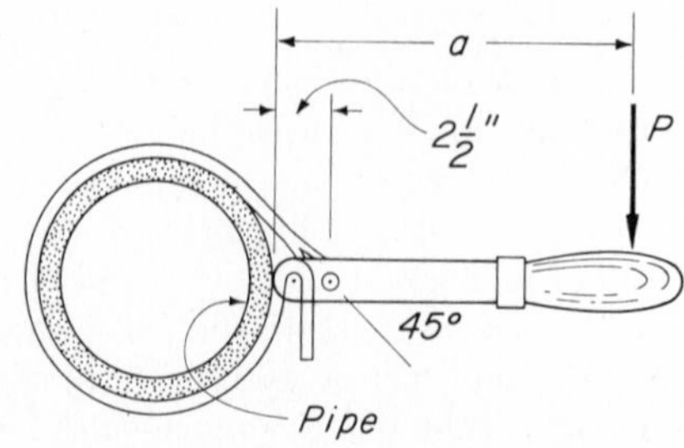

FIG. 9-29 Prob. 9-46. Strap wrench.

REVIEW QUESTIONS

9-1. What is meant by (*a*) frictional resistance and (*b*) friction force?

9-2. Explain the meaning of normal pressure.

9-3. What is meant by impending motion?

9-4. Define limiting friction.

9-5. What is generally meant by angle of inclination?

9-6. What is the coefficient of friction?

9-7. Define the angle of friction.

9-8. What relation exists between the coefficient of friction and the tangent of the angle of friction?

9-9. Explain the meaning of angle of repose.

9-10. Does the friction force at a dry surface depend upon (*a*) the degree of roughness of the contacting surfaces? (*b*) the normal pressure? (*c*) the areas of the contacting surfaces? (*d*) velocity (explain)? (*e*) temperature?

9-11. Does Eq. (9-2) relating the belt tensions on opposite sides of a pulley apply when slipping is not impending?

9-12. What is the advantage of a V-belt system?

9-13. What is the common unit of the coefficient of rolling resistance?

9-14. What is the approximate coefficient of rolling resistance for rigid surfaces of contact?

CHAPTER 10

Centroids and Centers of Gravity

10-1 Introduction

In the study of parallel force systems in space, in Chap. 7, the *center of a parallel force system* was found to be the point through which passed the resultant force of the system. This center was located by the principle of moments: about any point, the moment of a resultant force equals the algebraic sum of the moments of the separate forces of the system. *Centroids of areas* and *centers of gravity of bodies* are located in similar manner, as is explained in the following articles.

10-2 Centers of Gravity

A body is composed of a system of particles each of which is subjected to the gravitational pull of the earth. The resultant of these vertical, parallel,[1] pulls (forces) on all the particles comprising the body is the **force of gravity,** also called the *weight W of the body*. The point in the body through which the gravity force W acts is called its **center of gravity.**

In the building of airplanes, ships, and automobiles, each made up of numerous separate parts, the position of the center of gravity must be such that proper stability is achieved. In these three instances, in fact, the center of gravity must lie in a vertical plane also containing the central longitudinal

[1]Since all gravity forces are directed toward the center of the earth's attraction, they cannot, strictly speaking, *be* parallel, but, of course, for practical purposes, they are considered to be so.

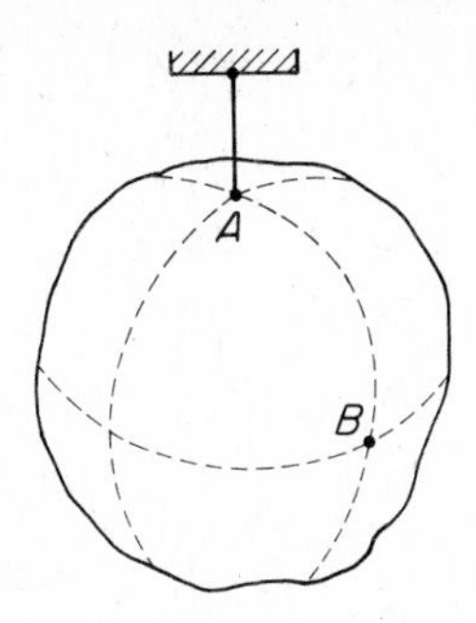

FIG. 10-1 Center of gravity by experiment.

axis, and in ships and automobiles it must lie low enough to ensure desired lateral stability. Therefore, the various parts must be so placed that their separate weights, as well as the moments of these weights, will balance about the central vertical plane.

The center of gravity of an irregular body may be approximately located *experimentally* as follows: Let the body be suspended from a flexible cord attached at any point A, as in Fig. 10-1. When it is brought to rest, the body is in equilibrium under the action of two forces W and T_A, the tension in the cord, whose action lines therefore are collinear and pass through the center of gravity of the body. If the cord is then attached at a second point B and the body is allowed to come to rest, the action lines of W and T_B will be collinear. Evidently, then, the action lines of T_A and T_B intersect at the center of gravity of the body. When the body is in position A, two vertical planes, each containing the extension of the cord, may be marked on its surface, as is indicated in Fig. 10-1. The center of gravity of the body lies on the line of intersection of these two planes. A third vertical plane, marked on the surface of the body when the cord is attached at B, will intersect the line of intersection of the first two planes at the center of gravity.

Other methods of determining the center of gravity of a body *experimentally* are (1) to *balance* it on a point support, (2) to balance it on a knife-edge support in at least two positions, and (3) when its weight is known, to place it on two knife-edge supports, one placed on a scale that measures that reaction, after which the distance $\bar{x}$ from the other support to the vertical plane containing the center of gravity may be computed.

To determine *mathematically* the center of gravity of a flat plate of irregular shape but of uniform thickness and of homogeneous material, the plate may be divided into a number of small elements of equal size, as shown in Fig. 10-2. If the weight of each element is w, concentrated at its center, a parallel force system is formed whose resultant W, the weight of the entire plate, passes through the center of gravity of the plate.

Let the weights of these elements be designated w_1, w_2, etc., let their coordinates be (x_1 and y_1), (x_2 and y_2), etc., and let the coordinates of the resultant weight W be $\bar{x}$ and $\bar{y}$ (*bar x* and *bar y*). Now, about any axis the

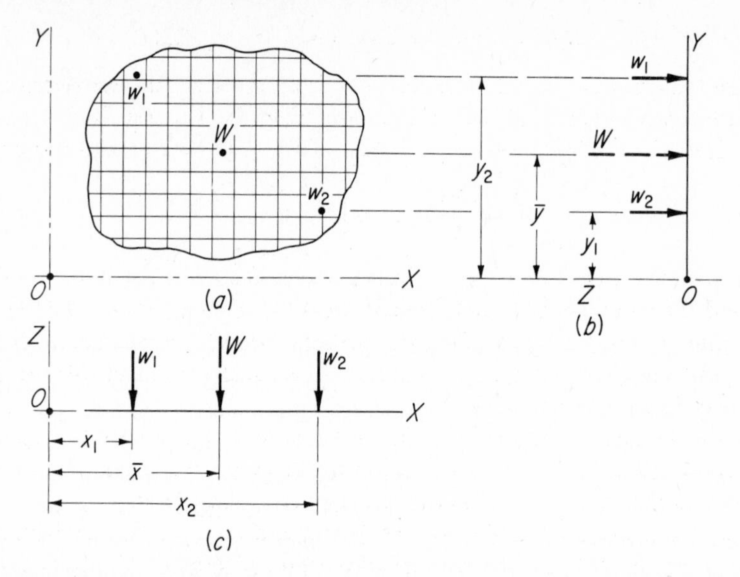

FIG. 10-2 Center of gravity of homogeneous flat plate. (*a*) Plate divided into equal-size elements. (*b*) Forces projected into *YZ* plane. (*c*) Forces projected into *XZ* plane.

moment of the resultant W equals the algebraic sum of the moments of the separate weights. Consequently, by moments about O in the XZ plane,

$$W\bar{x} = w_1x_1 + w_2x_2 + \cdots = \Sigma wx \qquad (a)$$

and, by moments about O in the YZ plane,

$$W\bar{y} = w_1y_1 + w_2y_2 + \cdots = \Sigma wy \qquad (b)$$

or

$$\bar{x} = \frac{\Sigma wx}{W} \quad \text{and} \quad \bar{y} = \frac{\Sigma wy}{W} \qquad (\textbf{10-1})$$

thus locating the center of gravity (*c.g.*).

10-3 Centroids

Let us imagine now that the flat plate in Fig. 10-2 gradually reduces in thickness. When finally its thickness becomes zero, it no longer has weight and only a surface or an area remains. The point in the plate which formerly was the center of gravity of the body now is the **centroid** of the area. If a is the surface area of one small element in Fig. 10-2 and t is the thickness of the plate, its volume is ta and its weight w is δta, where δ is the density. If δta is substituted for w in Eq. (*a*) of Art. 10-2, and A is the total surface area, $W = \delta tA$, and

$$\delta t A\bar{x} = \delta t a_1 x_1 + \delta t a_2 x_2 + \cdots = \delta t \Sigma ax \qquad (a)$$

When the constants δ and t are canceled from the first and last terms of this equation, we find that $A\bar{x} = \Sigma ax$. Similarly, from Eq. (*b*), Art. 10-2, we would find that $A\bar{y} = \Sigma ay$. Consequently, Eq. (10-1), Art. 10-2, would become

$$\bar{x} = \frac{\Sigma ax}{A} \quad \text{and} \quad \bar{y} = \frac{\Sigma ay}{A} \qquad \textbf{(10-2)}$$

Apparently, then, *about any axis* **the moment of an area equals the algebraic sum of the moments of its component areas,** wherein *the moment of an area is defined as the product of the area multiplied by the perpendicular distance from the moment axis to the centroid of the area.* By means of this principle, we may locate the centroid of any area.

In the design of beams and columns, we constantly find it necessary to locate the centroids of their cross-sectional areas. Also, in the design of beams, we often find it necessary to compute the moment of an area.

The terms centroid and center of gravity are often used interchangeably. Also, a *gravity axis* passing through the center of gravity of a body is often called a *centroidal axis.*

10-4 Centroids of Simple Geometric Areas

By geometry and by integration (see Appendix A), the centroids of some simple geometric areas are found to be located as illustrated in Fig. 10-3.

The centroid of a rectangle, or of any parallelogram, lies at the intersection C of its two diagonals. This point is also the intersection of the two lines bisecting the two pairs of opposite sides. The truth of this statement can be deduced as follows: Let the rectangle be divided into narrow strips running parallel to the two opposite sides. The centroid of each strip lies at its

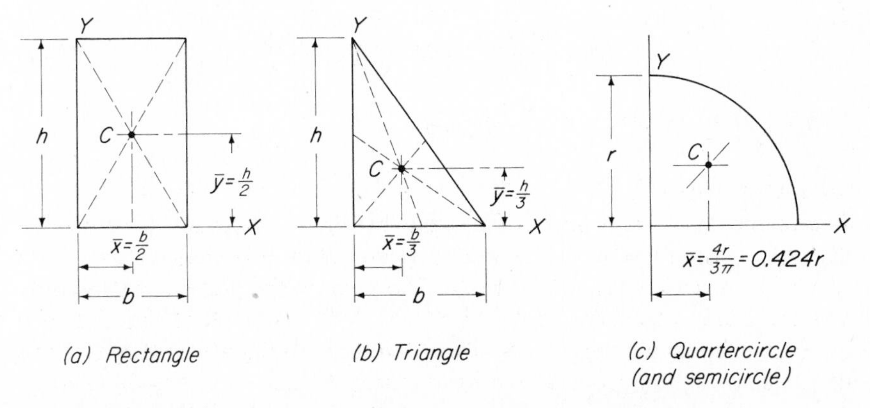

FIG. 10-3 Centroids of simple geometric areas. (*a*) Rectangle. (*b*) Triangle. (*c*) Quartercircle (and semicircle).

mid-point, the centroids of all the parallel strips will lie on a line bisecting the two opposite sides, and the centroid C of the entire area will lie at the intersection of the two bisecting lines which is at their mid-points.

The centroid of a triangle lies at the intersection of its medians. If the triangle is divided into infinitesimally narrow strips running parallel to any base, the centroid of each strip is at its mid-point, the centroids of all the parallel strips will lie on the median of that base, and the centroid C of the entire area lies at the intersection of the three medians, at a distance of one-third of the length of the median from its base.

The centroid of a semicircle is most easily determined by integration, as is shown in Prob. A-3. By symmetry, the centroids of a quarter circle and of a semicircle, shown in Fig. 10-3, are seen to lie at equal distances from the X axis.

10-5 Centroids of Composite Areas

A composite area is one made up of a number of simple areas. To determine its centroid, a composite area is generally divided into two or more simple component areas. The centroid of the composite area is then found by the previously stated principle: about any axis, *the moment of an area equals the algebraic sum of the moments of its component areas*. The following problem illustrates the application of this principle.

ILLUSTRATIVE PROBLEMS

10-1. Locate the centroid of the composite area shown in Fig. 10-4 with respect to the X and Y axes.

Solution: Let the X and Y axes intersect at the lower left corner of the area. Then all moments are positive. This composite area divides naturally into three simple geometric areas, a rectangle, a triangle, and a semicircle, whose centroids we obtain from Fig. 10-3. To aid in computing $\bar{x}$ and $\bar{y}$, we represent these areas as forces concentrated at their respective centroids and shown in two edge views of the area similar to a parallel force system in space. The total area is their resultant which passes through the centroid C of the composite area. From the given dimensions we find that

$$A_1 = (12)(6) = 72 \qquad A_2 = (\tfrac{1}{2})(12)(6) = 36 \qquad A_3 = (\tfrac{1}{2})(3.1416)(6)^2 = 56.6$$

Hence, $$A = 72 + 36 + 56.6 = 164.6 \text{ sq in.}$$

Then, in the front view, to find $\bar{x}$,

$[M_O]$ $$164.6\bar{x} = (72)(6) + (36)(8) + (56.6)(14.54)$$

or $164.6\bar{x} = 432 + 288 + 823 = 1{,}543$ and $\bar{x} = 9.37$ in. *Ans.*

Likewise, in the side view, to find $\bar{y}$,

$[M_O]$ $$164.6\bar{y} = (72)(3) + (56.6)(6) + (36)(8)$$

or $164.6\bar{y} = 216 + 340 + 288 = 844$ and $\bar{y} = 5.13$ in. *Ans.*

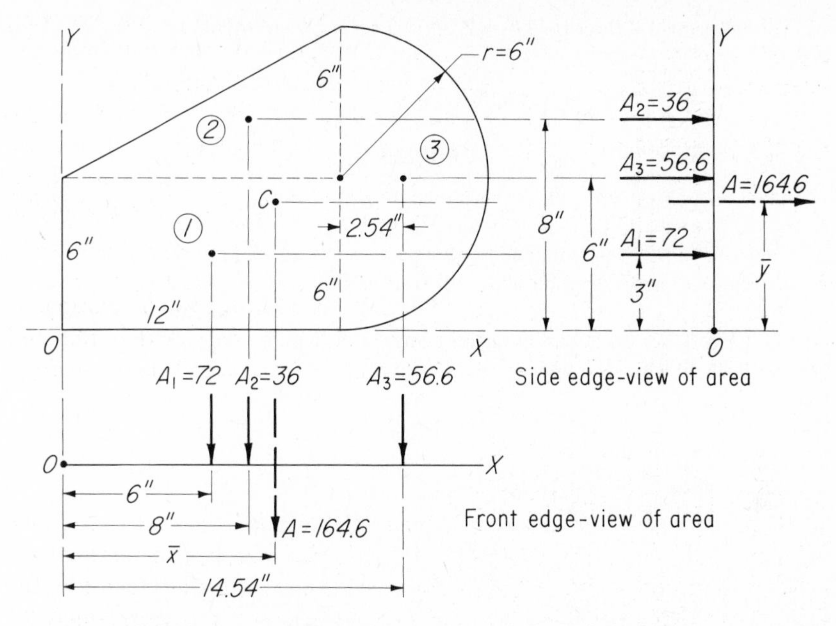

FIG. 10-4 Location of centroid of a composite area.

10-2. In Fig. 10-5 is shown the cross-sectional area of a composite steel beam made up of an American standard I-beam, 12 by $5\frac{1}{4}$ in. by 50 lb per ft, with a $\frac{1}{2}$- by 8-in. cover plate riveted to its top flange. Locate the centroid of the composite area.

Solution: This area is symmetrical about the Y axis. Hence, we need determine only the distance $\bar{y}$. The cross-sectional area A_1 of the plate is $(\frac{1}{2})(8)$ or 4 sq in. From a handbook we find the cross-sectional area A_2 of the I-beam to be 14.6 sq

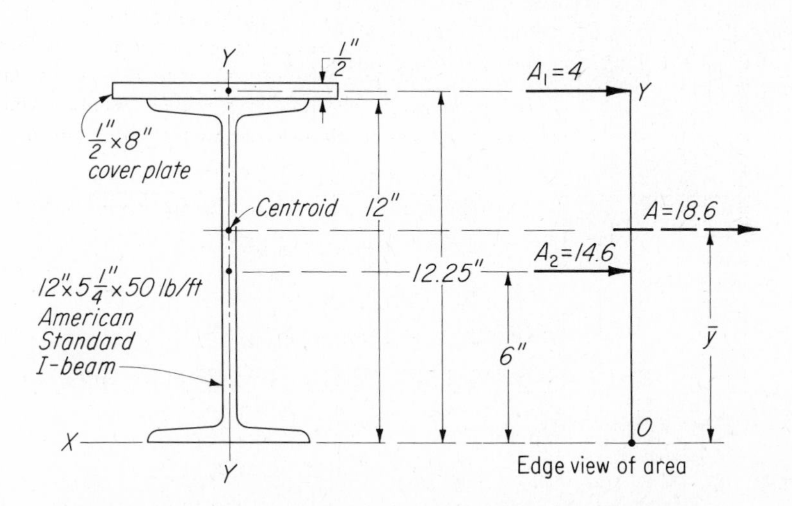

FIG. 10-5 Location of centroid of cross-sectional area of beam.

in. The total area A then is 14.6 + 4 or 18.6 sq in. Then, by moments about O, we have

$[M_O]$ $18.6\bar{y} = (4)(12.25) + (14.6)(6) = 136.6$ and $\bar{y} = 7.35$ in. *Ans.*

PROBLEMS

10-3. Locate the centroid, with respect to the X and Y axes, of the area shown in Fig. 10-6. (HINT: The moment of an area with a part removed equals the moment of that area *minus* the moment of the part removed. Hence, represent the part removed by an arrow applied at its centroid and oppositely directed.)

Ans. $\bar{x} = 6.82$ in.; $\bar{y} = 5.23$ in.

10-4. Determine the location of the centroid of the area shown in Fig. 10-7. (See hint in Prob. 10-3.)

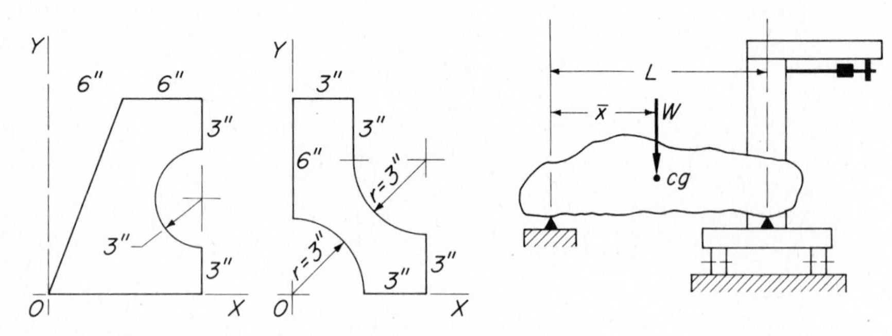

FIG. 10-6 Prob. 10-3.

FIG. 10-7 Prob. 10-4.

FIG. 10-8 Prob. 10-5.

10-5. The irregular body shown in Fig. 10-8 is placed upon two knife-edge supports, one resting on a scale that indicates a load of 225 lb. If the weight W of the body is 350 lb and if L is 7 ft, compute the distance $\bar{x}$ from the left support to the vertical plane through the body containing its center of gravity. *Ans.* $\bar{x} = 4.5$ ft

10-6. The timber beam, shown in cross section in Fig. 10-9, is made of two planks, 2 by 8 in., and one plank 2 by 10 in., all securely nailed together. Compute the

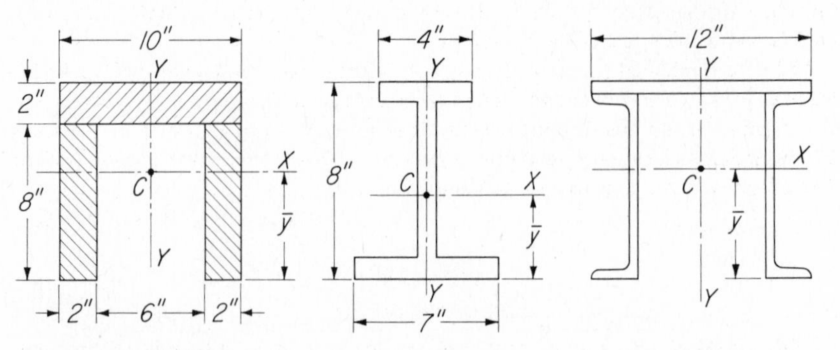

FIG. 10-9 Prob. 10-6.

FIG. 10-10 Prob. 10-7.

FIG. 10-11 Prob. 10-8 and 11-21.

distance $\overline{y}$ from the bottom of the beam to its centroid. YY is an axis of symmetry.

Ans. $\overline{y} = 5.92$ in.

10-7. The cast-iron beam, shown in cross section in Fig. 10-10, is symmetrical about the YY axis. The two flanges and the web are 1 in. thick. Compute the distance $\overline{y}$ from the bottom of the beam to its centroid.

10-8. In Fig. 10-11 is shown the cross section of a steel beam built up of two American standard channels, 10 in. by 20 lb per ft, to whose top flanges are riveted a ½- by 12-in. plate. The area of each channel is 5.86 sq in. YY is an axis of symmetry. Compute the distance $\overline{y}$ from the bottom of the beam to its centroid.

Ans. $\overline{y} = 6.78$ in.

10-9. The structural-steel angle shown in Fig. 10-12 measures 8 by 4 in. and is 1 in. thick. Compute the distances $\overline{x}$ and $\overline{y}$ from the heel of the angle to its centroid. (Check answers against values given in a handbook.)

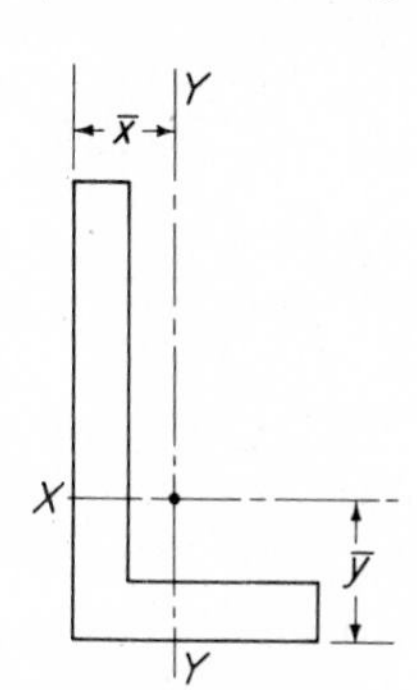

FIG. 10-12 Prob. 10-9.

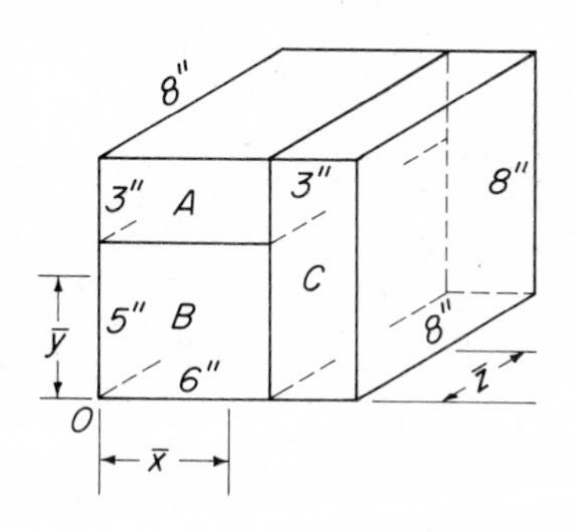

FIG. 10-13 Prob. 10-10.

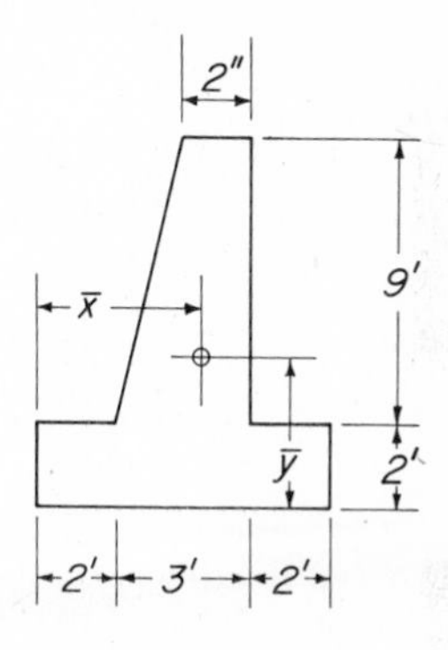

FIG. 10-14 Prob. 10-11.

10-10. The three rectangular wood blocks *A*, *B*, and *C*, shown in Fig. 10-13, are bonded together to make a single block 8 by 9 by 8 in. Block *A* weighs 0.01 pci, *B* weighs 0.015 pci, and *C* weighs 0.02 pci. Calculate the three dimensions $\overline{x}$, $\overline{y}$, and $\overline{z}$ which locate the center of gravity *c.g.* of the entire block. (NOTE: The object used in this problem is admittedly an impractical one. However, it does serve well to introduce the quantities of volume, density, weight, and center of gravity in an easily understood manner, using simple values.)

10-11. A reinforced-concrete dam, whose cross section is shown in Fig. 10-14, extends across a small river and diverts water into an irrigation canal. To determine the stability of the dam against overturning due to water pressure, it is necessary to determine the location of its center of gravity *c.g.* Calculate the horizontal distance $\overline{x}$ and vertical distance $\overline{y}$, using a 1-ft length of the dam. The concrete weighs 150 pcf.

Ans. $\overline{x} = 3.61$ ft; $\overline{y} = 4.22$ ft

SUMMARY

(By article number)

10-1. *Centroids of areas* and *centers of gravity of bodies* are located in a manner similar to that used to locate the *center* of a system of parallel forces.

10-2. The gravitational pull of the earth on a body is its *weight*, also called the

force of gravity. The point in the body through which this gravity force acts is called its **center of gravity.**

10-3. The **centroid of an area** occupies the position of the center of gravity of a homogeneous thin plate whose thickness approaches zero.

10-5. A **composite area** is one made up of a number of simpler areas. The *centroid of a composite area* is located by use of the following principle: about any axis, *the moment of an area equals the algebraic sum of the moments of its component areas.*

REVIEW PROBLEMS

10-12. The timber T-beam shown in cross section in Fig. 10-15 is made of two planks, 2 by 6 in., securely spiked together. *Compute the distance $\bar{y}$ to its centroid C.*

Ans. $\bar{y} = 5$ in.

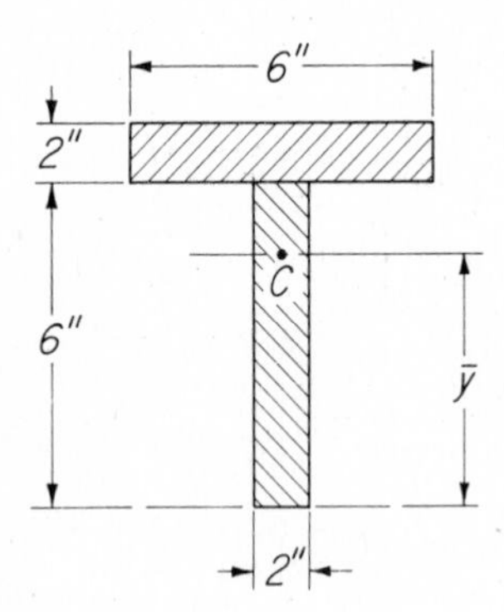

FIG. 10-15 Prob. 10-12.

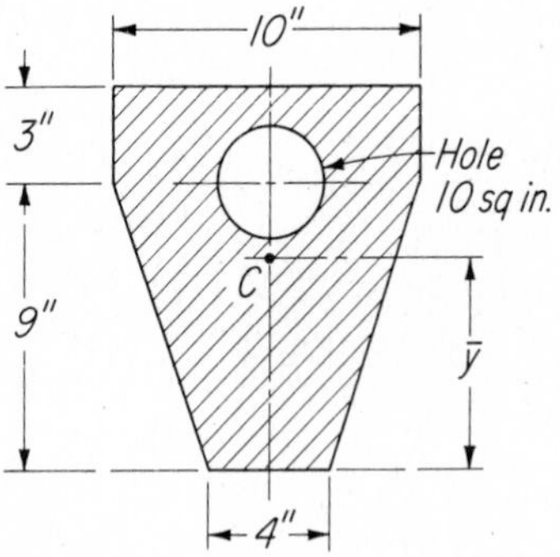

FIG. 10-16 Prob. 10-13.

10-13. The surface shown in Fig. 10-16 is symmetrical about a vertical center line through the circular hole whose area is 10 sq in. *Compute the distance $\bar{y}$ to its centroid C.* *Ans.* $\bar{y} = 6.61$ in.

10-14. The cross section of a steel beam built up of a wide-flange section, with a channel section riveted to its top flange, is shown in Fig. 10-17. The actual depth of the wide-flange section is 18 in. and its area is 14.71 sq in. The area of the channel is 6.03 sq in., the thickness of its web is 0.28 in., and the distance from the back to its center of gravity is 0.70 in. *Compute $\bar{y}$ to the centroid C of the composite section.*

10-15. In Fig. 10-18 is shown the cross section of a steel beam, built up of one web plate, $\frac{3}{4}$ by 18 in., one cover plate, 1 by 12 in., and four equal-leg angles, 5 by 5 by $\frac{3}{4}$ in., assembled as shown. The distance from the heel of each angle to its centroid is 1.52 in. *Compute $\bar{y}$ of the composite section.*

10-16. Determine $\bar{y}$ for the composite section shown in Fig. 10-19. The area of each 10-in. American standard channel is 4.47 sq in. The channels are 10 in. high measured along the back (or long side) of the channels. *Ans.* $\bar{y} = 8.15$ in.

10-17. Locate the centroid of the entire sail area of the boat shown in Fig. 10-20. The area of the jib is 90 sq ft and the area of the mainsail is 310 sq ft.

Ans. $\bar{x} = 13.75$ ft; $\bar{y} = 14.65$ ft. Both measurements are from point *A*

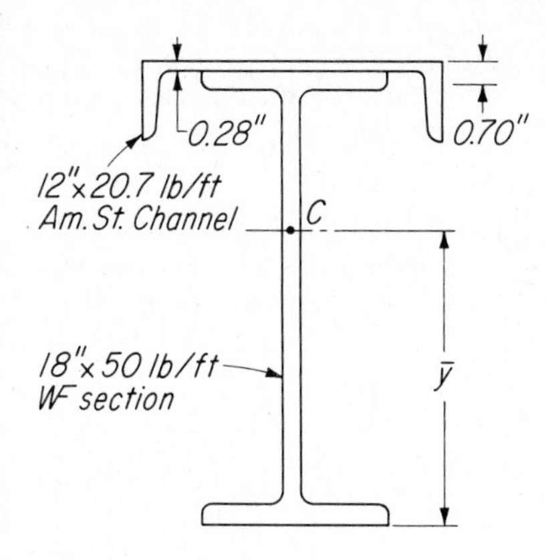

FIG. 10-17 Prob. 10-14.

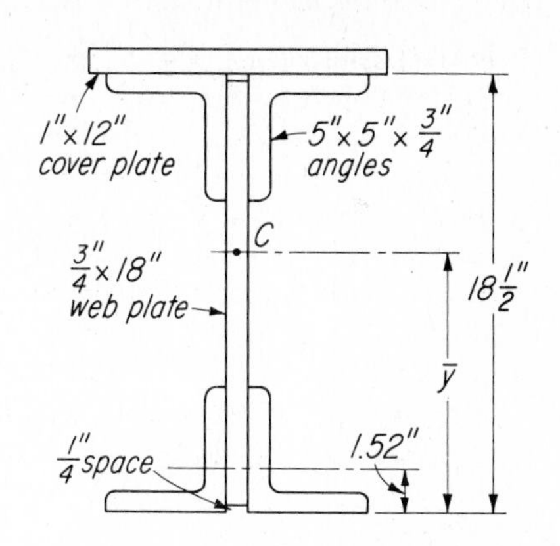

FIG. 10-18 Probs. 10-15 and 11-27.

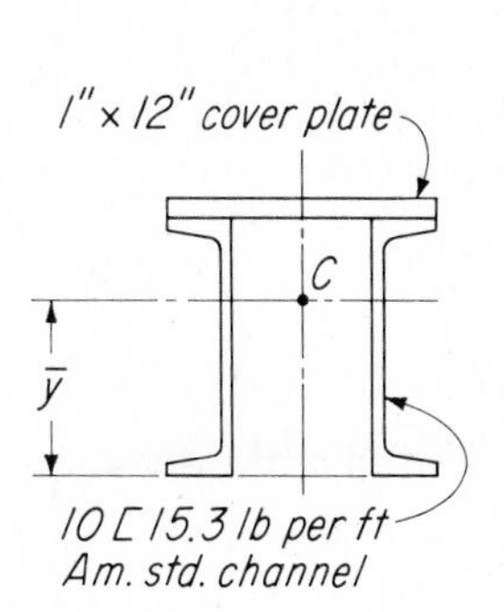

FIG. 10-19 Prob. 10-16.

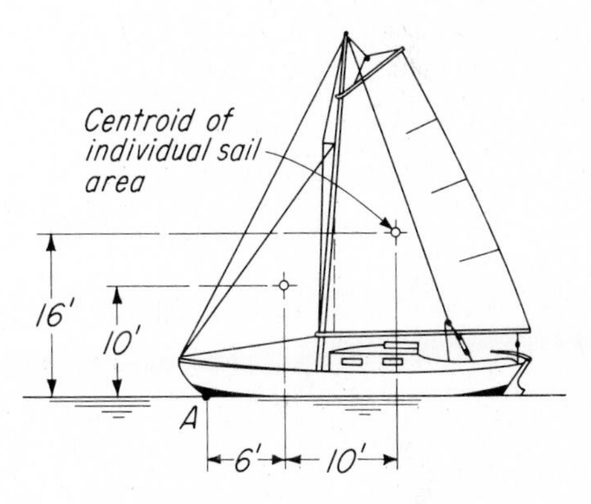

FIG. 10-20 Prob. 10-17.

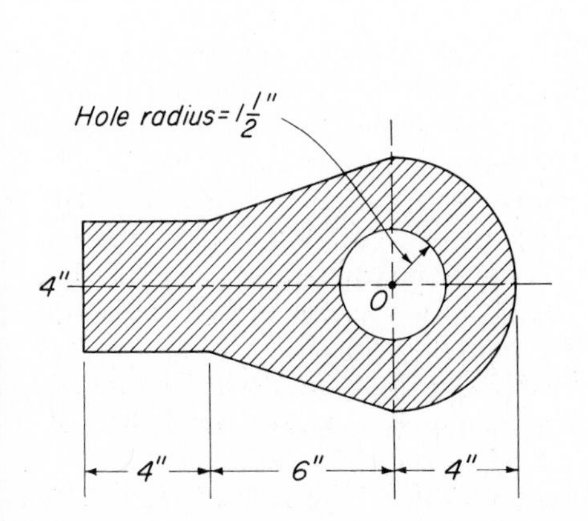

FIG. 10-21 Prob. 10-18.

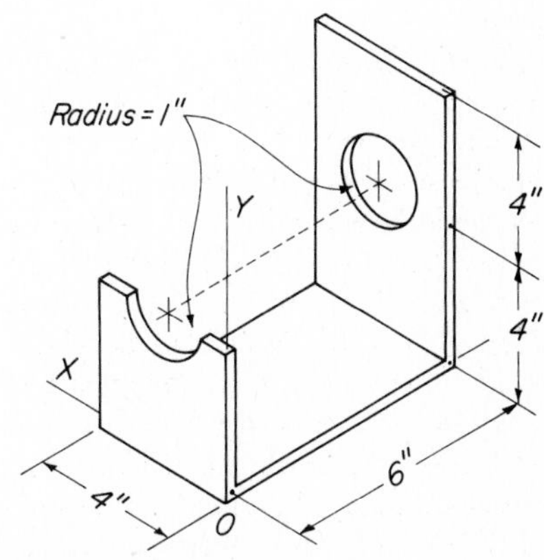

FIG. 10-22 Prob. 10-19.

10-18. Locate the centroid of the area shown in Fig. 10-21.

10-19. Calculate $\overline{y}$ and $\overline{z}$ (i.e., locate the center of gravity) for the sheet metal bracket shown in Fig. 10-22. All thicknesses are small and equal.

Ans. $\overline{y} = 2.135$ in.; $\overline{z} = 3.74$ in.

REVIEW QUESTIONS

10-1. What is the force of gravity and how may it be measured?

10-2. What is meant by center of gravity? By what means may its location in a body be determined?

10-3. Explain the meaning of centroid of an area.

10-4. By what principle may the centroid of a composite area be located? Explain.

10-5. In what phase of engineering practice is the location of the centroid of an area required?

CHAPTER 11

Moments of Inertia of Areas

11-1 Introduction and Definitions

To determine certain *shearing stresses* in beams, a quantity must be determined involving the *first moment of an area* Q, which, with respect to a reference axis, is the product of the area and the perpendicular distance from the axis to the centroid of the area. That is, $Q = A\bar{y}$.

Now, to determine certain *bending stresses* in beams, and also in the design of columns, a quantity must be determined which involves the *second moment of an area,* commonly called the **moment of inertia** and denoted by the symbol I. With respect to a reference axis in the plane of the area, **the second moment of an area, or its moment of inertia, is the sum of the products of the elemental areas, each multiplied by the square of the distance from the reference axis to its centroid.** The moment of inertia is, therefore, the product of the entire area multiplied by the mean value of the squares of the distances from the reference axis to the centroids of the elemental areas. When the number of areas is finite, $I = \Sigma ay^2$, by which equation the *approximate moment of inertia* may be obtained, as outlined in Art. 11-2. When each small area is infinitesimal, or dA, and the number of areas is infinite, $I = \int y^2\, dA$, by which equation the *exact moment of inertia* may be obtained, as is shown in Appendix B. The exact moment of inertia of a composite area is obtained as illustrated in Art. 11-4.

In practical engineering language, the moment of inertia of the cross-sectional area of a beam or a column of a given material, with respect to a reference axis, may be said to represent the relative capacity of the section to resist bending or buckling in a direction perpendicular to the reference axis.

When bending stresses are involved, the moment of inertia required is with respect to an axis lying in the plane of the cross-sectional area and is called the **rectangular moment of inertia.** In shafts that transmit power, and are therefore subjected to twisting or torsional stresses, the moment of inertia required is with respect to an axis perpendicular to the plane of the cross-sectional area; it is then called the **polar moment of inertia** of the area.

Units of Moment of Inertia. Moment of inertia is the product of an area and the square of a distance. If the area is in square inches and the distance in inches, the product will be in *inches to the fourth power.* That is, $(\text{in.}^2)(\text{in.})^2 = \text{in.}^4$ This unit has no physical conception.

11-2 Approximate Determination of Moment of Inertia

The moment of inertia of an area may be obtained approximately by dividing the total area A into a finite number of smaller areas a, then multiplying each small area a by the square of the distance (y^2) from the reference axis to its centroid. The moment of inertia then is the sum of these products. Consequently,

Approximate moment of inertia:

$$I = \Sigma a y^2 \tag{11-1}$$

The approximate method is used chiefly with irregular areas. The following example illustrates the method. A rectangular area was chosen here in order that its approximate and exact moments of inertia might be compared and thus show the probable error of the approximate method.

ILLUSTRATIVE PROBLEM

11-1. Determine the approximate moment of inertia of the area shown in Fig. 11-1 with respect to axis XX passing through its centroid.

Solution: Let the area arbitrarily be divided into 10 strips of equal width running parallel with the XX axis, each area then containing (2)(10) or 20 sq in. We find that, of the distances to the respective centroids, $y_1 = 9$ in., $y_2 = 7$ in., $y_3 = 5$ in., $y_4 = 3$ in., and $y_5 = 1$ in. Clearly, with respect to axis XX, the moment of inertia of the lower half equals that of the upper half; therefore we determine only the latter and then multiply it by 2 to obtain I of the entire area. That is,

$$[I = \Sigma a y^2] \qquad I_X = (2)(a_1 y_1^2 + a_2 y_2^2 + a_3 y_3^2 + a_4 y_4^2 + a_5 y_5^2)$$

But each of the areas a_1, a_2, etc., equals 20 sq in. Hence,

$$I_X = (2)(20)(81 + 49 + 25 + 9 + 1) \qquad \text{or} \qquad I_X = 6{,}600 \text{ in.}^4 \qquad \textit{Ans.}$$

The exact moment of inertia of a rectangular area is $(\frac{1}{12})bh^3$ which, for the area in this example, is 6,667 in.^{4}, thus indicating an error in the approximate result of only 1 percent.

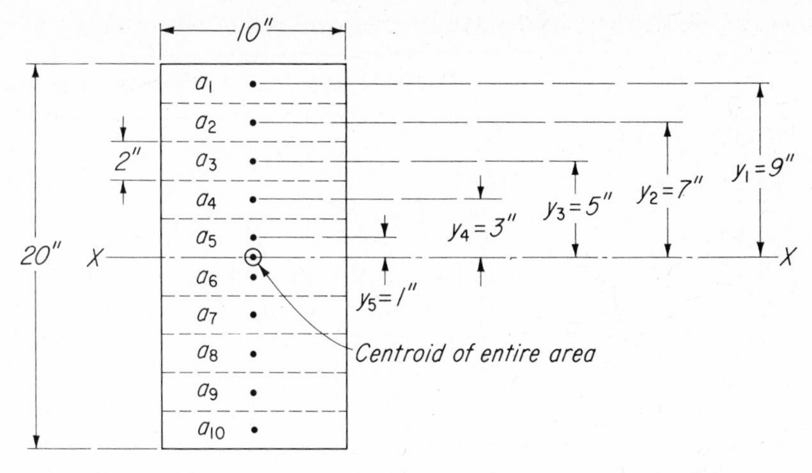

FIG. 11-1 Moment of inertia of an area. Approximate method.

11-3 Moments of Inertia of Simple Areas

The moments of inertia of many simple geometric areas are readily determined by integration, as is illustrated in Appendix B. These areas and their moments of inertia are generally expressed in terms of simple dimensions which completely define them.

The moments of inertia and the radii of gyration of a number of simple areas are given in Table 11-1. These and many others are usually found in handbooks.

The moments of inertia of other simple areas *from which some parts have been removed,* such as those shown in Fig. 11-2, are easily obtained from the following principle: *With respect to any axis, the moment of inertia of an area with a part removed equals the moment of inertia of the area minus the moment of inertia of the part removed.* However, if the thickness is very small,

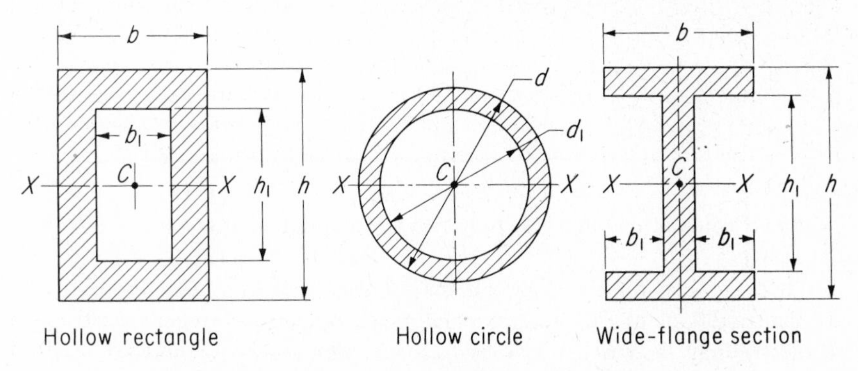

FIG. 11-2 Special equations for obtaining moments of inertia.

Table 11-1 MOMENTS OF INERTIA AND RADII OF GYRATION OF SIMPLE AREAS

Area	Moment of inertia	Radius of gyration
Rectangle (or parallelogram) 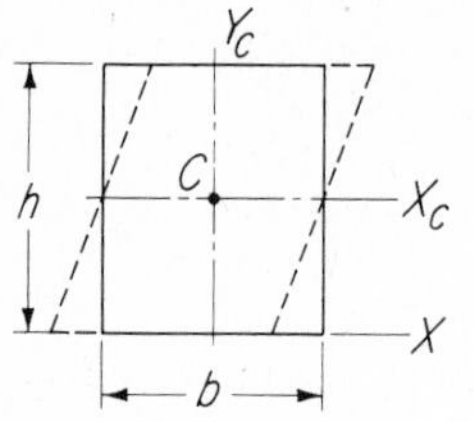	$I_{X_C} = \frac{bh^3}{12}$ $I_X = \frac{bh^3}{3}$	$k_{X_C} = \frac{h}{\sqrt{12}} = 0.289h$ $k_X = \frac{h}{\sqrt{3}} = 0.577h$
Triangle 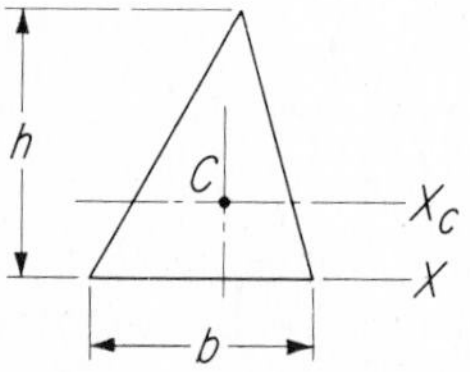	$I_{X_C} = \frac{bh^3}{36}$ $I_X = \frac{bh^3}{12}$	$k_{X_C} = \frac{h}{\sqrt{18}} = 0.236h$ $k_X = \frac{h}{\sqrt{6}} = 0.408h$
Circle 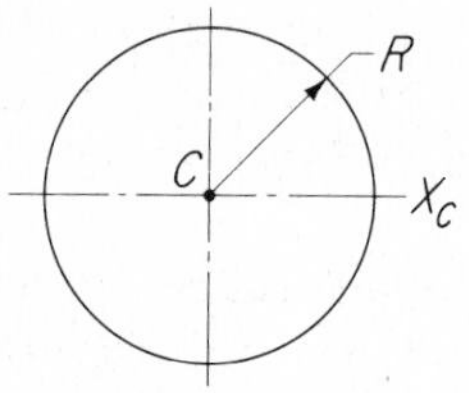	$I_{X_C} = \frac{\pi R^4}{4} = 0.7854R^4$	$k_{X_C} = \frac{R}{2} = 0.5R$
Semicircle 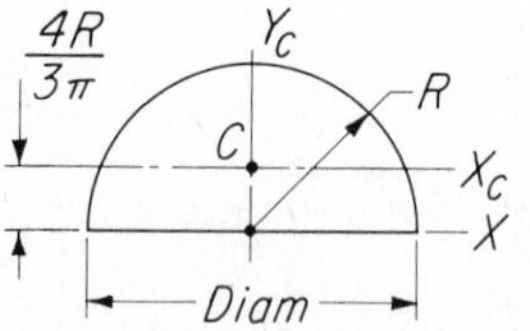	$I_{X_C} = \left(\frac{\pi}{8} - \frac{8}{9\pi}\right)R^4$ $= 0.110R^4$ $I_X = I_{Y_C} = \frac{\pi R^4}{8}$ $= 0.3927R^4$	$k_{X_C} = \frac{\sqrt{9\pi^2 - 64}}{6\pi}R$ $= 0.264R$ $k_X = k_{Y_C} = \frac{R}{2} = 0.5R$

longhand calculations may be required to maintain reasonable precision in the results.

When the axes for the area and for the part removed coincide and both are centroidal, as in Fig. 11-2, special simple equations giving the moment of inertia may be written as shown above. Otherwise the general method outlined in the following article must be used.

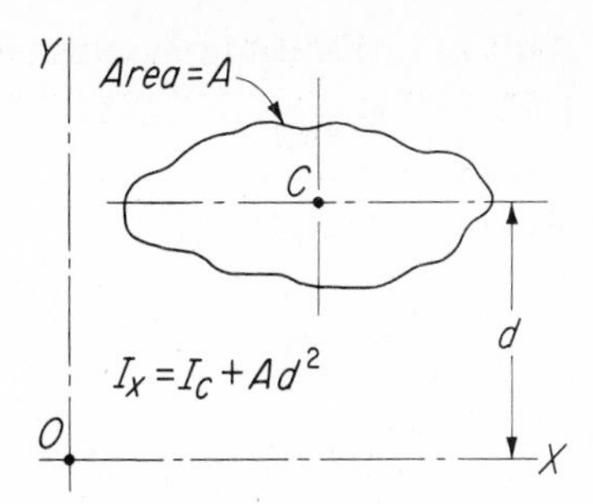

FIG. 11-3 Moment of inertia of area with respect to a noncentroidal axis.

The moments of inertia of standard sizes of structural-steel shapes and of structural timbers, commonly used as beams and columns, are given in various handbooks. The moments of inertia of nonstandard sizes and of built-up sections must be computed by the methods here outlined.

11-4 Moments of Inertia of Composite Areas

In modern structural practice, and especially in aircraft construction, relatively few members used to resist bending are of simple geometric cross-sectional area such as rectangular or circular. Many sections, however, are so shaped that they are readily divided into two or more simple *component areas*. The entire area is then called a *composite area*. The moment of inertia of each component area, with respect to a centroidal axis of the composite area, is then determined by the following principle, illustrated in Fig. 11-3 and more fully explained in Appendix B:

The moment of inertia of an area with respect to any axis not through its centroid is equal to the moment of inertia of that area with respect to its own parallel centroidal axis plus the product of the area and the square of the distance between the two axes. When this principle is stated in symbols, we have

$$I_X = I_C + Ad^2 \tag{11-2}$$

where I_X = moment of inertia of an area with respect to any noncentroidal axis, generally, in.^4

I_C = moment of inertia of the area with respect to its parallel centroidal axis, in.^4

A = area, sq in.

d = perpendicular distance between the parallel axes, in.

Equation (11-2) is called the **transfer formula,** since it "transfers" the moment of inertia of an area from its centroidal axis to any other parallel axis. *The moment of inertia of a composite area* then is the sum of the moments of inertia of the component areas, all with respect to a common axis. Close study of this equation reveals that the moment of inertia of an area is always least with respect to a centroidal axis. The following problems will illustrate the method.

ILLUSTRATIVE PROBLEMS

11-2. Locate the centroid of the composite T-beam section shown in Fig. 11-4 and determine its moment of inertia I_X with respect to the axis XX.

Solution: The composite area is divided into two rectangular areas A and B, and the centroid of each area is located by inspection, as in Fig. 11-4. Next, we draw the edge view of the entire area and apply the separate areas, represented by arrows, at their respective centroids. Then, by moments about O,

$$76\bar{y} = (36)(11.5) + (40)(5) = 614 \qquad \text{and} \qquad \bar{y} = 8.08 \text{ in.} \qquad \textit{Ans.}$$

Having thus located the centroid of the composite area, we then determine distances d_A and d_B, from axis XX to the centroids of areas A and B; that is, $d_A = 11.5 - 8.08 = 3.42$ in. and $d_B = 8.08 - 5 = 3.08$ in. Now, the moment of inertia of each of areas A and B, with respect to axis XX, is given by Eq. (11-2). That is, $I_X = I_C + Ad^2$. The total moment of inertia of the composite area then is $I_X = \Sigma(I_C + Ad^2)$. Since $A_A = 36$ sq in., $A_B = 40$ sq in., $d_A = 3.42$ in., and $d_B = 3.08$ in., we have

$$I_X = \Sigma[I_C + Ad^2] \qquad \text{in which} \qquad I_C = \frac{bh^3}{12}$$

Area A: $I_X = \dfrac{(12)(3)^3}{12} + (36)(3.42)^2 = 27 + 421$ . . . . 448 in.^{4}

Area B: $I_X = \dfrac{(4)(10)^3}{12} + (40)(3.08)^2 = 333 + 380$ . . . . 713 in.^{4}

Moment of inertia of the composite area . . . . 1,161 in.^{4}

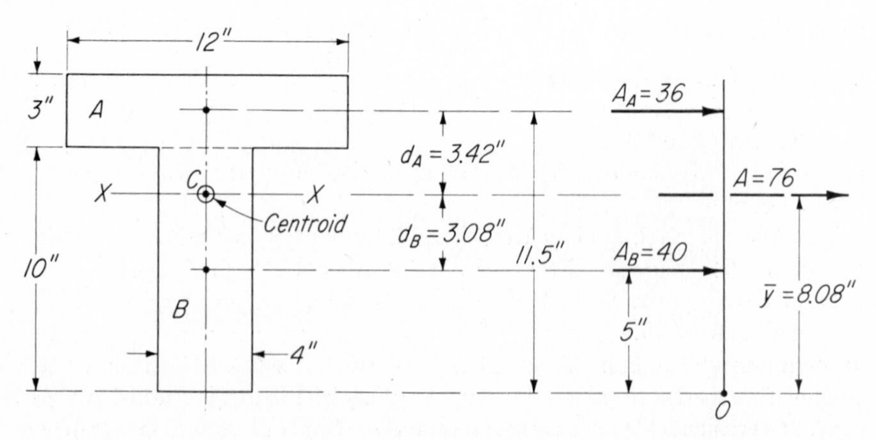

FIG. 11-4 Centroid and moment of inertia of a composite area.

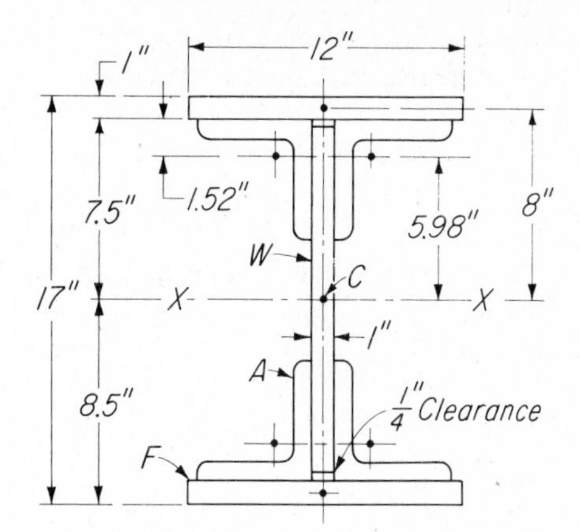

FIG. 11-5 Moment of inertia of a composite area.

11-3. In Fig. 11-5 is shown the cross-sectional area of a steel beam, built up of one web plate 1 by 14.5 in., four angles 5 by 5 by $\frac{3}{4}$ in., and two cover plates 1 by 12 in. Compute the moment of inertia of this composite section with respect to axis XX passing through its centroid.

Solution: This section is symmetrical about the X axis which, then, is located 8.5 in. above the bottom. From a handbook we find that the moment of inertia I_C of each of the four angles about its own centroidal X axis is 15.7 in.^{4}, that the area A of each angle is 6.94 sq in., and that the distance y from the back of the angle to its centriodal axis is 1.52 in. Considering separately the two flanges F, the four angles A, and the web W, we have

$$I_X = \Sigma[I_C + Ad^2]$$

2 flanges: $\quad I_X = 2\left[\frac{(12)(1)^3}{12} + (12)(8)^2\right] = 2(1 + 768) \ldots\ldots\ldots\ldots 1{,}538 \text{ in.}^4$

4 angles: $\quad I_X = 4[15.7 + (6.94)(5.98)^2] = 4(15.7 + 248.2) \ldots\ldots\ldots 1{,}055 \text{ in.}^4$

1 web plate: $I_X = \frac{(1)(14.5)^3}{12} = \frac{(1)(3{,}048)}{12} \ldots\ldots\ldots\ldots\ldots\ldots\ldots\ldots 254 \text{ in.}^4$

Moment of inertia of the composite area $\ldots\ldots\ldots\ldots\ldots\ldots\ldots\ldots\ldots$ 2,847 in.^{4}

11-5 Radius of Gyration

In the design of columns a term called the **radius of gyration** is very useful. It is denoted by the symbol k.

The radius of gyration of an area is that distance from its moment-of-inertia axis at which the entire area could be considered as being concentrated without changing the numerical value of its moment of inertia.

That is, if all the elemental areas dA are placed side by side in a strip parallel to the moment-of-inertia axis as in Fig. 11-6, and if the thickness of this strip is infinitesimal, the moment of inertia I_C of each elemental area dA would become zero, and that term would therefore drop out of Eq. (11-2). The moment of inertia I of each small area then would be $y^2\,dA$, and of the entire

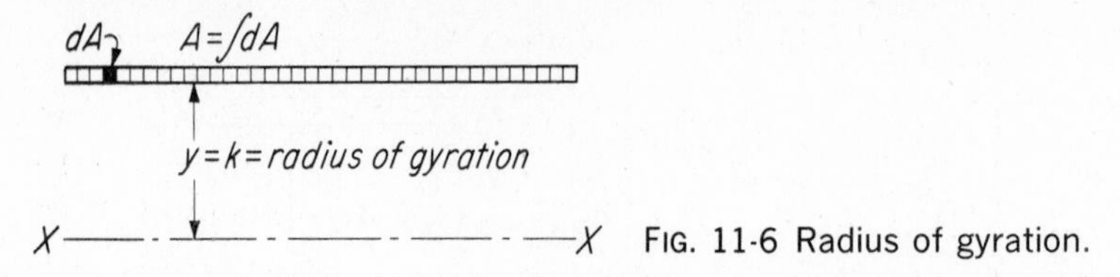

FIG. 11-6 Radius of gyration.

area I would be the sum of these, or $\int y^2\, dA$ (see also Appendix B). But, since y now has the same value for all of the small areas, $y^2 = k^2$ and $\int dA$ (the sum of all the small areas) $= A$. Therefore

$$I_X = Ak^2 \qquad \text{or} \qquad k = \sqrt{\frac{I}{A}} \tag{11-3}$$

In many instances, the moments of inertia of the cross-sectional area of a column with respect to its rectangular, centroidal X and Y axes are not equal. Therefore, one of its two major radii of gyration will have a smaller numerical value than the other. **A compression member tends to buckle in the direction of its least radius of gyration.**

PROBLEMS

11-4. Change the 3-, 4-, and 12-in. dimensions of Prob. 11-2 to 2, 2, and 5 in., respectively, and solve.

11-5. In Fig. 11-7 is shown the cross-sectional area of a beam which is symmetrical about its centroidal X axis. Using the transfer formula, compute its moment of inertia I_X and check it by the special equation given in Art. 11-3. *Ans.* $I_X = 2.656$ in^4

11-6. Locate the centroidal XX axis and compute the moment of inertia I_X of the angle shown in cross section in Fig. 11-8. (Check answers against values given in handbook.)

11-7. Determine the location of the centroidal XX axis of the area shown in Fig. 11-9 and compute the moment of inertia I_X of the area.

Ans. $\bar{y} = 4$ in.; $I_X = 576$ in.^{4}

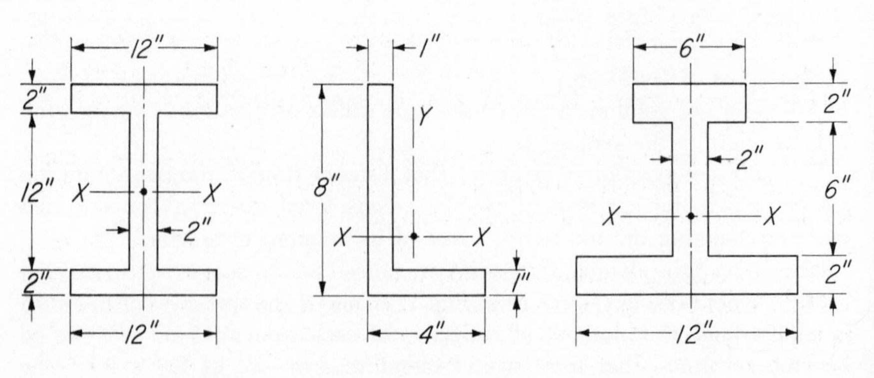

FIG. 11-7 Prob. 11-5. FIG. 11-8 Prob. 11-6. FIG. 11-9 Prob. 11-7.

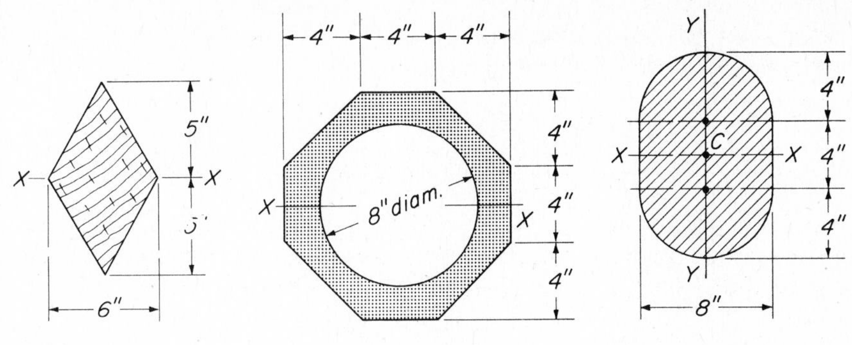

FIG. 11-10 Prob. 11-9.

FIG. 11-11 Prob. 11-10.

FIG. 11-12 Prob. 11-11.

11-8. Show that the moment of inertia of a triangular area about an axis through its apex and parallel to the base is $I = bh^3/4$.

11-9. The cross section of a timber beam is shown in Fig. 11-10. Compute the moment of inertia I_X of this section.

11-10. Determine the moment of inertia I_X of the area shown in Fig. 11-11.

11-11. Calculate the moments of inertia I_X and I_Y of the strut shown in Fig. 11-12. Also compute the least radius of gyration of the section.

Ans. $I_X = 786$ in.^{4}; $I_Y = 372$ in.^{4}; $k_Y = 2.13$ in.

11-12. Locate the centroidal axis and calculate I_X of the area shown in Fig. 11-13 using each method of subdivision indicated.

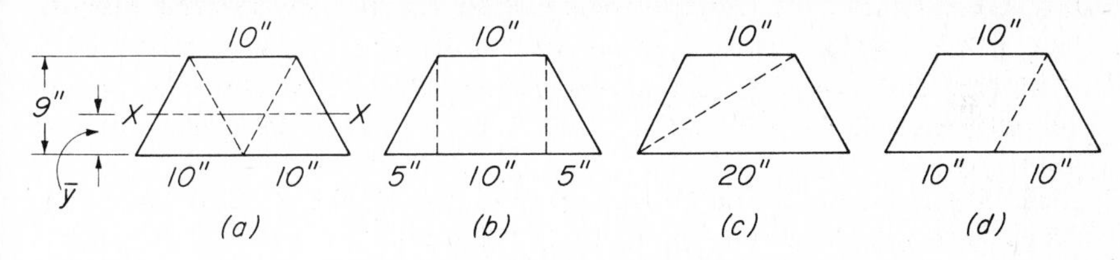

FIG. 11-13 Prob. 11-12.

11-13. Locate the centroidal axis and calculate I_X of the section shown in Fig. 11-14.

Ans. $\bar{y} = 6\frac{1}{4}$ in. (from top); $I_X = 1{,}170$ in.^{4}

11-14. A timber beam is fabricated from four separate pieces of lumber as shown in Fig. 11-15. Calculate I_X.

11-15. The timber cross section shown in Fig. 11-15 will be used as a column. The least radius of gyration will be needed before the load capacity of the column is calculated. Compute the least radius of gyration of the section.

Ans. $k = 2.75$ in.

11-16. Calculate the radius of gyration k_X of the area shown in Fig. 11-16.

11-17. The I-beam in Fig. 11-17 is 20 in. deep and its cross-sectional area is 24.8 sq in. The moment of inertia with respect to its centroidal X axis is 1,502 in.^{4} A 12- by $\frac{1}{2}$-in. cover plate is riveted to its top flange. Locate the centroidal X axis of the composite area and compute the moment of inertia I_X.

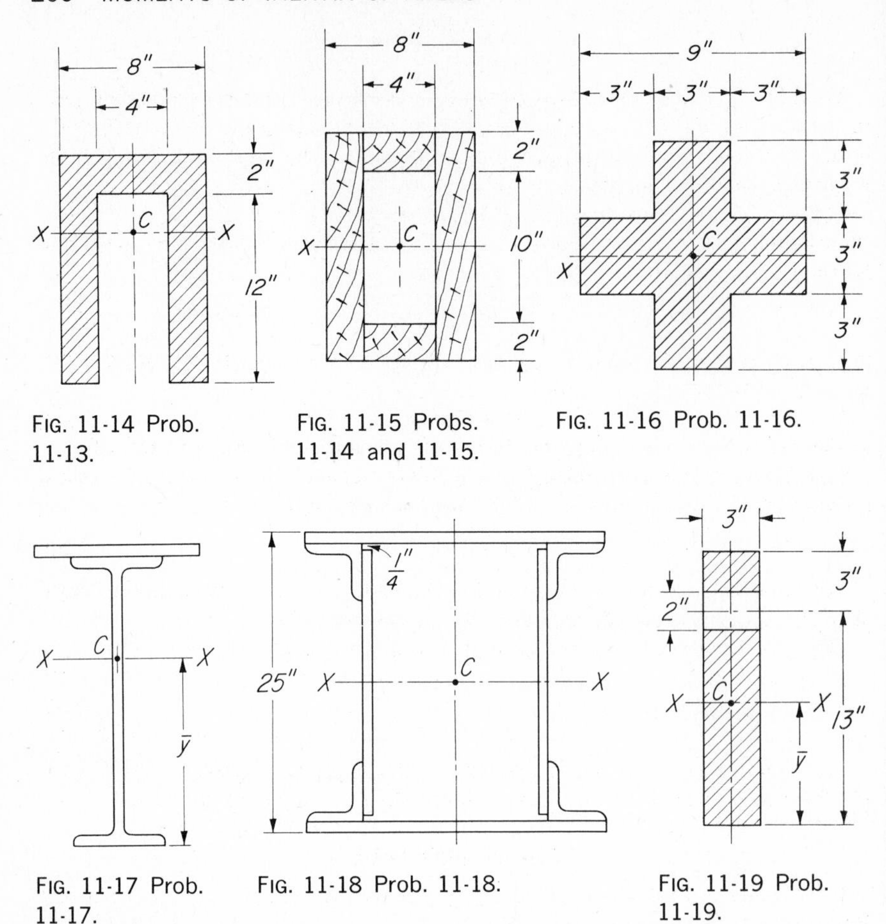

FIG. 11-14 Prob. 11-13.

FIG. 11-15 Probs. 11-14 and 11-15.

FIG. 11-16 Prob. 11-16.

FIG. 11-17 Prob. 11-17.

FIG. 11-18 Prob. 11-18.

FIG. 11-19 Prob. 11-19.

11-18. The box girder, whose cross-sectional area is shown in Fig. 11-18, is built up of four angles 4 by 4 by ½ in., two web plates 23.5 by ½ in., and two cover plates 25 by ½ in., assembled to provide a section 25 in. deep. The centroidal moment of inertia of each angle is 5.60 in.^{4}, its cross-sectional area is 3.75 sq in., and the distance $\bar{y}$ from the back of the angle to its own centroid is 1.18 in. Compute the moment of inertia I_X of the entire section.

11.19 A 2-in. diam hole has been bored in the upper half of the 3- by 16-in. timber joist shown in Fig. 11-19. Locate the centroidal X axis. By what percentage has the moment of inertia of the solid joist been weakened by this hole?

Ans. 16.9 percent; ($\bar{y} = 7.29$ in.; $I_X = 851$ in.^{4})

11-20. By what percentage will the solid joist in Prob. 11.19 be weakened, if the hole is bored through its middle? Compare with answer to Prob. 11.19 and draw conclusions.

SUMMARY

(By article number)

11-1. The quantity **moment of inertia,** also called *second moment of area,* is used in the analysis and design of beams and columns. It represents the capacity of a beam or a column to resist bending or buckling. The general expression for the moment of inertia of an area is $I = \int y^2 \, dA$.

11-2. To obtain its moment of inertia, an irregular area is divided into narrow strips *running parallel to the reference axis.* **The approximate moment of inertia** then is found by the equation

$$I = \Sigma a y^2 \tag{11-1}$$

in which the area a of each strip is multiplied by the square of the distance from its centroid to the reference axis.

11-3. The exact moment of inertia of a simple geometric area such as a rectangle, a triangle, and a circle is obtained by integration as described in Appendix B.

11-4. The moment of inertia of an area with respect to a noncentroidal axis is given by the following equation, called **the transfer formula:**

$$I_X = I_C + Ad^2 \tag{11-2}$$

The moment of inertia of a composite area is the sum of the moments of inertia of the component areas, all with respect to a common axis.

11-5. The **radius of gyration k** of an area is given by the equation

$$k = \sqrt{\frac{I}{A}} \tag{11-3}$$

This quantity is useful in the design of all compression members, since such members tend to buckle in the direction of their *least radius of gyration.*

REVIEW PROBLEMS

11-21. Refer to Prob. 10-8. Calculate I_X. The moment of inertia of each channel about its own centroidal axis is 78.5 in.^{4} *Ans.* $I_X = 266.5$ in.^{4}

11-22. In Fig. 10-5, the I-beam is 12 in. deep, its cross-sectional area is 14.57 sq in., and its moment of inertia with respect to its own centroidal X axis is 301.6 in.^{4} *Compute the moment of inertia I_X of the beam with cover plate with respect to the* centroidal X axis of the composite section. *Ans.* $I_X = 424$ in.^{4}

11-23. The beam section shown in Fig. 11-20 is built up of a 24-in. wide-flange section with a 15-in. channel riveted to its top flange. The area of the wide-flange section is 29.43 sq in. and its centroidal moment of inertia is 2,987 in.^{4} The area of the channel is 9.9 sq in., the thickness t of its web is 0.4 in., the distance y from its back to its horizontal centroidal axis is 0.79 in., and its moment of inertia about that axis is 8.2 in.^{4} *Locate the centroidal XX axis of the composite area, and compute the moment of inertia I_X.*

11-24. In Fig. 11-21 is shown the cross-sectional area of a latticed column in which two 15-in. channels are held together by lattice bars riveted to their flanges. The area of each channel is 14.64 sq in., the distance x from the back to its centroidal

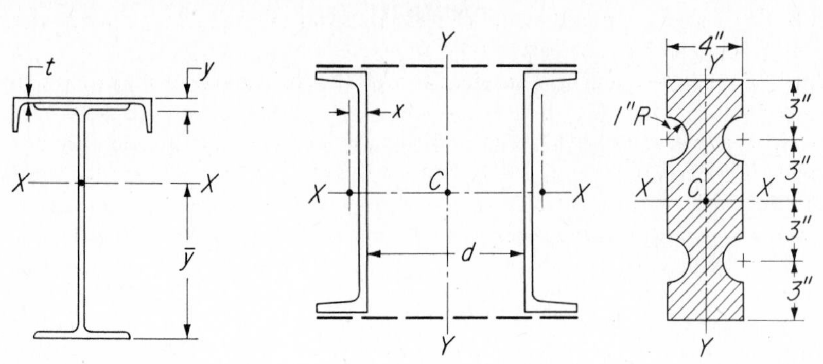

FIG. 11-20 Prob. 11-23. FIG. 11-21 Prob. 11-24. FIG. 11-22 Probs. 11-25 and 11-26.

axis is 0.80 in., and its moment of inertia about that axis is 11.2 in.^{4} The moment of inertia with respect to the X axis of each channel is 401.4 in.^{4} *Compute the distance* d *required to make* I_X *and* I_Y *equal.* (The lattice bars, indicated by dashed lines, are neglected, since they serve merely to hold the two channels in position.)

Ans. $d = 8.72$ in.

11-25. The 4- by 12-in. solid section in Fig. 11-22 has four semicircular grooves cut in its sides as shown. *Compute the moment of inertia of this section about the X axis.* *Ans.* $I_X = 518$ in.^{4}

11-26. Determine the moment of inertia of the area shown in Fig. 11-22 about the Y axis through the centroid C.

11-27. Compute the moment of inertia I_X with respect to the centroidal X axis of the section shown in Fig. 10-18 and described in Prob. 10-15. The area of each angle is 6.94 sq in., the distance from the back of the angle to its centroidal axis is 1.52 in., and its moment of inertia with respect to that axis is 15.7 in.^{4} The distance $\bar{y}$ is 11.44 in.

REVIEW QUESTIONS

11-1. What practical use is made of the quantity moment of inertia?

11-2. Define the moment of inertia of an area in terms of its elemental areas and with respect to an axis in the plane of the area.

11-3. In practical engineering language, what does the moment of inertia of the cross-sectional area of a beam or a column represent?

11-4. In obtaining the rectangular moment of inertia of an area, what is the position of the reference axis with respect to the area?

11-5. In obtaining the polar moment of inertia of an area, what is the position of the reference axis with respect to the area?

11-6. By what method may the approximate moment of inertia of an area be determined?

11-7. How is the exact moment of inertia of a simple geometric area obtained?

11-8. State the principle sometimes used to obtain the moment of inertia of an area with some part removed.

11-9. Give the transfer formula and a complete word statement of the meaning of its separate terms.

11-10. Explain briefly how to obtain the moment of inertia of a composite area.

11-11. What is the radius of gyration of an area, and what practical use is made of it?

CHAPTER 12

Miscellaneous Problems

12-1 Introduction

Some common engineering applications of the principles of statics will be presented in this chapter of miscellaneous problems. Hydrostatic pressure and some forms of hydrostatic loads will be discussed. Granular materials cause lateral pressures similar to those of true fluids. Therefore, the principles of hydrostatics can and will be used to evaluate the loads against and the stability of some simple retaining walls. Finally, flexible cables and arches will be considered.

12-2 Intensity and Direction of Hydrostatic Pressures

When a liquid is contained and restrained by the surface of a solid object, force is exerted on that surface by the liquid. The intensity of this force per unit area is called *pressure*. A characteristic of liquid pressure is that it is exerted equally in all directions. In engineering practice, pressure is usually expressed in pounds per square foot, abbreviated to psf, or in pounds per square inch, abbreviated to psi. For fluids at rest, the pressure is always exerted perpendicular to the resisting surface (see Fig. 12-1*a*).

The pressure p at any point in a liquid is

$$p = wh \tag{12-1}$$

in which w is the weight per unit volume of the liquid and h is the vertical distance from the free surface. That is, liquid pressure is zero at the free surface and increases at a uniform rate from there downward (see Fig. 12-1*b*).

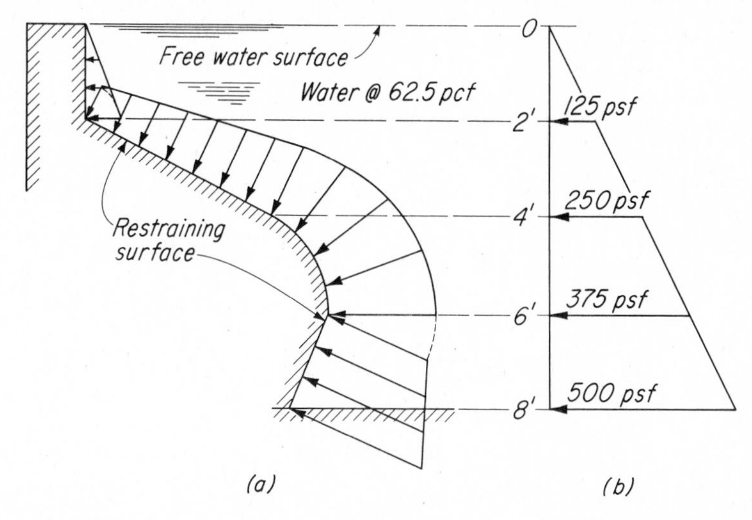

FIG. 12-1 Intensity and direction of hydrostatic pressure. (*a*) Direction of pressure is always perpendicular to restraining surface. (*b*) Intensity of pressure $p = wh$.

12-3 Buoyancy

The weight w per unit volume of a fluid is called its *weight density* and, in most engineering practice, is generally expressed in pounds per cubic foot, abbreviated pcf. An object of lower weight density than the fluid in which it is immersed or floats will rise or float because the upward pressure of the fluid on the object exceeds the downward pressure of the object on the fluid. The total upward force exerted upon such a body by the fluid is called its **buoyancy** and is equal to the weight of the fluid displaced. The load required to barely submerge a floating body will herein be referred to as its **load capacity;** it is equal to the difference between the weight of object and the weight of the fluid it can displace by complete submergence.

12-4 Hydrostatic Loads

The total hydrostatic force P acting against a submerged plane area is equal to the *average pressure* $\overline{p}$ against the area multiplied by the area A over which the pressure acts. That is,

$$\boldsymbol{P = \overline{p}(A)} \tag{12-2}$$

The average pressure p is always found at the centroid of the submerged plane area. The method of computing the hydrostatic loads and the resultant reacting forces on some simple structures will be illustrated in the following three problems.

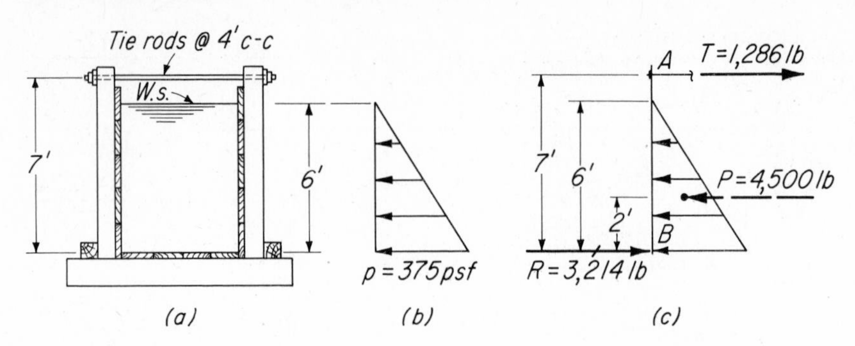

FIG. 12-2 Hydrostatic pressure and load against vertical surface. (*a*) Water flume. (*b*) Pressure diagram. (*c*) Load diagram.

ILLUSTRATIVE PROBLEMS

12-1. The sides of the water flume whose cross section is shown in Fig. 12-2*a* are supported by steel rods passing through opposite vertical timber posts spaced 5 ft on centers along the flume. Calculate the tension T in one steel rod, and the reacting force R at the bottom of one post.

Solution: The *pressure diagram* shown in Fig. 12-2*b* illustrates the uniform variation of hydrostatic pressure from the free water surface to the bottom. The hydrostatic pressure varies uniformly from zero at the top of the wall to 375 psf at the bottom. The average pressure is

$$\bar{p} = \frac{0 + 375}{2} = 187.5 \text{ psf}$$

The hydrostatic force acting against a section of wall 6 ft high and extending 2 ft each side of a post and rod is

$$[P = \bar{p}A] \qquad P = 187.5(6)(4) = 4{,}500 \text{ lb}$$

This resultant hydrostatic force acts through the centroid of the pressure prism as shown in Fig. 12-2*c*. Taking moments about the bottom of the wall (point B in the load diagram)

$$[\Sigma M_B = 0] \qquad T(7) = 4{,}500(2) \quad \text{or} \quad T = 1{,}286 \text{ lb} \qquad \textit{Ans.}$$

The reacting force R at the bottom of the post should now be found by a summation of moments about A. That is,

$$[\Sigma M_A = 0] \qquad R(7) = 4{,}500(5) \quad \text{or} \quad R = 3{,}214 \text{ lb} \qquad \textit{Ans.}$$

A check on some of the numerical work may be obtained by a force summation.

$$[\Sigma H = 0] \qquad 1{,}286 + 3{,}214 = 4{,}500 \qquad \textit{Check.}$$

12-2. In the cross section of the water flume shown in Fig. 12-3*a*, the sloping timber posts with their supporting steel rods are spaced 4 ft on centers along the flume. Calculate the tension T in one rod and the V and H components of the reaction forces at the bottom of one post.

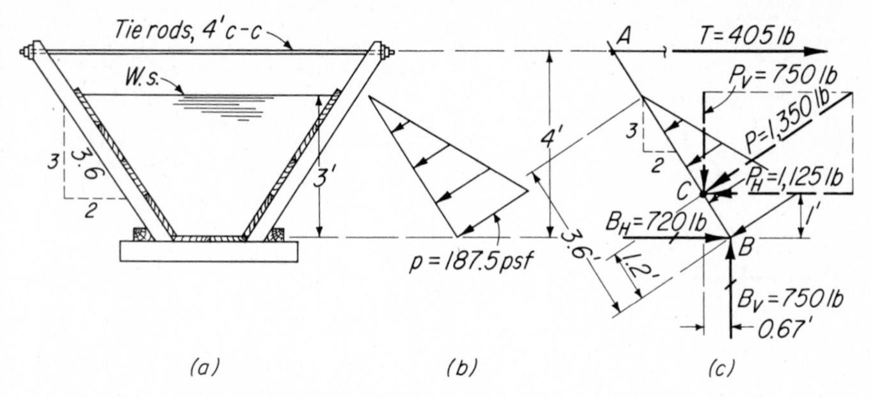

FIG. 12-3 Hydrostatic pressure and load against sloping surface. (*a*) Water flume. (*b*) Pressure diagram. (*c*) Load diagram.

Solution: The hydrostatic pressure p at the bottom of the flume is $3(62.5) = 187.5$ psf, and acts perpendicular to the sloping restraining surface, as shown in Fig. 12-3*b*. The total hydrostatic force P acting against the contributing area A for one post is the average pressure $\bar{p}$ multiplied by the area A on which it acts. Hence,

$$[P = \bar{p}A] \qquad P = \left(\frac{187.5}{2}\right)(4)(3.6) = 1{,}350 \text{ lb}$$

The line of action of this force passes through the centroid of the pressure prism as shown in Fig. 12-3*c* and intersects the wall at a distance of 3.6/3 or 1.2 ft from the bottom. The tensile force T in the rod may be found by a moment summation about B.

$$[\Sigma M_B = 0] \qquad T(4) = 1{,}350(1.2) \quad \text{or} \quad T = 405 \text{ lb} \quad \textit{Ans.}$$

The force P may be resolved into its vertical and horizontal components. By similar triangles

$$\frac{P}{3.6} = \frac{P_V}{2} = \frac{P_H}{3} \quad \text{or} \quad P_V = 750 \text{ lb} \quad \text{and} \quad P_H = 1{,}125 \text{ lb}$$

Summing forces in the vertical and horizontal directions gives

$$[\Sigma V = 0] \qquad B_V - P_V = 0 \quad \text{or} \quad B_V = 750 \text{ lb} \quad \textit{Ans.}$$
$$[\Sigma H = 0] \qquad B_H + 405 - 1{,}125 = 0 \quad \text{or} \quad B_H = 720 \text{ lb} \quad \textit{Ans.}$$

A check may be obtained by a moment summation about C.

$$[\Sigma M_C = 0] \qquad (405)(3) = 720(1) + (750)(0.67)$$

or

$$1{,}215 = 720 + 500 = 1{,}220$$

This checks within one-half of 1 percent. It might be noted that the hypotenuse of the 2:3 triangle is not precisely 3.6.

12-3. An opening in the vertical side of a wood-stave water tank is closed by means of a door hinged at the bottom and held in place by a bolted crossbar A,

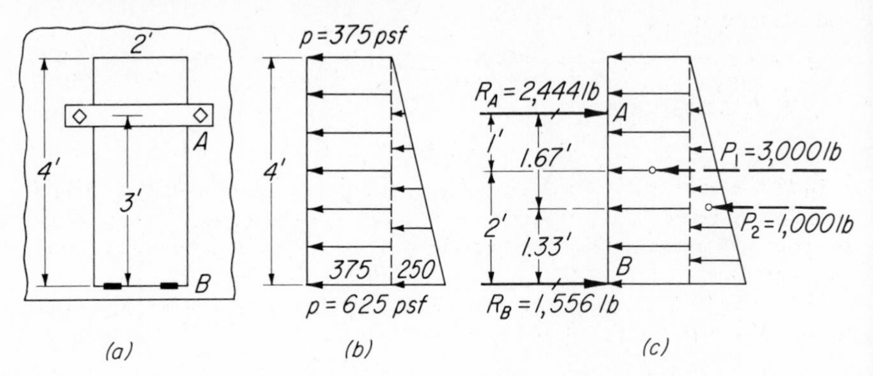

FIG. 12-4 Hydrostatic pressure and load against submerged surface. (*a*) Full view of door. (*b*) Pressure diagram. (*c*) Load diagram.

as illustrated in Fig. 12-4*a*. Calculate the force on each hinge and on each bolt due to the hydrostatic pressure caused by a depth of 10 ft of water in the tank, if the hinges are at the bottom of the tank.

Solution: With 10 ft of water in the tank, the top of the 4-ft-high door is 6 ft below the free water surface, and the pressure at that point is 6(62.5) or 375 psf. At the bottom of the door, the pressure is (62.5)(10) or 625 psf, as shown in the pressure diagram, Fig. 12-4*b*. The two bolts will have equal tensions, and the forces on the two hinges will be alike. An edge view of the door is shown in Fig. 11-4*c*. The total hydrostatic force acting against the door is

$[P = \bar{p}A]$ $$P = \frac{375 + 625}{2}(2)(4) = 4{,}000 \text{ lb}$$

Since the line of action of this force is not self-evident, it will be expedient to break the trapezoidal pressure prism into two prisms, one a rectangular prism representing a uniform pressure of 375 psf against the door and the other a triangular prism under which the pressure varies uniformly from zero at the top of the door to 250 psf at the bottom.

Considering each pressure prism separately,

$[P = \bar{p}A]$ $$P_1 = 375(2)(4) = 3{,}000 \text{ lb}$$

and $$P_2 = \tfrac{1}{2}(250)(2)(4) = 1{,}000 \text{ lb}$$

To find the total reaction force at A, take moments about B

$[\Sigma M_B = 0]$ $$3R_A - 3{,}000(2) - 1{,}000(1.333) = 0 \quad \text{or} \quad R_A = 2{,}444 \text{ lb}$$

Similarly,

$[\Sigma M_A = 0]$ $$3R_B - 3{,}000(1) - 1{,}000(1.67) = 0 \quad \text{or} \quad R_B = 1{,}556 \text{ lb}$$

Then the tension in each bolt is

$$T = \frac{2{,}444}{2} = 1{,}222 \text{ lb} \qquad \textit{Ans.}$$

and the force in each hinge is

$$F = \frac{1{,}556}{2} = 778 \text{ lb} \qquad Ans.$$

PROBLEMS

12-4. The sides of the water flume shown in Fig. 12-5 are supported by steel tie rods through vertical timber posts spaced 6 ft on centers along the flume. Calculate the tension T in one rod and the reaction R at the bottom of one post, due to the water pressure.

12-5. Solve Prob. 12-4 when the depth of water is 6 ft, the distance from the bottom to the rod is 8 ft, and the tie-rod spacing is 8 ft.

Ans. $T = 2{,}250$ lb; $R = 6{,}750$ lb

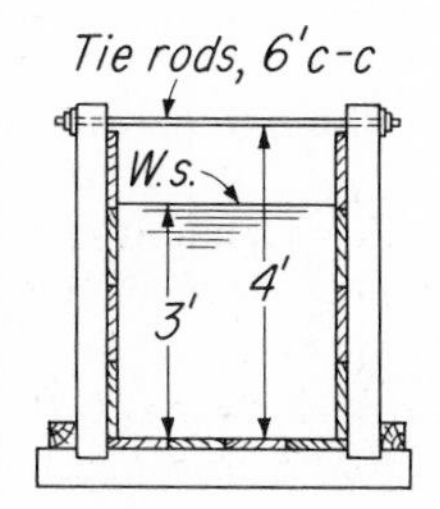

FIG. 12-5 Probs. 12-4 and 12-5.

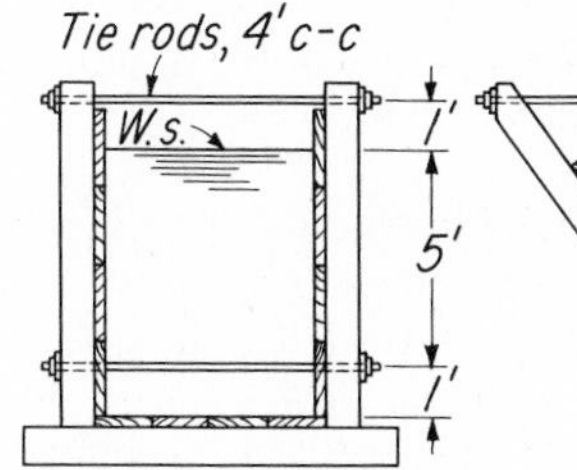

FIG. 12-6 Prob. 12-6.

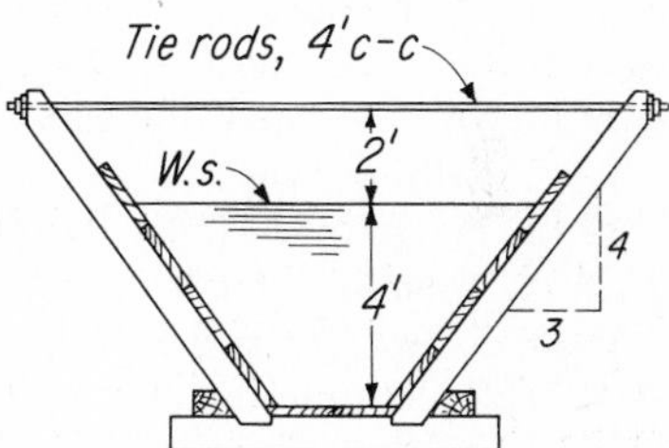

FIG. 12-7 Probs. 12-7 and 12-8.

12-6. Calculate the tensions T_1 and T_2 in the upper and lower tie rods, due to water pressure against one side of the flume shown in Fig. 12-6. Neglect any attachment of the bottom of the post to the horizontal beam.

12-7. Determine the tension T in the tie rod of the flume shown in Fig. 12-7, caused by the water pressure against the sloping side, and the V and H components of the reaction at the bottom B of the post.

Ans. $T = 695$ lb; $B_V = 1{,}500$ lb; $B_H = 1{,}305$ lb

12-8. Solve Prob. 12-7 when the depth of water is 4.5 ft, the distance from the bottom to the rod is 6 ft, the slope of the post is 3/2 (3 units vertical to 2 horizontal), and the tie-rod spacing is 5 ft.

12-9. In Fig. 12-8 is shown a temporary timber flashboard used on a concrete diversion dam to raise the water level. Planks are laid sloping from the bottom to the horizontal beam at the top, which in turn is supported by square timber posts spaced 6 ft on centers along the dam. Calculate the compressive force C in one post when depth of water h is 4 ft and θ is 45°. *Ans.* $C = 1{,}417$ lb

12-10. Solve Prob. 12-9 when the slope of the planks is 4/3 (4 units vertical to 3 horizontal), the spacing of the posts is 8 ft on centers, and h is 6 ft.

12-11. A simple form of self-opening gate, used for automatic control of the water level behind a small timber dam, is shown in Fig. 12-9. The gate is hinged at its upper edge and is kept closed by the weight W on the bracket. The opening in the dam covered by the gate is 2 ft wide and 3 ft high. If W is 1,800 lb, calculate the

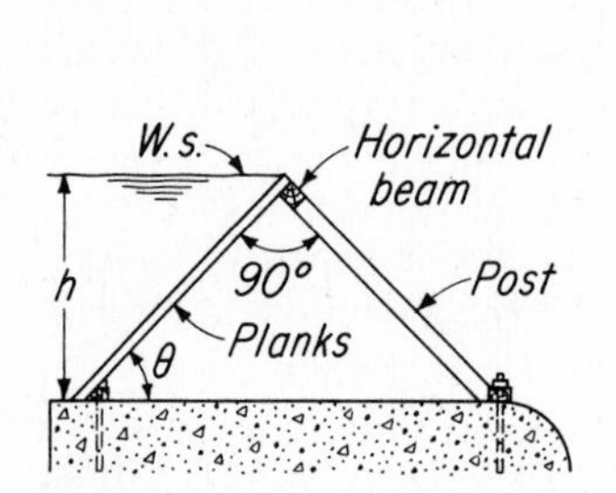

FIG. 12-8 Probs. 12-9 and 12-10.

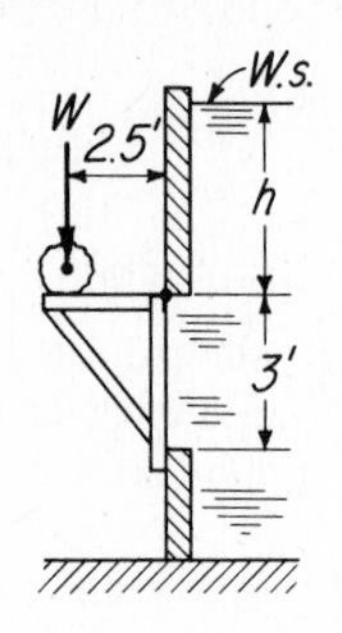

FIG. 12-9 Probs. 12-11 and 12-12.

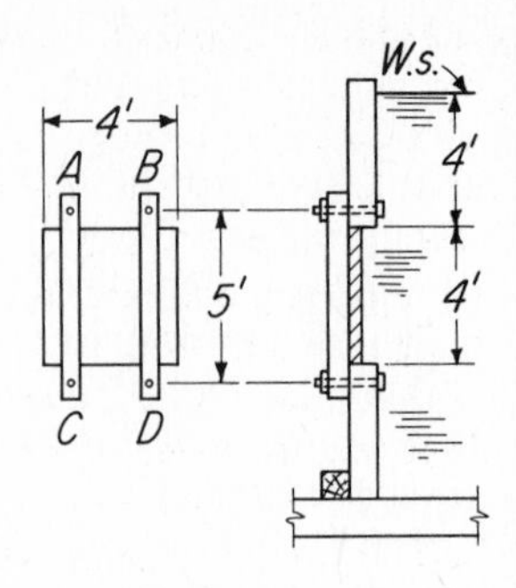

FIG. 12-10 Prob. 12-13.

level h above the top of the gate to which the water may rise before the pressure against it will open it. *Ans.* $h = 6$ ft

12-12. In Prob. 12-11, let the gate be 3 ft wide and 3 ft high, and let W be 2,025 lb. Then solve for h.

12-13. An opening in the side of a tank is temporarily closed as illustrated in Fig. 12-10. Calculate the tensions T produced in each of the four bolts by the water pressure. *Ans.* $T_A = T_B = 1{,}367$ lb; $T_C = T_D = 1{,}633$ lb

12-14. The concrete dam whose cross section is shown in Fig. 12-11 weighs 950 kips per linear foot along the dam. This weight may be considered as concentrated at the center of gravity C of the section. To prevent tensile stresses between the concrete and bedrock, the resisting moment M_R of the weight of the dam about point B at the base must be greater than the overturning moment M_O of the water pressure. (*a*) Calculate these moments for a 1-ft section. (*b*) To what maximum height h can the water behind the dam rise before tensile stresses will be indicated at the upstream face of the dam?

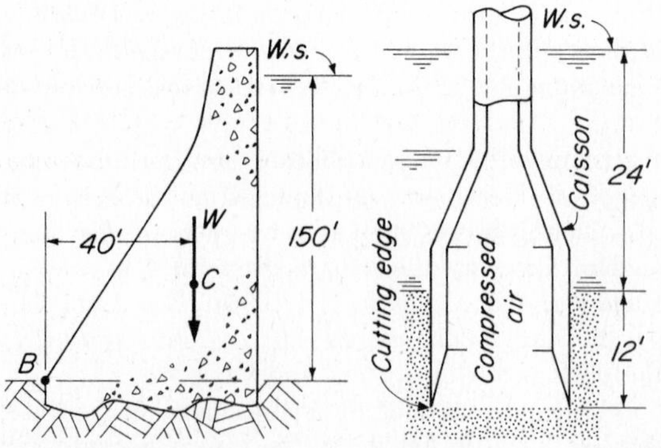

FIG. 12-11 Prob. 12-14.

FIG. 12-12 Probs. 12-15 and 12-16.

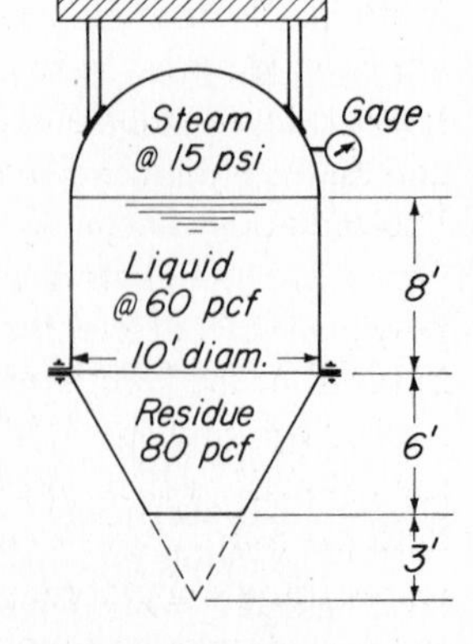

FIG. 12-13 Probs. 12-17 and 12-18.

12-15. Underwater excavation is accomplished by use of a heavily constructed concrete caisson, a cross section of which is shown in Fig. 12-12. By maintaining an air pressure inside the caisson equal to the upward or lateral pressure of the mud at its bottom, the mud is prevented from creeping under the cutting edge. Assuming that the mud acts as a fluid weighing 90 pcf, calculate the minimum air pressure p (in psi) required to maintain this equilibrium when the cutting edge is 36 ft below the free water surface. *Ans.* $p = 17.95$ psi

12-16. Solve Prob. 12-15 when the cutting edge has been lowered a total of 24 ft into the mud.

12-17. The pressure vessel shown in Fig. 12-13 is cylindrical in cross section. The bottom portion collects residue and is in the form of a truncated cone. Calculate the pressure intensity on the bottom. Assume that the residue acts as a true fluid weighing 80 pcf. *Ans.* $p = 21.65$ psi

12-18. Refer to Prob. 12-17. The bottom section is bolted to the cylindrical portion with 96 equally spaced bolts. The empty bottom weighs 2,400 lb. To effect a tight seal, the bolts are to be tensioned an additional 500 lb beyond that required to overcome the loads. Calculate the required total force per bolt.

12-19. A log weighing 40 pcf floats in still water. What percentage of its volume is above the water surface?

12-20. A lightly constructed houseboat is to rest on several 3-ft diam logs, 20 ft long, floating in water. The houseboat is estimated to weigh 11,300 lb, and the logs weigh 42.5 pcf. Let the load required to barely submerge a free-floating log be known as its load capacity. If only two-thirds of the load capacity of the logs may be used, how many are required? *Ans.* 6 logs

12-21. At a large construction job, a light steel pipeline carrying air runs horizontally through a water reservoir. The outside diameter of the pipe is 12 in. and it weighs 6 lb per linear foot. Every 10 ft along its length the pipe is fastened to a concrete-block anchor by means of a short, light steel cable. Calculate the tension T in each cable. *Ans.* $T = 432$ lb

12-5 Stability of Retaining Walls

Materials such as loose sand or gravel, granular soil, or mud, for example, are considered to cause pressures against retaining walls or other restraining surfaces in a manner *similar* to those of true fluids. Since the retained materials may be (1) dry and well compacted producing low lateral pressures, (2) semidry like ordinary soil, or (3) wet and fluid like mud or freshly poured concrete, producing high lateral pressures, the pressures resulting are not always proportional to the actual weights of the materials and are therefore expressed in terms of the weight of an "equivalent fluid" which will produce the given hydrostatic pressure. This equivalent weight w' is called the **fluid weight** of the material and the pressure p' that it will produce is called the **equivalent fluid pressure.** These equivalencies are used to determine the forces acting on retaining walls, sheet piling, caissons, and other similar structures subjected to various types of pressures.

A retaining wall is usually analyzed to determine (1) the factor of safety against sliding, (2) the factor of safety against overturning, and (3) the maximum soil pressure under the foundation. The problems of this chapter will be limited to a study of the resistance of the wall to overturning. The factor of safety against overturning may be defined as the moment of the resisting forces divided by the moment of the overturning forces, both with respect to the axis about which the wall would overturn.

ILLUSTRATIVE PROBLEM

12-22. The retaining wall shown in Fig. 12-14 is of reinforced concrete weighing 150 pcf. Lateral pressure is exerted against the retaining wall by earth having an actual weight of 100 pcf and a *fluid weight* of 25 pcf. (*a*) Determine the factor of safety against overturning of the wall about an axis through the toe *A*. (*b*) Does soil pressure exist over the entire area of the base?

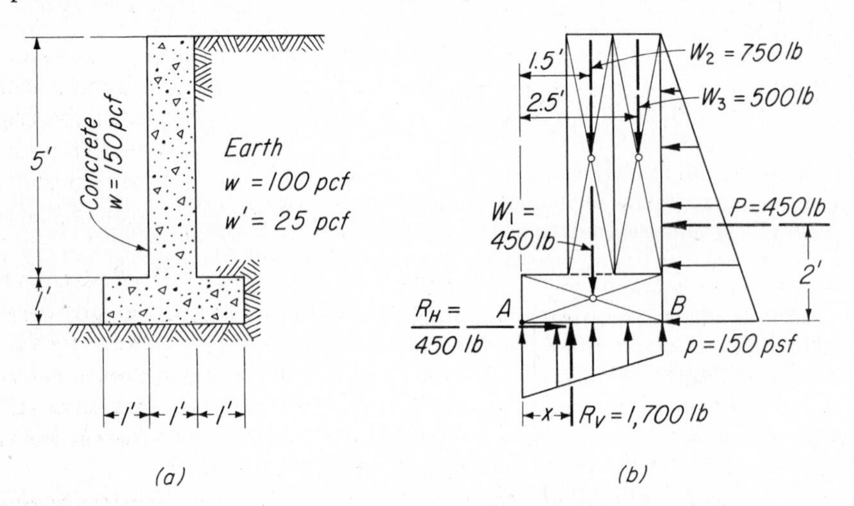

FIG. 12-14 Analysis of retaining wall against overturning. (*a*) Retaining wall. (*b*) Free-body diagram.

Solution: For convenience the retaining wall will be broken into two sections, the base and the wall. Consider a 1-ft length of wall. The weight of the footing is $W_1 = (150)(1)(3) = 450$ lb. The weight of the wall is $W_2 = (150)(1)(5) = 750$ lb. The weight of soil over the footing is $W_3 = (100)(1)(5) = 500$ lb. The lateral earth pressure varies from zero at the top of the wall to $(6)(25) = 150$ psf at the base of the footing. The lateral force P is

$$[P = \bar{p}A] \qquad P = \tfrac{1}{2}(150)(1)(6) = 450 \text{ lb}$$

If the wall overturns, it will do so about point A of Fig. 12-14*b*. The moment of the forces above the base that resist overturning about point A is

$$\text{R.M.} = (1.5)W_1 + (1.5)(W_2) + (2.5)(W_3)$$

or

$$\text{R.M.} = (1.5)(450) + (1.5)(750) + (2.5)(500) = 3{,}050 \text{ lb-ft}$$

The moment of the overturning forces above the base, with respect to A, is

$$\text{O.M.} = (2)(450) = 900 \text{ lb-ft}$$

The factor of safety against overturning is

$$\text{F.S.} = \frac{\text{R.M.}}{\text{O.M.}} = \frac{3{,}050}{900} = 3.39 \qquad \textit{Ans.}$$

The soil pressure under the base will not be uniform. The pressure will be greatest at point A of Fig. 12-14*b* and will be least at point B. If the resultant vertical force caused by this soil pressure falls within the middle one-third of the base, soil pressure will exist over the entire base as indicated in Fig. 12-14*b*. By a force summation in the vertical direction

$$[\Sigma V = 0] \qquad R_V = 450 + 750 + 500 = 1{,}700 \text{ lb}$$

The line of action of this force can be located by a moment summation about A

$$[\Sigma M_A = 0] \qquad (1.5)(450) + (1.5)(750) + (2.5)(500) - 2(450) - (x)(1{,}700) = 0$$

or

$$x = \frac{675 + 1{,}125 + 1{,}250 - 900}{1{,}700} = \frac{2{,}150}{1{,}700} = 1.265 \text{ ft}$$

Since the resultant vertical force R_V falls in the middle one-third of the base, compressive soil pressure will exist over the entire base. *Ans.*

PROBLEMS

12-23. The retaining wall shown in Fig. 12-15 is of concrete weighing 150 pcf. The wall is subjected to a lateral pressure due to soil having an equivalent *fluid weight* of 30 pcf. (*a*) Find the factor of safety against overturning and (*b*) determine if the foundation pressure will extend over the entire base.

Ans. (*a*) F.S. = 1.53; (*b*) the foundation soil pressure will *not* exist over the entire base (compressive pressure will exist on 63.5 percent of the base)

12-24. The retaining wall shown in Fig. 12-16 is constructed of reinforced concrete weighing 150 pcf. It is subjected to a lateral pressure due to soil with an actual weight w of 120 pcf and a *fluid weight* w' of 30 pcf. (*a*) Determine the factor of

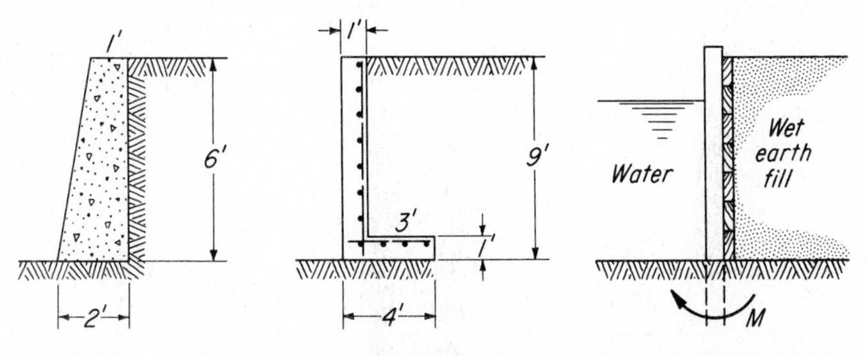

FIG. 12-15 Prob. 12-23. FIG. 12-16 Prob. 12-24. FIG. 12-17 Prob. 12-25.

safety against overturning and (*b*) comment on the pressure distribution on the base.

Ans. F.S. = 2.47; compressive soil pressure will *not* exist over entire base

12-25. A timber retaining wall supporting water on one side and loose wet soil with an equivalent *fluid weight* of 90 pcf on the other is shown in Fig. 12-17; it has planking nailed solidly to deeply driven round timber piles spaced 6 ft on centers. If the depth of water to the solid ground line is 6 ft and the depth of the supported soil is 9 ft, compute the reacting moment M in the pile at the ground line due to the lateral pressures of these materials. *Ans.* M = 52,100 lb-ft

12-6 Flexible Cables; Rigid Arches; Concentrated Vertical Loads

A cable is considered to be perfectly flexible when its resistance to bending is relatively so small that it may be disregarded in practical problems. In such a cable the internal force at any point, due either to its own weight or to externally applied loads, always parallels a tangent to the cable axis at that point. Telephone wires, power wires, and suspension-bridge cables are, for practical purposes, considered to be of this type. A cable supporting traffic lights is a common example of a cable subjected to concentrated forces.

A flexible cable is a nonrigid "structure" whose configuration is determined largely by its length and the loads it supports; when the loads are changed, the configuration of the cable will immediately change until equilibrium is established. An arch, on the other hand, is a rigid structure whose configuration, or shape, is predetermined, depending upon the loads it is to support. However, because the arch is rigid, a change in loads will not change its shape except to a very minute extent. The graphical solution of both types of problems is somewhat similar, since a flexible cable may be regarded as an "upside-down" arch.

When supports are not at the same level or when the loading is "unsymmetrical," as illustrated in Prob. 12-26, both types of problems become somewhat more complex. When supports are at the same level and the loading is "symmetrical," about a vertical center line, as illustrated in Prob. 12-31, the problems are rather simple.

In Fig. 12-18*a* a flexible cable supporting concentrated loads is shown. The supports are at different levels, the loads have different magnitudes, as do the horizontal dimensions between their lines of action. This case, then, would represent the most general type of problem and is the most difficult to solve.

Because a flexible cable is not a rigid body, its final shape under given loads cannot be predetermined, as in the case of the rigid arch. However, we do know beforehand the locations of the end supports A and E (dimensions L and y), and usually we also know the horizontal dimensions a, b, c, and d. This leaves four unknown quantities to be determined: the vertical and horizontal reaction components at A and E. The end reactions at either end may be found if we know the cable slope at that end. For example, if we know v (i.e., the location of point b) the reaction components at A may be found.

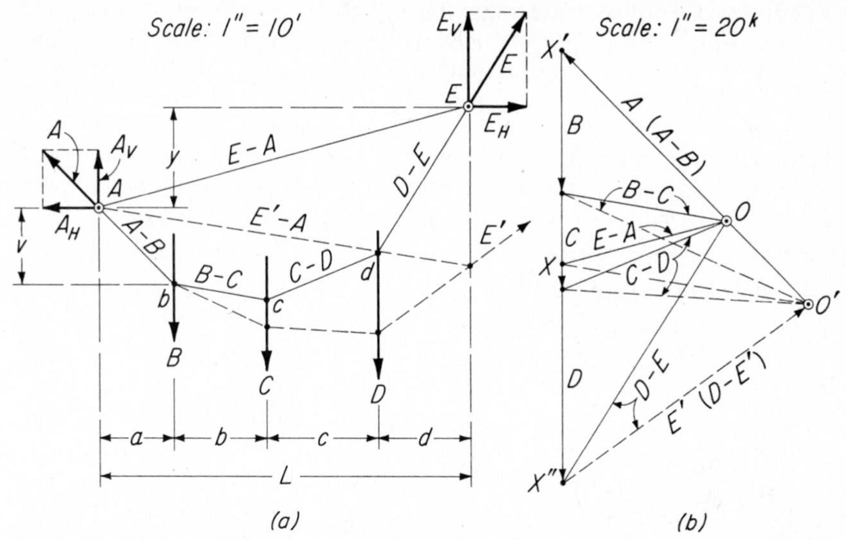

FIG. 12-18 Flexible cable, concentrated loads, graphical solution. (*a*) Space diagram. (*b*) Force diagram. Force polygon.

The unknown reaction components A_V and A_H may be determined analytically by two moment summations, one about E and the other about b, both involving A_V and A_H, which may then be solved simultaneously. A force summation of the entire cable system in both the vertical and horizontal directions will now yield E_V and E_H. A step-by-step graphical solution is outlined below.

PROCEDURE FOR GRAPHICAL SOLUTION

(A study of Prob. 5-17 is recommended.)

1. In space diagram, from given dimensions, locate supports *A* and *E*, point *b*, and the vertical action lines of loads *B*, *C*, and *D*. Then draw known string *AB*.
2. In force diagram, lay off known loads *B*, *C*, and *D*. Draw ray *AB* (whose direction is known), and select trial pole O' at random but lying on ray *AB*. Next draw trial rays (dashed) *BC*, *CD*, and DE'.
3. In space diagram, draw trial strings (dashed) *BC*, *CD*, and DE', parallel respectively to the corresponding trial rays. Then draw trial closing string $E'A$ (dashed). Also draw final closing string *EA*, which must pass through supports *E* and *A*.
4. In force diagram, draw ray $E'A$ (dashed) parallel to trial closing string $E'A$, thus locating point *X*. From *X* draw final ray *EA* parallel to final string *EA*, thus locating final pole *O*. Then draw final rays *BC*, *CD*, and *DE* (solid lines).

5. The stresses in cable segments *AB*, *BC*, *CD*, and *DE* are now found by scaling the corresponding rays. Note that the scaled value of ray *AB* gives the stress in cable segment *AB* as well as the reaction *A* and that ray *DE* gives the stress in *DE* as well as reaction *E*.

NOTE: The following facts should be observed: (*a*) The configuration of the cable can be controlled by the arbitrary location of point *b*. (*b*) When dimension v is least, the cable stresses are greatest; when v is greatest, the stresses are least. (*c*) The scaled values XX' and XX'' would be, respectively, the vertical reactions at supports *A* and *E*, if *AE* were a compression strut and if the resulting "frame" (strut plus cable) were then hinged at *A* and supported by rollers on a horizontal surface at *E* (in which case the reactions at those supports would be vertical).

ILLUSTRATIVE PROBLEM

12-26. In Fig. 12-18*a* let dimensions $L = 20$ ft, $y = 5$ ft, $v = 4$ ft, $a = 4$ ft, $b = 5$ ft, $c = 6$ ft, and $d = 5$ ft; and let loads $B = 15$ kips, $C = 10$ kips, and $D = 20$ kips. Solve (*a*) graphically for the four stresses in cable segments *AB*, *BC*, *CD*, and *DE* (scales on $8\frac{1}{2}$ by 11: 1 in. = 5 ft, 1 in. = 10 kips); and (*b*) analytically for the support reactions *A* and *E* which equal, respectively, cable stresses *AB* and *DE*. (NOTE: A complete graphical solution by students is recommended, as well as the analytical solution to determine cable stresses *BC* and *CD* which is called for in Prob. 12-29.)

a. Graphical solution: When the above procedure is followed, the following values may be scaled from the force diagram, Fig. 12-18*b*:

Reaction A = stress AB = 25.4 kips Reaction E = stress DE = 32.5 kips
Stress BC = 18.2 kips Stress CD = 19.3 kips

b. Analytical solution: Clearly, reaction *A* is parallel to cable segment *AB*. Therefore, a relation between A_V and A_H may be established either by a moment equation about point *b* or by proportion. The latter relation is by proportion

$$\frac{A_V}{4} = \frac{A_H}{4} \qquad \text{or} \qquad A_V = A_H$$

Then, by moments about *E*, we obtain

$$[\Sigma M_E = 0] \qquad 20A_V + 5A_V - 20(5) - 10(11) - 15(16) = 0$$

$$\text{or} \qquad 25A_V = 100 + 110 + 240 = 450 \qquad \text{and} \qquad A_V = A_H = 18 \text{ kips}$$

$$\text{from which} \quad A = \sqrt{(18)^2 + (18)^2} = \sqrt{648} \qquad \text{or} \qquad A = 25.4 \text{ kips} \qquad \textit{Ans.}$$

Then, to obtain E_V, E_H, and E, we have

$$[\Sigma V = 0] \qquad E_V + 18 - 15 - 10 - 20 = 0 \qquad \text{or} \qquad E_V = 27 \text{ kips}$$

$$[\Sigma H = 0] \qquad E_H = 18 \text{ kips}$$

and

$$E = \sqrt{(27)^2 + (18)^2} = \sqrt{729 + 324} = \sqrt{1{,}053} \qquad \text{or} \qquad E = 32.5 \text{ kips} \qquad \textit{Ans.}$$

PROBLEMS

12-27. Solve Prob. 12-26 when $L = 33$ ft, $y = 14$ ft, $v = 6$ ft, $a = 8$ ft, $b = 12$ ft, $c = 7$ ft, $d = 6$ ft, and loads $B = 42$ kips, $C = 21$ kips, and $D = 12$ kips. (Scales on 8½ by 11: 1 in. = 10 ft, 1 in. = 20 kips.)

12-28. Solve Prob. 12-26 when $L = 20$ ft, $y = 5$ ft, $v = 2$ ft, $a = 4$ ft, $b = 5$ ft, $c = 6$ ft, $d = 5$ ft, and loads $B = 15$ kips, $C = 20$ kips, and $D = 10$ kips. (Scales on 8½ by 11: 1 in. = 4 ft, 1 in. = 10 kips.)

Partial Ans. $A = 38$ kips, $E = 44$ kips

12-29. Complete the analytical solution of Prob. 12-26 by solving for the stresses in cable segments *BC* and *CD*. Calculate also the vertical distances y_c and y_d from

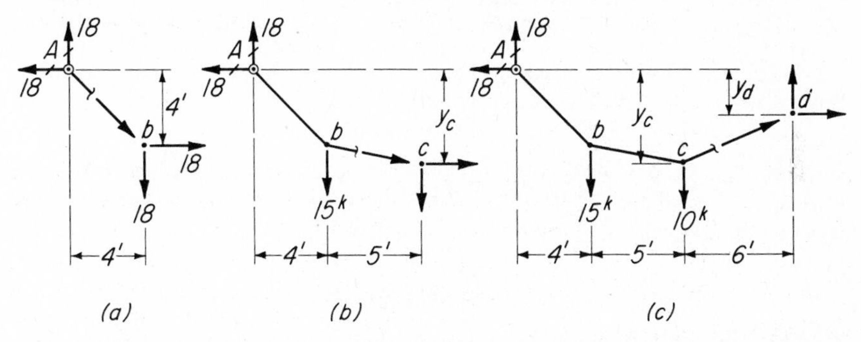

FIG. 12-19 Flexible cable, concentrated loads, analytical solution. (*a*) Free-body diagram. Segment *AB*. (*b*) Free-body diagram. Segment *ABC*. (*c*) Free-body diagram. Segment *ABCD*.

support *A* to point *c* and *d*. Check all results against the graphical solution. (Use free-body diagrams as shown in Fig. 12-19.)

Ans. $BC = 18.2$ kips, $y_c = 4.83$ ft; $CD = 19.3$ kips; $y_d = 2.5$ ft

12-30. Solve analytically for cable stresses *BC* and *CD* in Prob. 12-28, and for the vertical distances y_c and y_d from support *E* to points *c* and *d*.

ILLUSTRATIVE PROBLEM

12-31. In Fig. 12-20 is shown a typical tied, sectional roof-arch rib supporting loads *B*, *C*, and *D*. (*a*) Determine graphically the rise *r* and the stresses in sections *AB*, *BC*, *CD*, and *DE*, when $B = C = D = 5$ kips, $a = 10$ ft, and $\theta = 30°$. (*b*) Determine the shape of the arch and the stresses in all sections when $r = 6$ ft. (NOTE: It is recommended that students complete the solution as explained below. Scales on 8½ by 11: 1 in. = 10 ft, 1 in. = 5 kips.)

Solution: (*a*) The string and force polygons for this solution are shown in solid lines in Fig. 12-20. Note that for an upward-curving arch the pole must be located at the left of the load line.

(*b*) When the rise *r* is given a fixed value, an indirect graphical solution can be made by several trial locations of the pole *O*′. A simpler and direct analytical-graphical solution is to make a vertical cut through section *BC* of the rib and horizontal

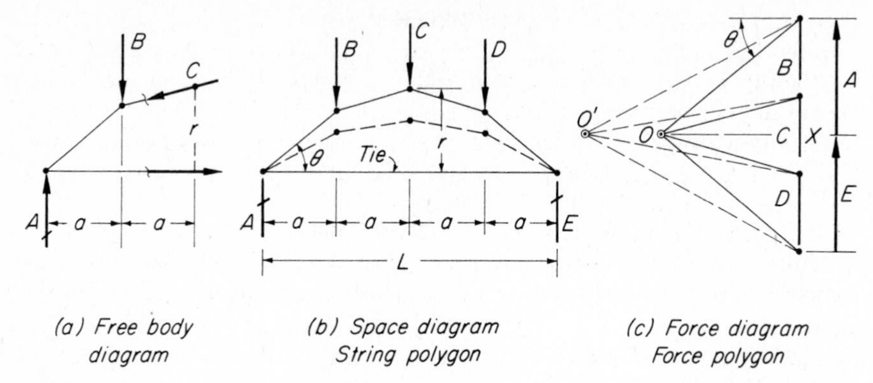

FIG. 12-20 Tied, sectional roof-arch rib. (*a*) Free-body diagram. (*b*) Space diagram. String polygon. (*c*) Force diagram. Force polygon.

tie, as shown in Fig. 12-20*a*, and to calculate the stress in the tie by moments about point *C*. When this stress is then laid off in the force polygon (line *O'X*), both string and force polygons (dash lines) are readily completed.

PROBLEMS

12-32. The cable shown in Fig. 12-21 supports four equal loads as shown. The cable is to be inclined at an angle of 45° to the horizontal at the supports. Determine graphically the maximum cable sag *s* and the maximum cable tension *T* if $W = 2$ kips. *Ans.* $s = 7.5$ ft; $T = 5.66$ kips

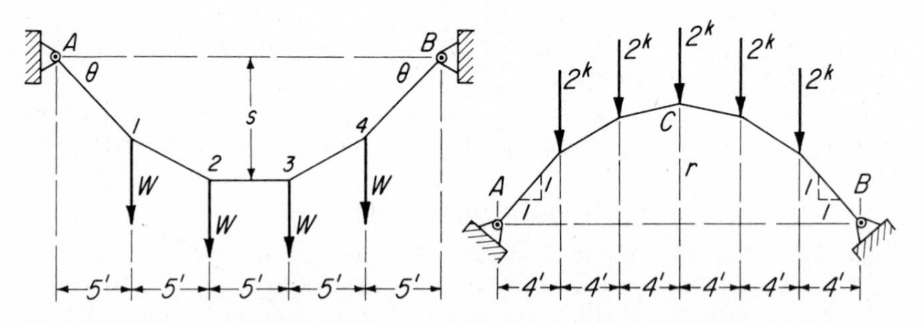

FIG. 12-21 Probs. 12-32 to 12-34. FIG. 12-22 Probs. 12-35 and 12-36.

12-33. The weights *W* in Fig. 12-21 are 50 lb each. What is the sag *s* if the horizontal pull at the supports is limited to 120 lb? *Ans.* $s = 15$ ft

12-34. The cable shown in Fig. 12-21 supports four equal loads. Determine graphically the allowable cable sag *s* if $W = 800$ lb and the maximum cable tension *T* is limited to 4,000 lb. Check solution analytically.

12-35. An arch is to support the five 2-kip loads, as shown in Fig. 12-22. Determine graphically the rise *r* of the arch and the reactions at *A* and *B* if the arch is constructed along the load line *ACB*. The load line is to have a 1 to 1 slope at supports. Check analytically. *Ans.* $r = 7.2$ ft; $R_A = R_B = 7.07$ kips

12-36. An arch is constructed along the load line *ACB*, shown in Fig. 12-22. Determine the reactions *A* and *B* and the compressive force *C* in the segment adjacent to the crown point *C*, if the load line makes an angle of 30° with the horizontal at *A* and *B*. Check analytically.

12-7 Flexible Cables, Horizontally Uniform Loads

A common problem is that of determining the maximum internal force *T* in a cable, due either to its own weight or to an externally applied load uniformly and horizontally distributed between the ends. A uniform horizontal distribution of load is generally assumed in telephone wires, and in power and trolley wires in which the ends are lying at, or near, the same level and in which the maximum sag does not exceed about 10 per cent of the distance between supports. The weight of the roadway of a suspension bridge is another example of uniform horizontal distribution of load on the supporting cables. If the suspension cable weighs little in comparison with the roadway it supports, its weight may be disregarded, and the distribution of loads is then horizontal, regardless of the amount of cable sag. The shape assumed by such cables is parabolic. A cable under only its own weight or a load that is uniformly distributed along the axis of the cable will assume the shape of a catenary. Only the case of uniform horizontal loading will be considered here.

Consider the flexible cable shown in Fig. 12-23*a*. Let *L* be the horizontal distance (span) between supports *A* and *B*. When these supports are at the same elevation, as in this case, the maximum vertical sag *s* of the cable will occur at the center *C*. At this point *C* of maximum sag, a tangent to the axis of the cable will always be horizontal, and so will the stress *H* within the cable.

If the uniform horizontal load is *w* lb per ft, the total resultant load on section *CD* is *wx*, which is considered to be concentrated at the centroid of

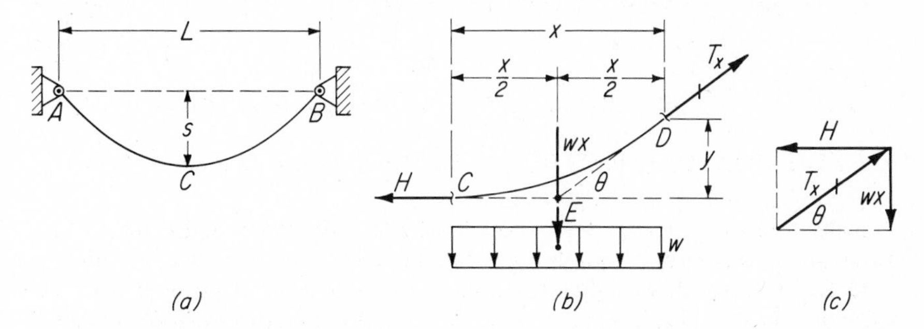

FIG. 12-23 Flexible cable, horizontally uniform load. Supports at same level. (*a*) Flexible cable. Supports at same level. (*b*) Free-body diagram of portion of loaded cable. (*c*) Force parallelogram.

the load area. The tension T_X in the cable at any point D is always directed parallel to the cable axis at D, as shown in Fig. 12-23*b*.

Now, since only three forces, H, wx, and T_X act on section CD of the cable, they must be in equilibrium and also concurrent (see Art. 2-15). Forces H and wx, whose action lines are known, intersect at point E, through which point the action line of T_X then also must pass.

The magnitude of stress H for any given section of cable CD is established by a summation of moments about D. That is,

$$[\Sigma M_D = 0] \qquad Hy - wx\left(\frac{x}{2}\right) = 0 \qquad \text{or} \qquad H = \frac{wx^2}{2y}$$

From this equation, we note that the maximum value of H occurs when $x = L/2$ and $y = s$. Then

$$\mathbf{H = \frac{wL^2}{8s}} \tag{12-3}$$

From Fig. 12-23*c* we note that T_X is equal and opposite to the resultant of H and wx. Hence, the cable tension at any horizontal distance x from the center C is

$$T_X = \sqrt{H^2 + (wx)^2}$$

From the above equation, we note than when $x = 0$ (at the center of the span) $T_X = H$, and that as the value of x increases, so does the value of stress T_X, which becomes greatest at the supports A and B.

The maximum value of T_X then also occurs when $x = L/2$ and $y = s$. That is,

$$T_{max} = \sqrt{H^2 + \left(\frac{wL}{2}\right)^2} = \sqrt{\frac{w^2L^4}{64s^2} + \frac{w^2L^2}{4}}$$

or

$$\mathbf{T_{max} = \frac{wL}{2}\sqrt{\frac{L^2}{16s^2} + 1}} \tag{12-4}$$

From Fig. 12-23*c*, the angle θ of the cable at any point is given by

$$\tan\theta = \frac{wx}{H}$$

At the supports $x = L/2$. Therefore

$$\mathbf{\tan\theta_{max}} = \frac{wL}{2H} = \frac{wL}{2(wL^2/8s)} = \mathbf{\frac{4s}{L}} \qquad \text{or} \qquad \mathbf{\theta = \tan^{-1}\frac{4s}{L}} \tag{12-5}$$

ILLUSTRATIVE PROBLEM

12-37. A power wire weighing 2 lb per ft is stretched between two supports at equal elevations, 100 ft apart. If the sag at the center is 10 ft, compute the tension H at the center and the maximum tension T at one support.

Solution: From Eq. (12-3)

$$\left[H = \frac{wL^2}{8s}\right] \qquad H = \frac{2(100)^2}{8(10)} = 250 \text{ lb} \qquad \textit{Ans.}$$

From Eq. (12-4) the maximum tension is

$$\left[T_{\max} = \frac{wL}{2}\sqrt{\frac{L^2}{16s^2} + 1}\right] \qquad T_{\max} = \frac{2(100)}{2}\sqrt{\frac{(100)^2}{16(10)^2} + 1} = 269 \text{ lb} \qquad \textit{Ans.}$$

PROBLEMS

12-38. Compute the maximum tension in a flexible wire weighing 0.3 lb per ft stretched between two supports 160 ft apart and at equal levels if the sag is 16 ft.

12-39. Determine the sag in the wire of Prob. 12-38 if it is stretched until the tension reaches the maximum allowable value of 480 lb. *Ans.* $s = 2.01$ ft

12-40. In order to control the sag in a radio aerial wire, one end A is fastened to the top of a tower, as shown in Fig. 12-24, while the other passes over a small sheave B and supports a weight W. If the wire weighs 0.05 lb per ft, and if A and B are 80 ft apart and at the same level, compute the weight W required to keep the sag at 2 ft. Neglect the diameter of the sheave B and the weight of the wire between B and W.

12-41. If the maximum allowable tension in the aerial wire of Prob. 12-40 is 120 lb, compute the minimum permissible sag.

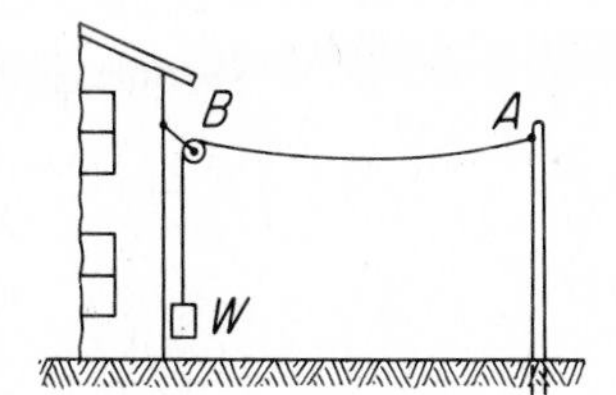

FIG. 12-24 Probs. 12-40 and 12-41.

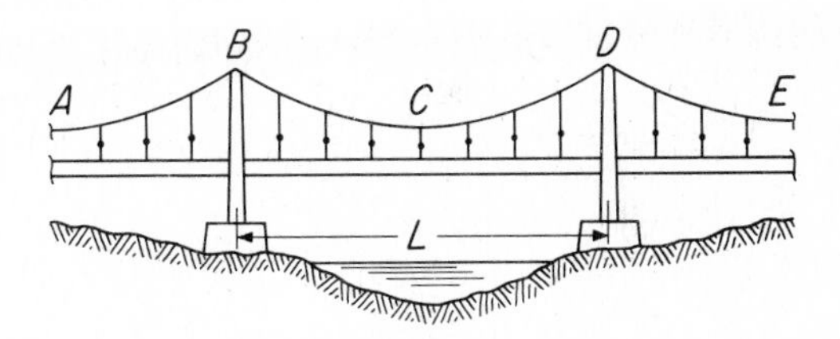

FIG. 12-25 Probs. 12-42 and 12-56.

12-42. A footbridge is supported by two like suspension cables, shown in Fig. 12-25. The tops of the towers are at the same level and are 300 ft apart. The weight of the bridge on each cable is 200 lb per horizontal foot, and the sag at the center is 50 ft. (The weight of the cable would probably be less than 4 lb per ft and may, therefore, be neglected.) Determine graphically the tensions in the middle, C, and at the ends, B and D, of the cable, and the angle θ the cable makes with the horizontal at the tower. Check the results analytically using the equations of Art. 12-7.

When the end supports of flexible wires and cables are *not at the same level,* which generally is the case with communication and power wires in rolling or rough terrain, the point of greatest sag will *not* be midway between supports, but can be located as shown in Prob. 12-43. The maximum stress in the cable will then occur at the higher of the two supports. When the difference in elevation of the end supports does not exceed about 20 percent of the horizontal span, and the maximum sag does not exceed about 10 percent of this span, measured vertically from a straight line between the end supports, the weight of the cable may, without appreciable error, be assumed to be a horizontally uniformly distributed load. The problem may then be solved as follows.

ILLUSTRATIVE PROBLEM

12-43. A power cable spans a river and is supported at towers A and B, as shown in Fig. 12-26. Support A is 120 ft below B, and the lowest point of the cable is $s' = 30$ ft below A. The horizontal span L is 600 ft. During winters, when coated with ice, the cable weighs 10 lb per ft. Solve for the distance x from A to the point of lowest sag C, and for cable stresses H, T_A, and T_B under these conditions.

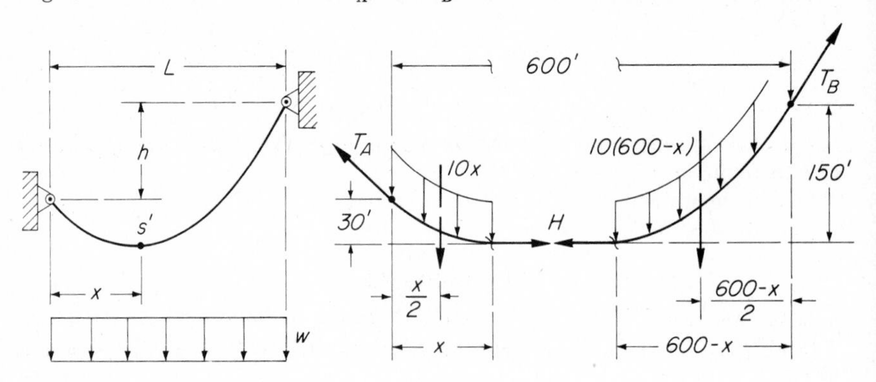

FIG. 12-26 Flexible cable, horizontally uniform load. Supports at different levels. (*a*) Flexible cable, horizontally uniform load. (*b*) Free-body diagrams.

Solution: The cable is cut at the point of maximum sag C, and free-body diagrams of both portions are drawn, as in Fig. 12-26*b*. The load on the left portion is clearly $10x$, and on the right portion is $10(600 - x)$, both concentrated at the mid-points (horizontally) of the respective sections. To solve for x, two equations of moment equilibrium may now be written, one about A and the other about B, thus:

$$[\Sigma M_A = 0] \qquad 30H = 10x\left(\frac{x}{2}\right) = 5x^2 \quad \text{or} \quad H = \frac{x^2}{6} \tag{a}$$

$$[\Sigma M_B = 0] \qquad 150H = (6{,}000 - 10x)\left(300 - \frac{x}{2}\right)$$

$$\text{or} \qquad 150H = 5x^2 - 6{,}000x + 1{,}800{,}000 \tag{b}$$

Substituting Eq. (*a*) for H in Eq. (*b*), we obtain

$$150\left(\frac{x^2}{6}\right) = 25x^2 = 5x^2 - 6{,}000x + 1{,}800{,}000$$

from which $20x^2 + 6{,}000x = 1{,}800{,}000$ or $x^2 + 300x = 90{,}000$

Completing the square gives

$$x^2 + 300x + (150)^2 = 90{,}000 + (150)^2 = 112{,}500$$

or $x + 150 = \pm\sqrt{112{,}500} = 335$ and $x = 185$ ft *Ans.*

Substituting this value of x in Eq. (a), we obtain

$$H = \frac{x^2}{6} = \frac{(185)^2}{6} = \frac{34{,}225}{6} = 5{,}704 \text{ lb} \qquad \textit{Ans.}$$

The vertical component of T_A is clearly $10x = 10(185) = 1{,}850$ lb, and the vertical component of T_B is $10(600 - 185) = 4{,}150$ lb. The horizontal component of both T_A and T_B is $H = 5{,}704$ lb. Hence,

$$T_A = \sqrt{(1{,}850)^2 + (5{,}704)^2} = 6{,}000 \text{ lb} \qquad \textit{Ans.}$$

and

$$T_B = \sqrt{(4{,}150)^2 + (5{,}704)^2} = 7{,}054 \text{ lb} \qquad \textit{Ans.}$$

If desired, the slopes of the cable at A and B are easily determined, since $\cos \theta_A = H/T_A$ and $\cos \theta_B = H/T_B$.

PROBLEMS

12-44. Solve Prob. 12-43 when $L = 400$ ft, $h = 80$ ft, and $s' = 20$ ft (see Fig. 12-26).

12-45. Solve Prob. 12-43 (see Fig. 12-26 when $L = 300$ ft, $h = 60$ ft, and $s' = 10$ ft.

Still another type of flexible-cable problem is that of the cableway, often used to transport people and/or materials short distances over irregular terrain, or to transport material such as freshly mixed concrete from the mixing plant to the point of placement in a large concrete dam. For large spans, and where supports are not at the same level, such problems become rather complex, and will not be dealt with here. However, for spans not exceeding a few hundred feet where supports are at, or near, the same level, and the carriage load does not exceed a few thousand pounds, the approximate method shown in Prob. 12-46 will give results sufficiently accurate for practical purposes. The maximum cable tension occurs at the supports when the carriage is at the mid-point. In this type of cableway, however, it is customary to use a cable that is continuous between the points of anchorage (D and E in Fig. 12-27) over the supporting towers to which the cable is completely fastened to prevent slipping. Under these conditions, the greatest stress occurs in the backstay sections of the cable, being greatest there when dimension d is comparatively small, and decreasing as d increases.

ILLUSTRATIVE PROBLEM

12-46. At a forest camp a cableway spans a river, as shown in Fig. 12-27. The tower supports A and B are at the same elevation. The horizontal span L is 200 ft; the weight w of the cable is 4 lb per ft; the height h of the towers is 20 ft; and the weight W of the fully loaded carriage is 800 lb. For convenient operation, the maximum sag s' of the cable must exceed 6 ft when the loaded carriage is at mid-span. (*a*) Determine the maximum tension T in section AB of the cable under these conditions. (*b*) If the cable is continuous from D to E, calculate the tension in the backstay portion BE, and the vertical compression in tower B, when θ is 45° (disregard weight of cable in backstay). (*c*) If the maximum allowable safe tension in the cable is 12,000 lb, calculate the required horizontal dimension d from base of tower B to anchorage E.

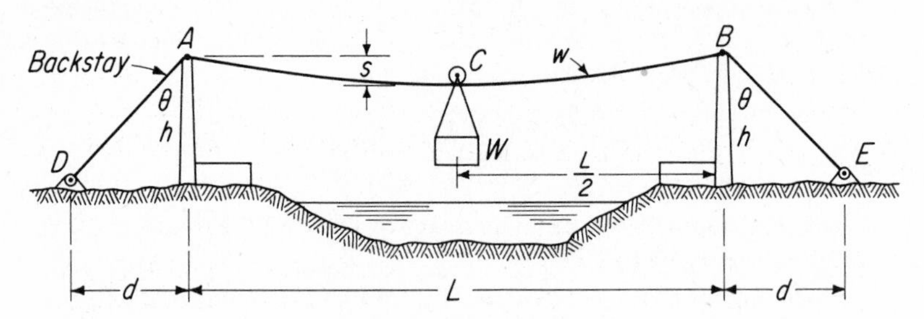

FIG. 12-27 Simple cableway supporting moving concentrated load.

Solution: (*a*) The cable is cut at the midpoint C and a free-body diagram of either half is drawn, as shown in Fig. 12-28*a*. The cable weight of 400 lb is applied at the mid-point of the right half as shown, and one-half of the carriage weight, equaling

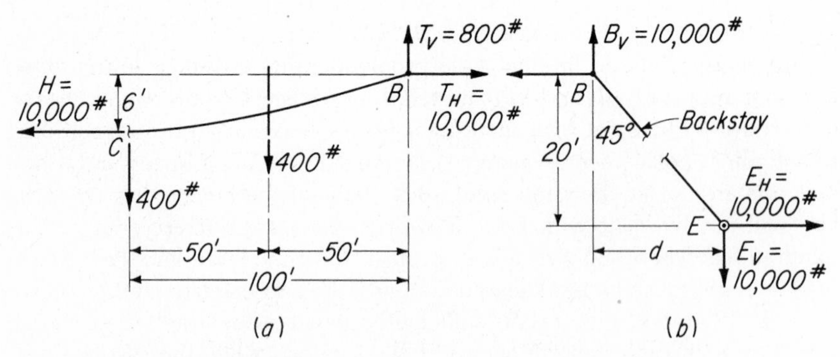

FIG. 12-28 Free-body diagrams. (*a*) Right half of cable AB. (*b*) Right backstay BE.

400 lb, is applied at the mid-point C of the entire cable, the other half of the carriage weight being supported by the left half of the cable not shown. Clearly, then,

$$[\Sigma V = 0] \qquad T_V = 400 + 400 = 800 \text{ lb}$$

A moment equation about C gives T_H as follows:

$[\Sigma M_C = 0]$ $\qquad 6T_H + 400(50) = 800(100)$

from which $6T_H = 80{,}000 - 20{,}000 = 60{,}000$ and $T_H = 10{,}000$ lb

Then, $\qquad T_B = \sqrt{(800)^2 + (10{,}000)^2} = 10{,}032$ lb $\qquad$ *Ans.*

(*b*) The free-body diagram of the backstay portion *BE* of the cable is shown in Fig. 12-28*b*. Clearly,

$[\Sigma H = 0]$ $\qquad T_H = E_H = 10{,}000$ lb

and, when $\theta = 45°$, $\qquad E_H = E_V = 10{,}000$ lb

Hence, $BE = \sqrt{(10{,}000)^2 + (10{,}000)^2} = \sqrt{200{,}000{,}000} = 14{,}140$ *lb* $\qquad$ *Ans.*

From Fig. 12-28 it is clear that the vertical compressive stress in the tower is

$$T_V + B_V = 800 + 10{,}000 = 10{,}800 \text{ lb} \qquad \textit{Ans.}$$

(*c*) The stress in *BE* exceeds the allowable 12,000 lb, but can be reduced to 12,000 lb by moving anchorage *E* to the right, thereby reducing the vertical component of *BE* while the horizontal component remains constant. The calculations may be as follows:

The square of the allowable stress = $(12{,}000)^2 = 144{,}000{,}000$
The square of the component E_H = $(10{,}000)^2 = 100{,}000{,}000$
The square of the component E_V = $(E_V)^2 = 44{,}000{,}000$

from which $\qquad E_V = \sqrt{44{,}000{,}000} = 6{,}630$ lb

Then, by proportion $\qquad \dfrac{E_V}{E_H} = \dfrac{6{,}630}{10{,}000} = \dfrac{20}{d}$

or $\qquad 6{,}630d = 200{,}000$ and $d = 30.2$ ft $\qquad$ *Ans.*

PROBLEMS

12-47. Solve parts (*a*), (*b*), and (*c*) of Prob. 12-46 when $L = 400$ ft, $w = 8$ lb per ft, $h = 30$ ft, $\theta = 60$, $W = 2{,}000$ lb, $s' = 10$ ft, and the maximum allowable tension in the cable is 38,000 lb. *Ans.* $T_B = 34{,}000$ lb; $BE = 39{,}400$ lb; $d = 60$ ft

12-48. Solve parts (*a*), (*b*), and (*c*) of Prob. 12-46 when $L = 300$ ft, $w = 6$ lb per ft, $h = 50$ ft, $\theta = 45°$, $W = 600$ lb, $s' = 25$ ft, and the maximum allowable tension in the cable is 5,200 lb.

SUMMARY

(*By article number*)

12-2. The intensity of **hydrostatic pressure** at any point in a fluid is *wh*. This pressure is equal in all directions. *h* is the vertical distance below the *free surface*.

12-3. The **buoyancy** of a body is equal to the weight of the fluid displaced.

12-4. The *hydrostatic force* acting on a submerged plane area is equal to the product of the average pressure against the plane and the area of the submerged plane.

12-5. Lateral pressures caused by loose earth, sand, or mud against retaining walls and surfaces are similar to those caused by fluids, and are expressed in terms of the weight of an "equivalent fluid" which would produce these pressures, called the **equivalent fluid weight.**

12-6. The stress in a flexible cable or in an arch designed to be free of bending stresses under load is always parallel to the axis of the cable or arch. The stresses can be found graphically by drawing a force polygon.

12-7. A flexible cable under uniform horizontal load will hang as a parabola. The maximum cable tension is given by

$$T_{max} = \frac{wL}{2}\sqrt{\frac{L^2}{16s^2} + 1} \tag{12-4}$$

and the angle of the cable at the support is given by

$$\theta = \tan^{-1}\left(\frac{4s}{L}\right) \tag{12-5}$$

REVIEW PROBLEMS

12-49. Figure 12-29 shows a partial section of a timber flume subjected to water pressure. The planking is supported by sloping posts *A* spaced 4 ft on centers which are held in place partly by the struts *B*. Compute the total *hydrostatic load P* against one post, and the compressive force *C* in one strut.

Ans. $P = 10{,}000$ lb; $C = 6{,}670$ lb

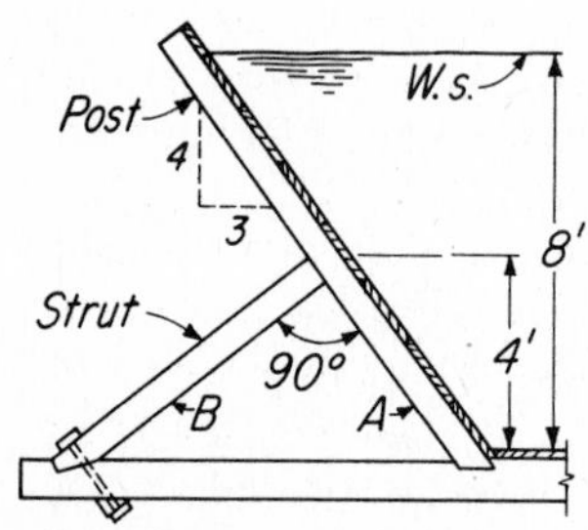

FIG. 12-29 Prob. 12-49.

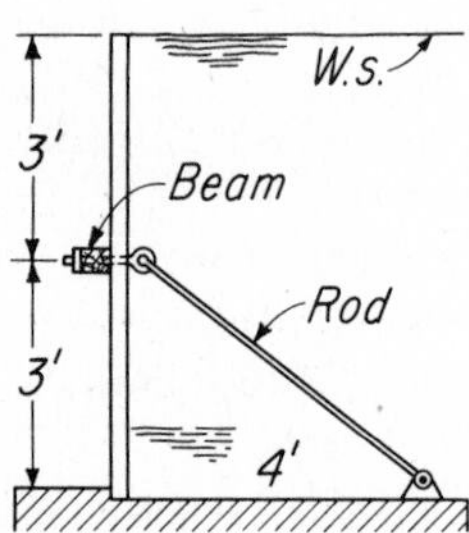

FIG. 12-30 Prob. 12-50.

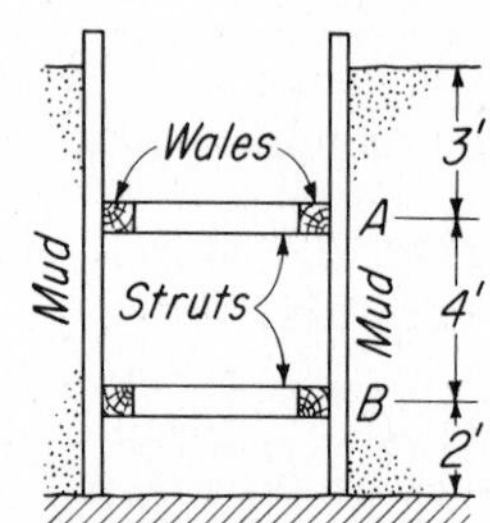

FIG. 12-31 Prob. 12-51.

12-50. The vertical planking of the timber flashboard shown in Fig. 12-30 is supported partly by a horizontal beam at mid-height bolted to inclined steel rods. If the rods are spaced 4 ft on centers, compute the tension *T* in one rod.

12-51. The trench shoring shown in Fig. 12-31 supports mud with an *equivalent fluid weight* of 80 pcf. Horizontal struts are wedged between the wales and are spaced 3 ft on centers along the trench. Compute the compressive stresses, in each of struts *A* and *B*, due to the pressure of the mud. *Ans.* $A = 2{,}430$ lb; $B = 7{,}290$ lb

12-52. A temporary timber cofferdam will be subjected to the mud and water loads shown in Fig. 12-32. Assuming that the water weighs 62.5 pcf, the mud 120 pcf, and that the fluid weight of the mud is 120 pcf, determine the compressive force in the timber struts behind the dam. The struts are spaced at 6-ft centers. Neglect the weight of the timber structure in these calculations.

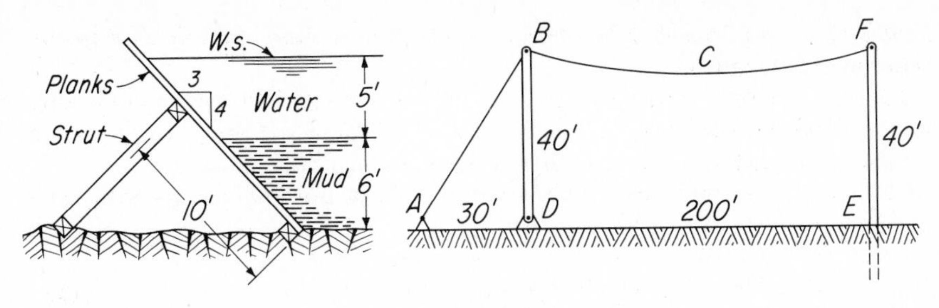

FIG. 12-32 Prob. 12-52.

FIG. 12-33 Prob. 12-54.

12-53. A pontoon bridge uses welded steel pontoons, 4 ft deep by 6 ft wide and 20 ft long, supporting a timber deck. Each pontoon weighs 5,000 lb and the timber deck weighs 300 lb per ft of bridge. The pontoons are spaced 16 ft on centers. (*a*) How deep in the water will the pontoons float when they support only the dead weight of the bridge? (*b*) If the top of the pontoons must remain 1 ft out of the water at maximum load, how much *additional* total deck load W can be supported by each pontoon? *Ans.* (*a*) $d = 1.31$ ft; (*b*) $W = 12{,}700$ lb

12-54. A *flexible cable* is stretched between the tops of two poles which are 200 ft apart, as shown in Fig. 12-33. Pole BD is pin-connected at the bottom, and its top is supported by the guy wire AB, lying in the plane $ABDE$. The lower end of the pole FE is buried solidly in the ground. If the weight of the cable is 0.3 lb per ft and the sag at the center C is 5 ft, compute the tension in the guy wire AB, and the reacting moment M at E in the pole FE. Neglect the weight of the guy wire.
Ans. $AB = 500$ lb; $M_E = 12{,}000$ lb-ft

12-55. A pipeline is supported over a river by means of wire hangers suspended from a *flexible cable*, similar to the main span of Fig. 12-25. The backstays are unloaded and make an angle of 45° with the ground. The tops of the towers are at the same height and are 160 ft apart. The cable sag is 25 ft. The pipeline weighs 500 lb per ft, including an allowance for the weight of the cable and hangers. Determine graphically the tension at the center of the span and in the backstays. Check all results analytically.
Ans. $H = 64{,}000$ lb; $T_{max} = 75{,}500$ lb; $T_{BSTY} = 90{,}500$ lb

12-56. In Fig. 12-25 the span L of the footbridge is 160 ft, the maximum sag s of the cable at C is 20 ft, and the weight of the cable and deck is 250 lb per horizontal foot. Determine graphically the tension in the cable at C and D. Check analytically.

REVIEW QUESTIONS

12-1. Define *pressure* and state the usual units of pressure.

12-2. What is the direction of the pressure of a fluid at rest on a submerged surface?

12-3. How is the intensity of liquid pressure at any point below the free surface obtained?

12-4. What is meant by (*a*) buoyancy and (*b*) load capacity?

12-5. How may the hydrostatic force against a submerged plane area be obtained?

12-6. What is meant by (*a*) fluid weight and (*b*) equivalent fluid pressure?

12-7. The outside rays in the force polygon for a flexible cable or arch give us what information?

12-8. What is the effect of moving the pole of the force polygon of a flexible cable farther from the loads?

12-9. What is the shape of a cable under uniformly distributed load if the load is (*a*) uniform per horizontal foot of span and (*b*) uniform per foot along the cable axis?

12-10. What is true of the horizontal component H of the stress T at any point in a flexible cable supported at its ends?

PART 2

DYNAMICS

CHAPTER 13

Basic Principles of Dynamics

13-1 Introduction

In the study of the motion of a body, we can observe that it moves a certain distance in a given interval of time and that equal distances may not be covered in equal intervals of time. Certain relationships exist between displacement, time, velocity, and rate of change of velocity. These relationships comprise the study of *kinematics*.

If we measure the force that produces the motion of a body, we may study the effects of the force on the motion of the body. This study is called *kinetics*.

13-2 Definitions

Statics deals with the forces acting on rigid bodies at rest. **Dynamics** deals with the motion of rigid bodies and with the forces that produce or change their motion.

A particle is a quantity of matter so small, *in comparison with its range of motion*, that its dimensions may be disregarded. Thus, stars, planets, and projectiles are commonly considered as particles. In many problems in dynamics, the dimensions of the body are immaterial. In such problems, we regard the body as dimensionless, and treat it as a particle.

13-3 Kinematics and Kinetics

The study of dynamics is generally divided into two main branches: *kinematics* and *kinetics*. **Kinematics** deals with motions in themselves, entirely apart from the forces that produce or change these motions. Kinematics, then, concerns only the study of displacement, velocity, and acceleration, usually with respect to time. **Kinetics** is the study of the changes in the motion of a rigid body produced by an *unbalanced* force system, and of the means of determining the forces required to produce any desired change in its motion.

13-4 Types of Motion

The motion of particles and bodies in space is governed by definite laws and principles. Such motion may be along any imaginable path, either in a plane or in space. The simplest path is a straight line.

Rectilinear motion, or **rectilinear translation,** is motion along a straight-line path. Thus, a train traveling along a straight track and a freely falling body are examples of rectilinear motion.

Curvilinear motion is motion along a curved path and may be either in a plane or in space. An automobile rounding a curve on a level road and the path of a projectile in still air are examples of *plane curvilinear motion.*

Rotation is a special case of plane curvilinear motion in which all particles travel in parallel circular paths of constant radii about a fixed *axis of rotation,* which is perpendicular to the planes of motion. A pulley rotating about a fixed shaft and the flyball of a steam-engine governor, rotating about a fixed vertical shaft, are examples of rotation.

Plane motion is the motion of a rigid body in which all particles move in parallel planes. Rectilinear translation, plane curvilinear motion, and rotation are all special cases of plane motion. *The general case of plane motion* is a combination of rectilinear motion and plane curvilinear motion. The wheels of a locomotive, rolling on a straight track, and the connecting rod between the piston and the driving wheels are examples of the general case of plane motion.

The types of motion to be considered in this text are shown in Table 13-1.

13-5 Displacement

The displacement of a body in motion is its change of position with respect to some arbitrary fixed point. In engineering practice, **linear displacement,** for which the symbol is s, is generally given in feet. Large displacements, such as of automobiles and trains, are often given in miles. **Angular displacement** θ, such as that of a rotating pulley, is usually given in revolutions (rev) or in radians (rad). (One revolution equals 360 deg or 2π rad. $\pi = 3.1416$.) Displacement has direction as well as magnitude and is, therefore, a vector quantity.

13-6 Velocity

The velocity of a body is its rate of change of position, or its displacement with respect to time. The commonly used unit of time is the second (sec). Other units are the minute and the hour. **Linear velocity** v is then expressed

Table 13-1 TYPES OF PLANE MOTION ILLUSTRATED

Type of motion	Definition sketch	Example
Rectilinear translation	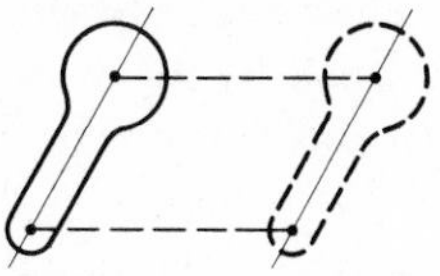	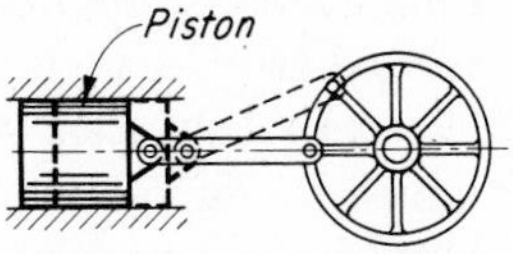 Piston
Rotation	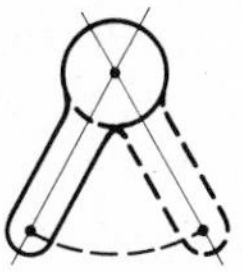	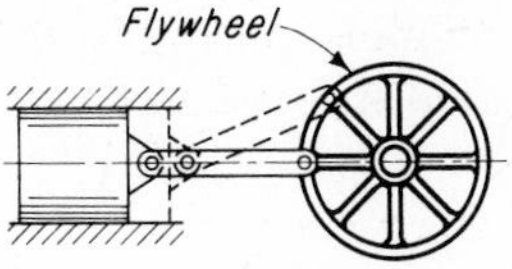 Flywheel
General case of plane motion	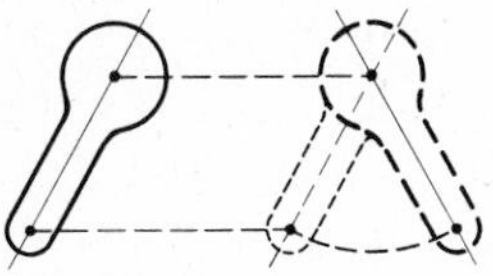	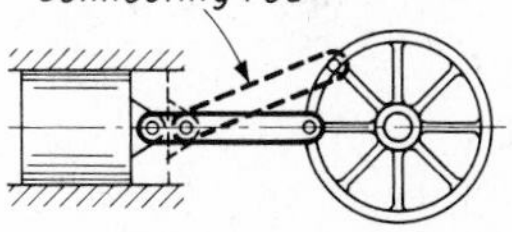 Connecting rod

in feet per second (ft per sec), feet per minute (ft per min), miles per hour (mph), etc. **Angular velocity** ω is expressed either in revolutions per second (rps) or per minute (rpm) or in radians per second (rad per sec) or per minute (rad per min). Like displacement, velocity has direction as well as magnitude, and is therefore a vector quantity. **Speed** *is the magnitude of velocity only,* without regard to direction, and is, therefore, a scalar quantity.

13-7 Acceleration

If the velocity of a body is changing, the body is said to be *accelerated.* Hence, **acceleration is the rate of change in velocity with respect to time.** If linear velocity is expressed in feet per second (ft per sec), the change in velocity, per unit of time, which is the **linear acceleration** a, is expressed in feet per second per second (ft per sec^2). Also, if angular velocity ω is expressed in revolutions per second (rps), **angular acceleration** α will be in revolutions per second per second (rps per sec). Like velocity and displacement, acceleration is a vector quantity. The acceleration associated with a *decreasing velocity* is sometimes called a **deceleration.**

13-8 Sign Convention

Difficulty is often experienced in the matter of signs. Such difficulty is largely eliminated if we *let the initial direction of displacement determine the positive sense of accelerations, velocities, displacements, forces, and moments of forces.*

13-9 Newton's Laws of Motion

The three laws of motion, formulated by Sir Isaac Newton, follow:

First Law: A body will remain at rest or in uniform motion in a straight line unless acted upon by an externally unbalanced force.

Second Law: A body acted upon by an externally unbalanced force will receive an acceleration which is proportional to the force and in the direction in which the force acts.

Third Law: To every action there is an equal and opposed reaction.

These laws form the basis for predicting the motions of bodies, and for the determination of the forces involved. A thorough discussion of their application is found in Chap. 15.

SUMMARY

(By article number)

13-1 through 13-3. Divisions of Engineering Mechanics

	Dynamics	
Statics	**Kinematics**	**Kinetics**
The study of effects of forces on a rigid body at rest	The study of the motion of a body without consideration of the forces involved	The study of the forces that produce or change the velocity of a rigid body

13-4. The three types of motion are (1) **rectilinear motion** (or translation), or motion along a straight path; (2) **curvilinear motion,** or motion along a curved path; and (3) **the general case of plane motion,** which is a combination of rectilinear and plane curvilinear motion.

13-5. Displacement is the change of position of a body with respect to some arbitrary fixed point.

13-6. Velocity is the rate of change of position of a body with respect to time.

13-7. Acceleration is the rate of change of velocity with respect to time.

13-8. Sign convention. The initial direction of the displacement determines the positive sign of accelerations, velocities, displacements, forces, and moments of forces.

REVIEW QUESTIONS

13-1. Define the following terms: (*a*) dynamics, (*b*) kinematics, (*c*) kinetics.

13-2. Name three types of motion.

13-3. Define the following terms: (*a*) displacement, (*b*) velocity, and (*c*) acceleration.

13-4. Give the symbols denoting linear displacement, velocity, and acceleration, and the engineering units in which these quantities are commonly expressed.

13-5. Give the symbols denoting angular displacement, velocity, and acceleration, and the engineering units in which these quantities are commonly expressed.

13-6. Define Newton's first law of motion.

13-7. Define Newton's second law of motion.

13-8. Define Newton's third law of motion.

CHAPTER 14

Kinematics of Rectilinear Motion

14-1 Equations of Rectilinear Motion. Uniform Acceleration

If a body moves with *uniform velocity* v during an interval of time t, the displacement s is the product of the velocity and the time. Hence

$$s = v \cdot t \qquad (a)$$

Acceleration is the rate of change of velocity with respect to time. If a body moves with an **initial velocity** v_0 and is then given a uniform acceleration a, the *increase in velocity* during an interval of time t is the product of the acceleration a and time t or $a \cdot t$. Its **final velocity** v at the end of any time period t will then be its initial velocity v_0 plus the increase in velocity at, or

$$\mathbf{v = v_0 + at} \qquad \text{(14-1)}$$

This relation is shown diagrammatically in Fig. 14-1, in which the vertical ordinates v_0, v_{avg}, and v, between the x axis and the velocity curve, represent, respectively, the initial, the average, and the final velocities.

When the acceleration of a body is constant, its velocity increases at a uniform rate, and the displacement of the body during a time interval t is the product of its *average velocity* v_{avg} and the time t. From Fig. 14-1, we see that the average velocity is one-half the sum of the initial and final velocities. Therefore, since $v = v_0 + at$,

$$v_{avg} = \frac{1}{2}(v_0 + v) = \frac{1}{2}v_0 + \frac{1}{2}(v_0 + at) = v_0 + \frac{1}{2}at \qquad (b)$$

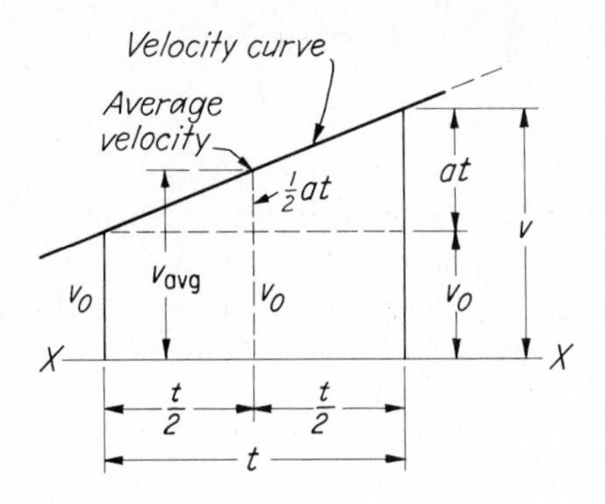

FIG. 14-1 Initial, final, and average velocity.

The displacement s, therefore, is the product of the average velocity v_{avg} and the time t. That is, $s = (v_0 + \frac{1}{2}at)t$, or

$$s = v_0 t + \frac{1}{2}at^2 \tag{14-2}$$

Table 14-1 EQUATIONS OF UNIFORMLY ACCELERATED RECTILINEAR MOTION

To find	When given	Use equation		Derived from
t	$a \quad v_0 \quad v$	$t = \dfrac{v - v_0}{a}$	(A)	Eq. (14-1)
a	$t \quad v_0 \quad v$	$a = \dfrac{v - v_0}{t}$	(B)	Eq. (14-1)
	$t \quad v_0 \quad s$	$a = \dfrac{2s - 2v_0 t}{t^2}$	(C)	Eq. (14-2)
	$v_0 \quad v \quad s$	$a = \dfrac{v^2 - v_0^2}{2s}$	(D)	Eq. (14-3)
v_0	$t \quad a \quad v$	$v_0 = v - at$	(E)	Eq. (14-1)
	$t \quad a \quad s$	$v_0 = \dfrac{s}{t} - \dfrac{at}{2}$	(F)	Eq. (14-2)
	$a \quad v \quad s$	$v_0 = \sqrt{v^2 - 2as}$	(G)	Eq. (14-3)
v	$t \quad a \quad v_0$	$v = v_0 + at$	(14-1)	
	$a \quad v_0 \quad s$	$v = \sqrt{v_0^2 + 2as}$	(H)	Eq. (14-3)
s	$t \quad a \quad v_0$	$s = v_0 t + \frac{1}{2}at^2$	(14-2)	
	$a \quad v_0 \quad v$	$s = \dfrac{v^2 - v_0^2}{2a}$	(I)	Eq. (14-3)
	$t \quad v_0 \quad v$	$s = \frac{1}{2}t(v_0 + v)$	(J)	Eq. (14-2)

From Eq. (14-1), $t = (v - v_0)/a$. This value substituted in Eq. (14-2) gives

$$v^2 = v_0^2 + 2as \qquad (14\text{-}3)$$

Equations (14-1) *to* (14-3) *express the basic relations between time, acceleration, velocity, and displacement of a body with uniformly accelerated motion.*

In the solution of problems, any one of the quantities t, a, v_0, v, and s may be the unknown to be solved for. For easy reference the basic equations, and others which are readily derived from them, are tabulated in Table 14-1.

Units.[1] To avoid confusion in applying these equations, the following engineering units should always be used: a in ft per sec^2; v in ft per sec; s in ft; t in sec. Velocities are often given in miles per hour (mph). 60 mph = 88 ft per sec; 1 mph = 1.467 ft per sec. To convert miles per hour to feet per second, multiply by 88/60, or 1.467. 1 mile = 5,280 ft.

Table 14-2 UNITS OF MOTION. ABBREVIATIONS

Unit of motion	Moderate	Extreme
Linear acceleration a	ft per sec^2	fpsps
Angular acceleration α	rad per sec^2	rad psps
	rev per sec^2	rpsps
Linear velocity v	ft per sec	fps
Angular velocity ω	rad per sec	rad ps
Linear displacement s	ft	ft
Angular displacement θ	rad	rad

[1] The matter of suitable abbreviations for the units of motion is not yet completely solved. The common abbreviation of feet per second to ft per sec is *moderate* and the abbreviation to fps is *extreme*. In some parts of teaching, the moderate system is desirable, since it lends itself readily to common algebraic processes which the extreme system does not. In professional practice, however, and in student problem work, the greater brevity of the extreme system has considerable advantage. Table 14-2 shows the two systems applied to the most commonly used units of motion.

ILLUSTRATIVE PROBLEM

14-1. A car starting from rest ($v_0 = 0$) is given a constant acceleration a until it reaches a final velocity v of 64 ft per sec after having traveled a distance s of 512 ft. Calculate the acceleration a and the required time t.

Solution: Given $v_0 = 0$, $v = 64$ ft per sec; $s = 512$ ft. Find a and t. Using Eq. (D) to find a, we have

$$\left[a = \frac{v^2 - v_0^2}{2s}\right] \qquad a = \frac{(64)^2 - 0}{2(512)} = 4 \text{ ft per sec}^2 \qquad \textit{Ans.}$$

Then, from Eq. (A), we obtain

$$\left[t = \frac{v - v_0}{a}\right] \qquad t = \frac{64 - 0}{4} = 16 \text{ sec} \qquad \textit{Ans.}$$

PROBLEMS

14-2. Calculate the constant acceleration a required to give a train starting from rest a velocity of 44 ft per sec in 22 sec.

14-3. An automobile traveling at 60 mph is slowed to a stop (decelerated) at a uniform rate in 16 sec. Compute the deceleration a. *Ans.* $a = 5.5$ ft per sec^2

14-4. Compute the time t required to increase the speed of a car from 15 to 60 mph, if its acceleration is 4.4 ft per sec^2. Also compute the distance traveled s.

14-5. An airplane attains a velocity of 90 mph after traveling 66 ft from rest on a catapult. Assuming uniform acceleration, compute the time required to travel the first 66 ft on the catapult and the acceleration of the plane.

Ans. $t = 1$ sec; $a = 132$ ft per sec^2

14-6. What distance s does the car in Prob. 14.4 travel in the 10 sec? Calculate by Eq. (14-2) and check by Eq. (C).

14-7. A car being tested passes point A with an initial velocity v_0 of 30 mph, and reaches point B, 770 ft distant, in 10 sec. (a) What must be its constant acceleration a and (b) what is its final velocity v?

Ans. (a) $a = 6.6$ ft per sec^2; (b) $v = 110$ ft per sec

14-8. A car is traveling at 30 mph on an acceleration lane to a freeway. What acceleration is required to obtain a speed of 60 mph in a distance of 440 ft? What time is required to travel this distance?

14-9. What must be the initial velocity v_0 of a body being accelerated 6 ft per sec^2 in order to travel a distance of 400 ft in 10 sec? *Ans.* $v_0 = 10$ ft per sec

14-10. A helicopter rises with a constant vertical acceleration of 2 ft per sec^2. What will be its vertical velocity when it has risen 16 ft above the point where its upward velocity was 8 ft per sec?

14-11. A car is accelerated 5.5 ft per sec^2. Calculate the initial velocity v_0 the car must have if it is to attain a final velocity v of 45 mph in a distance of 352 ft. Compute also the time t required to attain that final velocity.

Ans. $v_0 = 22$ ft per sec; $t = 8$ sec

14-12. The retrorockets on a space vehicle can provide a deceleration of 25 ft per sec^2. How long must the rockets be fired to change the velocity of the vehicle from 12,000 to 10,000 ft per sec? How far will the vehicle travel during this time?

14-13. A body starting from rest is accelerated 8 ft per sec^2. What final velocity v will it have after traveling 225 ft? *Ans.* $v = 60$ ft per sec

14-14. A speed boat has a velocity of 30 mph as it comes out of a turn. It accelerates at the rate of 1 ft per sec^2 for the first $\frac{1}{2}$ mile of the straightaway and then maintains this final velocity for the next $\frac{1}{2}$ mile. How long did it take to cover the mile?

14-15. An automobile starting from rest is accelerated 6 ft per sec^2 for 10 sec and is then brought to a stop in the next 15 sec with a uniformly decreasing velocity. Calculate the total distance s traveled in the 25 sec. *Ans.* $s = 750$ ft

14-16. The elevator in an office building, starting from rest at the first floor, is accelerated 2.5 ft per sec^2 for 5 sec. It continues at constant velocity for 12 sec more and is then stopped in 3 sec with a constant deceleration. If the floors are 12 ft-6 in. apart, at what floor does the elevator stop?

14-2 Freely Falling Bodies

The gravitational attraction of the earth is evidenced by a pull or force on every material body, the intensity of which is measured by its weight W. A body falling in space encounters air resistance which varies with its velocity, and with its size and shape. Because of air resistance, a falling body quickly reaches a maximum or **terminal velocity,** ranging from a few feet per second for a feather to about 30 ft per sec for a drop of water, about 175 ft per sec for a man dropping with an unopened parachute, and about 18 ft per sec with an open parachute.

For comparatively low velocities and for small, heavy bodies at somewhat higher velocities, we may neglect air resistance without appreciable error. Experiments have shown that **the acceleration given a freely falling body by the gravitational pull W is constant and equals approximately 32.2 ft per sec^2,** which is denoted by the symbol g. The value of g is slightly higher at the poles than at the equator and decreases slightly with altitude, being about $\frac{1}{4}$ of 1 percent lower at an altitude of 25,000 ft than at sea level.

For freely falling bodies, a equals g and Eqs. (14-1) to (14-3) of Art. 14-1 then become

$$v = v_0 + gt \qquad (14\text{-}4)$$

$$s = v_0 t + \frac{1}{2} g t^2 \qquad (14\text{-}5)$$

$$v^2 = v_0^2 + 2gs \qquad (14\text{-}6)$$

In all the equations in Table 14-1, a now equals g, or 32.2 ft per sec^2. *The direction of the initial displacement should be considered as positive.* Velocity and acceleration in the same direction are then also positive, while those oppositely directed are negative.

ILLUSTRATIVE PROBLEM

14-17. A small steel ball is shot vertically upward from the top of a building 80 ft above the street, with an initial velocity of 60 ft per sec. (*a*) How high above the building will the ball rise and (*b*) in what time will it reach the maximum height? (*c*) Compute the velocity with which it will strike the street and (*d*) the total time elapsed until it strikes.

Solution: At the point of maximum height, the final velocity of the ball is zero. On its upward journey, v_0 is positive but a is negative. Then, $v_0 = 60$ ft per sec; $v = 0$; and $a = -32.2$ ft per sec^2. From Eq. (14-6), or from Eq. (*I*), Table 14-1,

$$\left[s = \frac{v^2 - v_0^2}{2a}\right] \qquad s = \frac{0 - (60)^2}{2(-32.2)} = 55.9 \text{ ft} \qquad \textit{Ans.}$$

From Eq. (*A*), Table 14-1 the time required to reach maximum height is

$$\left[t = \frac{v - v_0}{a}\right] \qquad t = \frac{0 - 60}{-32.2} = 1.86 \text{ sec} \qquad \textit{Ans.}$$

From its maximum height, the ball must fall 55.9 + 80, or 135.9 ft. Its velocity upon striking the street is found by Eq. (*H*) of Table 14-1. Now $v_0 = 0$; $a = -32.2$ ft per sec^2; and $s = -135.9$ ft. Therefore to obtain its final velocity, we have

$$[v = \sqrt{v_0^2 + 2as}] \qquad v = \sqrt{0 + 2(-32.2)(-135.9)} = \pm 93.6$$

or

$$v = -93.6 \text{ ft per sec} \qquad \textit{Ans.}$$

The negative root was chosen because the velocity we seek is in the opposite direction to the initial displacement.

Then, the time to drop the 135.9 ft is

$$\left[t = \frac{v - v_0}{a}\right] \qquad t = \frac{-93.6 - 0}{-32.2} = 2.91 \text{ sec}$$

The total elapsed time is

$$t = 1.86 + 2.91 = 4.77 \text{ sec} \qquad \textit{Ans.}$$

Alternate Solution: The final velocity and the total elapsed time could have been found by applying the same equations to the entire flight. The initial velocity was 60 ft per sec and the ball ended up 80 ft below its initial position. Therefore $s = -80$ ft.

$$[v = \sqrt{v_0^2 + 2as}] \qquad v = \sqrt{(60)^2 + 2(-32.2)(-80)} = \pm 93.6$$

or

$$v = -93.6 \text{ ft per sec} \qquad \textit{Ans.}$$

The negative root should be chosen because the velocity we seek is in the opposite direction to the initial direction of motion. The total elapsed time is

$$\left[t = \frac{v - v_0}{a}\right] \qquad t = \frac{-93.6 - 60}{-32.2} = 4.77 \text{ sec} \qquad \textit{Ans.}$$

PROBLEMS

14-18. In Prob. 14-17 change the initial velocity to 80 ft per sec and solve.

14-19. Parachute jumpers land with an average vertical velocity of 18 ft per sec. With what height of ordinary jump does this compare? *Ans.* $h = 5.03$ ft

14-20. A workman's hammer is accidentally dropped from the twentieth floor of a building under construction. With what velocity does it strike the pavement 304 ft below, and what time t is required?

14-21. A ballast bag filled with sand is dropped from a balloon. The observer noted that the bag struck the ground 6 sec after its release. If air friction is disregarded, how high was the balloon above the ground? *Ans.* $h = 580$ ft

14-22. A workman drops a wrench from the roof of a tall building 300 ft above the street and at the same instant shouts a warning to men below. If sound travels 1,120 ft per sec, how long after the warning is heard below does the wrench strike?

14-23. A stone is dropped in still air from the deck of a high bridge. The sound of the splash reaches the deck 3 sec later. If sound travels 1,120 ft per sec in still air, how high is the deck above the water? (This requires the solution of a quadratic equation.) *Ans.* $s = 134$ ft

14-24. A rocket fired vertically burns its fuel for 20 sec. At burnout (instant at which the rocket runs out of fuel) the rocket has an upward velocity of 439 mph and has reached an altitude of 4 miles. Determine the maximum altitude reached by the rocket and the total time of free flight (burnout to ground). Neglect air resistance.

14-3 Rectilinear Motion. Variable Acceleration

When a body moves with constant acceleration, definite equations may be written from which its state of motion at any time is readily obtained, as was shown in Arts. 14-1 and 14-2. When the acceleration is variable, no such general equations can be written, because innumerable variations are possible. But, if the displacement, or the velocity, or the acceleration of the body can be expressed in the form of a simple mathematical equation, the various other conditions of its motion may be determined by elementary calculus, as illustrated in Appendix D. A number of problems in which simple variable accelerations occur are, however, readily solved by means of motion diagrams, as is explained in the following article.

14-4 Motion Diagrams

The relations among acceleration, velocity, and displacement are generally expressed in the form of algebraic equations, from the solution of which desired results are then obtained, as was illustrated in the preceding articles.

Corresponding results may be obtained by means of so-called *motion diagrams*, constructed in accordance with the two laws given in the following article. These laws are word statements of the fundamental mathematical relations of motion, as expressed by the following differential equations: $v = ds/dt$ and $a = dv/dt = d^2s/dt^2$, which are more fully explained in Appendix E.

14-5 Laws of Motion Diagrams

We shall now develop a series of three related diagrams, constructed in accordance with the following laws, by means of which we may determine unknown magnitudes of acceleration, velocity, and displacement.

The first law is used to determine the shape of a diagram. The second law is used to determine values of ordinates representing either acceleration, velocity, or displacement. The wording and meaning of these laws should be memorized.

First law of motion diagrams:

[The slope of the curve at any point in any diagram] = [the length of the ordinate at the corresponding point in the next higher diagram]

Second law of motion diagrams:

$$\left[\begin{array}{l}\textbf{The difference in length of any}\\ \textbf{two ordinates in any diagram}\end{array}\right] = \left[\begin{array}{l}\textbf{the area between corresponding}\\ \textbf{ordinates in the next higher dia-}\\ \textbf{gram}\end{array}\right]$$

Definitions. Six terms appearing in these two laws will now be defined:

Curve. Any line that may be precisely defined by a mathematical equation is called a *curve*. (Thus, a straight line may, mathematically speaking, be referred to as a curve.)

Slope. The *slope* of a line is the ratio of y to x, as indicated in Fig. 14-2. The slope of a curve at any point is the slope of a tangent to the curve at that point. The slope of a curve may be positive or negative. A horizontal line has *zero* slope. A straight line has a *uniform* slope. A curved line has a *varying* slope, and the slope may be either *increasing* or *decreasing* for successive points along the curve (see Fig. 14-3).

Ordinate. With reference to a system of rectangular axes (X and Y), an *ordinate* is a line drawn parallel to the Y axis and extending from the X axis to the curve. *Principal ordinates* are those giving initial and final values in any time period.

Diagram. The framework of a diagram consists of two vertical Y axes spaced a given time period t apart; a horizontal X axis; and a *curve* extending between the two Y axes. The intersection of the X axis with the left Y axes is called the *origin*.

The seven slopes of curves shown in Fig. 14-3 are the only ones that will

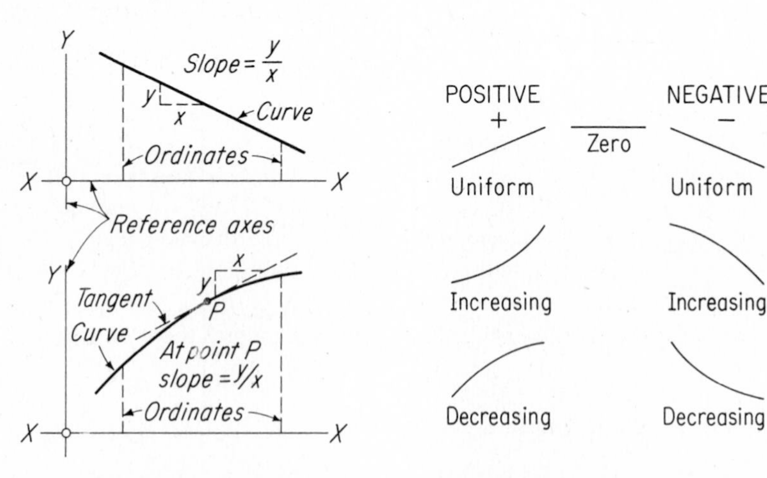

FIG. 14-2 Slopes of curves.

FIG. 14-3 Lines having different slopes or changes in slopes.

be encountered in motion diagrams. They are easily memorized. Zero slope is represented by a horizontal line. The three positive slopes always *rise* from left to right. The three negative slopes always *fall* from left to right.

If the quantity to be expressed at the beginning of the time period is zero, the *initial ordinate* at the origin will, of course, be zero, and the curve will then start at the origin. Conversely, if the quantity to be expressed at the beginning of the time period has a finite value, positive or negative, this value is then shown as an ordinate, positive (above the X axis) or negative (below the X axis), extending from the origin to the beginning of the curve. From this beginning point the curve is then drawn, always from left to right with its correct slope (see Fig. 14-3), proportionately correct but preferably *not to scale,* until it terminates at the right Y axis. The ordinate extending from this point of termination to the X axis is called the *final ordinate* and represents the value of the expressed quantity at the end of the given time period.

To assist students in memorizing the first and second laws of motion diagrams and the seven possible slopes, the four sets of practice problems appearing on the next page should be copied and the work called for completed. By this method students can, within a few hours, learn to solve seemingly complex problems, quickly and with comparative ease.

First Law of Motion Diagrams

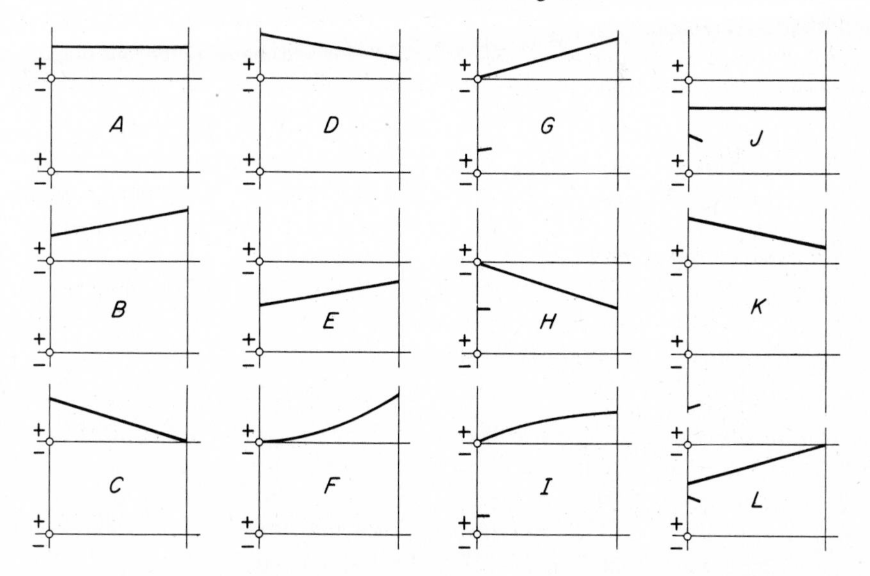

Draw curve in lower diagram. Start curve at origin.

Draw curve in lower diagram. Start curve where indicated.

Second Law of Motion Diagrams

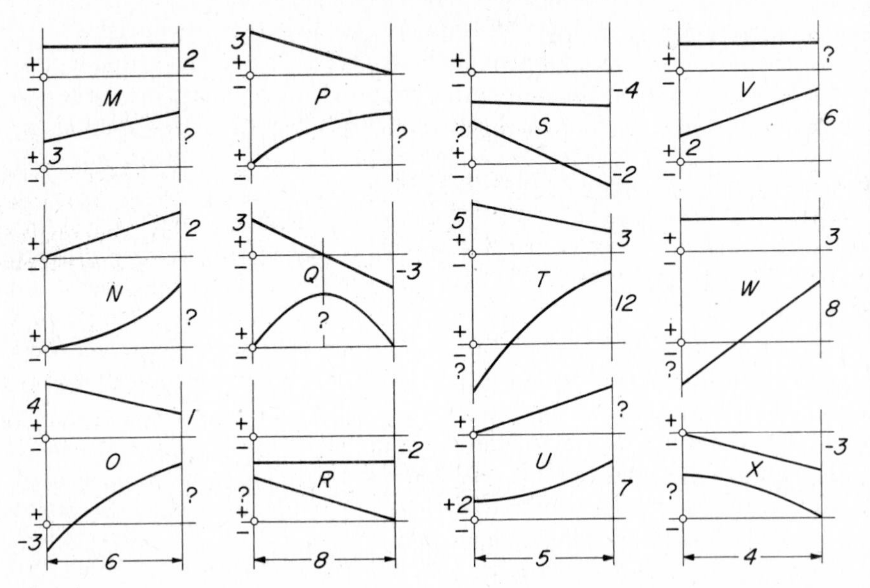

Calculate values of principal ordinates. Calculate values of principal ordinates.

Practice Problems. In order to gain proficiency in the application of the first and second laws of motion diagrams, students should copy and complete the four practice sheets shown above. Draw curves proportionately correct, but not to scale.

Areas under Curves. By *area under a curve* is meant that area lying between the curve and the horizontal reference axis, and bounded on the left and right by vertical ordinates. Areas under the curves most commonly encountered are given in Table 14-3.

Next Higher Diagram. It is understood that the diagrams are to be drawn in the following order from top to bottom of the sheet: top, acceleration; middle, velocity; and bottom, displacement.

14-6 Applications of Motion Diagrams

Memorization of the two laws of motion diagrams, the seven slopes, and the areas under curves given in Table 14-3 is an important aid in their application. The procedure involved in the solution of the following example is indicated by numerous notes in Fig. 14-4. The final simplified solution, as it should appear on paper, is shown in Fig. 14-5. *Motion diagrams are not generally drawn to scale,* but reasonable proportions should be maintained.

Table 14-3 AREAS UNDER CURVES

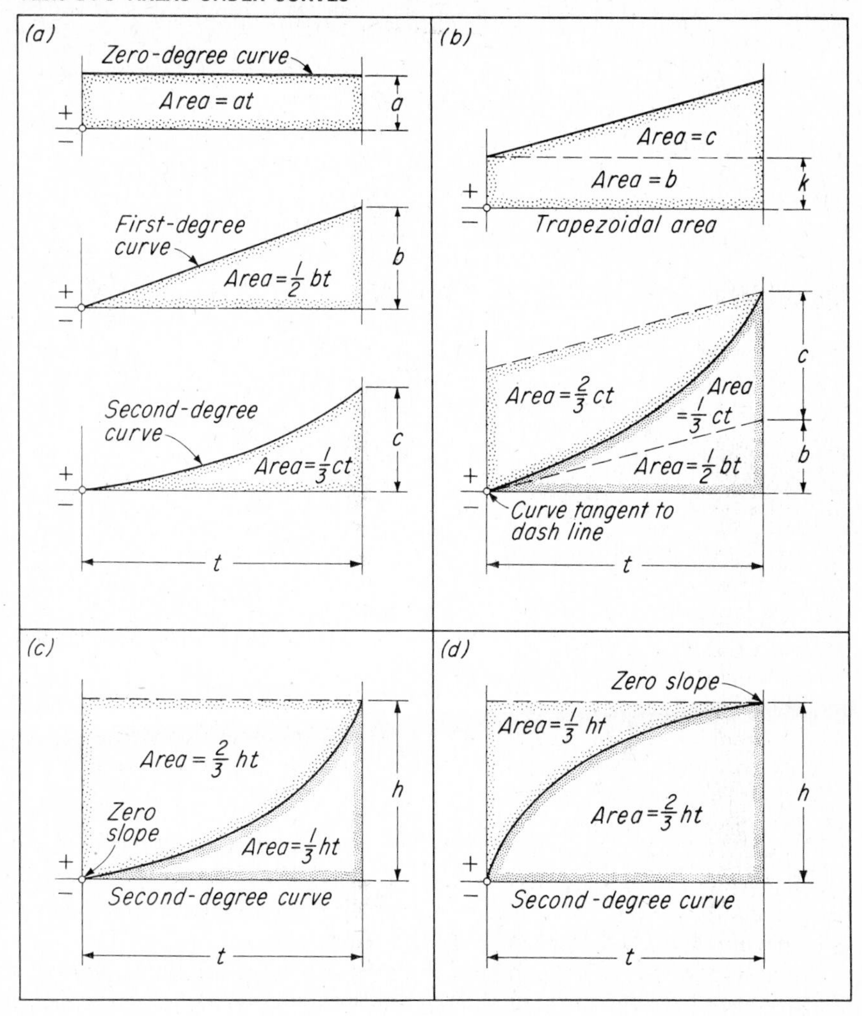

ILLUSTRATIVE PROBLEM

14-25. A car starting from rest at A is given an acceleration which increases uniformly from 0 at A to 12 ft per sec^2 in 4 sec. At B, the brakes are applied with constant force and the car is decelerated uniformly at -8 ft per sec^2 for 3 sec until it stops at C. Draw the acceleration, velocity, and displacement diagrams. Compute and indicate the velocities and displacements at B and C.

Solution: In Fig. 14-4, the letters A to J denote *principal ordinates*. The encircled letters K to P denote areas under curves. The various notes indicate the relations

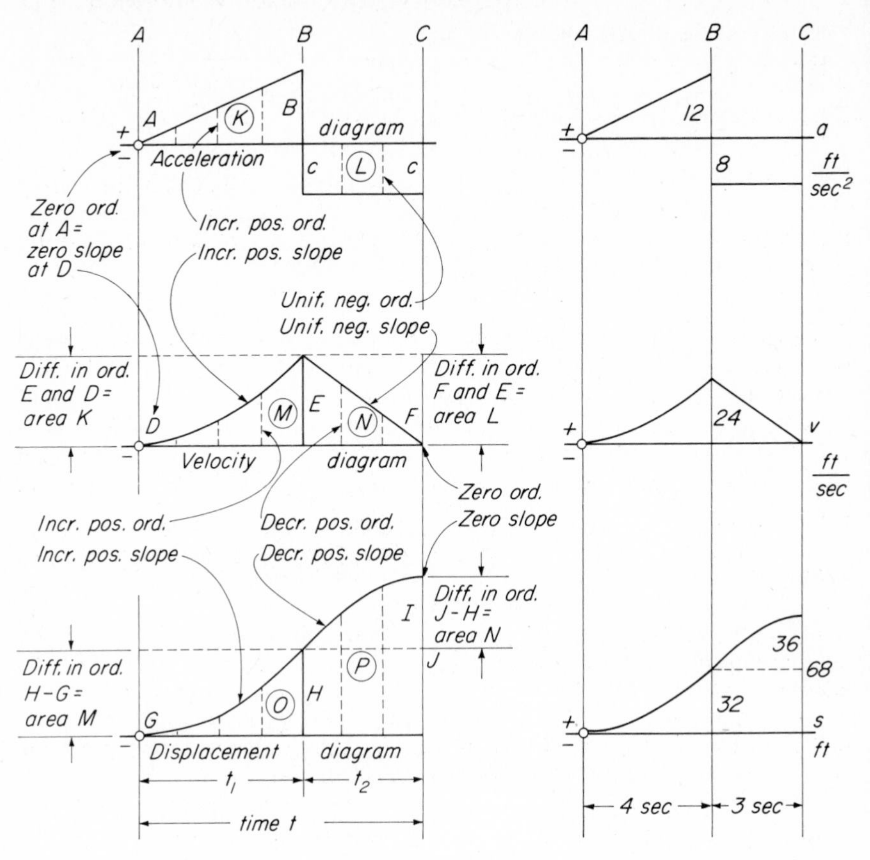

FIG. 14-4 Construction of motion diagrams.

FIG. 14-5 Completed motion diagrams.

that govern the construction of the diagrams. These notes and diagrams should be carefully studied in connection with the statement and solution of the problem.

In Fig. 14-5 is shown the completed problem, stripped of all but the necessary details. The shape of the acceleration diagram and the values of its principal ordinates are obtained from the statement. During the first 4 sec, the acceleration (ordinates) is positive and uniformly increasing from 0 to $+12$. At B, a sudden change in acceleration (ordinates) is made to -8, at which value it remains constant for 3 sec. Also obtained from the statement are the facts that $v_A = 0$, $v_C = 0$, $s_A = 0$. Thereafter, the velocity and displacement diagrams are completed in accordance with the laws of motion diagrams, and the values of all principal ordinates are computed and are noted on the diagrams. Frequent reference to Fig. 14-4 should be made.

PROBLEMS

14-26. The diagram in Fig. 14-6 shows the acceleration of a body to be uniform at 4 ft per sec². Assume the body to start from rest. Draw the acceleration, velocity,

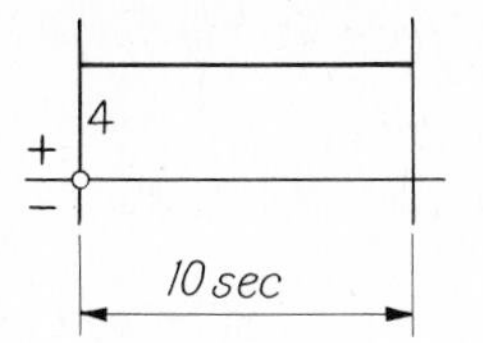

FIG. 14-6 Prob. 14-26.

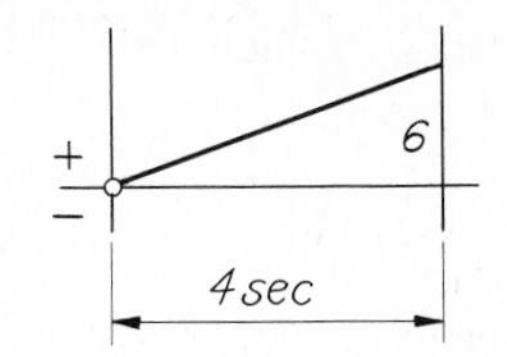

FIG. 14-7 Probs. 14-28 and 14-31.

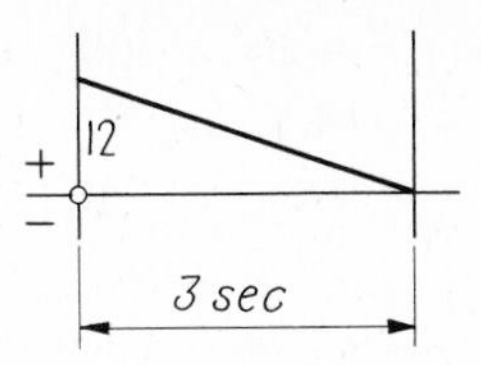

FIG. 14-8 Prob. 14-32.

and displacement diagrams, and determine the velocity and displacement at the end of 10 sec. *Ans.* $v = 40$ ft per sec; $s = 200$ ft

14-27. In Prob. 14.26 assume an initial velocity v_0 of 20 ft per sec and solve.

14-28. A body starting from rest is given a *uniformly increasing acceleration* as indicated by the diagram in Fig. 14-7. Complete the acceleration, velocity, and displacement diagrams, and compute the velocity and displacement at the end of 4 sec.

14-29. Assume the body in Prob. 14.28 to have an initial velocity v_0 of 10 ft per sec and solve. *Ans.* $v = 22$ ft per sec; $s = 56$ ft

14-30. Solve Prob. 14.4 by motion diagrams.

14-31. A body starting from rest has the velocity shown in the diagram of Fig. 14-7. Complete the motion diagrams and find the acceleration and the distance traveled at the end of 4 sec. *Ans.* $a = 1.5$ ft per sec^2; $s = 12$ ft

14-32. The diagram in Fig. 14-8 shows the *uniformly decreasing acceleration* of a body starting from rest. Draw the acceleration, velocity, and space diagrams, and indicate its velocity and the distance traveled at the end of 3 sec.

Ans. $v = 18$ ft per sec; $s = 36$ ft

14-33. In Prob. 14-32, assume an initial velocity of 12 ft per sec and solve.

14-34. Solve Prob. 14-14 by motion diagrams.

14-35. Solve Prob. 14-1 by motion diagrams.

14-36. Solve Prob. 14-7 by motion diagrams.

14-37. Solve Prob. 14-24 by motion diagrams.

14-38. In Fig. 14-9 is shown the acceleration diagram of an automobile which starts from rest and is given a *uniformly increasing acceleration* reaching 8 ft per sec^2 in 6 sec. The brakes are then suddenly applied, and the car is brought to a stop in 8 sec with a *uniform deceleration* of -3 ft per sec^2. Complete the accelera-

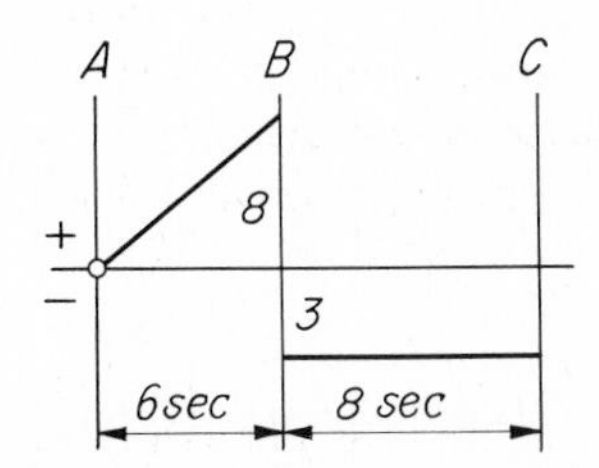

FIG. 14-9 Probs. 14-38 and 14-39.

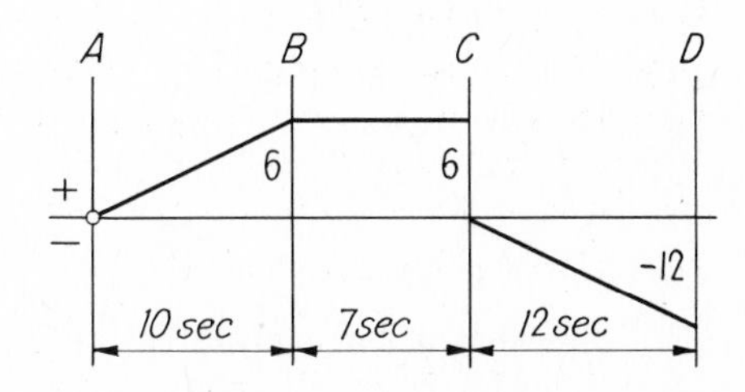

FIG. 14-10 Prob. 14-40.

tion, velocity, and displacement diagrams, and compute the values of all principal ordinates.

14-39. If the car in Prob. 14-38 has an initial velocity at A of 24 ft per sec, how much longer than 8 sec is required to stop it with the given deceleration, and what total distance is traveled between A and C? *Ans.* 8 sec longer; 576 ft

14-40. The diagram shown in Fig. 14-10 indicates the acceleration given a car during a continuous time period of 29 sec. The car starts from rest at A and stops at D. Draw complete acceleration, velocity, and displacement diagrams. Compute and indicate the values of all principal ordinates.

Ans. $v_B = 30$ ft per sec; $v_C = 72$ ft per sec; $s_D = 1{,}033$ ft

14-41. A switch engine accelerates 2 ft per sec^2 from rest until reaching a velocity of 15 mph, at which speed it continues until it is 561 ft from its starting point. It is then brought to a stop in 5.5 sec with a constant deceleration. Draw complete a, v, and s diagrams, and determine the total time t elapsed and the distance s traveled.

14-42. An automobile can accelerate or decelerate at 6 mph per sec and has a maximum speed of 90 mph. If it starts from rest, what is the shortest time that it can travel 1 mile if it is to come to rest at the end of the mile run? Draw motion diagrams. *Ans.* $t = 55$ sec

SUMMARY

(By article number)

14-1. The three basic **equations of uniformly accelerated rectilinear motion** follow:

$$v = v_0 + at \tag{14-1}$$

$$s = v_0 t + \frac{1}{2}at^2 \tag{14-2}$$

$$v^2 = v_0{}^2 + 2as \tag{14-3}$$

14-2. When air resistance is neglected, **freely falling bodies move with a constant downward acceleration of 32.2 ft per sec^2**, which is denoted by the symbol g. In Eqs. (14-1) to (14-3), then, g replaces a and is 32.2 ft per sec^2.

14-5. The laws governing the construction of **motion diagrams** are as follows:

First law of motion diagrams:

[The slope of the curve at any point in any diagram] = [the length of the ordinate at the corresponding point in the next higher diagram]

Second law of motion diagrams:

[The difference in length of any two ordinates in any diagram] = [the area between corresponding ordinates in the next higher diagram]

REVIEW PROBLEMS

14-43. Calculate the *uniform acceleration* a required to increase the velocity of an airplane from rest to 90 mph in 8 sec. What total distance s is traveled in this time?

14-44. An automobile being tested for accelerating power passes point A with a velocity of 15 mph and 5 sec later passes point B, which is 300 ft distant. Compute (a) the *average (uniform) acceleration* a, and (b) the final velocity v at B.

Ans. (a) $a = 15.2$ ft per sec^2; (b) $v = 98$ ft per sec

14-45. Cars A and B travel in parallel lanes along a straight highway at 30 mph. B is 40 ft behind A and decides to pass A. What *constant acceleration* must be given car B for it to reach a point 60 ft ahead of A in 10 sec?

Ans. $a = 2$ ft per sec^2

14-46. An elevator ascends 64 ft from rest in 8 sec. Assuming uniform acceleration, what was the maximum velocity? What was the acceleration?

14-47. A switch engine starting from rest is uniformly *accelerated* 2 ft per sec^2 for 10 sec. Then it continues at uniform velocity for the next 20 sec, after which it is brought to a stop, with constant deceleration, in 8 sec. Determine (a) the velocity v at the end of the first 10 sec, and the following total distances traveled: (b) s_{10} at the end of 10 sec; (c) s_{30} at the end of 30 sec, and (d) s_{38} at the end of 38 sec.

Ans. (a) $v = 20$ ft per sec; (b) $s_{10} = 100$ ft; (c) $s_{30} = 500$ ft; (d) $s_{38} = 580$ ft

14-48. When an elevator cage is 120 ft above the bottom of the shaft and is going up at the rate of 20 ft per sec, a bolt suddenly detaches from under the floor of the cage. (a) With what *velocity* will the bolt strike the bottom of the shaft, and (b) what time elapses? *Ans.* (a) $v = 90.2$ ft per sec; (b) $t = 3.42$ sec

14-49. The displacement of a point on a moving body was measured in each frame of a motion-picture film. The displacements determined from each successive frame measured with respect to the initial position of the body were as follows: 0.25 in., 1.5 in., 4.0 in., 9.5 in., 16.0 in., 24.5 in., and 34.5 in. The camera was 20 frames per second. Plot the displacement, velocity, and acceleration curves. What was the maximum acceleration of the point?

14-50. A small stone is dropped down a well. The splash is heard 4 sec later. If sound travels 1,120 ft per sec, how deep is the well? (Solution of a quadratic equation is required.)

14-51. Draw *motion diagrams* for the conditions of motion described in Prob. 14-15 and solve. *Ans.* $s = 750$ ft

14-52. Solve Prob. 14-17 by *motion diagrams*. Let upward acceleration, velocity, and displacement be positive.

14-53. Draw *motion diagrams* for Prob. 14-16 and solve.

14-54. A rocket-driven test vehicle runs on a level track. It is braked by dropping a scoop into a water trough. The vehicle starts from rest and accelerates at a uniform rate of 40 ft per sec^2 for 20 sec. The vehicle then decelerates at the rate of 15 ft per sec^2 while coasting for 20 sec. What deceleration rate is required if the vehicle is braked to a stop in the next 2,500 ft? What total length of track is required?

14-55. A rocket is launched vertically upward. The fuel lasts 20 sec and during this time the rocket has a uniform upward acceleration of 30 ft per sec^2. Neglect air friction. Calculate (*a*) the velocity at the instant the fuel burned out; (*b*) the maximum altitude reached by the rocket; and (*c*) the velocity of the rocket on striking the earth. (*d*) How long was the rocket in the air?

14-56. A body starts from rest with an acceleration that increases uniformly from 8 to 16 ft per sec^2 in 6 sec. Draw the *motion diagrams*, and determine the final velocity and total distance traveled by the end of 6 sec. [See diagrams (*b*) of Table 14-3.]

Ans. $v = 72$ ft per sec; $s = 192$ ft

14-57. A parachutist jumps from a balloon. He falls 5 sec before his chute opens. If he reaches his terminal velocity of 18 ft per sec with the opened chute in 3 sec, what was his maximum deceleration? Assume that the deceleration varies uniformly from a maximum when the chute first opens to zero when terminal velocity is reached.

14-58. Trains run at 60 mph on a certain stretch of track. How far back of a stopped train should a warning torpedo be placed to signal an oncoming train, if the brakes are applied at once and the train can be decelerated at a uniform rate of 2.5 ft per sec^2?

14-59. The acceleration of a streetcar starting from rest increases uniformly from 2 to 9 ft per sec^2 in 8 sec. Then the car continues at uniform velocity for the next 20 sec, and in 11 sec more is brought to a stop with a constant deceleration. Draw complete *motion diagrams*, and determine the total distance traveled. [See diagrams (*b*) of Table 14-3.]

Ans. $s = 1{,}260.7$ ft

REVIEW QUESTIONS

14-1. Give the three basic equations that express the rectilinear motion of a uniformly accelerated body.

14-2. What is the acceleration of a freely falling body, when air resistance is neglected, and by what symbol is it denoted?

14-3. State the first law of motion diagrams, and explain its use.

14-4. State the second law of motion diagrams, and explain its use.

14-5. Define the following terms: (*a*) curve, (*b*) slope of curve, (*c*) ordinate, (*d*) area under curve.

14-6. State the names of the three motion diagrams in order, starting with the top diagram.

14-7. Sketch from memory the seven possible slopes of curves.

14-8. Draw to scale lines (about 2 in. long) having the following slopes: $+\frac{1}{4}$; $+\frac{1}{2}$; $+1$; $+2$; -2; -1.6; -0.6; $-\frac{1}{3}$.

CHAPTER 15

Kinetics of Rectilinear Motion

15-1 Introduction

In this analysis of the rectilinear motion of a rigid body, all forces are presumed to lie in a plane also containing the center of gravity of the body, and all particles of the body are presumed to move in paths parallel to the straight-line path of its center of gravity. Consequently, all particles have the same acceleration.

15-2 Accelerating Force

According to Newton's second law of motion, **a body acted upon by an unbalanced force will receive an acceleration that is proportional to the force and is in the direction in which the force acts.** This law, also referred to as the **acceleration law,** enables us to determine the acceleration that a known force will give a body, or the magnitude of the force required to impart a definite acceleration to the body.

For example, in Fig. 15-1 are shown three bodies, each acted upon by an unbalanced external force F, producing an acceleration a, which is proportional to the force and in the same direction. (F may be the resultant of several forces acting on the body.) In the case of the freely falling body (Fig. 15-1*c*), where air resistance is neglected, *the unbalanced force is its weight W, and the acceleration a produced by W has been found by experiment to equal approximately* 32.2 *feet per second per second* (ft per sec^2) usually denoted by g. Since F is always proportional to a, we may write

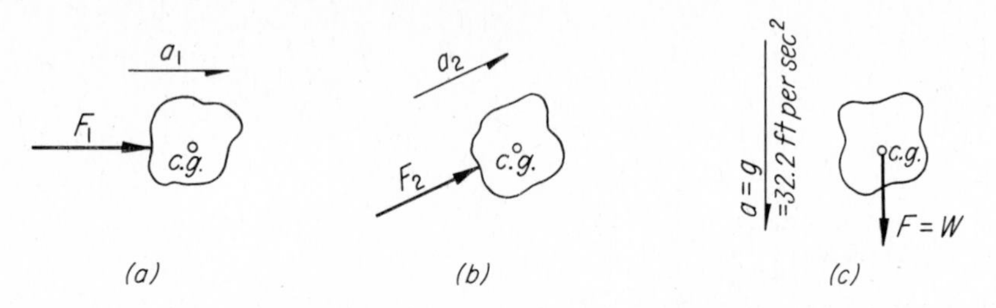

FIG. 15-1 Accelerating forces and accelerations produced. (*a*) Horizontal. (*b*) Angular. (*c*) Freely falling body.

$$\frac{F_1}{a_1} = \frac{F_2}{a_2} = \frac{F}{a} = \frac{W}{g} \tag{a}$$

from which

Accelerating force:

$$\boldsymbol{F = \frac{Wa}{g} = Ma} \tag{15-1}$$

where F = accelerating force, lb
W = weight of the body, lb
a = acceleration, ft per sec^2 produced by F
g = 32.2 ft per sec^2

The quantity W/g is commonly referred to as the **mass** of the body and is denoted by the symbol M. In engineering practice, *length, time,* and *force* are generally considered to be the *fundamental quantities,* and the system of units based on these quantities is called the *gravitational system.* Mass, therefore, is a derived quantity. Use of this term, and of the concept of *mass,* often simplifies discussions of bodies in motion. However, *in the solution of practical engineering problems, the derived quantity M must always be replaced by the fundamental quantities* W/g, in which W represents *force* (pounds), and g *length* and *time* (feet per second per second).

Equation (15-1) shows that the force F required to give a body a definite acceleration a is proportional to its mass ($M = W/g$). For example, a light wooden door requires less force to slam it shut quickly than does a heavy steel door whose mass is relatively high.

15-3 Inertia Force

According to *Newton's first law of motion,* **a body will remain at rest or in uniform motion in a straight line unless acted upon by an unbalanced force.** This law suggests that a body cannot of itself change its state of motion. This resisting property of a body is known as **inertia.** The first law, therefore, is also referred to as the **inertia law.** For example, a large ship increases its speed

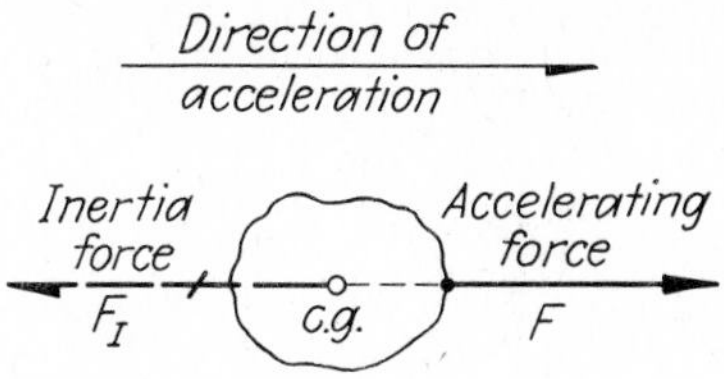

FIG. 15-2 Action and reaction.

very slowly even after the application of many thousands of horsepower to its propellers; likewise, it will continue its motion long after the power has been shut off, thus demonstrating its *inertia,* or resistance to changes in motion.

In the first and second of Newton's laws, mention is made of an *unbalanced force,* which must be external, but the third law states that *to every action there is an equal and opposed reaction.* This would suggest that the reaction to the unbalanced *external* force or action might be an *internal* force or reaction, which could only be the *inertia of the body.* Inertia is a *property* of a body only, and is not generally regarded as a force. However, **the inertia of a body manifests itself as a force** whenever a change in its motion is produced by the action of an accelerating force. Therefore we refer to it as an **inertia force.** The inertia force F_I is always equal in magnitude to the accelerating force F. Consequently,

Inertia force:

$$F_I = F = \frac{Wa}{g} \qquad \textbf{(15-2)}$$

from which

Acceleration:

$$a = \frac{F_I g}{W} \qquad \textbf{(15-3)}$$

As is indicated in Fig. 15-2, *the accelerating and inertia forces are always collinear, have opposite senses, and act through the center of gravity of the body.*

15-4 Friction and Other Forces

Bodies in motion are generally acted upon by several external forces, the *resultant* of which then is the accelerating force. These forces are considered to act in a single plane, called the *reference plane,* passing through the center of gravity of the body. Consider a body resting on a level surface, as shown in Fig. 15-3*a*. Its weight W is counteracted by the reacting force N of the supporting surface. If a force P is applied parallel to the supporting surface (Fig. 15-3*b*), the frictional resistance or friction force F is developed as a reaction to P, hence is oppositely directed. As long as P is less than the greatest frictional resistance F which can be developed, P will equal F and the body will remain at rest in *static equilibrium.*

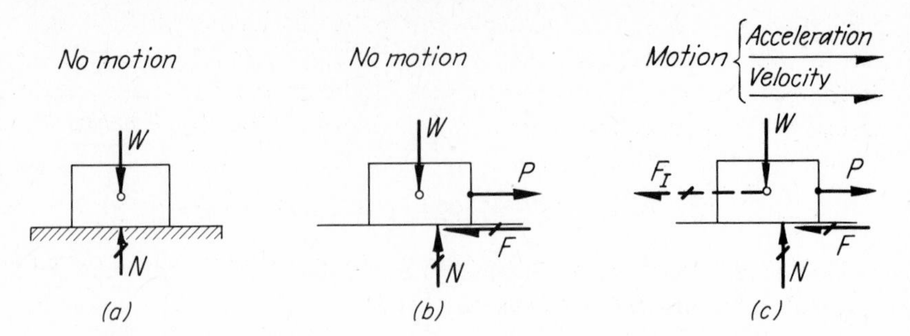

FIG. 15-3 Friction forces. (*a*) Body at rest. (*b*) $P = F$. (*c*) $P > F$.

If now we increase force P until it exceeds the maximum possible frictional resistance F, the difference $P - F$, acting to the right, will be *the unbalanced external accelerating force* which will cause the body to be *accelerated* to the right, as indicated by the a and v **motion arrows** in Fig. 15-3*c*. The equal and opposite reaction to this accelerating force will be the inertia force F_I. As may be seen from this simple illustration, *the accelerating force is seldom a single force, but is generally the resultant of a number of forces acting on the body.* The small bars across the shanks of the arrows representing forces N, F, and F_I indicate that they are *reacting forces* as distinguished from *acting forces*. N and F_I are always reacting forces; the friction force F may be either an acting or a reacting force, depending upon the conditions under which motion takes place.

In Fig. 15-4*a* is shown a body moving up an incline with decreasing velocity. The motion arrow v indicates displacement and velocity up the incline, here considered as positive. The fact that the velocity decreases indicates a negative acceleration directed down the incline. *The inertia force is always directed oppositely to the acceleration,* and therefore acts up the incline, as is shown in the free-body diagram, Fig. 15-4*c*.

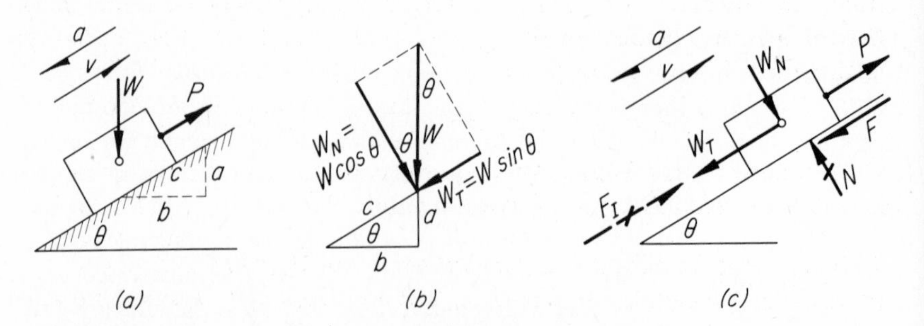

FIG. 15-4 Body moving up incline with decreasing velocity. (*a*) Space diagram. (*b*) Components of W. (*c*) Free-body diagram.

Solution of problems of this type is simplified by having *all forces acting parallel and perpendicular to the incline,* as in Fig. 15-4c. This method necessitates resolving W into two components, W_T tangent to the incline, and W_N normal to the incline. The diagram in Fig. 15-4b will be found useful. When the slope of the incline is given by an angle θ,

$$W_T = W \cdot \sin\theta \qquad \text{and} \qquad W_N = W \cdot \cos\theta \tag{a}$$

When slope is given as a fraction or in percentage, the hypotenuse $c = \sqrt{a^2 + b^2}$ and, by similarity of triangles,

$$\frac{W_T}{a} = \frac{W_N}{b} = \frac{W}{c} \tag{b}$$

or

$$W_T = \frac{a}{c}W \qquad \text{and} \qquad W_N = \frac{b}{c}W \tag{c}$$

Clearly,

$$\frac{a}{c} = \sin\theta \qquad \text{and} \qquad \frac{b}{c} = \cos\theta \tag{d}$$

For slopes of 10 percent or less, the values of $\cos\theta$ (or b/c) are practically unity, and $\sin\theta$ equals practically the percent slope expressed in decimal form. When so considered, $W_N = W$, and $W_T = W\sin\theta$, or $W \cdot$ (percent of slope).

15-5 Dynamic Equilibrium. The Force Method

Since the *accelerating force,* which is the resultant of all the external forces, is balanced by the internal *inertia force,* equilibrium of forces exists, similar to static equilibrium. That is, *the resultant of the external impressed forces is equated to the resultant of the internal inertia forces acting on all the particles of the body,* thus establishing equilibrium. This statement is known as **d'Alembert's principle.**

Because an inertia force exists only as a result of *dynamic* conditions, the state of equilibrium thus created by its use is referred to as **dynamic equilibrium.** The equations of static equilibrium may then be applied also to conditions of dynamic equilibrium, thus greatly simplifying the solution of problems in dynamics. These equations are

$$\Sigma F_X = 0 \qquad \Sigma F_Y = 0 \qquad \text{and} \qquad \Sigma M = 0$$

This method of solution, based on dynamic equilibrium of forces, will hereafter be referred to as the **force method.**

15-6 Dynamic Equilibrium of Concurrent Force Systems

In studying the rectilinear motion of a body, we are, as a rule, interested primarily in one of two things: (1) the motion of the body and the magnitudes of the external forces influencing its motion, regardless of their actual points

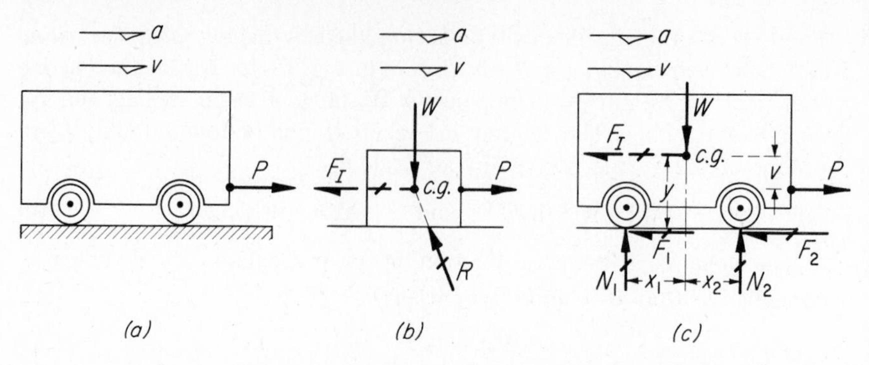

FIG. 15-5 Free-body diagrams of moving trailer. (*a*) Space diagram. (*b*) Concurrent force system. (*c*) Nonconcurrent force system.

of application; or (2) the magnitudes and actual points of application of each of the forces influencing its motion.

Consider the moving trailer shown in Fig. 15-5*a*. If the problem comes under case (1), the body may conveniently be shown as a dimensionless rectangle with all forces acting through its center of gravity, as shown in Fig. 15-5*b*. Then we have a *concurrent force system*, completely solved by the equations $\Sigma F_X = 0$ and $\Sigma F_Y = 0$. The reaction R, however, is usually resolved into its vertical and horizontal components, N the normal reaction of the supporting surface, and F the friction force, as in Fig. 15-6*b*. If the problem comes under case (2), a fully dimensioned free-body diagram must be used, as shown in Fig. 15-5*c*. Now we have a *nonconcurrent force system*, solved by the equations $\Sigma F_X = 0$, $\Sigma F_Y = 0$, and $\Sigma M = 0$.

ILLUSTRATIVE PROBLEMS

15-1. A body weighing 200 lb is pulled with increasing velocity along a level surface by a constant force P of 60 lb, as shown in Fig. 15-6*a*. The coefficient of friction

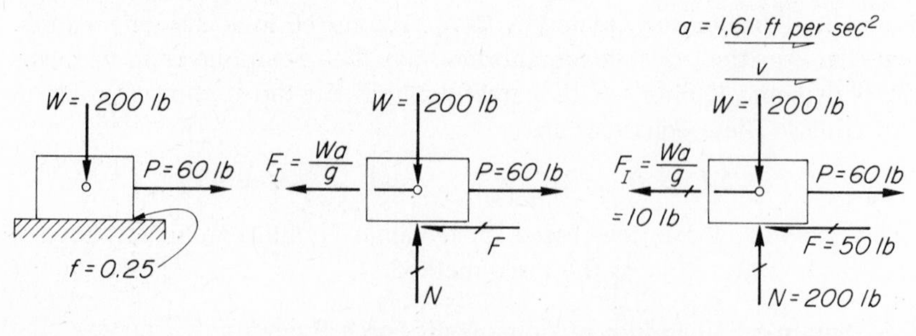

FIG. 15-6 Body in accelerated motion. Dynamic equilibrium of concurrent force system.

f is 0.25, between the body and the supporting surface. Compute the acceleration given the body by force P.

Solution: The initial direction of motion (velocity and displacement) obviously will be to the right. Since force P causes the motion, the acceleration also will be to the right, and we so indicate by motion arrows on the free-body diagram in Fig. 15-6*b*. The direction of the friction force F is opposite to that of the velocity v, and the direction of the inertia force F_I is opposite to that of the acceleration a. The value of N we obtain from $\Sigma V = 0$. Then $F = fN$. Finally, from $\Sigma H = 0$, we obtain the value of F_I, the inertia force, from which a may then be computed, since $a = F_I g/W$. Then, since $W = 200$ lb, $f = 0.25$, and $g = 32.2$ ft per sec^2, we have

$$[\Sigma V = 0] \qquad N = 200 \text{ lb}$$

$$[F = fN] \qquad F = (0.25)(200) = 50 \text{ lb}$$

$$[\Sigma H = 0] \qquad F_I + 50 = 60 \qquad \text{or} \qquad F_I = 10 \text{ lb}$$

$$\left[a = \frac{F_I g}{W}\right] \qquad a = \frac{(10)(32.2)}{200} = 1.61 \text{ ft per sec}^2 \qquad \textit{Ans.}$$

The checking of results thus obtained is of extreme importance. To verify that the forces calculated are actually in equilibrium, their values should be recorded on the free-body diagram, Fig. 15-6*b*, *as soon as found*. So to record the values of the forces in the examples in a textbook might cause some confusion[1] as to which forces are originally known and which are to be determined. When all forces are thus recorded, Fig. 15-6*b* will appear as shown in Fig. 15-6*c*, which is *the complete free-body diagram*. A glance, or a simple algebraic summation of forces, will then verify the required conditions of equilibrium.

15-2. A box weighing 644 lb is hoisted vertically out of a ship's hold with an upward acceleration a of 10 ft per sec^2. Calculate the tension T in the hoisting cable.

Solution: The free-body diagram is drawn as in Fig. 15-7. Because the acceleration is upward, the inertia force F_I acts downward. Since W and a are known, F_I may be computed. T is then found by a summation of vertical forces. The equations are

[1] Although the separate steps in the solution of a problem can be *progressively* explained in the text, the gradual evolution of the free-body diagram, as it unfolds on the student's paper, unfortunately cannot readily be shown. Therefore, the free-body diagrams accompanying solved examples are generally shown as they appear at the beginning of the solution, thus clearly indicating the known quantities and the unknown to be determined.

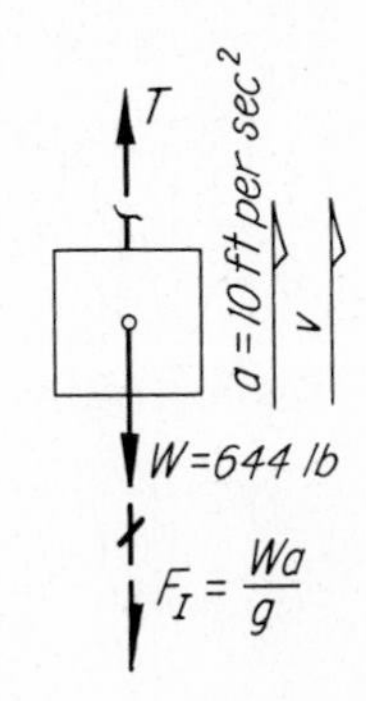

FIG. 15-7 Free-body diagram.

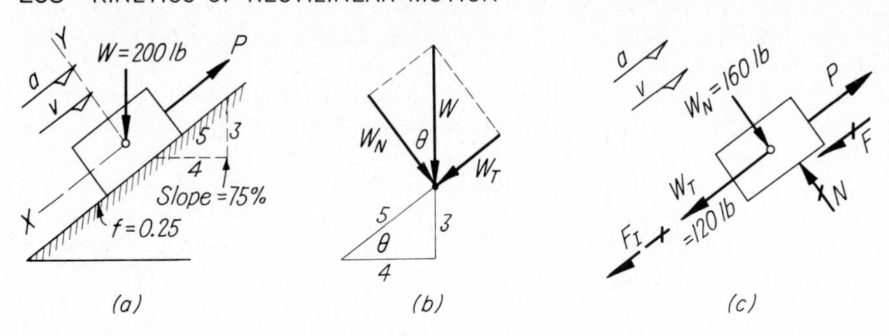

FIG. 15-8 Body moving up incline with increasing velocity. (*a*) Space diagram. (*b*) Components of *W*. (*c*) Free-body diagram.

$$\left[F_I = \frac{Wa}{g}\right] \qquad F_I = \frac{(644)(10)}{32.2} = 200 \text{ lb}$$

$$[\Sigma V = 0] \qquad T = W + F_I = 644 + 200 \qquad \text{or} \qquad T = 844 \text{ lb} \qquad \textit{Ans.}$$

15-3. The body shown in Fig. 15-8*a* weighs 200 lb and is pulled up a 75 percent incline by a force *P*. The coefficient of friction *f* is 0.25 between contacting surfaces. (*a*) Determine the constant value of *P* required to bring the body from rest to a velocity of 20 ft per sec in 10 sec. (*b*) What value of *P* will then suffice to maintain that velocity?

Solution: From the statement of the problem we see that a and v are directed up the incline. After the free-body diagram is drawn as in Fig. 15-8*c*, the components W_T and W_N are computed. Next we may determine the acceleration a, since v_0, v, and t are known. Then $F_I = Wa/g$. The force P is then obtained by a summation of forces parallel to the incline. Note that F is *opposed* to the direction of motion, and F_I is *opposed* to the direction of the acceleration. Computing the components of W, we obtain

$$W_T = \frac{a}{c}W = \frac{3}{5}(200) = 120 \text{ lb} \qquad \text{and} \qquad W_N = \frac{b}{c}W = \frac{4}{5}(200) = 160 \text{ lb}$$

To obtain the acceleration, when $v_0 = 0$, $v = 20$ ft per sec, and $t = 10$ sec, we use Eq. (*B*), Table 14-1. That is,

$$\left[a = \frac{v - v_0}{t}\right] \qquad a = \frac{20 - 0}{10} = 2 \text{ ft per sec}^2$$

Then, to obtain F_I,

$$\left[F_I = \frac{Wa}{g}\right] \qquad F_I = \frac{(200)(2)}{32.2} = 12.4 \text{ lb}$$

We may now evaluate forces N, F, and P as follows:

$$[\Sigma F_Y = 0] \qquad N = W_N = 160 \text{ lb}$$
$$[F = fN] \qquad F = (0.25)(160) = 40 \text{ lb}$$
$$[\Sigma F_X = 0] \qquad P = 40 + 120 + 12.4 = 172.4 \text{ lb} \qquad \textit{Ans.}$$

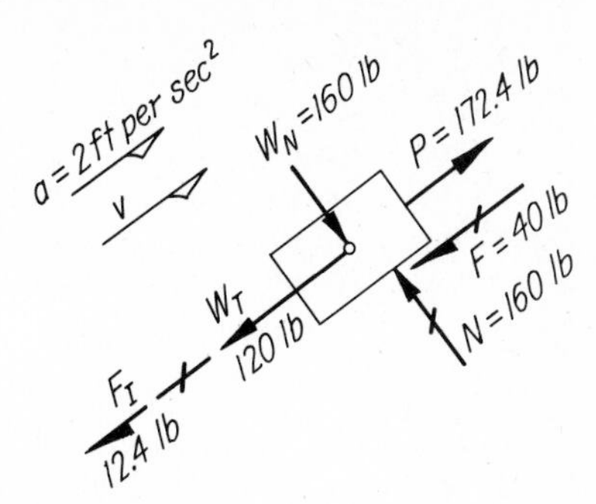

FIG. 15-9 *Complete* free-body diagram.

When all values have been recorded on the free-body diagram (Fig. 15-8*c*), it will appear as in Fig. 15-9. A glance, or a simple algebraic summation of forces, will then quickly verify the required condition of dynamic equilibrium.

If the velocity remains constant after reaching 20 ft per sec, the acceleration becomes zero, and $F_I = 0$. Then, the force required to maintain that velocity is

$$P = F + W_T = 40 + 120 \qquad \text{or} \qquad P = 160 \text{ lb} \qquad \textit{Ans.}$$

15-4. The block shown in Fig. 15-10*a* weighs 100 lb and is acted upon by a horizontal force *P*. The coefficient of friction *f* is 0.4 between the body and the incline. Calculate the magnitude of *P* required (*a*) barely to prevent the block from slipping down the incline, (*b*) to push the block up the incline with uniform velocity, and (*c*) to give the block an acceleration of 8.05 ft per sec² up the incline.

Solution: The components of *W* and *P*, parallel (tangent) and perpendicular (normal) to the incline, are readily computed with the aid of the diagrams shown in Fig. 15-10*b* and *c*. From Fig. 15-10*b*, the components of *W* are

$$W_T = \frac{3}{5}(100) = 60 \text{ lb} \qquad \text{and} \qquad W_N = \frac{4}{5}(100) = 80 \text{ lb}$$

Since *P* is unknown, its components can be expressed only in terms of *P*. That is, from Fig. 15-10*c*,

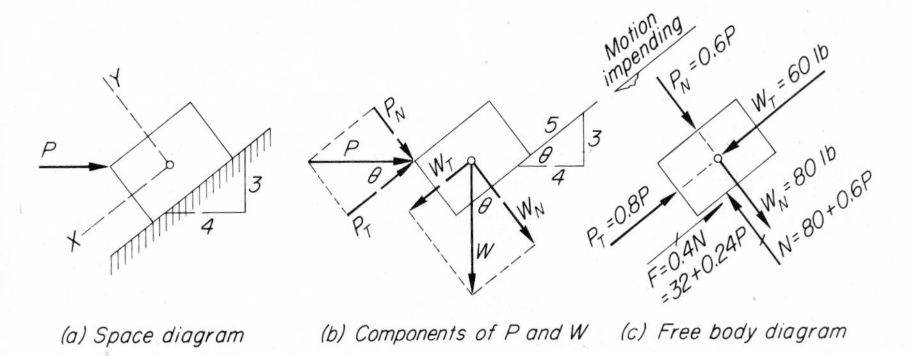

FIG. 15-10 Body on incline. Motion impending down incline.

$$P_T = \frac{4}{5}P = 0.8P \qquad \text{and} \qquad P_N = \frac{3}{5}P = 0.6P$$

Part (a): When the block is barely kept from sliding, motion impends *down* the incline. The maximum frictional resistance ($F = 0.4N$) is then developed, acting *up* the incline, as shown in Fig. 15-10*c*. The values of N, F, and P are now determined in that order. Hence,

$[\Sigma F_Y = 0]$ $$N = 80 + 0.6P$$

Then

$[F = fN]$ $$F = (0.4)(80 + 0.6P) = 32 + 0.24P$$

We may now obtain the force P from a summation of forces parallel to the incline. Hence,

$[\Sigma F_X = 0]$ $$0.8P + (32 + 0.24P) = 60$$

or $$1.04P = 60 - 32 = 28$$

and $$P = 26.9 \text{ lb} \qquad \textit{Ans.}$$

Part (b): When the body moves up the incline with uniform velocity, the acceleration, and hence F_I, is zero. The maximum frictional resistance ($F = 0.4N$) is developed acting down the incline against the direction of motion, as shown in Fig. 15-11.

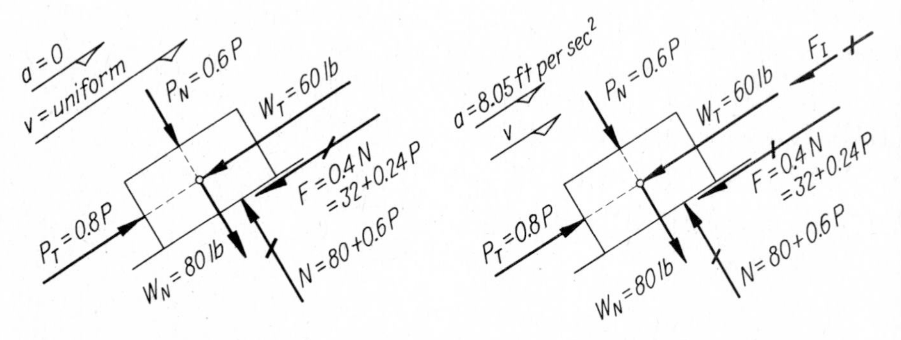

FIG. 15-11 Uniform motion up the incline.

FIG. 15-12 Accelerated motion up the incline.

The values of all forces except P are the same as in part (*a*). We may now obtain P by a summation of forces parallel to the incline. That is,

$[\Sigma F_X = 0]$ $$0.8P = 60 + (32 + 0.24P)$$

or $$0.56P = 60 + 32 = 92$$

and $$P = 164 \text{ lb} \qquad \textit{Ans.}$$

Part (c): The acceleration of 8.05 ft per sec² *up* the incline merely introduces an inertia force F_I acting *down* the incline, as shown in Fig. 15-12. All other forces remain the same as in part (*b*). The value of F_I is

$$\left[F_I = \frac{Wa}{g}\right] \qquad F_I = \frac{(100)(8.05)}{32.2} = 25 \text{ lb}$$

We find P from a summation of forces parallel to the incline, which gives

$$[\Sigma F_X = 0] \qquad 0.8P = 60 + 25 + (32 + 0.24P)$$

or $$0.56P = 60 + 25 + 32 = 117 \quad \text{and} \quad P = 209 \text{ lb} \qquad \textit{Ans.}$$

The value of each force solved for should be recorded on the proper free-body diagram as soon as determined. When all forces are so recorded, the summations, $\Sigma F_X = 0$ and $\Sigma F_Y = 0$, should be made to verify that force equilibrium exists.

PROBLEMS

15-5. A man, standing in the aisle of a streetcar, faces the front of the car. To maintain his balance, at what angle θ with the vertical must he lean, forward or backward, when the car (*a*) accelerates 3.22 ft per sec^2 and (*b*) decelerates 6.44 ft per sec^2? *Ans.* (*a*) $\theta = 5°43'$; (*b*) $\theta = 11°19'$

15-6. Solve Prob. 15-1 when $f = 0.2$, $P = 50$ lb, and $W = 100$ lb.

15-7. Calculate the force P required to give the body shown in Fig. 15-3 an acceleration to the right of 5 ft per sec^2, if $W = 322$ lb and $f = 0.3$. If the initial velocity v_0 of this body is 20 ft per sec, what distance s will it be moved in 10 sec? *Ans.* $P = 146.6$ lb; $s = 450$ ft

15-8. A box slides down an incline and thence across a level floor. If the box slides 15 ft across the floor before coming to rest and the coefficient of friction f is 0.12, what was the velocity of the box at the bottom of the incline?

15-9. A man weighing 161 lb enters an elevator about to descend. What pressure N does he exert on the floor of the elevator (*a*) when it is still at rest, (*b*) when it descends with an acceleration of 4 ft per sec^2, and (*c*) when it is brought to a stop with a deceleration of 4 ft per sec^2? (This problem verifies a common experience.) *Ans.* (*a*) $N = 161$ lb; (*b*) $N = 141$ lb; (*c*) $N = 181$ lb

15-10. In Fig. 15-13 is shown a materials hoist such as is commonly used on construction jobs. The hoisting platform weighs 420 lb; the platform must carry a

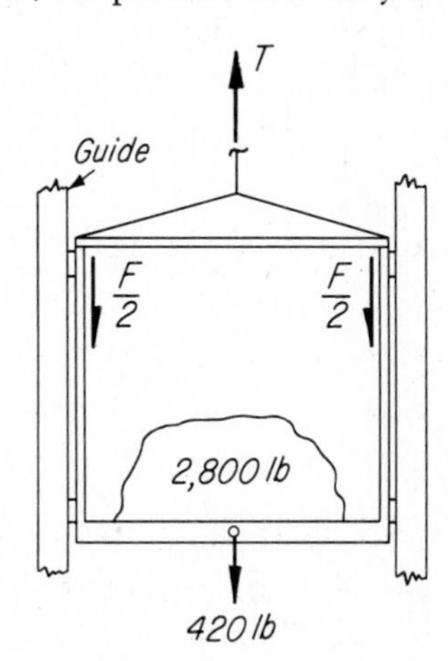

FIG. 15-13 Prob. 15-10.

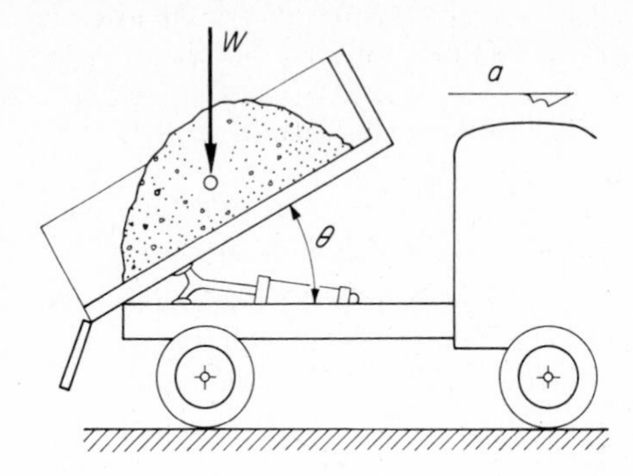

FIG. 15-14 Prob. 15-11.

maximum load of 2,800 lb. The total friction resistance F against both side guides is 50 lb. The hoisting cable must provide strength equal to twice the maximum tension T caused by an upward acceleration of 30 ft per sec^2. Calculate T and the required cable strength S.

15-11. The dump box on the truck shown in Fig. 15-14 makes an angle of 30° with the horizontal. Calculate the required acceleration of the truck to the right if the weight W is to slide down the box. Assume $f = 0.7$.

Ans. $a = 2.82$ ft per sec^2

15-12. A body slides down a 30° incline. The coefficient of friction f is 0.25. (*a*) How long will it take the body to slide 60 ft starting from rest? (*b*) What should be the inclination θ of the plane if the body is to slide down the incline with constant speed?

15-13. The body shown in Fig. 15-15 weighs 100 lb and is pulled along a level surface by the sloping force P. The coefficient of friction $f = 0.4$. (*a*) Calculate the acceleration a given the body when P is 50 lb. (*b*) What distance s will it travel from rest in 10 sec?

15-14. Calculate the distance s the block shown in Fig. 15-16 will be moved from rest in 10 sec, if $P = 39$ lb, $W = 100$ lb, and $f = 0.2$. *Ans.* $s = 209$ ft

15-15. A pile-driver hammer weighing 644 lb is to be raised 48 ft in 4 sec. The total friction on the guides is constant at 100 lb. What constant pull must be exerted on the cable? *Ans.* $T = 864$ lb

15-16. In Prob. 15-14 (and Fig. 15-16), reverse the sense of P and solve. (P will then act upward and to the left. Note the effect of the decrease in frictional resistance.)

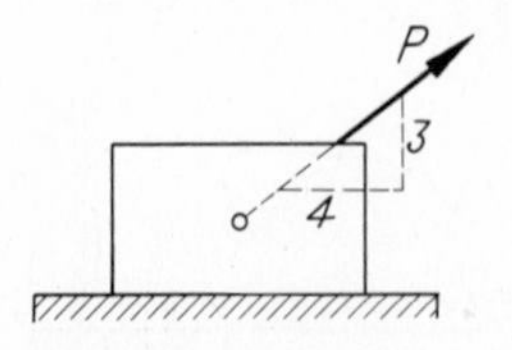

FIG. 15-15 Probs. 15-13 and 15-22.

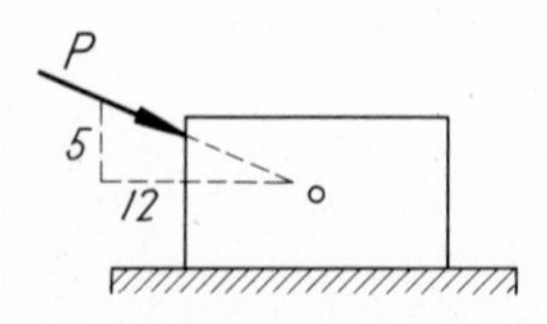

FIG. 15-16 Probs. 15-14, 15-16, and 15-23.

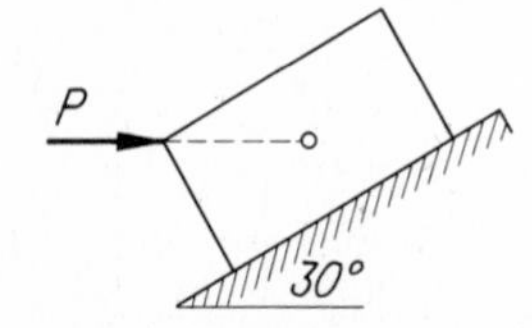

FIG. 15-17 Probs. 15-21 and 15-24.

15-17. An automobile weighing 4,000 lb travels at 30 mph. When the brakes are applied with constant force, calculate the total frictional resistance F that must be developed between the tires and the roadbed, if the car is stopped in a distance of 55 ft. *Ans.* $F = 2{,}180$ lb

15-18. Solve Prob. 15-3 if W is 100 lb and $f = 0.5$.

15-19. A trailer weighing 40,000 lb is given a uniform acceleration while traveling on a level road. The velocity is increased from 10 to 30 mph in a distance of $\frac{1}{2}$ mile. If the total frictional resistance (bearings, rolling resistance of wheels, air friction, etc.) is 20 lb per ton, what was the drawbar pull? *Ans.* $P = 804$ lb

15-20. A car skids 88 ft in coming to rest from a velocity of 30 mph. What was the coefficient of friction f between the tires and the road?

15-21. The body shown in Fig. 15-17 weighs 50 lb. If P is 70 lb and f is 0.2, what acceleration will be given the body?

15-22. If the block shown in Fig. 15-15 weighs 100 lb and f is 0.4, what minimum force P, acting in the direction shown, will move it with uniform velocity? (See Prob. 15-4.) *Ans.* $P = 38.5$ lb

15-23. In Fig. 15-16 is shown a block weighing 80 lb. The coefficient of sliding friction f is 0.3. Calculate the minimum constant force P required to move it 90 ft from rest in 6 sec.

15-24. The block shown in Fig. 15-17 weighs 50 lb. If the coefficient of friction f is 0.4, what minimum force P, acting in the direction shown, is required (*a*) barely to keep it from sliding down the incline and (*b*) to move it up the incline with uniform velocity? *Ans.* (*a*) $P = 7.2$ lb; (*b*) $P = 63.5$ lb

15-25. In Fig. 15-18 A and B represent, respectively, a loaded truck and a trailer, connected with a bar. The trailer weighs 12,880 lb, and its road resistance F (friction force F = rolling friction plus bearing friction and air resistance) is 150 lb. Calculate the force P in the bar when the truck (*a*) runs with uniform velocity, (*b*) accelerates 3 ft per sec^2, and (*c*) decelerates 5 ft per sec^2. (Draw a separate free-body diagram of the trailer for each part.)
Ans. (*a*) $P = 150$ lb; (*b*) $P = 1{,}350$ lb; (*c*) $P = -1{,}850$ lb

15-26. In Prob. 15-25, the truck weighs 16,100 lb and its road resistance is 200 lb. Calculate the force that is exerted by the truck in moving itself and trailer when it (*a*) runs at uniform velocity, (*b*) accelerates 3 ft per sec^2, and (*c*) decelerates 5 ft per sec^2. In parts (*a*) and (*b*) represent the *tractive force TF* by a forward-acting force pulling on the truck, and in (*c*) the *retarding force RF* as a backward-acting force. Note that the conditions of motion in this problem are the same as in Prob. 15-25. Hence, use the forces in the bar as obtained in that problem.

15-27. A 1,610-lb sled is brought from rest to 30 mph in a horizontal distance of

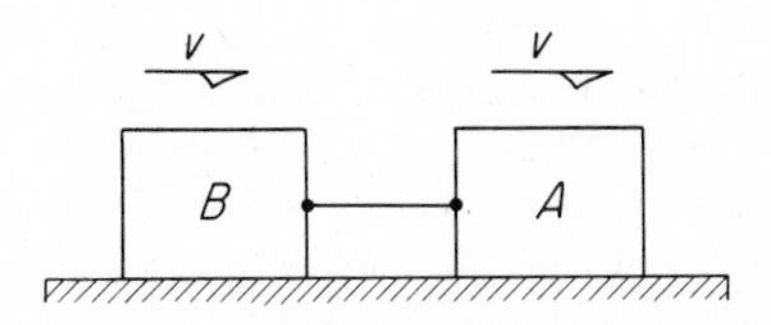

FIG. 15-18 Probs. 15-25 and 15-26.

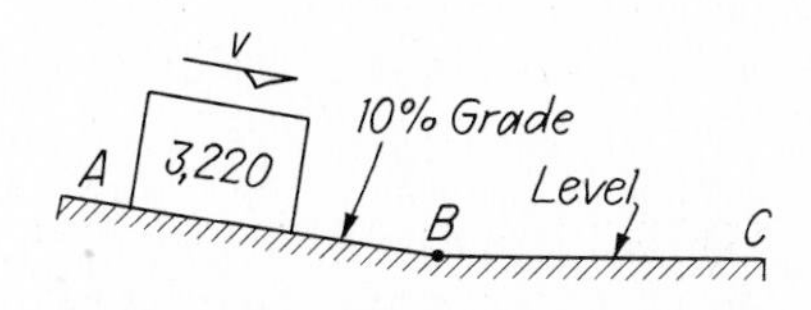

FIG. 15-19 Prob. 15-28.

440 ft. If the tow rope makes an angle of 30° with the horizontal, find the required constant tension in the rope. Assume the coefficient of friction f to be 0.1.

15-28. A car weighing 3,220 lb coasts down a 10 percent grade from A to B, a distance of 100 ft, and then coasts along a level road until stopping at C (Fig. 15-19). Assume the road resistance to remain constant at 40 lb for the entire distance A to C. If the initial velocity of the car at A is 15 mph, compute (*a*) its velocity at B and (*b*) the distance s it coasts from B to C.

Ans. (*a*) $v_B = 32.4$ ft per sec $= 22.1$ mph; (*b*) $s = 1{,}310$ ft

15-7 Systems of Bodies in Motion

Two or more bodies connected together in such manner that their motions are dependent upon one another are referred to as a **system of bodies.** The motions of the bodies in a system may differ in direction only, as in Fig. 15-21, where A moves up the incline while B moves vertically downward, or the motions may have the same direction but differ in sense and in magnitude, as in Fig. 15-20, in which both A and B move vertically (direction), but A moves upward (sense) while B moves downward with a velocity equal to one-half that of A.

The weight and frictional resistance of cords and sheaves in such systems are herein considered to be small in comparison with other forces involved and will therefore be neglected. When this is done, *the tension in the cord is regarded as constant throughout its entire length.* Such terms as *weightless cord* and *weightless and frictionless sheave* are useful to indicate that these items, being negligible, are disregarded.

Difficulty is sometimes encountered in determining which way the bodies in a system will move. To aid in visualizing this, *let all but one body in turn be held immovable while the static tension in the connecting cord is determined. The motion, if any, will take place in the direction of the external pull that produces the greatest static tension in the cord.*

In the following problems all forces, or their resultants, are considered to act through the centers of gravity of the bodies; the force systems acting upon them are therefore concurrent.

ILLUSTRATIVE PROBLEMS

15-29. The two weights A and B, shown in Fig. 15-20*a*, are supported by an arrangement of light cords and sheaves whose negligible weights and friction may be disregarded. If A weighs 60 lb and B weighs 180 lb, calculate the acceleration a of each of blocks A and B after release, and the tension T in the main cord.

Solution: A study of Fig. 15-21*a* shows that the displacement of B will be one-half that of A. The acceleration and velocity of B at any instant will then be one-half that of A, or $a_B = \frac{1}{2}a_A$, and $v_B = \frac{1}{2}v_A$, as noted in Fig. 15-20*c*. We may now determine the acceleration a of body A by considering A and B separately as free bodies.

The main cord is now cut just above A and also above the sheave supporting B,

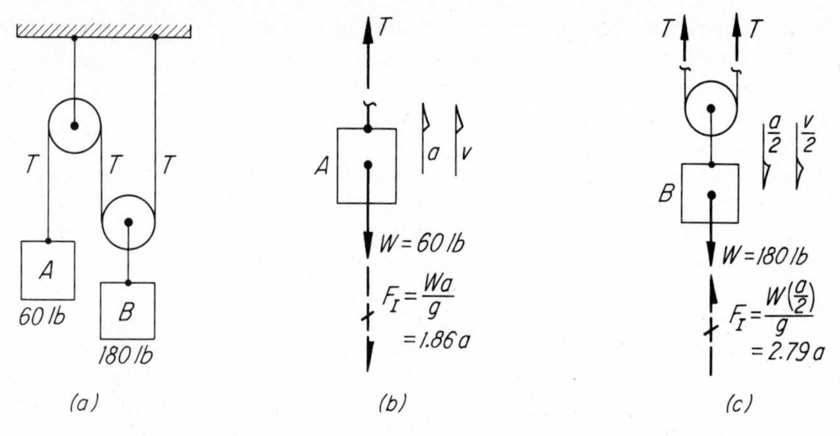

FIG. 15-20 A system of connected bodies. (*a*) Space diagram. (*b*) Free body *A*. (*c*) Free body *B*.

and the free-body diagrams shown in Fig. 15-20*b* and *c* are drawn. Next we must determine which way each body will move. Let each body in turn be held immovable. When *B* is held, the pull of *A* produces a static tension of 60 lb in the cord. When *A* is held, the pull of *B* produces a static tension of 90 lb in the cord on each side of the supporting sheave. Since *B* produces the greater tension, the motion (a, v, and s) of *B* will be downward and that of *A* upward, which directions we then indicate on the free-body diagrams. The inertia force for each body may now be computed in terms of its acceleration a. Hence,

$$\left[F_I = \frac{Wa}{g}\right] \qquad \text{For body } A\text{:} \quad F_I = \frac{60a}{32.2} = 1.86a \qquad (a)$$

$$\left[F_I = \frac{Wa}{g}\right] \qquad \text{For body } B\text{:} \quad F_I = \frac{90a}{32.2} = 2.79a \qquad (b)$$

We may now determine a by two summations of the forces acting on each of the two bodies. That is,

$$[\Sigma V = 0] \qquad \text{For body } A\text{:} \quad T = 60 + 1.86a \qquad (c)$$
$$[\Sigma V = 0] \qquad \text{For body } B\text{:} \quad 2T = 180 - 2.79a \qquad (d)$$

When Eq. (*c*) is multiplied by 2, Eqs. (*c*) and (*d*) become equalities, and their right sides may then be equated to obtain a. That is,

$$120 + 3.72a = 180 - 2.79a \qquad \text{or} \qquad 6.51a = 180 - 120 = 60$$

from which

$$a_A = 9.22 \text{ ft per sec}^2 \qquad \text{and} \qquad a_B = 4.61 \text{ ft per sec}^2 \qquad \textit{Ans.}$$

The values of F_{IA} and F_{IB} then are as follows:

$$F_{IA} = (1.86)(9.22) = 17.2 \text{ lb}$$

and

$$F_{IB} = (2.79)(9.22) = 25.8 \text{ lb}$$

We may now find the tension T from either Fig. 15-20*b* or *c*. As a check, both should be used. The equations are

$[\Sigma V = 0]$ For body *A*: $T = 60 + 17.2$ or $T = 77.2$ lb *Ans.*

$[\Sigma V = 0]$ For body *B*: $2T = 180 - 25.8$ or $T = 77.2$ lb *Check.*

15-30. The two bodies *A* and *B* in Fig. 15-21*a* are connected with a weightless cord passing over a weightless and frictionless sheave. The coefficient of friction f is 0.25 between body *A* and the incline. Determine the acceleration a of the system, when released from rest, and the tension T in the connecting rope.

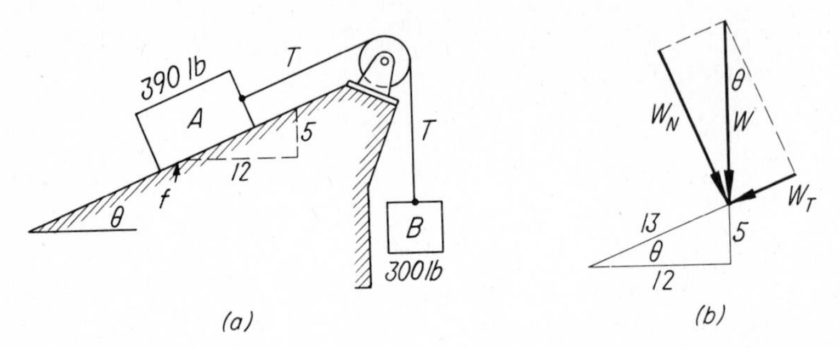

FIG. 15-21 A system of connected bodies. (*a*) Space diagram. (*b*) Components of *W*.

Solution: Clearly, the two bodies, being directly connected, have the same motion (a, v, and s) except for direction. The rope is cut near each body, and free-body diagrams are drawn, as in Fig. 15-22. The components W_T and W_N and the force N are then determined. By similarity of triangles (Fig. 15-21*b*),

$$W_T = \frac{5}{13}(390) = 150 \text{ lb} \qquad \text{and} \qquad W_N = \frac{12}{13}(390) = 360 \text{ lb}$$

To obtain N, using Fig. 15-23*a*,

$$[\Sigma F_Y = 0] \qquad N = 360 \text{ lb}$$

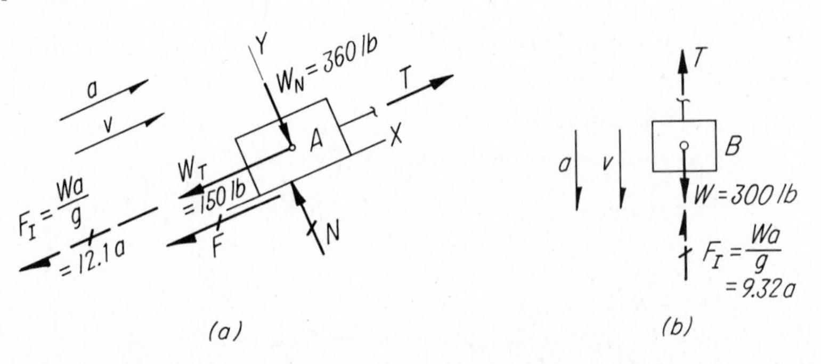

FIG. 15-22 A system of connected bodies. (*a*) Free-body diagram of *A*. (*b*) Free-body diagram of *B*.

The direction of the motion must now be determined. When A is held immovable, the pull on the cord by B is 300 lb downward. When B is held, the component W_T pulls 150 lb down the incline. Therefore, if motion occurs, B will move downward and A will move up the incline, as indicated by the motion arrows in Fig. 15-22*a* and *b*. Hence, for body A, F and F_{IA} both act down the incline; for body B, F_{IB} acts upward. We may then find the acceleration a of the system by writing two equations of force equilibrium, $\Sigma F_X = 0$ for body A, and $\Sigma F_Y = 0$ for body B. When a has thus been found, F_I for each body and T are readily calculated.

Since $f = 0.25$ and $N = 360$ lb, the friction force is

$[F = fN]$ $$F = (0.25)(360) = 90 \text{ lb}$$

In terms of a, the inertia forces are

$\left[F_I = \frac{Wa}{g}\right]$ For body A: $$F_{IA} = \frac{390a}{32.2} = 12.1a \qquad (e)$$

$\left[F_I = \frac{Wa}{g}\right]$ For body B: $$F_{IB} = \frac{300a}{32.2} = 9.32a \qquad (f)$$

We may then write two equations of force equilibrium. That is,

$[\Sigma F_X = 0]$ For body A: $$T = 150 + 90 + 12.1a \qquad (g)$$
$[\Sigma F_Y = 0]$ For body B: $$T = 300 - 9.32a \qquad (h)$$

Since Eqs. (*g*) and (*h*) are equalities,

$$150 + 90 + 12.1a = 300 - 9.32a$$

or $21.42a = 300 - 240 = 60$ and $a = 2.8$ ft per sec^2 *Ans.*

To obtain T, we first evaluate F_{IA} and F_{IB} as follows:

$$F_{IA} = 12.1a = (12.1)(2.8) = 33.9 \text{ lb}$$
$$F_{IB} = 9.32a = (9.32)(2.8) = 26.1 \text{ lb}$$

We then calculate T as follows:

For body A: $T = 150 + 90 + 33.9$ or $T = 273.9$ lb *Ans.*
For body B: $T = 300 - 26.1$ or $T = 273.9$ lb *Check.*

All numerical values should be recorded on the two free-body diagrams, so that a *visual check* is obtained.

PROBLEMS

In the following problems, the weights of cords and sheaves are presumed to be negligible. Their frictional resistance and inertia will therefore be neglected.

15-31. Body A in Fig. 15-23 weighs 96.6 lb and B weighs 128.8 lb. Calculate (*a*) the acceleration a of the system after release, (*b*) the tension T_1 in the main cord, (*c*) the tension T_2 in the upper cord, and (*d*) the velocity v of the system after it has moved 10 ft from rest.

Ans. (*a*) $a = 4.6$ ft per sec^2; (*b*) $T_1 = 110$ lb; (*c*) $T_2 = 220$ lb; (*d*) $v = 9.59$ ft per sec

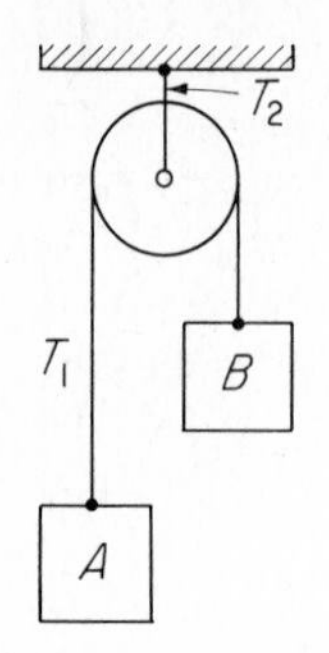

FIG. 15-23 Probs. 15-31, 15-32, 15-74, and 15-75.

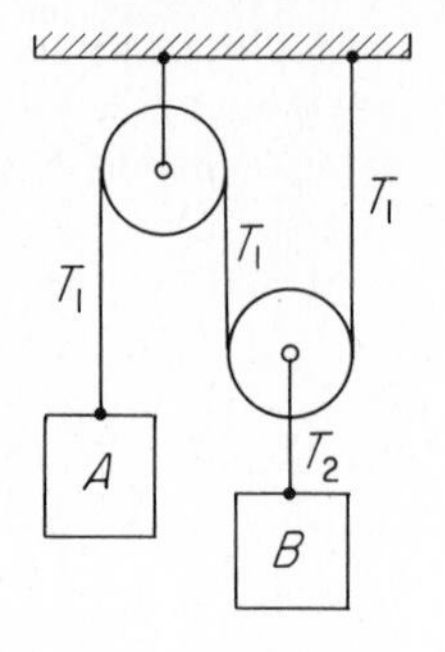

FIG. 15-24 Probs. 15-33 and 15-34.

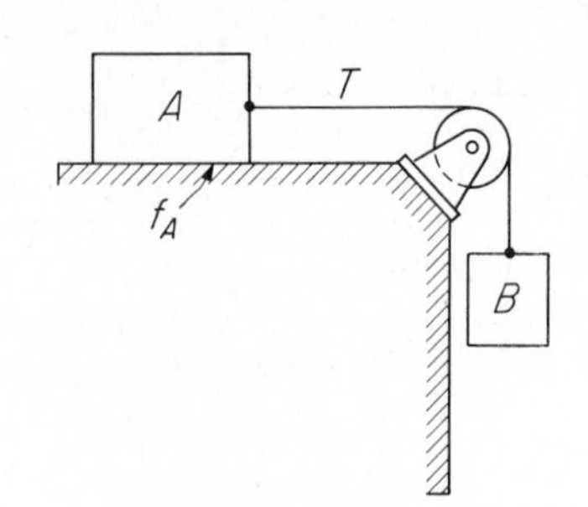

FIG. 15-25 Probs. 15-35 to 15-38 and 15-76.

15-32. In Fig. 15-23 body *A* weighs 80 lb. Calculate the weight W_B of body *B* required to accelerate the system 6.44 ft per sec^2, with *B* moving downward.

15-33. If body *A* in Fig. 15-24 weighs 112 lb, calculate the weight W_B of *B* required to give body *A* an acceleration of 8.05 ft per sec^2 with *B* moving downward. (Note that the displacement of *B* is one-half that of A. Hence, $a_B = \frac{1}{2}a_A$ and $v_B = \frac{1}{2}v_A$.)

Ans. $W = 320$ lb

15-34. Calculate the acceleration *a* of block *A* shown in Fig. 15-24, and the two cord tensions T_1 and T_2. Body *A* weighs 50 lb and *B* weighs 80 lb.

Ans. $a = 4.6$ ft per sec^2; $T_1 = 42.85$ lb; $T_2 = 85.7$ lb

15-35. In Fig. 15-25, body *A* weighs 40 lb. If f_A is 0.5, calculate the weight W_B of *B* required to accelerate the system 6 ft per sec^2.

15-36. Blocks *A* and *B* in Fig. 15-25 weigh 322 and 161 lb, respectively. When $f_A = 0.3$, (*a*) what distance *s* will the system move from rest in 3 sec and (*b*) what is the tension *T* in the rope?

15-37. The block *A* in Fig. 15-25 weighs 644 lb. What must be the weight of block *B* if block *A* is to move 12 ft from rest in 2 sec? $f_A = 0.25$.

Ans. $W_B = 346$ lb

15-38. In Fig. 15-25 block *A* weighs 322 lb and block *B* weighs 161 lb. If $f_A = 0.1$, find the total distance block *B* travels if the cord breaks 3 sec after the system starts from rest.

15-39. Body *A* in Fig. 15-26 weighs 100 lb and *B* weighs 260 lb. The coefficients of sliding friction, f_A and f_B, are 0.2. Determine (*a*) the acceleration of the system after it is released, (*b*) the tension *T* in the rope before *B* strikes the stop at *C*, (*c*) the velocity of *B* the instant it strikes the stop at *C*, and (*d*) the distance *s* body *A* will continue to the right after *B* stops.

Ans. (*a*) $a = 2.86$ ft per sec^2; (*b*) $T = 28.9$ lb; (*c*) $v = 7.56$ ft per sec; (*d*) $s = 4.44$ ft

15-40. Solve Prob. 15-39 if $W_A = 100$ lb, $W_B = 390$ lb, $f_A = 0.1$, and $f_B = 0.25$.

15-41. In the system shown in Fig. 15-27, *A* weighs 520 lb, *B* weighs 386 lb, and f_A is 0.25. Calculate (*a*) the acceleration *a* of the system, (*b*) the distance *s* it moves in 4 sec from rest, and (*c*) the tension *T* in the cord.

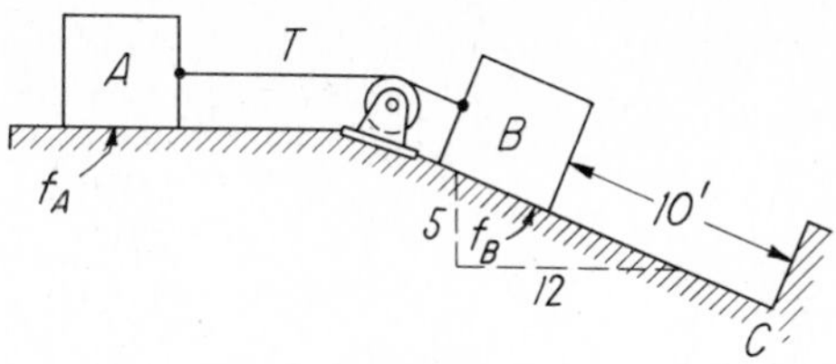

FIG. 15-26 Probs. 15-39 and 15-40.

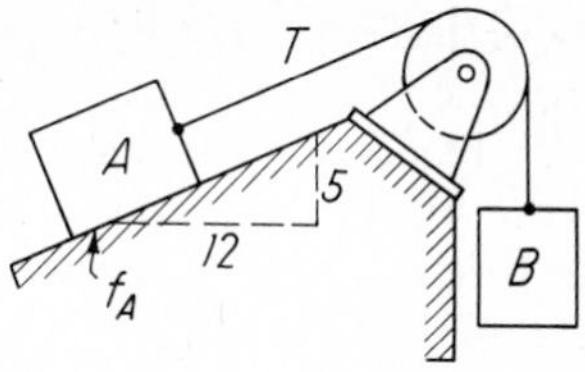

FIG. 15-27 Probs. 15-41, 15-42, and 15-73.

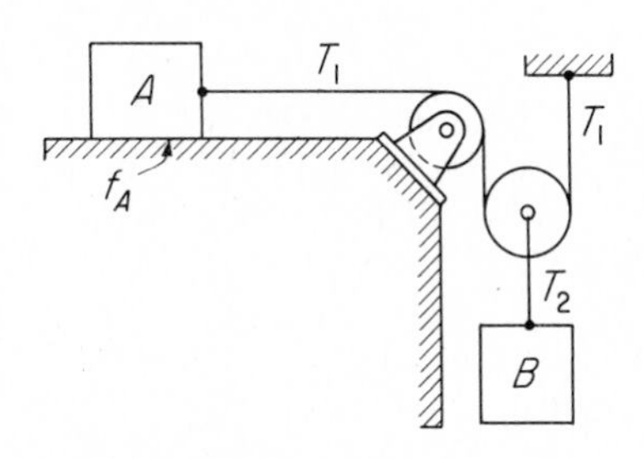

FIG. 15-28 Probs. 15-43 and 15-44.

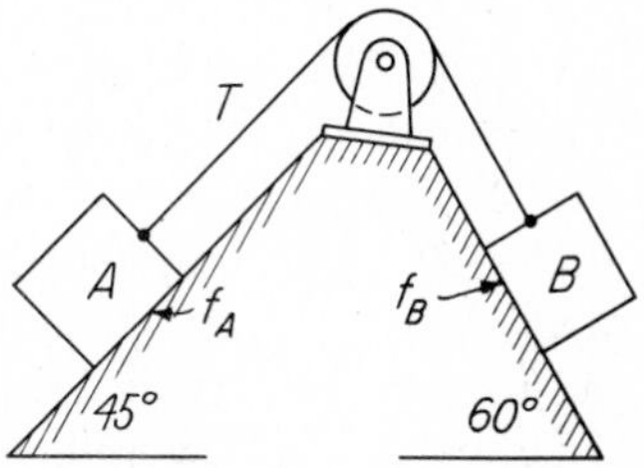

FIG. 15-29 Probs. 15-45 and 15-77.

15-42. Solve Prob. 15-41 if $W_A = 1{,}300$ lb, $W_B = 161$ lb, and $f_A = 0.2$.
Ans. (*a*) $a = 2.178$ ft per sec²; (*b*) $s = 39.25$ ft; (*c*) $T = 172$ lb

15-43. Block *A* in Fig. 15-28 weighs 100 lb, *B* weighs 80 lb, and f_A is 0.3. Find the acceleration *a* of the system and tensions T_1 and T_2.
Ans. $a = 2.68$ ft per sec²; $T_1 = 38.3$ lb; $T_2 = 76.7$ lb

15-44. Block *A* in Fig. 15-28 weighs 500 lb. What must be the weight of block *B* to give the system an acceleration of 16.1 ft per sec² if $f_A = 0.2$?

15-45. In Fig. 15-29, *A* weighs 100 lb, *B* weighs 200 lb, and f_A and f_B are 0.4. (*a*) How far and in which direction will *A* move in 3 sec and (*b*) what is the tension *T* in the cord?

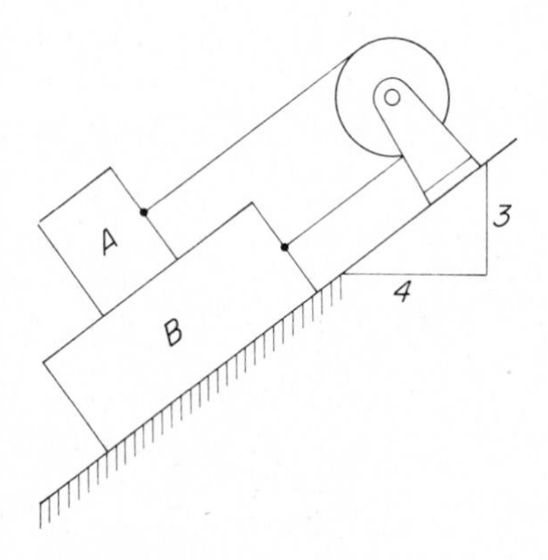

FIG. 15-30 Prob. 15-46.

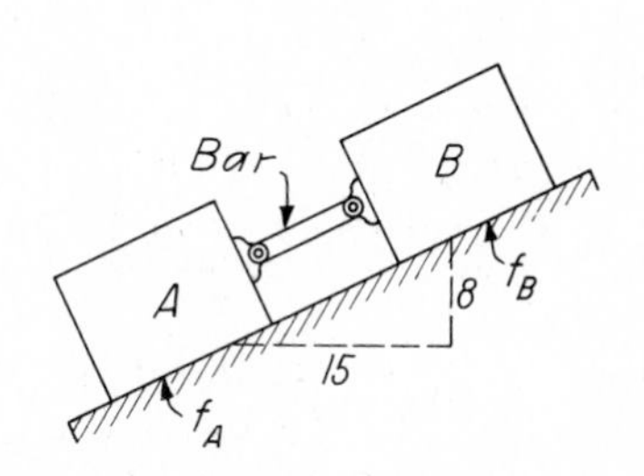

FIG. 15-31 Probs. 15-47 and 15-48.

15-46. Blocks A and B shown in Fig. 15-30 are released from rest. Calculate the acceleration of block B if $W_A = 100$ lb, $W_B = 200$ lb, and $f = 0.1$ for all surfaces.

15-47. Blocks A and B in Fig. 15-31 are connected with a bar whose small weight may be disregarded. If A and B each weigh 340 lb, and f_A is 0.2 and f_B is 0.4, calculate (*a*) the acceleration a of the system after release and (*b*) the force P in the bar. (HINT: First, treat A and B as one body, and find a for the system, then a for body A alone, if detached from the bar. The difference in the two accelerations is caused by the force P in the bar.) *Ans.* (*a*) $a = 6.63$ ft per sec^2; (*b*) $P = 30$ lb tension

15-48. Solve Prob. 15-47 when $A = 340$ lb, $f_A = 0.5$, $B = 680$ lb, and $f_B = 0.3$.

15-8 Dynamic Equilibrium of Nonconcurrent Force Systems

In many problems involving dynamic equilibrium, the *actual* points of application of the forces acting on the body, as well as their magnitudes, must be considered, as in the example below. These forces then generally form a **noncurrent force system in dynamic equilibrium,** the solution of which involves *moment equilibrium* as well as *force equilibrium*. The necessary equations of solution then are

$$\Sigma F_X = 0 \qquad \Sigma F_Y = 0 \qquad \text{and} \qquad \Sigma M = 0$$

ILLUSTRATIVE PROBLEM

15-49. The garage door shown in Fig. 15-32 weighs 200 lb. It moves on rollers which run on an overhead track. Total resistance to motion ($F_1 + F_2$) is 10 percent of the weight, or 20 lb. Calculate the acceleration a given the door by the force P, of 45 lb. Determine also the resulting roller reactions R_1 and R_2, and the friction forces F_1 and F_2 at each roller.

Solution: The door will be accelerated to the right. Therefore, the inertia force F_I acts to the left through its center of gravity.

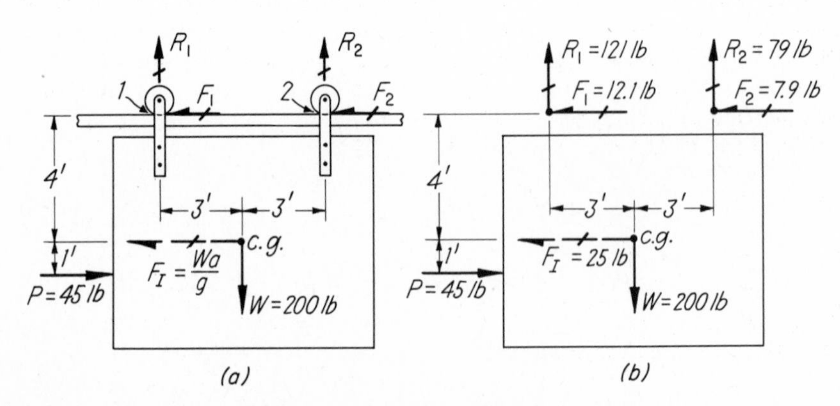

FIG. 15-32 Dynamic equilibrium of nonconcurrent force system. (*a*) Space diagram showing reactions. (*b*) Complete free-body diagram.

To evaluate the inertia force, we have

$[\Sigma H = 0]$ $\quad F_I + (F_1 + F_2) = P \quad$ or $\quad F_I = 45 - 20 = 25$ lb

from which we then obtain the acceleration:

$$\left[a = \frac{F_I g}{W}\right] \quad a = \frac{(25)(32.2)}{200} \quad \text{or} \quad a = 4.025 \text{ ft per sec}^2 \qquad \textit{Ans.}$$

Reactions R_1 and R_2 are now obtained from summations of moments about points 2 and 1, respectively. The equations are

$$[\Sigma M_2 = 0] \quad 6R_1 + (4)(25) = (5)(45) + (3)(200)$$
$$6R_1 = 225 + 600 - 100 = 725 \quad \text{and} \quad R_1 = 121 \text{ lb} \qquad \textit{Ans.}$$
$$[\Sigma M_1 = 0] \quad 6R_2 + (5)(45) = (3)(200) + (4)(25)$$
$$6R_2 = 600 + 100 - 225 = 475 \quad \text{and} \quad R_2 = 79 \text{ lb} \qquad \textit{Ans.}$$

The friction force at each roller is then 10 percent of the roller reaction. Hence,

$$F_1 = 0.1R_1 = (0.1)(121) \quad \text{or} \quad F_1 = 12.1 \text{ lb} \qquad \textit{Ans.}$$
$$F_2 = 0.1R_2 = (0.1)(79) \quad \text{or} \quad F_2 = 7.9 \text{ lb} \qquad \textit{Ans.}$$

When the values of all forces have been recorded on the free-body diagram (Fig. 15-32*b*), the following check equations will verify the required equilibrium:

$$[\Sigma H = 0] \quad P = F_I + F_1 + F_2 \quad \text{or} \quad 45 = 25 + 12.1 + 7.9 = 45 \qquad \textit{Check.}$$
$$[\Sigma V = 0] \quad W = R_1 + R_2 \quad \text{or} \quad 200 = 121 + 79 = 200 \qquad \textit{Check.}$$

PROBLEMS

15-50. Solve Prob. 15-49 when $W = 300$ lb, $P = 75$ lb, and the total resistance to motion $(F_1 + F_2)$ is 15 percent of W.

15-51. Block A in Fig. 15-33 weighs 644 lb and rests on the bed of a truck. The coefficient of friction f_A is 0.3, between block and truck. Assuming that the block will slide, or tip, if the truck is suddenly stopped with a high deceleration, determine (*a*) which will it do and why, and (*b*) in what shortest distance s can the truck be stopped from a speed of 30 mph to the right, for the block to remain in position. (Disregard truck in free-body diagram.) *Ans.* (*a*) The block will slide when F_I exceeds $F = 193.2$ lb (it cannot tip because the overturning moment is less than the counteracting moment of its weight); (*b*) $s = 100$ ft

15-52. Solve Prob. 15-51 when f_A is 0.6.

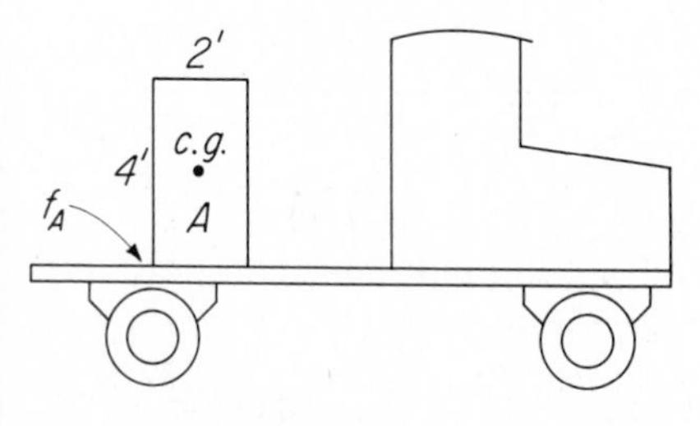

FIG. 15-33 Probs. 15-51 to 15-53.

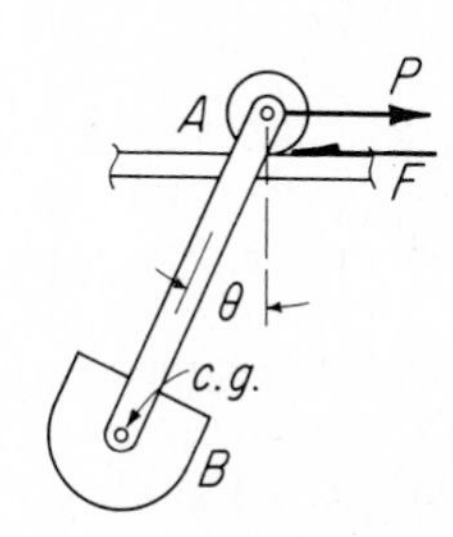

FIG. 15-34 Prob. 15-55.

15-53. In Prob. 15-51, what value of f_A will cause block A to be on the verge of tipping, and at what acceleration a will this occur?

Ans. $f_A = 0.5$; $a = 16.1$ ft per sec^2

15-54. An overhead crane in a foundry moves a 10-ton ladle horizontally. If the crane travels 10 ft while stopping and its initial velocity was 8.03 ft per sec, what is the tension in the cable supporting the ladle and what angle θ does it make with the vertical? Assume constant acceleration. *Ans.* $T = 20{,}100$ lb; $\theta = 5.71°$

15-55. The loaded ore bucket B, in Fig. 15-34 weighs 2,000 lb and is moved along a level overhead track by a force P. The total resistance to motion F is 100 lb. The weight of wheel A and of the bar may be neglected. When the bucket is accelerated 8.05 ft per sec^2, calculate (*a*) the required force P, (*b*) the angle θ of the bar with the vertical, and (*c*) the tension T in the bar.

15-56. The trolley bar shown in Fig. 15-35 is hinged at B to the top of a streetcar and is held against the overhead trolley wire by a spring at B exerting a moment M. The bar is 12 ft long and weighs 48 lb. The friction and weight of the wheel A are negligible. If the moment M is 400 lb-ft, calculate the vertical pressure between the wheel and the wire (*a*) when the car is at rest and (*b*) when it accelerates 8.05 ft per sec^2 to the left. *Ans.* (*a*) $R_A = 14.4$ lb; (*b*) $R_A = 11$ lb

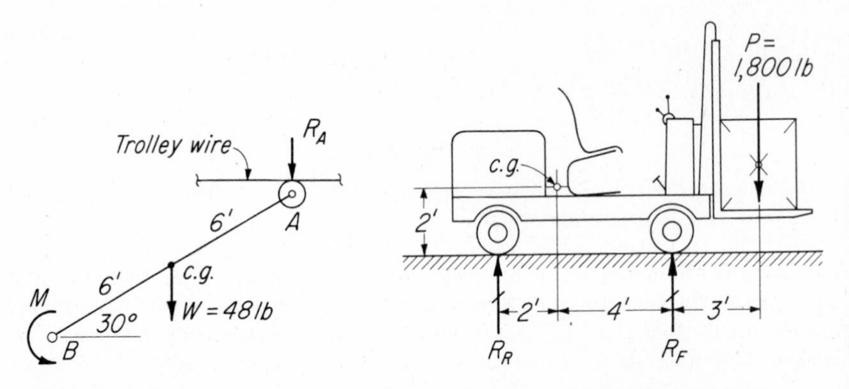

FIG. 15-35 Prob. 15-56.

FIG. 15-36 Probs. 15-57 and 15-58.

15-57. The fork-lift truck shown in Fig. 15-36 weighs 2,100 lb. What upward acceleration can be given the 1,800-lb load P and still maintain a 100-lb reaction force on each of the two rear wheels? *Ans.* $a = 10.7$ ft per sec^2

15-58. If the 1,800-lb load shown in Fig. 15-36 has been raised until its center of gravity is 15 ft above the floor, what rearward acceleration may be given the 2,100-lb fork-lift truck and still maintain a 100-lb reaction force on each of the two rear wheels? The center of gravity of the truck is 2 ft above the floor.

15-59. The airplane in Fig. 15-37 weighs 19,320 lb and is slowed to a stop after landing by a constant braking force applied to the two rear wheels of its tricycle landing gear. Calculate the maximum vertical component R_F of the front-wheel reaction if the plane lands at 90 mph and stops in 600 ft. (Because of the lifting action of the wings while the plane is in motion, R_F will be maximum just before the plane stops. A simple point free body is sufficient.) *Ans.* $R_F = 7{,}180$ lb

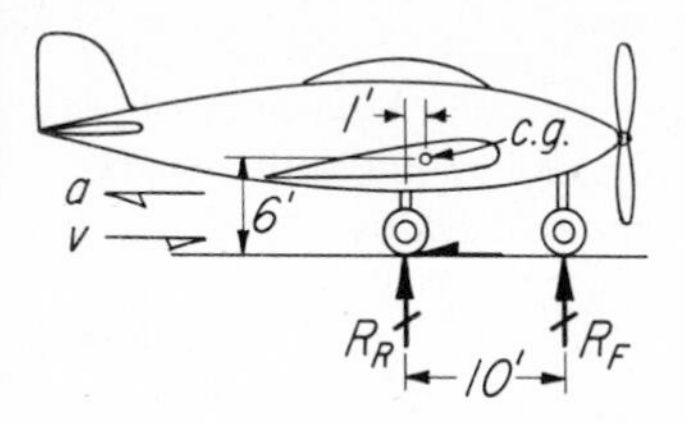

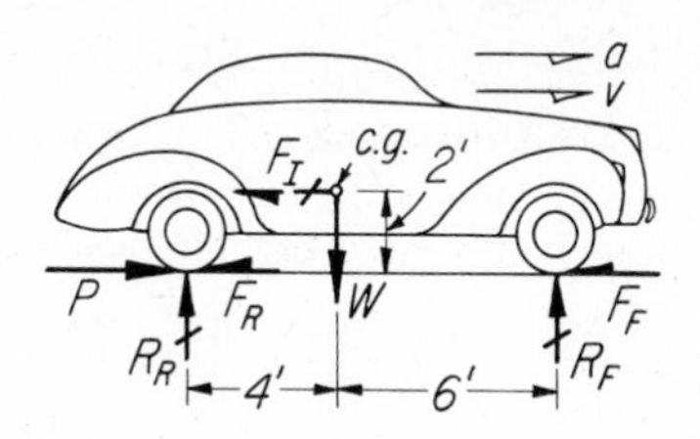

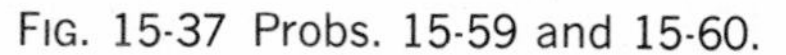

FIG. 15-37 Probs. 15-59 and 15-60.

FIG. 15-38 Probs. 15-61 and 15-90.

15-60. Solve Prob. 15-59 for the rear-wheel reaction R_R just before stopping and when at rest.

15-61. The car shown in Fig. 15-38 weighs 3,220 lb. The total average road resistance ($F = F_F + F_R$) at moderate speeds on the level or low grades is 2 percent of the weight, or 64.4 lb. The effort of the motor in driving the car is shown as a forward-acting tractive force P under the rear wheels. Calculate (*a*) the acceleration a given the car when P is 564.4 lb, (*b*) the resulting wheel reactions R_F and R_R, and (*c*) the friction forces F_F and F_R. (A simple point free-body diagram is sufficient.)

Ans. (*a*) $a = 5$ ft per sec^2; (*b*) $R_F = 1{,}188$ lb, $R_R = 2{,}032$ lb; (*c*) $F_F = 23.8$ lb, $F_R = 40.6$ lb

15-62. A 2- by 4- by 8-in. brick is standing on its end on the bed of a truck. The 2-in. dimension is parallel to the direction of motion. Will the brick tip or slide if the truck is given a large acceleration (*a*) if $f = 0.4$, and (*b*) if $f = 0.15$?

Ans. (*a*) Brick will tip; (*b*) brick will slide

SUMMARY

(By article number)

15-2. The resultant of an unbalanced force system acting on a body is called the **accelerating force** ***F***; this produces an acceleration a in line with the proportional to the force. The accelerating force acting on a freely falling body is its weight W, and its acceleration g is 32.2 ft per sec^2. Hence the proportionality

$$\frac{F}{a} = \frac{W}{g} \qquad \text{or} \qquad F = \frac{Wa}{g} \tag{15-1}$$

15-3. According to Newton's third law, *every action has an equal and opposed reaction.* The equal and opposed reaction to an acceleration force is the **inertia** of the body, manifesting itself as a force called the **inertia force.** Therefore,

$$\textbf{Accelerating force} \qquad F = \frac{Wa}{g} = F_I \qquad \textbf{Inertia force} \tag{15-2}$$

15-5. Use of the inertia force to balance the accelerating force creates **dynamic equilibrium** similar to *static equilibrium.*

15-6. When only the unknown magnitudes of forces acting on a moving body are desired, the body may be considered as a particle without size or shape. All forces then act through its center of gravity, and the problem involves a **concurrent force system in dynamic equilibrium** and may be solved by the equations of *static force equilibrium,*

$$\Sigma F_X = 0 \qquad \text{and} \qquad \Sigma F_Y = 0$$

15-8. When the actual points of application of the forces must be considered, the problem involves a **nonconcurrent force system in dynamic equilibrium** and may be solved by the equations of *static force and moment equilibrium.*

$$\Sigma F_X = 0 \qquad \Sigma F_Y = 0 \qquad \text{and} \qquad \Sigma M = 0$$

REVIEW PROBLEMS

15-63. A mine cage weighs 2,440 lb and must carry a maximum load of 4,000 lb. The average frictional resistance F of the side guides is 100 lb. *What constant cable tension T is required* to give the loaded cage an upward velocity of 15 ft per sec from rest, in a distance of 12.5 ft? *Ans.* $T = 8{,}340$ lb ($a = 9$ ft per sec^2)

15-64. Calculate the value of *the constant horizontal force P required to move a body* weighing 644 lb along a level floor a distance of 100 ft in 5 sec, if the coefficient of friction f is 0.2.

15-65. A train weighing 1,000 tons is pulled up a 1 percent grade with uniform velocity. The train resistance (bearing friction, rolling friction, air resistance, etc.) is 12 lb per ton of weight. *Calculate the constant drawbar pull which the locomotive must exert.* (Let W_T be 1 percent of W.) *Ans.* $P = 16$ tons

15-66. Determine the minimum distance s in which a car traveling at 30 mph can be stopped on the level, if the coefficient of friction f is 0.8 between the tires and the pavement. *Ans.* $s = 37.5$ ft

15-67. The body shown in Fig. 15-4 is pulled up the incline with decreasing velocity by the force P. Let $W = 100$ lb, $P = 60$ lb, $f = 0.2$, and $\theta = 30°$. (*a*) *Compute the deceleration a.* (*b*) If v_0 is 30 ft per sec, in what distance s will the body come to a stop? (*c*) What time t is required to stop it? (*d*) Will the body start down the incline again after stopping?

Ans. (*a*) $a = 2.35$ ft per sec^2; (*b*) $s = 191$ ft; (*c*) $t = 12.8$ sec; (*d*) No!

15-68. *Calculate the acceleration a* given body A in Fig. 15-39 when $W_A = 100$ lb, $P = 100$ lb, and $f_A = 0.4$. *Ans.* $a = 7.18$ ft per sec^2

15-69. In Fig. 15-39 determine *the force P required to maintain uniform velocity* of body A up the incline, if $W_A = 100$ lb and $f_A = 0.4$.

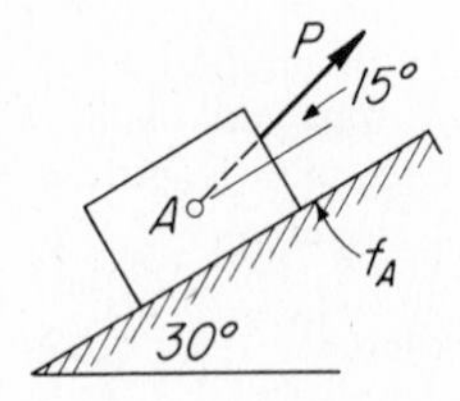

FIG. 15-39 Probs. 15-68 and 15-69.

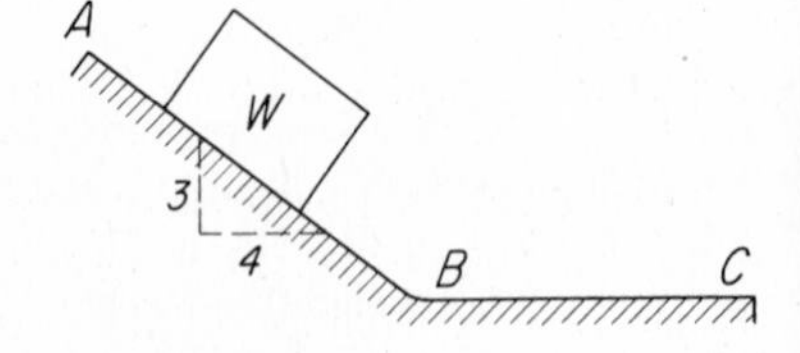

FIG. 15-40 Probs. 15-70 and 15-71.

15-70. The diagram in Fig. 15-40 illustrates a chute in a factory used for sliding boxes to a lower floor. The boxes start from rest at A; they slide a distance of 20 ft down to B and then along the level to C, where they enter an endless conveyer belt. *Calculate the distance BC* required, if f is 0.2 and the boxes are to enter the belt with a velocity of 4 ft per sec. *Ans.* $s = 42.8$ ft ($v_B = 23.8$ ft per sec)

15-71. If the incline AB in Fig. 15-40 is 10 ft long and the flat portion BC is 12 ft long, what will be the velocity of a 100-lb box at C if it starts from rest at A. Assume $f = 0.1$.

15-72. What horizontal force is required to move a 500-lb weight up a 30° incline with an acceleration of 3.22 ft per sec^2? Assume $f = 0.2$. *Ans.* $P = 504$ lb

15-73. In Fig. 15-27, body A weighs 520 lb, and f_A is 0.25. If the cord snaps while A is moving up the incline with a velocity of 20 ft per sec, (*a*) how far will it continue before stopping and (*b*) what time does it require to slide 30 ft back down the incline after stopping? *Ans.* (*a*) $s = 10.1$ ft; (*b*) $t = 3.48$ sec

15-74. Block A in Fig. 15-23 weighs 120 lb. *Calculate the weight* W_B of block B required to accelerate the *system* 8.05 ft per sec^2 with B moving downward.
Ans. $W_B = 200$ lb

15-75. In Fig. 15-23, A weighs 60 lb and B weighs 80 lb. Determine (*a*) *the acceleration a of the system* after release and (*b*) the tension T in the cord.

15-76. Bodies A and B in Fig. 15-25 weigh 100 and 80 lb, respectively, and f_A is 0.4. Calculate (*a*) *the acceleration a of the system* and (*b*) the tension T in the cord.

15-77. Block A in Fig. 15-29 weighs 100 lb and block B weighs 40 lb. Find the acceleration of the system if $f_A = 0.2$ and $f_B = 0.5$.

15-78. Blocks A, B, and C in Fig. 15-41 weigh 60, 80, and 100 lb, respectively; f_A is 0.2 and f_B is 0.4. Compute (*a*) *the acceleration a of the system* and (*b*) the tensions T_1 and T_2 in the two cords. (HINT: To obtain a, consider the inertia force of the entire system when $W = 60 + 80 + 100 = 240$ lb. Then $a = F_I g/240$.)
Ans. (*a*) $a = 7.51$ ft per sec^2; (*b*) $T_1 = 26$ lb, $T_2 = 76.7$ lb

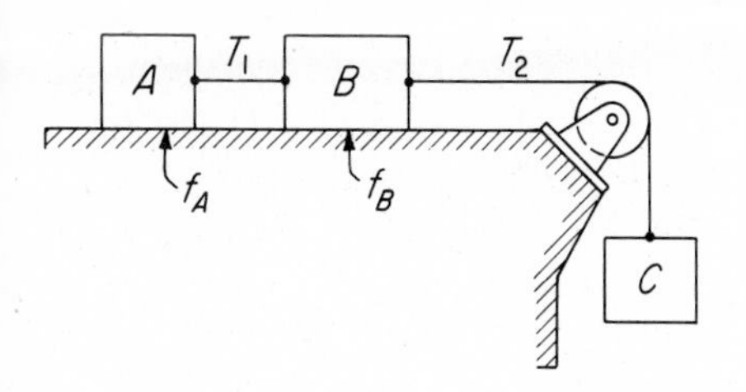

FIG. 15-41 Prob. 15-78.

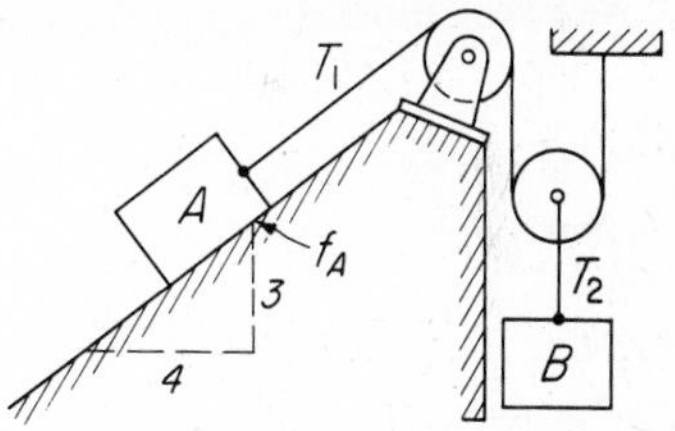

FIG. 15-42 Probs. 15-79 to 15-81.

15-79. In Fig. 15-42, A weighs 100 lb, and f_A is 0.2. *Calculate the weight* W_B of B required to accelerate body A 8.05 ft per sec^2 up the incline.
Ans. $W_B = 231$ lb

15-80. *Calculate the distance s* that body A in Fig. 15-42 will move from rest in 4 sec, if A weighs 100 lb, B weighs 180 lb, and f_A is 0.2. Compute also the tensions T_1 and T_2 in the cords.

15-81. *Determine the weight* W_A of body A in Fig. 15-42 required to accelerate it 8.05 ft per sec^2 down the incline, if B weighs 80 lb and f_A is 0.2.
Ans. $W_A = 237$ lb

15-82. Blocks A and B in Fig. 15-43 slide down the incline in (frictionless) contact with each other. A weighs 200 lb, B weighs 100 lb, $f_A = 0.4$, and $f_B = 0.2$. Compute (*a*) *the acceleration a of the system* and (*b*) the pressure P between the blocks. (See Prob. 15-47.) *Ans.* (*a*) $a = 10.73$ ft per sec^2; (*b*) $P = 10.6$ lb

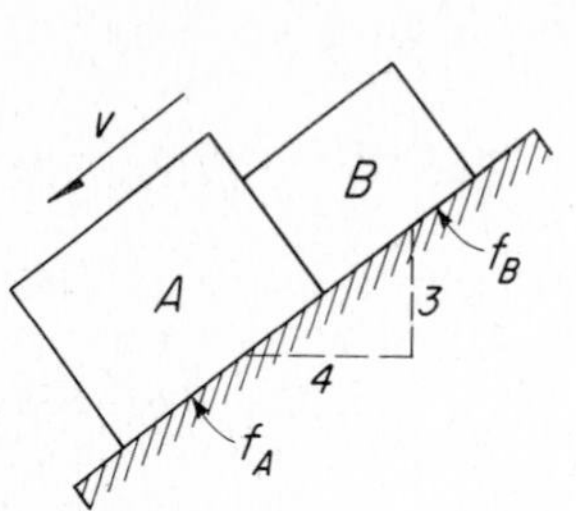

FIG. 15-43 Prob. 15-82.

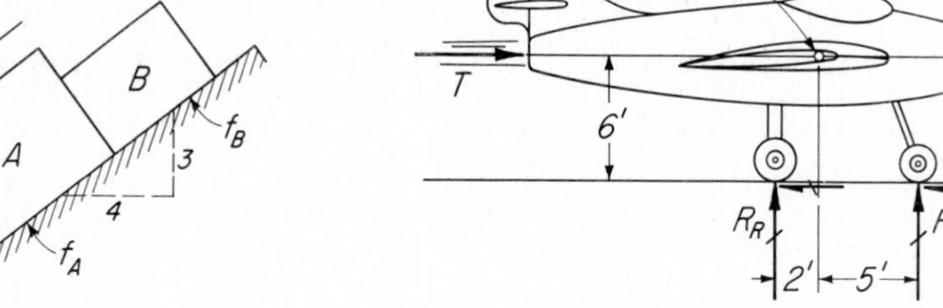

FIG. 15-44 Prob. 15-83.

15-83. The jet airplane shown in Fig. 15-44 weighs 19,300 lb and its engine develops a 2,165-lb thrust during take-off. The rolling resistance is 5 percent of the weight. (*a*) What is the vertical component of the front-wheel reaction R_F just after the plane starts to roll down the runway? (*b*) What is the acceleration of the plane at this time?

15-84. If the coefficient of friction between the brick and the truck bed of Prob. 15-62 is 0.2, will the brick slide or topple? *Ans.* It will slide

15-85. Solve Prob. 15-62 if the 4-in. dimension is parallel to the direction of motion.

15-86. A car with a rear-wheel drive has a 10-ft wheel base. The center of gravity of the car is 3 ft above the pavement and 6 ft ahead of the rear wheels. Determine the maximum acceleration the car can be given if the f between the tires and the pavement is 0.6. Neglect air friction and rolling friction of front tire.

Ans. $a = 9.44$ ft per sec^2

15-87. The center of gravity of a two-wheeled trailer is directly over the wheels and 4 ft above the pavement. The tongue is horizontal and 2 ft above the pavement. What acceleration must be given to the trailer if the pull on the tongue is to remain horizontal. Assume rolling friction under the wheels is 10 percent of the trailer weight. *Ans.* $a = 3.22$ ft per sec^2

15-88. Block A in Fig. 15-45 weighs 600 lb and must be transported upright on a trailer. The coefficient of friction f is 0.25, between A and the trailer. The block is prevented from sliding to the right by stop B, and from tipping to the right by rope C, which is slightly slack. (*a*) *What maximum acceleration may be given the trailer* to the right without causing the block to tip or slide to the left? (*b*) What tension T is produced in the rope when the trailer is stopped with constant deceleration from 45 mph in 150 ft? *Ans.* (*a*) $a = 8.05$ ft per sec^2; (*b*) $T = 75.6$ lb

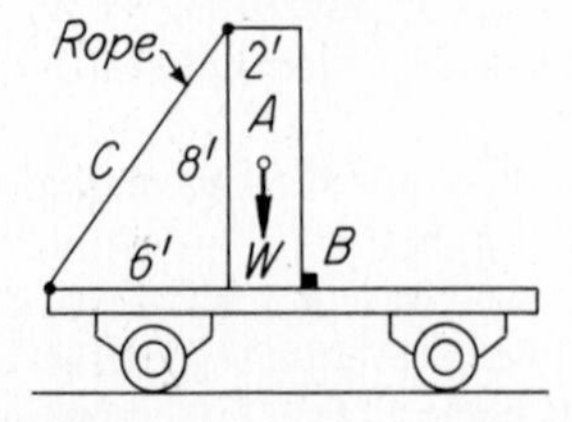

FIG. 15-45 Probs. 15-88 and 15-89.

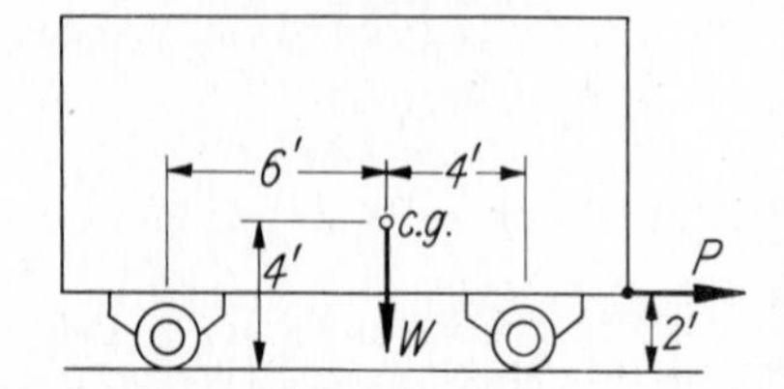

FIG. 15-46 Prob. 15-92.

15-89. In Fig. 15-45, block A weighs 600 lb. It rests on the bed of a trailer and is prevented from sliding and tipping to the right by the block B and the slightly slack rope C. If the strength of the rope is 100 lb, *in what minimum distance may the trailer be stopped* from a speed of 30 mph to the right without snapping the rope?

15-90. In Prob. 15-61 change the tractive force P to 864.4 lb and solve.

15-91. A truck weighs 4,830 lb and has a wheel base of 11 ft. Its center of gravity is 5 ft in front of the rear axle and is 3 ft above the ground. *Determine the vertical wheel reactions* R_F and R_R, when the truck is stopped by a constant braking force in 110 ft, from a speed of 30 mph to the right. (When the truck decelerates, the road friction is due largely to braking action, and F then is not dependent entirely upon the magnitudes of the wheel reactions. The friction forces are therefore merely indicated on the free-body diagram, without values.)

Ans. $R_F = 2{,}555$ lb; $R_R = 2{,}275$ lb

15-92. The loaded trailer shown in Fig. 15-46 weighs 5 tons and is pulled along a level road. The average road resistance is 2 percent of the weight. *Calculate the force P, the wheel reactions* R_F *and* R_R*, and the friction forces* F_F and F_R, when the trailer (*a*) is pulled with uniform speed and (*b*) is accelerated 4.025 ft per sec^2.

Ans. (*a*) $P = 200$ lb, $R_F = 6{,}040$ lb, $R_R = 3{,}960$ lb, $F_F = 121$ lb, $F_R = 79$ lb;
(*b*) $P = 1{,}450$ lb, $R_F = 5{,}790$ lb, $R_R = 4{,}210$ lb, $F_F = 116$ lb, $F_R = 84$ lb

REVIEW QUESTIONS

15-1. Define the term kinetics.

15-2. When the external forces acting on a body do not balance, what is their resultant called, and what is the effect of that resultant?

15-3. What accelerating force acts on a freely falling body and what acceleration does it produce?

15-4. According to Newton's third law, to every action there is an equal and opposed reaction. What is the reaction to an accelerating force?

15-5. Explain the meaning of inertia of a body. When does inertia manifest itself?

15-6. Why is the inertia of a body represented as a force, and through what point in a body does its inertia force always act?

15-7. Give a few illustrations of manifestations of inertia of bodies.

15-8. What is meant by dynamic equilibrium? On what principle is dynamic equilibrium based?

15-9. If, in problems involving dynamic equilibrium, we are interested only in the magnitudes of the forces without regard to their actual points of application, what type do we assume the force system to be and what equations will solve it?

15-10. When, in problems involving dynamic equilibrium, the actual points of application of the forces must be considered, what type of force system do we have and what equations will solve it?

15-11. In dynamics, what is meant by a system of bodies?

15-12. Do all bodies in a system necessarily have the same motion? How may their motions differ?

15-13. By what method may we readily determine the way a system of bodies will move?

CHAPTER 16

Curvilinear Motion

16-1 Introduction

Curvilinear motion is motion along any path which is not a straight line. This motion may be in a plane, as when a train rounds a curve on a level roadbed, or may be in space, as when a car travels a curving roadbed over the rounded top of a hill. Only *plane curvilinear motion* is considered here.

Suppose one whirls overhead a heavy weight fastened to the end of a cord. Everyone is familiar with the pull exerted on the cord in such a case. If the cord breaks or is released, the weight will move off on a tangent to the circular path. A constant pull on the cord must be maintained if the weight is to be forced into a circular path. This inward-directed pull is necessary to cause the continuous change in the direction of the velocity. Such a force will be shown to be an accelerating force and the opposed outward-directed centrifugal force will be shown to be an inertia force.

16-2 Normal Acceleration in Curvilinear Motion

Recall that velocity has *direction* as well as magnitude and is therefore a *vector* quantity. It is this directional characteristic that distinguishes velocity from the scalar quantity speed. Acceleration is the rate of change of velocity with respect to time. The acceleration may be produced by a change in the direction of the velocity, a change in the magnitude of the velocity, or both. Consider the motion of a particle traveling in a circular path with constant speed. The velocity at any point is tangent to the path of motion of the particle. Since the direction of motion is constantly changing, the velocity

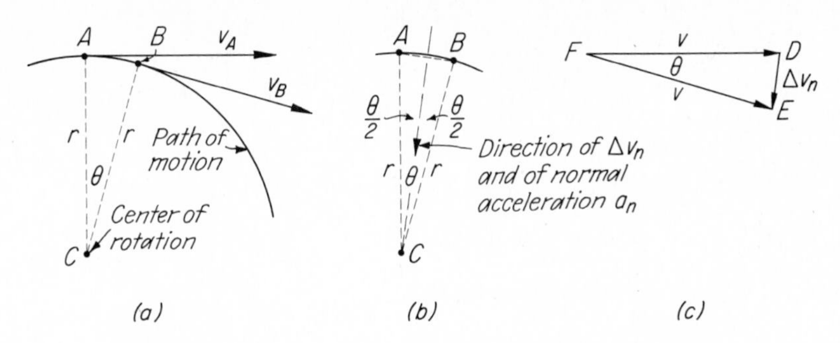

FIG. 16-1 Normal acceleration in curvilinear motion.

is constantly changing, and hence the particle is undergoing a continuous and constant acceleration directed toward the center of rotation. We may evaluate this normal acceleration as follows.

Let the body move from A to B along the horizontal circular path indicated in Fig. 16-1*a*. Let v_A and v_B represent its tangential velocities at A and B. These velocities are again shown as vectors in Fig. 16-1*c*. For uniform motion, $v_A = v_B = v$. The vector Δv_n represents the *normal* change in velocity of the body resulting from the constant *normal* acceleration a_n, both of which are directed toward the center of rotation C.

Let us now assume that the angle θ is very small. Then we may consider that *arc AB* equals *chord AB*. The two triangles ABC and DEF are similar, since both are isosceles triangles with mutually perpendicular sides. Hence, we have

$$\frac{AB}{r} = \frac{\Delta v_n}{v} \tag{a}$$

When the speed along the path is uniform, $AB = vt$ and the normal acceleration a_n is constant; then $\Delta v_n = a_n t$. When these substitutions are made in Eq. (a), we obtain

$$\frac{vt}{r} = \frac{a_n t}{v} \tag{b}$$

from which

$$\boldsymbol{a_n = \frac{v^2}{r}} \tag{16-1}$$

Since a_n, v, and r are instantaneous quantities, Eq. (16-1) *holds for variable as well as for uniform conditions of plane curvilinear motion.*

Should the speed of the body be changing along its curved path of motion, as the result of a tangential acceleration a_t, its true acceleration would be $a = \sqrt{a_n^2 + a_t^2}$, as indicated in Fig. 16-2.

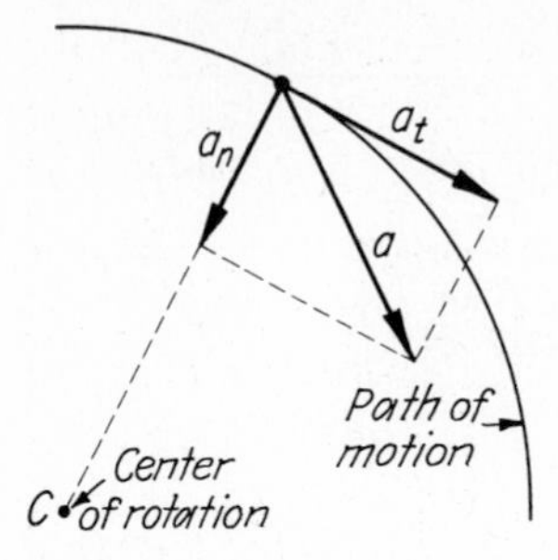

FIG. 16-2 Normal and tangential acceleration components.

16-3 Centrifugal Force

According to Newton's first law, a change in the direction of motion of a body can be caused only by an unbalanced force, and such a force, according to Newton's second law, is an accelerating force.

Equation (16-1) reveals that, in uniform circular motion, there exists a constant acceleration normal to the path of motion; that is, directed toward the center of rotation C, as shown in Figs. 16-1*b* and 16-2. This acceleration must be, and is, caused by a constant accelerating force, also directed toward the center of rotation C, called the **centripetal force.** Its equal and opposed reaction (Newton's third law), which must be an inertia force, is called the **centrifugal force,** and is defined by the usual symbols $F_I = Wa_n/g$.

The small body of weight W in Fig. 16-3*a* is suspended from a weightless arm supported at O, and thus forms a pendulum. When this body moves in a horizontal circle with uniform speed v, the arm generates a cone, from which is derived the name **conical pendulum.** The vertical **axis of rotation** passes through O. The intersection C of this axis with the plane in which the body moves is called the **center of rotation.** The free-body diagram in Fig. 16-3*b*

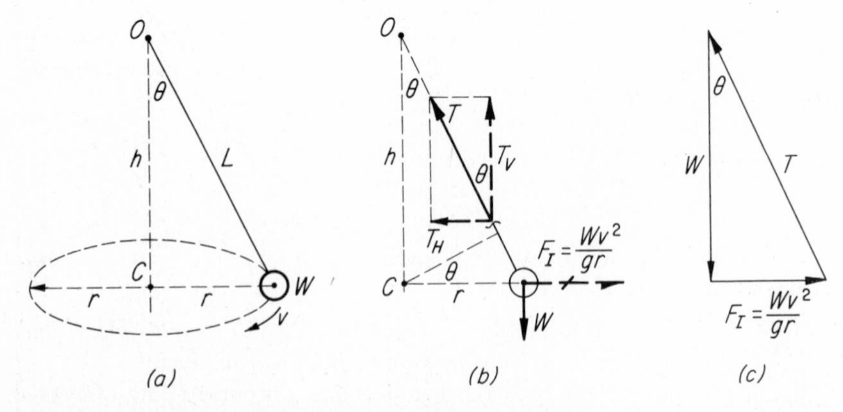

FIG. 16-3 Centrifugal force. Body moving in a horizontal circle. (*a*) Conical pendulum. (*b*) Free-body diagram. (*c*) Force diagram.

shows the tension T in the arm resolved into components T_V and T_H. Clearly, T_V supports the weight W of the body and T_H then is the unbalanced *centripetal accelerating force*, the reaction to which is the *centrifugal inertia force* $F_I = Wa_n/g$. But, from Eq. (16-1), $a_n = v^2/r$. Hence,

Centrifugal force: $$F_I = \frac{Wa_n}{g} \quad \text{or} \quad \boldsymbol{F_I = \frac{Wv^2}{gr}} \qquad \textbf{(16-2)}$$

where v = speed, ft per sec
r = radius, ft
g = 32.2 ft per sec^2

The centrifugal inertia force is made use of in the design of governors to control the speed of machines. Its existence also necessitates the banking of railroad and highway curves, as is explained in later articles.

The angle θ made by the arm with the vertical axis of rotation is obtained from the relation shown in Fig. 16-3c as follows:

$$\tan\theta = \frac{F_I}{W} = \frac{Wv^2/gr}{W} \qquad (c)$$

or

$$\boldsymbol{\tan\theta = \frac{v^2}{gr}} \qquad \textbf{(16-3)}$$

In Eq. (16-3), v is in feet per second. Multiplying v by 60 changes it from feet per second to feet per minute. In one revolution the ball travels $2\pi r$ ft. Hence, if N is the number of revolutions per minute,

$$\boldsymbol{N = v \cdot \frac{60}{2\pi r}} \qquad \textbf{(16-4)}$$

and

$$\boldsymbol{v = \frac{2\pi r N}{60}} \qquad \textbf{(16-5)}$$

where N = number of revolutions per minute, rpm
v = tangential velocity, ft per sec
r = radius of curvature of path, ft

ILLUSTRATIVE PROBLEM

16-1. The vertical shaft of the flyball governor in Fig. 16-4 is driven by the steam engine whose speed it controls. Because of centrifugal forces, increases and decreases in this speed cause the flyballs to move outward or inward, respectively, thus raising or lowering sleeve A, which in turn operates other controlling mechanisms. (a) What number N of revolutions per minute (rpm) must the governor turn, if the desired speed is maintained when $\theta = 30°$ and $L = 1$ ft? (b) If each ball B weighs 15 lb, what is the tension T in each main arm caused by ball B only? Assume the weight of the arm and of link b to be negligible. (NOTE: In an actual design these weights

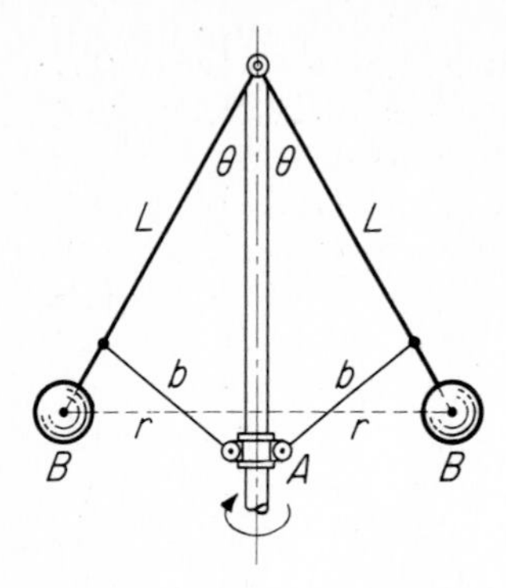

FIG. 16-4 Flyball governor.

could not be neglected. Also, a force would act in link b, and the solution below would be somewhat modified.)

Solution: To obtain N, we must first evaluate v from Eq. (16-3), in which $\tan\theta = v^2/gr$. Since $\tan 30° = 0.577$, $g = 32.2$ ft per sec^2, and $r = L\sin\theta = (1)(0.5) = 0.5$ ft, we have

$$\left[v = \sqrt{gr\tan\theta}\right] \qquad v = \sqrt{(32.2)(0.5)(0.577)} = 3.05 \text{ ft per sec}$$

N is obtained from Eq. (16-4): when $v = 3.05$ ft per sec and $r = 0.5$ ft,

$$\left[N = v \cdot \frac{60}{2\pi r}\right] \qquad N = \frac{(3.05)(60)}{(2)(3.1416)(0.5)} \qquad \text{or} \qquad N = 58.2 \text{ rpm} \qquad \textit{Ans.}$$

We now calculate tension T in the arm as follows: from Fig. 16-3c, $T = \sqrt{F_I^2 + W^2}$ and from Eq. (c) $F_I = W \cdot \tan\theta$, Hence, when $W = 15$ lb and $\tan\theta = 0.577$,

$$[F_I = W \cdot \tan\theta] \qquad F_I = (15)(0.577) = 8.66 \text{ lb}$$

$$\left[T = \sqrt{F_I^2 + W^2}\right] \quad T = \sqrt{(8.66)^2 + (15)^2} = \sqrt{300} \qquad \text{or} \qquad T = 17.3 \text{ lb} \qquad \textit{Ans.}$$

PROBLEMS

16-2. A sample of a slurry is to be divided into its liquid and solid components by means of a centrifuge. The sample container travels in an 18-in. circle. What rpm is necessary to maintain a normal acceleration equal to 600 times that of gravity?

16-3. A 12-lb ball attached to a weightless cord is swung around in a horizontal circle with a radius of 5 ft. The cord generates a cone. If the strength of the cord is 30 lb, at what speed N (rpm) will it break? *Ans.* $N = 36.7$ rpm

16-4. A conical pendulum consists of a 32.2-lb weight suspended on a 3-ft string. The weight travels in a circular path with a velocity v such that the string makes an angle of 30° with the vertical. Determine the velocity of the weight and the tension in the string.

16-5. One blade of a two-blade airplane propeller is 0.1 lb heavier than the other blade. Consider this difference in weight to be concentrated at a point 2 ft from the center of the shaft. Compute the radial side thrust F on the shaft caused by this unbalanced weight when the propeller turns 1,800 rpm. *Ans.* $F = 221$ lb

16-6. Solve Prob. 16-1 when $L = 9$ in., $\theta = 45°$, and $W_B = 20$ lb.

16-7. A 60-lb boy has a velocity of 20 ft per sec as he passes the lowest position on a swing whose ropes are 25 ft long. Find the tension in each rope if the distance from the center of rotation to the boy's center of gravity is 24 ft.

16-8. The balls B of the flyball governor in Fig. 16-4 are adjustable on the main arms. If the weight of the arm and the pull from link b are negligible, at what length of arm L should they be adjusted so as to revolve with a speed of 36 rpm while maintaining angle θ at 40°? What is the tension T in the main arms, if each ball weighs 12 lb? *Ans.* $L = 0.384$ ft; $T = 13.05$ lb

16-9. The central portion of the circular plate shown in Fig. 16-6 is horizontal, and its outer flange is shaped in the form of a cone. The plate rotates in a horizontal plane about the center of the vertical shaft. Calculate the angular speed N (in rpm) beyond which (*a*) body B will slide, if r_B is 1 ft and f_B is 0.2. (*b*) Beyond what speed N will body A slide or tip if r_A is 2 ft and f_A is 0.4?

Ans. (*a*) $N = 24.3$ rpm; (*b*) $N = 22.1$ rpm

16-10. A schematic diagram of an apparatus used for aerial rides at an amusement park is shown in Fig. 16-5. What is the angular velocity in rpm if $\theta = 30°$, $B = 12$ ft, and $L = 30$ ft?

16-11. If the maximum angular velocity of the apparatus shown in Fig. 16-5 is to be 10 rpm, what should be the maximum value of the dimension B if the angle θ is not to exceed 60°? $L = 30$ ft. *Ans.* $B = 24.9$ ft

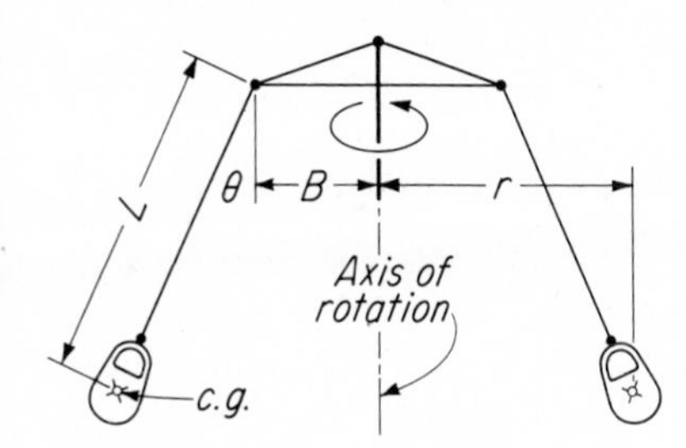

FIG. 16-5 Probs. 16-10, 16-11, and 16-61.

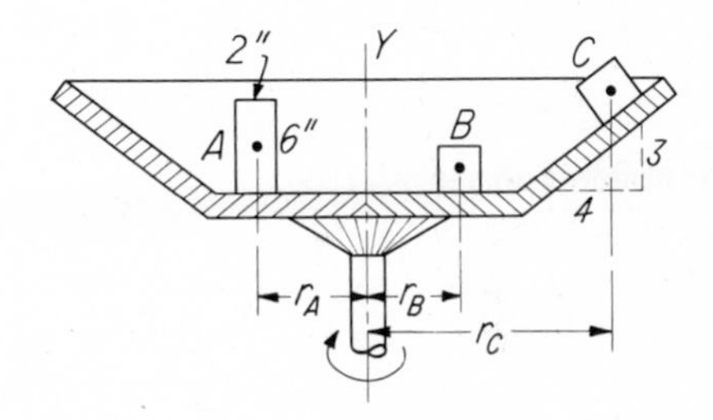

FIG. 16-6 Probs. 16-9 and 16-13.

16-12. Determine the maximum safe speed v at which a car can travel without skidding on a level highway curve, whose radius is 300 ft (*a*) when f is 0.2 for a slippery pavement and (*b*) when f is 0.6 for a dry pavement.

16-13. The plate shown in Fig. 16-6 rotates in a horizontal plane about the central axis Y. Its outer flange is shaped into the form of a cone. Calculate the angular speed N (in rpm) beyond which body C will slide outward off the plate, if r_C is 3 ft and f_C is 0.25. *Ans.* $N = 34.7$ rpm

16-4 Banking of Highway and Railway Curves

The outward pull exerted by some "unseen" force on a person in a car rounding a curve on a level, unbanked highway provides a physical demonstration of the existence of the centrifugal inertia force. The skidding and overturning of cars on curves give further demonstrations.

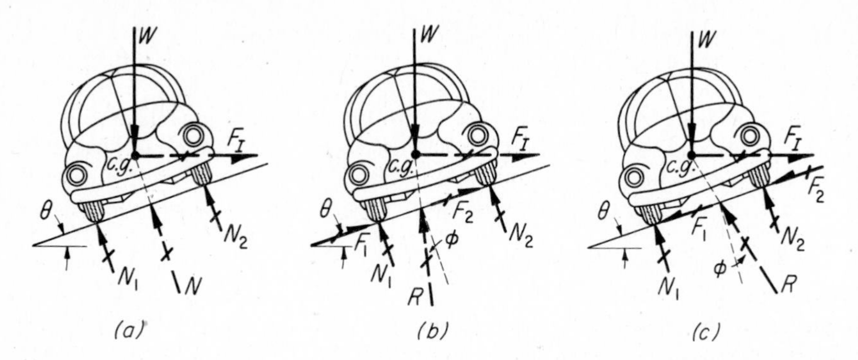

FIG. 16-7 Centrifugal force and banking of highway curves. (*a*) Normal speed. (*b*) Below normal speed. (*c*) Above normal speed.

To overcome these undesirable conditions, the outer rail on railway curves and the outer edge of the pavement on highway curves are elevated over the inner rail or edge by an amount e called the **superelevation.** The superelevation that should be given a curve depends upon its radius of curvature and upon the average speed of the vehicles.

When a car rounds a curve at the exact speed for which the curve was superelevated (Fig. 16-7*a*), no **side friction** against the tires is developed. This speed will be referred to as the **normal speed,** also referred to as *equilibrium speed* in railroad work. The total reaction N will then be *normal* to the pavement, because the opposing components of the weight W and of the centrifugal inertia force F_I, acting parallel to the road surface, now balance each other. Then $N_1 = N_2$, and $N = N_1 + N_2$. The three forces W, N, and F_I act through the center of gravity of the car and are in equilibrium.

At speeds *below the normal speed* (Fig. 16-7*b*), the tangential (parallel to the incline) component of F_I is insufficient to balance the tangential component of W. The car therefore tends to skid *inward* and *down* the incline, but this tendency is now resisted by the *side-friction forces* F_1 and F_2 acting *up* the incline against the tires. The total reaction R is the resultant of N_1, F_1, N_2, and F_2 and is now inclined *up* the incline *against* the direction of possible skidding. Forces W, R, and F_I are in equilibrium.

At speeds *above* the *normal speed* (Fig. 16-7*c*), the tangential component of F_I exceeds that of W; the car then tends to skid *outward* and *up* the incline. This tendency is resisted by the side-friction forces F_1 and F_2 acting *down* the incline. The reaction R is now inclined *down* the incline *against* the direction of possible skidding, and equilibrium of forces W, R, and F_I is established.

To obtain an equation giving the required superelevation e, let forces W, F_I, and N in Fig. 16-8*a* be those acting through the center of gravity of a

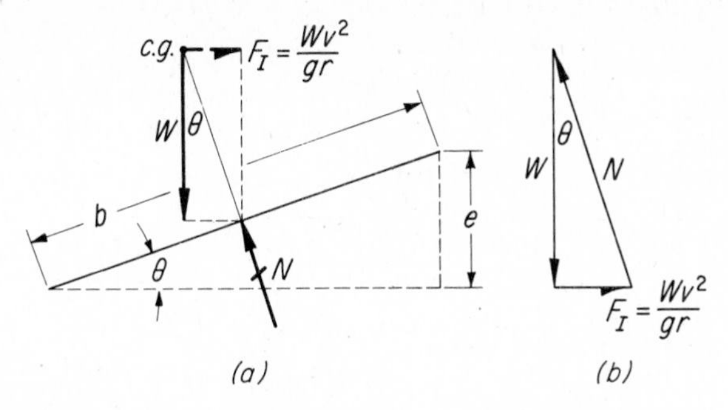

FIG. 16-8 Superelevation of highway curve. Vehicle moving at *normal speed.* (*a*) Free-body diagram. (*b*) Force diagram.

car traveling at *normal speed* around a curve of radius r and superelevation e. If b is the width of the pavement,

$$\sin\theta = \frac{e}{b} \tag{a}$$

For railway and highway curves, the angle θ is generally small. The sine and the tangent of any small angle are nearly equal. When so considered, $\sin\theta = \tan\theta = e/b$. From Fig. 16-8*b*, $\tan\theta = F_I/W = v^2/gr$, from Eq. (16-3). We then obtain

$$\tan\theta = \frac{v^2}{gr} = \frac{e}{b} \tag{b}$$

from which *the required superelevation e for a normal speed v is*

$$\boldsymbol{e = \frac{v^2 b}{gr}} \tag{16-6}$$

and *the normal speed v for a curve of radius r is*

$$\boldsymbol{v = \sqrt{gr\tan\theta}} \tag{16-7}$$

For *railway curves,* b is the distance from center to center of the *normal rail reactions* N_1 and N_2. This distance is generally considered to be 4.9 ft. For speeds lower or higher than the normal speed on any superelevated curve, pressure, called **flange pressure,** is exerted by the flanges of the wheels against the inner and outer rails, respectively, thus preventing the wheels from leaving the rails.

The most important highway-curve speed to determine is that at which skidding impends, or beyond which skidding will occur. Let forces W, F_I, and R in Fig. 16-9*a* be those (acting through its center of gravity) of a car rounding a curve at the maximum permissible speed without skidding outward. The instant skidding impends, the angle ϕ between R and a line normal to the

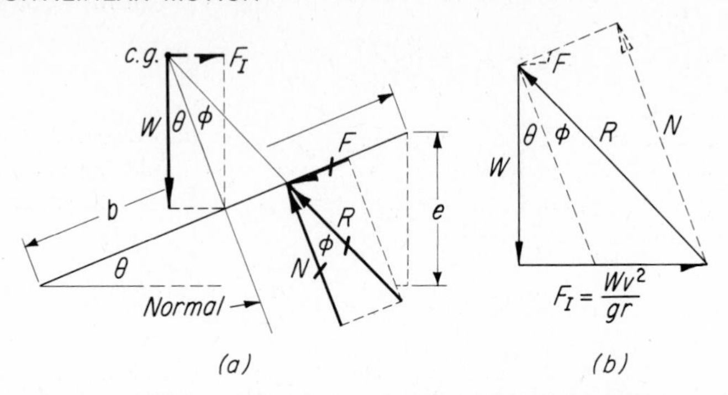

FIG. 16-9 Superelevation of highway curve. When skidding impends, tan $\phi = F$. (*a*) Free-body diagram. (*b*) Force diagram.

incline is *the angle of limiting static friction* (Art. 9-1) whose tangent is f, the coefficient of static friction ($\tan \phi = F/N = f$). Forces W, F_I, and R are in equilibrium. From the force diagram in Fig. 16-9*b*

$$\tan(\theta + \phi) = \frac{Wv^2/gr}{W} = \frac{v^2}{gr} \qquad (c)$$

from which *the maximum highway-curve speed without outward skidding is*

$$\boldsymbol{v_{\max} = \sqrt{gr \tan(\theta + \phi)}} \qquad \textbf{(16-8)}$$

The instant skidding impends inward, at slow speeds on highly superelevated roads, the angle ϕ is on the opposite side of a *normal* to the incline. Then the equation for *the minimum highway-curve speed to prevent inward skidding* is

$$\boldsymbol{v_{\min} = \sqrt{gr \tan(\theta - \phi)}} \qquad \textbf{(16-8a)}$$

but, in this equation, ϕ may not exceed θ. (Note that, when the side friction F is zero, ϕ is zero in Eqs. (16-8) and (16-8*a*) and the normal speed as given by Eq. (16-7) is obtained.)

ILLUSTRATIVE PROBLEM

16-14. The width b of the pavement of a highway curve is 20 ft, and its radius r is 400 ft. (*a*) Calculate the superelevation e required for a *normal speed* of 30 mph and (*b*) determine the maximum speed v at which a car may travel without skidding on this curve, when f is 0.25 for a slippery pavement.

Solution: We may obtain the desired results directly from Eqs. (16-6) and (16-8), but a free-body diagram is always helpful. The diagram in Fig. 16-9 is of the type required in this solution and will therefore be referred to. Given: $b = 20$ ft, $r = 400$ ft, $v = 30$ mph $= 44$ ft per sec. From Eq. (16-6), the superelevation e for the normal speed v is

$$\left[e = \frac{v^2 b}{gr}\right] \qquad e = \frac{(44)^2(20)}{(32.2)(400)} = 3 \text{ ft} \qquad \textit{Ans.}$$

The highest speed that may be traveled on any curve without skidding is given by Eq. (16-8): $v = \sqrt{gr \tan(\theta + \phi)}$. At maximum speed, skidding impends outward or *up* the incline. Hence R is inclined *down* the incline, as shown in Fig. 16-9*a*. From Fig. 16-9*a*,

$$\sin\theta = \frac{3}{20} = 0.15 \qquad \text{and} \qquad \theta = 8°38'$$

Also

$$\tan\phi = \frac{F}{N} = f = 0.25 \qquad \text{and} \qquad \phi = 14°02'$$

Then $\theta + \phi = 8°38' + 14°02' = 22°40'$, and $\tan(\theta + \phi) = \tan 22°40' = 0.4176$.
Equation (16-8) then gives

$$[v = \sqrt{gr \tan(\theta + \phi)}] \qquad v = \sqrt{(32.2)(400)(0.4176)}$$
$$= 73.2 \text{ ft per sec} \qquad \text{or} \qquad 50 \text{ mph} \qquad \textit{Ans.}$$

PROBLEMS

16-15. A highway curve with a radius of 900 ft is to be superelevated to accommodate a *normal speed* (no side friction) of 45 mph. Calculate the required superelevation e if the width b of the pavement is 30 ft. *Ans.* $e = 4.52$ ft

16-16. Calculate the *normal speed* v for a car on a highway curve whose radius is 600 ft, if b is 20 ft and e is 2 ft.

16-17. A railroad car weighing 60,000 lb rounds a level curve with a radius of 100 ft at a speed of 15 mph. Compute the total flange pressure P against the outer wheels. *Ans.* $P = 9{,}030$ lb

16-18. What is the required superelevation in inches for a standard-gage track if the normal speed is 60 mph and the radius of the curve is 1,760 ft? ($b = 4.9$ ft.) *Ans.* $e = 8.05$ in.

16-19. What is the maximum permissible speed on the superelevated curve of Prob. 16-18 if the total allowable flange force is 10 percent of the normal force?

16-20. An automobile rounds a level highway curve whose radius r is 200 ft. The tread of the car (distance center to center of wheels) is 4.8 ft, and its center of gravity is 2 ft above the ground. Assuming that side friction and roughness prevent skidding, at what speed v will the car tip over?

16-21. In Prob. 16-14, determine the maximum speed v without skidding if f is 0.6 for a dry pavement.

16-22. A 1,000-ft radius highway curve is to be superelevated so that skidding will not be impending at speeds below 60 mph. If the minimum f is assumed to be 0.2, what superelevation in inches is required for a 24-ft roadway? What will be the most comfortable speed for the passengers on this curve if it has the superelevation found above? *Ans.* $e = 11.1$ in.; $v = 24.1$ mph

16-23. A highway curve of 800-ft radius is to be banked to accommodate a maximum speed of 60 mph without skidding during wet weather, when f is 0.2. If the width b of the pavement is 40 ft, calculate the angle of bank θ and the required superelevation e. *Ans.* $\theta = 5°24'$; $e = 3.78$ ft

16-24. A boy running a foot race rounds a flat curve with a 44-ft radius. If his speed is 15 mph, his body will incline at what angle to the vertical?

16-25. A *level* concrete highway curve has a radius of 300 ft. At what top speed v may a car round this curve without skidding (*a*) when f is 0.2 for a slippery pavement and (*b*) when f is 0.8 for a dry pavement? *Ans.* (*a*) $v = 30$ mph; (*b*) $v = 60$ mph

16-26. Solve Prob. 16-25 when the pavement is 20 ft wide and the outer edge is superelevated 4 ft.

16-27. An automobile weighing 4,000 lb has a tread of 5 ft and its center of gravity is 27 in. above the ground. Assume that 60 percent of its weight is supported by the rear wheels. Calculate the normal wheel reactions (N_1 and N_2) and the side friction forces (F_1 and F_2) against each rear wheel when the car rounds a level highway curve of 800-ft radius at a speed of 60 mph.

Ans. $N_1 = 876$ lb; $F_1 = 263$ lb; $N_2 = 1{,}524$ lb; $F_2 = 457$ lb

16-28. In the circular motordrome shown in Fig. 16-10, the upper section C is cylindrical and sections B and A are cone-shaped. What are the *normal speeds* for riders on sections A and B when they follow paths of 30- and 40-ft radius, respectively? (Strictly speaking, the radius should be that of the center of gravity of the rider and motorcycle.) What are the highest and lowest speeds at which rider A may drive on the circular path of 30-ft radius without skidding upward or downward, if the coefficient of friction is 0.4?

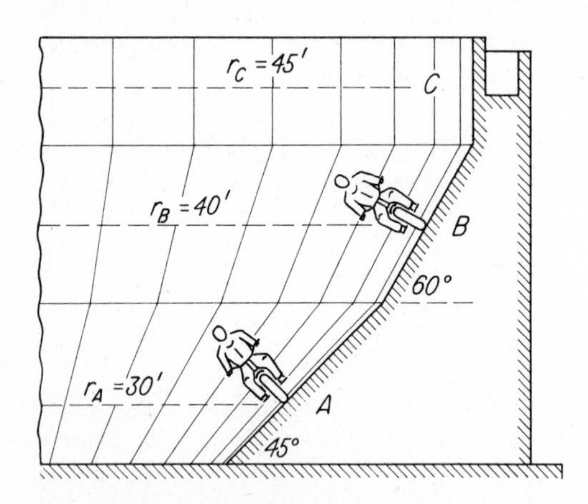

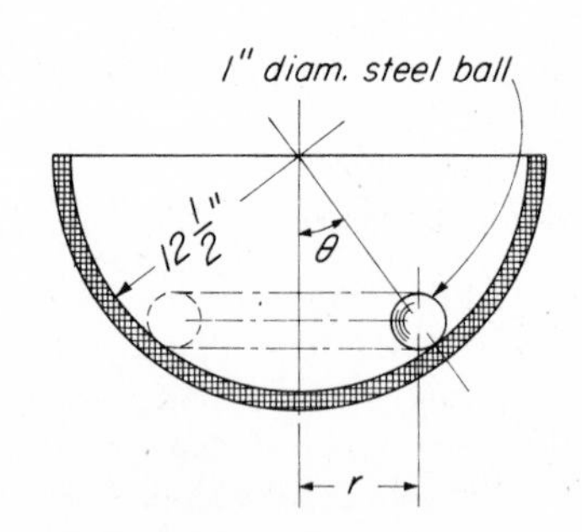

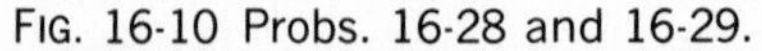
FIG. 16-10 Probs. 16-28 and 16-29.

FIG. 16-11 Prob. 16-30.

16-29. In Prob. 16-28, what pressure P do rider B and his motorcycle excert against the 60° incline while traveling at *normal speed*, if rider and motorcycle weigh 500 lb? *Ans.* $P = 1{,}000$ lb

16-30. The steel ball shown in Fig. 16-11 rotates in a horizontal plane inside the hemisphere-shaped container. What is the velocity of the ball when the angle θ is 30°?

16-5 Banking of Airplanes

An airplane in flight is sustained entirely by the air pressure against its wings and body and the uplift created by the partial vacuum on the upper wing

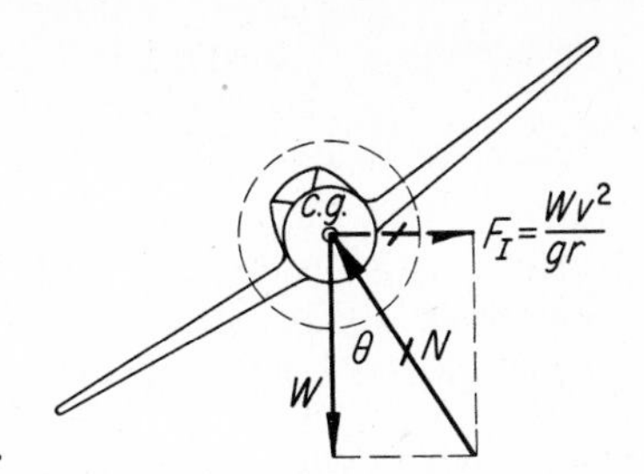

FIG. 16-12 Banking of airplanes.

surfaces. Because no appreciable resistance is offered by the air against any tendency of the plane to *sideslip*, the resultant N of the sustaining air pressure is considered to act normally to a transverse axis through the wings of the plane. This resultant N is called the **lift** and **θ is the angle of bank** *between N and the vertical,* indicated in Fig. 16-12. In level, straight-line flight, $\theta = 0$, $F_I = 0$, and the lift N equals the weight W of the plane. On steep banks N may be several times W. From Fig. 16-12 we see that, *for any angle of bank,*

$$\tan\theta = \frac{Wv^2/gr}{W} \qquad \text{or} \qquad \boldsymbol{\tan\theta = \frac{v^2}{gr}} \tag{16-3}$$

ILLUSTRATIVE PROBLEM

16-31. What angle of bank θ must the pilot of an airplane maintain if he wants to fly a horizontal circular path with a radius r of 2,000 ft at a speed v of 300 mph? Calculate the magnitude of the lift N under these flight conditions, if the plane weighs 16,100 lb.

Solution: Given: $r = 2{,}000$ ft; $v = 300$ mph $= 440$ ft per sec; $W = 16{,}000$ lb. We obtain θ, the angle of bank, from Eq. (16-3). That is,

$$\left[\tan\theta = \frac{v^2}{gr}\right] \qquad \tan\theta = \frac{(400)^2}{(32.2)(2{,}000)} = 3.01$$

and

$$\theta = 71°38' \qquad \textit{Ans.}$$

When $W = 16{,}100$ lb,

$$\left[F_I = \frac{Wv^2}{gr}\right] \qquad F_I = \frac{(16{,}100)(440)^2}{(32.2)(2{,}000)} = 48{,}400 \text{ lb}$$

and, from Fig. 16-11, the lift N is

$$[N = \sqrt{W^2 + F_I^2}] \quad N = \sqrt{(16{,}100)^2 + (48{,}400)^2} \qquad \text{or} \qquad N = 51{,}000 \text{ lb} \qquad \textit{Ans.}$$

PROBLEMS

16-32. Solve Prob. 16-31 when r is 3,520 ft and $v = 240$ mph.

16-33. An airplane flies in a horizontal circle at a constant speed of 120 mph. The instruments show the angle of bank θ to be 30°. Calculate the radius r of the circle and the lift N, if the plane weighs 8,660 lb. *Ans.* $r = 1{,}665$ ft; $N = 10{,}000$ lb

16-34. An airplane banks at an angle of 60° while making a horizontal turn at

360 mph. Determine the force exerted by a 180-lb pilot against the seat.

Ans. $F = 360$ lb

16-35. In Prob. 16-33, compute the angle of bank θ and the lift N when r is 500 ft.

16-36. An airplane weighing 12,000 lb is constructed to withstand a maximum lift N of 38,800 lb. (*a*) Calculate the maximum permissible angle of bank θ. (*b*) What is the radius of a horizontal curve flown at this angle when the speed is 180 mph?

Ans. (*a*) $\theta = 72°$; (*b*) $r = 704$ ft

16-37. If an airplane is so constructed that the upward-acting load on the wings N is not to exceed three times its weight, what is its maximum permissible angle of bank θ in horizontal flight?

16-38. An airplane makes a turn without change of altitude and at a constant speed of 420 mph. At what angle should the plane be banked if there is to be no sideslipping and the radius of the curve is 490 ft? If the pilot can stand an acceleration of 9 g for a short time, is this turn too sharp? *Ans.* $\theta = 87°36'$; yes

16-6 Curvilinear Motion in a Vertical Plane

Let us consider the motion of the small body in Fig. 16-13*a* along the *smooth curved plane* from A to B. If friction and air resistance are negligible, and hence are neglected, the change in the magnitude of its velocity $(v - v_0)$ may be shown to equal the change in the velocity of a freely falling body while being displaced the vertical distance h, measured between the initial and final positions of the center of gravity of the body. So long as the body follows the path, the direction of its motion at any instant is parallel to the path.

From Fig. 16-13*b*, we see that the tangential component W_T of the weight of the body now is the *accelerating force* producing a tangential acceleration a_t. Since the acceleration is proportional to the accelerating force, as illustrated in Fig. 16-14, we obtain

$$\frac{W_T}{a_t} = \frac{W}{g} \quad \text{or} \quad \frac{a_t}{g} = \frac{W_T}{W} = \sin\theta \quad \text{and} \quad a_t = g \cdot \sin\theta \qquad (a)$$

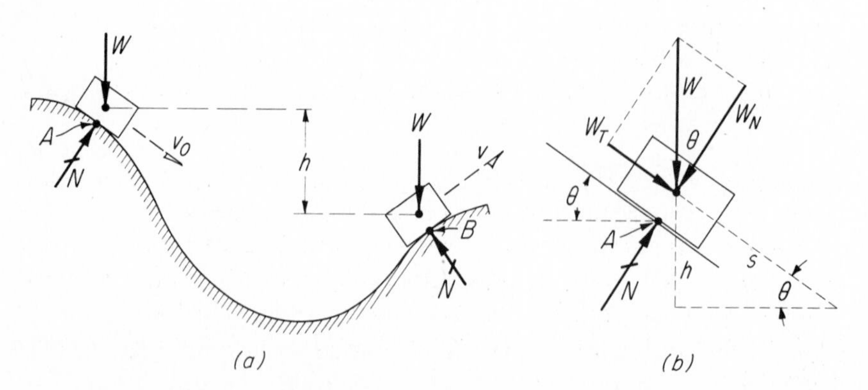

FIG. 16-13 Motion of a body along a frictionless curved path. (*a*) Frictionless curved path. (*b*) Forces acting on body at A.

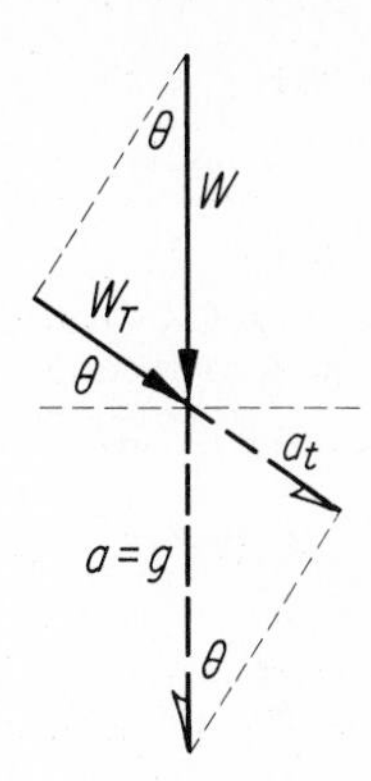

FIG. 16-14 Acceleration relationships.

If a very short time period is now considered, the small distance s moved by the body along the curved path (Fig. 16-13*b*) may be assumed to be straight. Consequently, in terms of the vertical displacement h,

$$\sin\theta = \frac{h}{s} \qquad \text{or} \qquad s = \frac{h}{\sin\theta} \tag{b}$$

We may now substitute Eqs. (*a*) and (*b*) in the basic equation of uniformly accelerated motion, $v^2 = v_0^2 + 2as$, because in so doing the variable, $\sin\theta$, is entirely eliminated. Then, since $a_t = g \cdot \sin\theta$ and $s = h/\sin\theta$, we have

$$v^2 = v_0^2 + 2(g \cdot \sin\theta)\left(\frac{h}{\sin\theta}\right) \tag{c}$$

from which we obtain

$$\boldsymbol{v^2 = v_0^2 + 2gh} \tag{16-9}$$

where v = final velocity parallel to the final path, ft per sec
v_0 = initial velocity parallel to the initial path, ft per sec
g = 32.2 ft per sec^2
h = vertical distance between the initial and final positions of the c.g., ft

Equation (16-9) is seen to be the same as Eq. (14-6), Art. 14-2, which is one of the basic equations of motion of freely falling bodies.

Therefore, *the change in velocity of a body moving between any two points along a frictionless path is equal to the velocity change it would undergo as a freely falling body while being displaced the vertical distance h between the horizontal planes containing the initial and final positions of its center of gravity.*

Signs. In Eq. (16-9), h is positive for upward displacements and is negative for downward displacements. Then v, v_0, and g are always positive.

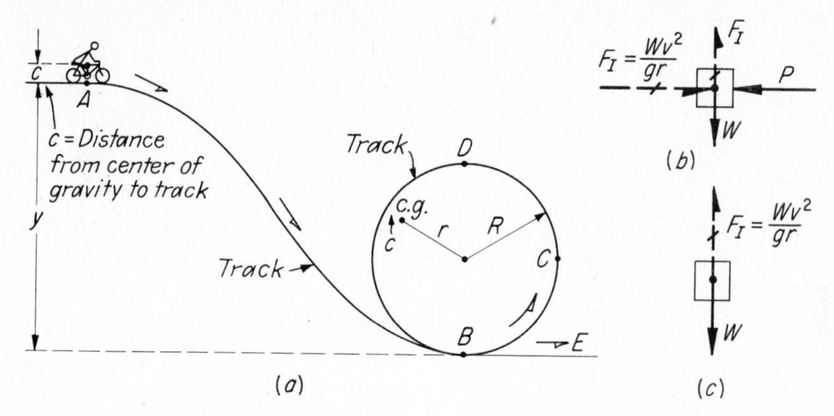

FIG. 16-15 Curvilinear motion in a vertical plane. (*a*) Body moving along frictionless path. (*b*) Pressure P at C. (*c*) Pressure at $D = 0$.

ILLUSTRATIVE PROBLEM

16-39. A circus performer rides a bicycle from rest at A in Fig. 16-15*a* down the frictionless incline to B, then around the circular track and out along the level E. The distance c from the track to the center of gravity of bicycle and rider is 3 ft. When y is 70 ft, R is 20 ft, and r of *c.g.* is 17 ft, (*a*) what is the rider's velocity at B and at D? (*b*) What total pressure P do bicycle and rider exert against the track at C, if their combined weight is 161 lb? (*c*) What minimum velocity at D is required for the bicycle to remain in contact with the track?

Solution: In each part to be solved we must recognize clearly the position of the center of gravity of rider and bicycle. Let the vertical distance between its initial and final positions be h. Then, in part (*a*) $h_B = 70$ ft at point B, and $h_D = 36$ ft at point D; in part (*b*) $h_C = 53$ ft; and in part (*c*) $h_D = 36$ ft. (NOTE: In many similar problems, where the distance c is small, it may be disregarded without appreciable error.)

a. To determine the velocities at B and D: when $v_A = 0$, $h_B = 70$ ft, and $h_D = 36$ ft:

$[v^2 = v_0{}^2 + 2gh]$ $v_B{}^2 = 0 + (2)(32.2)(70) = 4{,}508$ and $v_B = 67.1$ ft per sec *Ans.*
$[v^2 = v_0{}^2 + 2gh]$ $v_D{}^2 = 0 + (2)(32.2)(36) = 2{,}318$ and $v_D = 48.2$ ft per sec *Ans.*

b. At C, the pressure P is due entirely to the centrifugal force (Wv^2/gr), as is shown in Fig. 16-15*b*. Here, $h = 73 - 20 = 53$ ft. To compute P, we must first obtain v_C. Since $v_A = 0$ and $r = 17$ ft, we get

$[v^2 = v_0{}^2 + 2gh]$ $v_C{}^2 = 0 + 2(32.2)(53) = 3{,}413$ and $v_C = 58.2$ ft per sec

$$\left[P = F_I = \frac{Wv^2}{gr}\right] \qquad P = \frac{(161)(3{,}413)}{(32.2)(17)} = 1{,}000 \text{ lb} \qquad \textit{Ans.}$$

c. To maintain contact with the track at D, the centrifugal force F_I must at least equal the weight of the body, as is indicated in Fig. 16-15c. Then $W = Wv^2/gr$, or $v^2 = gr$. Since $r = 17$ ft, we obtain

$$[v^2 = gr] \qquad v_D{}^2 = (32.2)(17) = 547 \qquad \text{and} \qquad v_D = 23.4 \text{ ft per sec} \qquad \textit{Ans.}$$

PROBLEMS

16-40. The small body shown in Fig. 16-16 is attached to a weightless cord and swings in a vertical circle having a radius of 5 ft. If its velocity at A is 30 ft per sec, what is it at B and at C? What minimum velocity must the body have at C to remain in its circular path?

Ans. $v_B = 35$ ft per sec; $v_C = 24$ ft per sec; $v_C = 12.7$ ft per sec

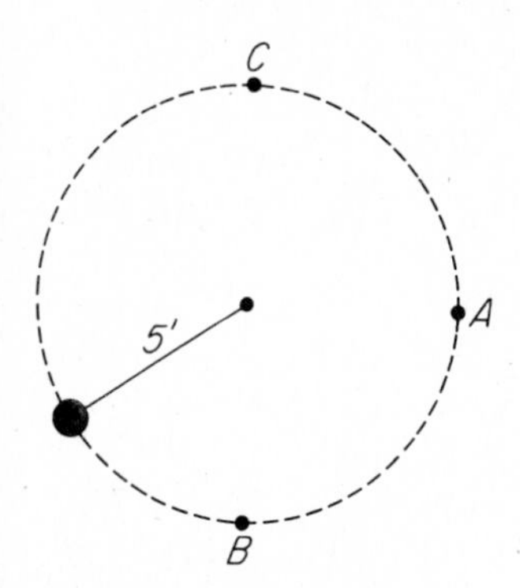

FIG. 16-16 Body swinging in a vertical circle.

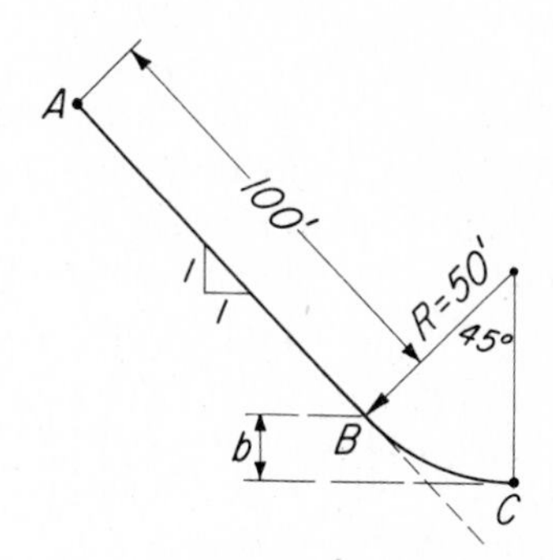

FIG. 16-17 Probs. 16-45 and 16-46.

16-41. If the body in Fig. 16-16 weighs 10 lb and the cord has a tensile strength T of 30 lb, what maximum speed may the body attain without breaking the cord?

16-42. If the weight of the body shown in Fig. 16-16 is 10 lb and it is released from rest at A, what will be the tension in the 5-ft cord when it has swung through an angle of 45°? *Ans.* 21.2 lb

16-43. Solve Prob. 16-42 if the initial velocity at point A was 10 ft per sec.

16-44. Solve Prob. 16-39 when $y = 60$ ft, $r = 15$ ft, and $c = 2$ ft.

16-45. A 200-lb ski jumper starts his run from point A of Fig. 16-17. (*a*) What is his velocity at C? (*b*) What normal force does the snow exert against his skis at C? Neglect all friction. *Ans.* $v = 74.2$ ft per sec; $F = 883$ lb

16-46. Solve Prob. 16-45 when distance AB equals 80 ft.

16-47. A body starts at rest and slides from A to B down a smooth, straight plane 10 ft long and inclined at 30° with the horizontal. Calculate (*a*) its velocity v_B at B, (*b*) its acceleration a between A and B, and (*c*) the time t_{AB} it takes to move from A to B.

Ans. (*a*) $v_B = 17.9$ ft per sec; (*b*) $a = 16.1$ ft per sec²; (*c*) $t_{AB} = 1.12$ sec

16-48. Compute the minimum height y from which the rider in Prob. 16-39 must start to remain in contact with track while circling the loop. (Draw a completely dimensioned diagram.)

16-7 Motion of Projectiles; Air Resistance Neglected

The flight of a projectile fired from a gun is influenced primarily by air resistance, which retards its motion, and by gravitational attraction, which eventually pulls the projectile down to earth. Other minor influences are rotation due to rifling of the gun barrel, windage, etc. The effects of air resistance are complex; they differ with the size and shape of the projectile. Consequently, this general study can consider only the effect of gravitational attractions.

At comparatively low velocities, air resistance is often negligible. *We may, therefore, consider as a projectile any body whose motion in space may be regarded as being affected only by gravity and by the magnitude and direction of its initial velocity.*

Let *ABCD* in Fig. 16-18 be the path (trajectory) of a projectile leaving point A at an angle θ with the horizontal and with an initial velocity v_0, whose vertical and horizontal components are $v_y = v_0 \sin\theta$ and $v_x = v_0 \cos\theta$. Since no horizontal forces act on the body, v_x *remains constant during flight.* But the vertical velocity component v_y is changed at the rate of 32.2 ft per sec^2 downward by gravitational attraction. Because $a_x = 0$ and $a_y = g$, we may use the equations of uniformly accelerated motion given in Table 14-1, Art. 14-1, *as they apply to the vertical motion of the projectile.*

Let us consider now the vertical motion of the projectile. In the basic equation ($s = v_0 t + \frac{1}{2}at^2$), $s = y$, $a = g$, and $v_0 \sin\theta$ replaces v_0. Then, *the vertical distance y between the starting point A and the striking point D is*

$$y = v_0 \sin\theta \cdot t + \frac{1}{2}gt^2 \tag{16-10}$$

The total time of flight may be obtained from Eq. (16-10), when y, v_0, and θ are known.

The horizontal distance x is called the range; it is maximum when $\theta = 45°$. Since the horizontal velocity component $v_0 \cos\theta$ is constant,

$$x = v_0 \cos\theta \cdot t \tag{16-11}$$

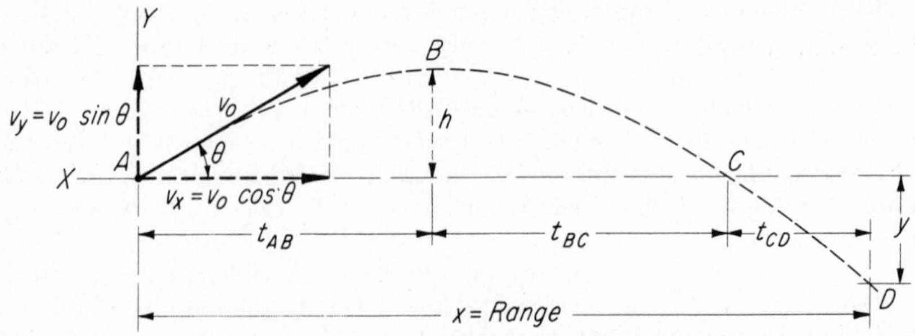

FIG. 16-18 The path of motion of a projectile.

Signs. The initial direction of v_0 determines the positive sign of y and g.

The use of Eq. (16-10) generally involves solution of a quadratic equation. This may be avoided by dividing the problem into several time elements, each of which may then be solved by such simpler relations as $v = at$, and $s = v_{avg} \cdot t$, as is illustrated in the following example.

ILLUSTRATIVE PROBLEM

16-49. A projectile leaves the muzzle A of a gun with an initial velocity of 300 ft per sec and at an angle θ of 30°, as illustrated in Fig. 16-18. It strikes the target at D, which is 200 ft lower than A. Compute the maximum height h to which it will rise, the actual velocity v_D with which it will strike the target, the total time of flight t, and the range x.

Solution: At B, v_y becomes 0. Therefore, h may be found from the basic equation $v^2 = v_0^2 + 2as$, in which $a = g = -32.2$ ft per sec² and $s = h$. The velocity v_D is determined from its vertical and horizontal components v_{yD} and v_{xD}. The total time t_{AD} is readily found by considering the motion in three separate time periods, t_{AB}, t_{BC}, and t_{CD}. Then $x = v_x \cdot t_{AD}$. At A, when $\theta = 30°$ and $v_0 = 300$ ft per sec, the velocity components are

$[v_y = v_0 \sin\theta]$ $\quad v_{yA} = (300)(0.5) = 150$ ft per sec

$[v_x = v_0 \cos\theta]$ $\quad v_{xA} = (300)(0.866) = 260$ ft per sec

To determine $h = s$, when $v_y = 0$ at B and $a = -g = -32.2$, we have $v^2 = v_0^2 + 2as$, or

$[v_{yB}^2 = v_{yA}^2 + 2as]$ $\quad 0 = (150)^2 + (2)(-32.2)h$ $\quad$ and $\quad h = 349$ ft $\quad$ *Ans.*

To find v_{yD}, the vertical velocity component at D, when $v_{yA} = 150$, $a = -g = -32.2$, and $s = -200$, we have $v^2 = v_0^2 + 2as$, or

$[v_{yD}^2 = v_{yA}^2 + 2as]$ $\quad v_{yD}^2 = (150)^2 + (2)(-32.2)(-200) = 35{,}300$

and $\quad v_{yD} = 188$ ft per sec

Since $v_{yD} = 188$ ft per sec and $v_{xD} = 260$ ft per sec, the actual velocity at D is

$$v_D = \sqrt{(188)^2 + (260)^2} \quad \text{or} \quad v_D = 321 \text{ ft per sec} \qquad \textit{Ans.}$$

The time required to rise the distance h is readily found since $h = v_{avg} \cdot t_{AB}$ and the average vertical velocity between A and B is ½(150) = 75 ft per sec. Therefore,

$$t_{AB} = \frac{349}{75} = 4.65 \text{ sec}$$

Similarly, the time required to drop from B to D, when $h = 349 + 200 = 549$ ft and $v_{avg} = \frac{1}{2}(0 + 188) = 94$ ft per sec, is

$$t_{BD} = \frac{549}{94} = 5.84 \text{ sec}$$

The total time of flight then is

$$t_{AD} = t_{AB} + t_{BD} = 4.65 + 5.84 \quad \text{or} \quad t_{AD} = 10.49 \text{ sec} \qquad \textit{Ans.}$$

which we could also obtain by solving Eq. (16-10), a quadratic, in which $y = -200$, $v_0 \sin\theta = 150$, and $g = -32.2$. That is,

$$\left[y = v_0 \sin\theta \cdot t + \frac{1}{2}at^2\right] \qquad -200 = 150t + \frac{1}{2}(-32.2)t^2$$

or $\qquad 16.1t^2 - 150t = 200 \qquad$ and $\qquad t = 10.49$ sec $\qquad$ *Check.*

The range is obtained from Eq. (16-11), in which $v_0 \cos\theta = 260$ ft per sec:

$$[x = v_0 \cos\theta \cdot t] \qquad x = (260)(10.49) = 2{,}729 \text{ ft} \qquad \textit{Ans.}$$

PROBLEMS

16-50. A projectile is fired with a muzzle velocity of 1,000 ft per sec from a gun aimed at an angle of 30° with the horizontal. How high will it rise?

Ans. $h = 3{,}875$ ft

16-51. Calculate the maximum horizontal distance x a stone can be thrown, if its initial velocity is 60 ft per sec.

16-52. A man throws a stone upward at an angle of 30° to the horizontal. It lands 173.2 ft measured horizontally and 6 ft below his arm measured vertically. Determine the time of flight and the initial velocity of the stone.

Ans. $t = 2.563$ sec; $v_0 = 78.0$ ft per sec

16-53. A nozzle on the end of a fire hose is 60 ft from a vertical wall and makes an angle of 45° with the horizontal. If the stream strikes the wall at a point 27.8 ft above the nozzle, what is the velocity of the water as it leaves the nozzle?

Ans. $v = 60$ ft per sec

16-54. Compute the instantaneous radius of curvature of the projectile of Prob. 16.49 as it passes point B.

16-55. A stunt man rides a motorcycle across the gap AB, as illustrated in Fig. 16-19. What minimum speed must he have at A to cross the gap when $d = 20$ ft, if the incline at A slopes 10° with the horizontal? A and B are at the same level.

Ans. $v_0 = 43.4$ ft per sec

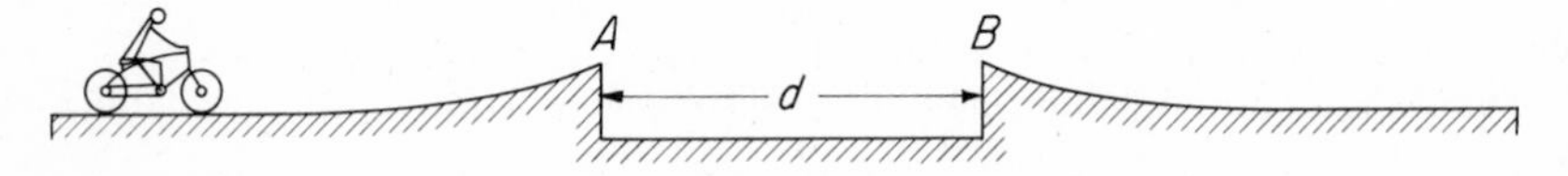

FIG. 16-19 Probs. 16-55 and 16-56.

16-56. If the slope of the incline at A in Fig. 16-19 is 15° and the speed of the rider is 45 mph, how wide a gap can he span? *Ans.* $d = 67.4$ ft

16-57. A stone is thrown upward at an angle of 20° and with an initial velocity of 50 ft per sec, from the top of a building 100 ft high. How high does it rise, how soon will it strike the ground, and at what horizontal distance from the starting point does it strike?

SUMMARY

(By article number)

16-3. When a weight, attached to a light cord, is swung with uniform speed in a horizontal circular path, it is accelerated toward the center of rotation by the horizontal component of the pull in the cord. The reaction to this accelerating pull or force is the *centrifugal inertia force,* which is *always directed outward from the center of rotation.* Its magnitude is determined by the equation

Centrifugal inertia force:

$$F_I = \frac{Wv^2}{gr} \tag{16-2}$$

16-4. *Railway and highway curves* are banked to overcome the tendency of the centrifugal inertia force to overturn passing vehicles. The proper *amount of bank* or *superelevation* e required for any speed v is

Superelevation:

$$e = \frac{v^2 b}{gr} \tag{16-6}$$

where r is the radius of curvature, and b is 4.9 ft for railway curves and is the width of pavement for highway curves.

When θ is the angle of bank, the *normal speed* at which no side-friction forces are exerted against the wheels is

Normal speed:

$$v = \sqrt{gr \tan \theta} \tag{16-7}$$

The maximum speed without outward skidding is

Maximum speed:

$$v = \sqrt{gr \tan (\theta + \phi)} \tag{16-8}$$

where ϕ is the angle whose tangent is f, the coefficient of static friction.

16-5. *The total normal air pressure* against the wings and body of an airplane in flight is called the **lift** N. When a plane banks,

$$N = \sqrt{W^2 + F_I^2}$$

where W is the weight of the plane and F_I is the centrifugal inertia force.

The angle of bank θ is the angle N makes with the vertical, and

$$\tan \theta = \frac{v^2}{gr} \tag{16-3}$$

where v is the velocity and r is the radius of curve.

16-5. In curvilinear motion in a vertical plane, the change in velocity of a body moving between any two points *along a frictionless path* is equal to the velocity change it would undergo as a freely falling body while being displaced the vertical distance h between those points, measured between the initial and final positions of the center of gravity. Consequently, $a = g$, and $s = h$; the basic equation, $v^2 = v_0 + 2as$, then becomes

$$v^2 = v_0^2 + 2gh \tag{16-9}$$

In this equation h is positive for downward displacements and is negative for upward displacements. Then, the signs of v, v_0, and g are always positive.

16-6. When air resistance is neglected, **the motion of a projectile** is influenced only by gravity and by the magnitude and direction of its initial velocity v_0, whose vertical and horizontal components are

$$v_y = v_0 \cdot \sin \theta \qquad \text{and} \qquad v_x = v_0 \cdot \cos \theta$$

in which θ designates the initial direction of flight. The force of gravity changes v_y at the rate of 32.2 ft per sec^2, while v_x remains constant. *The initial direction of v_0 determines the positive signs of acceleration, velocity, and displacement.*

The vertical distance y between the initial and final positions of the projectile is

$$y = v_0 \sin \theta \cdot t + \frac{gt^2}{2} \tag{16-10}$$

from which we may calculate *the total time of flight*, when y, v_0, and θ are known.

The *range* is obtained from the equation

$$x = v_0 \cos \theta \cdot t \tag{16-11}$$

in which $v_0 \cos \theta$ remains constant. The range x is maximum when $\theta = 45°$.

REVIEW PROBLEMS

16-58. A clamp weighing ½ lb is fastened to one spoke of a pulley at a point 9 in. from the axis of rotation. Calculate the radial side thrust F on the shaft caused by the *centrifugal inertia force* acting on the clamp, when the pulley turns at 400 rpm.

16-59. At what maximum speed v may a truck with a tread of 5 ft safely round a level highway curve, whose radius is 200 ft, without *skidding or tipping over*, if the center of gravity is 4 ft above the ground and f is 0.6? *Ans.* $v = 42.3$ mph

16-60. A level, circular platform in an amusement park increases its rotating speed slowly from rest to 9 rpm. If f is 0.2, at what maximum distance r from the axis of rotation could a person sit without sliding off? *Ans.* $r = 7.23$ ft

16-61. What maximum rotating speed N (rpm) may be given the swing shown in Fig. 16-5, if the radius r is limited to 34 ft? $B = 10$ ft, $L = 40$ ft.

16-62. Assume in Prob. 16-60 that a boy sits on the platform at a distance r of 4 ft from the axis of rotation. If $f = 0.2$, at what rpm N will he slide outward?

16-63. A projectile is fired at an angle of 30° with the horizon. The muzzle velocity is 2,000 ft per sec. What is the instantaneous radius of curvature of its path when it reaches the highest point of the trajectory? If the shell strikes the ground at the same elevation as the gun, what is the horizontal range?

Ans. $r = 92{,}900$ ft; $x = 107{,}500$ ft

16-64. A block starts from rest at the top of a smooth sphere. Neglect all friction. At what angle from the vertical will the block leave the surface of the sphere?

16-65. A bombing plane has a speed of 360 mph while flying 10,000 ft above the ground. How far ahead of the target should a bomb be released? Neglect air friction. Assume line of flight to be level. *Ans.* $s = 2.492$ miles

16-66. What *minimum rotating speed* N (rpm) must be maintained by the circular plate, in Fig. 16-6, to prevent body C from sliding down the conical flange, if r_C is 3 ft and f_C is 0.25? *Ans.* $N = 20.4$ rpm

16-67. A railroad curve has a radius of 2,000 ft, and the rails are 4.9 ft on centers.

(*a*) Calculate the required *superelevation e* of the outer rail to accommodate a *normal speed* of 45 mph. (*b*) What *flange pressure P* is exerted against the wheel flanges of a 64,400-lb car rounding the curve at 60 mph, if the tangential component of its weight is considered? *Ans.* (*a*) $e = 4$ in.; (*b*) $P = 3{,}375$ lb

16-68. What is the *maximum speed v* at which a car may round a highway curve, whose radius is 400 ft, without skidding, if the 30-ft-wide pavement is superelevated 3 ft, (*a*) when the pavement is slippery and f is 0.15 and (*b*) when the pavement is dry and f is 0.7? Assume that the car will not overturn.

16-69. A pilot observes that the speed of his plane is 180 mph horizontally and that its angle of bank is 45°. What is the radius of the curve he flies?

Ans. $r = 2{,}164$ ft

16-70. The pilot of an airplane wants to reverse his direction of flight from north to south as quickly as possible. (*a*) If his top speed on turns is 300 mph and the maximum permissible *angle of bank* for the airplane is 75°, on what minimum radius may he turn and how many seconds are required to travel the horizontal semicircle? (*b*) If the plane weighs 20,000 lb, what is the total *lift N*?

16-71. At a certain point in the trajectory of a wingless rocket the engine thrust is 1,000 lb, the remaining weight is 1,000 lb, and the path is inclined at an upward angle of 30° to the horizontal. Air resistance at its present speed is 200 lb. (*a*) What is the tangential acceleration of the rocket at this point? (*b*) What is the normal acceleration? (*c*) What is its true acceleration? (*d*) What angle does this true acceleration make with the horizon?

16-72. If the body shown in Fig. 16-16 has a velocity of 10 ft per sec as it passes point A, what is the true acceleration of the body at that time and what angle does the true acceleration vector make with the radius at A? See Prob. 16-40.

Ans. $a = 37.9$ ft per sec^2; $\theta = 58°$ measured downwardly from the radius

16-73. Calculate *the highest and lowest speeds* at which rider B may drive his motorcycle, without skidding, around the circular path in the motordrome shown in Fig. 16-10, where r_B is 40 ft and f_B is 0.4.

16-74. What *minimum speed* must be maintained by a man driving a motorcycle around the vertical section C of the motordrome in Fig. 16-10 in order to travel on the circular path of 45-ft radius, when f is 0.4? *Ans.* $v_{min} = 41$ mph

16-75. A section of track of a roller coaster lies in a *vertical plane* and has a radius of 20 ft. (*a*) *Calculate the maximum speed* at which a string of cars may pass over the top of this section without causing the wheels to exert an upward lift on the track. Assume the center of gravity to be 1 ft above the track. (*b*) How far above the top of this section of track must the cars start to attain that speed, if the velocity at the higher point is 10 ft per sec? *Ans.* (*a*) $v = 26$ ft per sec; (*b*) $h = 8.95$ ft

16-76. A stone is thrown horizontally with an initial velocity of 40 ft per sec from the deck of a bridge 200 ft above the water. (*a*) What is its actual velocity as it strikes the water and (*b*) how far away horizontally does it strike?

16-77. Water leaves the tip of a fire nozzle with a velocity of 60 ft per sec. If the nozzle is inclined up at an angle of 30° with the horizontal, how high above the nozzle will the jet strike a vertical wall 45 ft away? Disregard air friction.

16-78. If the nozzle of Prob. 16-77 is held at 45°, what is the highest window (measured vertically from the nozzle) that the jet can reach and what is the required horizontal distance? *Ans.* $h = 27.9$ ft; $x = 55.8$ ft

REVIEW QUESTIONS

16-1. What is meant by curvilinear motion and by plane curvilinear motion?

16-2. Explain the meaning of centrifugal force. When does it manifest itself, which way is it always directed, and what equation expresses it?

16-3. What is meant by normal acceleration? Toward which point is it always directed, and what equation expresses it?

16-4. Name some instances in which centrifugal forces play an important part.

16-5. What is meant by normal speed on highway and railway curves?

16-6. What equation gives the true (or total) acceleration of a body if its speed is changing along its curved path of motion?

16-7. Explain the meaning of superelevation. Why is it desirable?

16-8. Give the equation expressing the angle of bank of an airplane. What is likely to happen to an airplane if the tangent of its angle of bank θ does not equal v^2/gr?

16-9. When a body moves in a vertical plane along a frictionless curved path, what equation expresses the change in its velocity between any two points?

16-10. What two factors primarily influence the flight of a projectile? Which of these is constant and which is variable?

16-11. Under what conditions may other bodies be considered as projectiles?

16-12. When air resistance is neglected, what influences the vertical component of the initial velocity of a projectile, and what influences the horizontal component?

16-13. What is meant by the range of a projectile, and when is it (theoretically) maximum for any given initial velocity?

CHAPTER 17

Kinematics of Rotation

17-1 Rotational Motion

In rotational motion, as herein considered, every particle of a body moves in a path of constant radius about a fixed **axis of rotation,** usually *centroidal,* which is perpendicular to the plane of motion. Common examples of rotating bodies are flywheels on stationary engines, pulleys on fixed shafts, and rotors in motors and generators.

17-2 Angular Displacement

In Fig. 17-1 is shown a pulley of constant radius r rotating about its centroidal axis O. When a particle A on the rim of the pulley has moved to A', the angle swept through by the radius is the **angular displacement** θ of the particle. Another particle B at some other point on the pulley has moved to B', and its angular displacement θ is seen to be the same as that of particle A. From this we may conclude that **all particles of a rotating body have the same angular displacement.** Clearly, however, the *linear displacements* (AA' and BB') of particles A and B are not equal, but *are proportional to their distances away from the center of rotation O.*

The basic unit of angular displacement is the radian (rad). In engineering practice, however, the revolution is more commonly used. **A radian** *is the angle at the center subtended by an arc equal in length to the radius,* as shown in Fig. 17-2. The circumference of a circle is $2\pi r$. Since arc AB equals radius r, a complete circle, or one revolution, contains 2π radians (360 degrees); hence 1 radian $= 360/6.2832 = 57.3$ degrees (approximately). Because the length

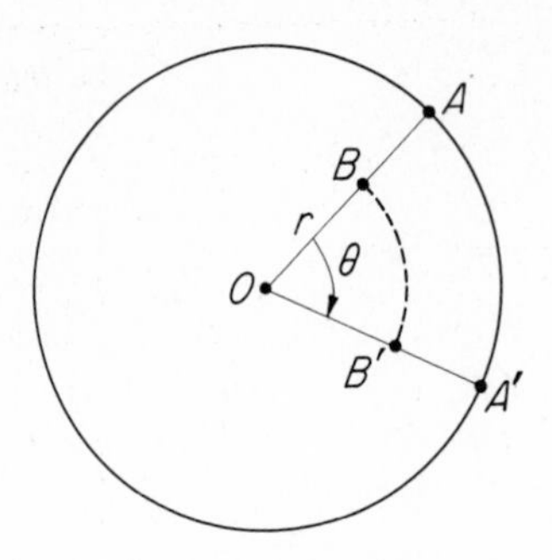

FIG. 17-1 Angular displacement.

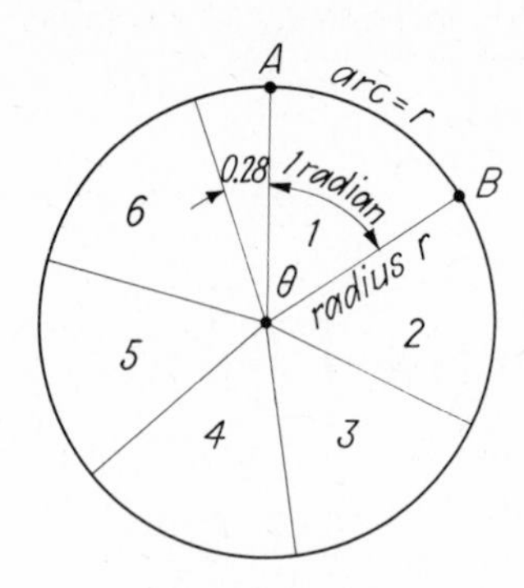

FIG. 17-2 One revolution = 360° = 2π rad.

of arc subtended by a radian is proportional to the radius (that is, arc/radius = constant) and because both radian and radius are measured in like units, *a radian is an abstract number without unit.* The choice of the radian greatly simplifies the relation between angular and linear units of motion, as indicated in Art. 17-8.

PROBLEMS

17-1. How many radians are in 1 rev? 50 rev? 360°?

Ans. 2π = 6.28 rad; 314 rad; 6.28 rad

17-2. Convert the following angular measurements to revolutions: 2π rad; 12.56 rad; 90°.

Ans. 1 rev; 2 rev, 0.25 rev

17-3. How many degrees are in 1 rad? 2π rad? 5 rev?

Ans. 57.3°; 360°; 1800°

17-4. In Fig. 17-1, what is the total angular displacement θ, in radians, of particle A during $\frac{1}{5}$ revolution? In degrees?

17-3 Angular Velocity

Angular velocity is the rate of change of angular displacement with respect to time and is denoted by the Greek-letter symbol ω (omega). Since the basic unit of time t is the second, *the basic unit of angular velocity is radians per second* (rad per sec). Other units are radians per minute (rad per min), revolutions per second (rps), and, most commonly used in engineering practice, revolutions per minute (rpm). Because all particles of a rotating body have equal *angular* displacements during any interval of time, their *angular* velocities and accelerations are likewise equal.

17-4 Angular Acceleration

Angular acceleration is the rate of change of angular velocity with respect to time and is denoted by the Greek-letter symbol α (alpha). The basic unit of angular acceleration is radians per second per second (rad per sec^2). Because

Table 17-1 UNIT CONVERSION FACTORS

Given →		α, angular acceleration				ω, angular velocity				θ, angular displacement	
To obtain ↓		$\frac{\text{rad per sec}}{\text{sec}}$	$\frac{\text{rad per min}}{\text{sec}}$	$\frac{\text{rps}}{\text{sec}}$	$\frac{\text{rpm}}{\text{sec}}$	$\frac{\text{rad}}{\text{sec}}$	$\frac{\text{rad}}{\text{min}}$	rps	rpm	rad	rev
		Multiply by									
α	$\frac{\text{rad ps}}{\text{sec}}$	1	$\frac{1}{60}$	2π	$\frac{2\pi}{60}$						
	$\frac{\text{rad pm}}{\text{sec}}$	60	1	120π	2π						
	$\frac{\text{rps}}{\text{sec}}$	$\frac{1}{2\pi}$	$\frac{1}{120\pi}$	1	$\frac{1}{60}$						
	$\frac{\text{rpm}}{\text{sec}}$	$\frac{60}{2\pi}$	$\frac{1}{2\pi}$	60	1						
ω	$\frac{\text{rad}}{\text{sec}}$					1	$\frac{1}{60}$	2π	$\frac{2\pi}{60}$		
	$\frac{\text{rad}}{\text{min}}$					60	1	120π	2π		
	rps					$\frac{1}{2\pi}$	$\frac{1}{120\pi}$	1	$\frac{1}{60}$		
	rpm					$\frac{60}{2\pi}$	$\frac{1}{2\pi}$	60	1		
θ	rad									1	2π
	rev									$\frac{1}{2\pi}$	1

the angular velocity ω may be expressed in radians per second, radians per minute, revolutions per second, or revolutions per minute, the angular acceleration α may be expressed in radians per second per second, radians per minute per second, revolutions per second per second, or revolutions per minute per second.

17-5 Conversions of Units of Angular Motion

In the study of angular motion, angular velocity is most conveniently expressed in radians per second. In practice, however, it is most generally expressed in revolutions per minute. Hence, conversion from one unit to the other is often necessary. For example, an angular velocity ω of 10 rad per sec is to be converted to revolutions per minute: One revolution equals 2π rad, and 1 min equals 60 sec. Hence,

$$\omega = 10\frac{\text{rad}}{\text{sec}} = 10\left(\frac{1}{2\pi}\right)\frac{\text{rev}}{\text{sec}} = 10\left(\frac{1}{2\pi}\right)(60)\ \text{rpm}$$

That is, to obtain ω in revolutions per minute, ω in radians per second is multiplied by a *conversion factor* of $60/2\pi$. For convenience, all necessary conversion factors are listed in Table 17-1.

PROBLEMS

17-5. In each of the following conversions, solve for X and/or Y: (*a*) $\theta = 1{,}257$ rad $= X$ rev; (*b*) $\omega = 40$ rad per sec $= X$ rps $= Y$ rpm.

Ans. (*a*) $\theta = 200$ rev; (*b*) $\omega = 6.37$ rps $= 382$ rpm

17-6. Convert the following quantities as indicated: (*a*) $\theta = 628$ rad $= X$ rev; (*b*) $\omega = 120$ rpm $= X$ rps $= Y$ rad per sec $= Z$ rad per min.

Ans. (*a*) $\theta = 100$ rev; (*b*) $\omega = 2$ rps $= 12.57$ rad per sec $= 754$ rad per min

17-7. Make the following conversions: $\alpha = 10$ rad per sec^2 $= X$ rad per min per sec $= Y$ rps per sec $= Z$ rpm per sec.

Ans. $\alpha = 600$ rad per min per sec $= 1.59$ rps per sec $= 95.4$ rpm per sec

17-8. Complete the following conversions: $\alpha = 270$ rpm per sec $= X$ rps per sec $= Y$ rad per sec^2 $= Z$ rad per min per sec.

17-6 Equations of Rotation. Uniform Acceleration

The equations of rotational motion with uniform acceleration are identical in form to those developed in Chap. 14 for rectilinear motion, except that the symbols α, ω, and θ, for acceleration, velocity, and displacement, respectively, now replace the corresponding symbols a, v, and s of rectilinear motion. The three basic equations follow:

Rectilinear motion		*Rotational motion*	
$v = v_0 + at$	(14-1)	$\omega = \omega_0 + \alpha t$	(17-1)
$s = v_0 t + \frac{1}{2}at^2$	(14-2)	$\theta = \omega_0 t + \frac{1}{2}\alpha t^2$	(17-2)
$v^2 = v_0^2 + 2as$	(14-3)	$\omega^2 = \omega_0^2 + 2\alpha\theta$	(17-3)

For easy reference, the basic equations and others that are readily derived from them are presented in Table 17-2. Except for the difference in symbols, these equations are identical to those in Table 14-1 of Art. 14-1.

Because t, α, ω, and θ may be given in various units containing seconds or minutes, radians or revolutions, great care must be exercised in use of the equations in Table 17-2. In general, the units of all quantities should be in

Table 17-2 EQUATIONS OF UNIFORMLY ACCELERATED ROTATIONAL MOTION

To find	When given	Use equation		Derived from
t	$\alpha \quad \omega_0 \quad \omega$	$t = \dfrac{\omega - \omega_0}{\alpha}$	(*A*)	Eq. (17-1)
α	$t \quad \omega_0 \quad \omega$	$\alpha = \dfrac{\omega - \omega_0}{t}$	(*B*)	Eq. (17-1)
	$t \quad \omega_0 \quad \theta$	$\alpha = \dfrac{2\theta - 2\omega_0 t}{t^2}$	(*C*)	Eq. (17-2)
	$\omega_0 \quad \omega \quad \theta$	$\alpha = \dfrac{\omega^2 - \omega_0^2}{2\theta}$	(*D*)	Eq. (17-3)
ω_0	$t \quad \alpha \quad \omega$	$\omega_0 = \omega - \alpha t$	(*E*)	Eq. (17-1)
	$t \quad \alpha \quad \theta$	$\omega_0 = \dfrac{\theta}{t} - \dfrac{\alpha t}{2}$	(*F*)	Eq. (17-2)
	$\alpha \quad \omega \quad \theta$	$\omega_0 = \sqrt{\omega^2 - 2\alpha\theta}$	(*G*)	Eq. (17-3)
ω	$t \quad \alpha \quad \omega_0$	$\omega = \omega_0 + \alpha t$	(17-1)	
	$\alpha \quad \omega_0 \quad \theta$	$\omega = \sqrt{\omega_0^2 + 2\alpha\theta}$	(*H*)	Eq. (17-3)
θ	$t \quad \alpha \quad \omega_0$	$\theta = \omega_0 t + \dfrac{1}{2}\alpha t^2$	(17-2)	
	$\alpha \quad \omega_0 \quad \omega$	$\theta = \dfrac{\omega^2 - \omega_0^2}{2\alpha}$	(*I*)	Eq. (17-3)
	$t \quad \omega_0 \quad \omega$	$\theta = \dfrac{1}{2}t(\omega_0 + \omega)$	(*J*)	Eq. (17-2)

seconds, and in either radians or revolutions. Proper conversions may be made either before or after the solutions of the appropriate equations, as required. The motion diagrams derived in Chap. 14 are as readily applied to angular motion as to linear motion, and are often very helpful in the solution of problems, as is shown in Prob. 17-10.

ILLUSTRATIVE PROBLEMS

17-9. The angular velocity ω of a flywheel increases at a uniform rate from 300 to 2,700 rpm, in 20 sec. Calculate the angular acceleration α in rpm per second and in rps per second, and the angular displacement θ in revolutions.

Solution: The appropriate equations are readily selected from Table 17-2, but care must be used in determining the proper units. Solutions in units only follow the solution of each equation. Given, $\omega_0 = 300$ rpm; $\omega = 2{,}700$ rpm; $t = 20$ sec. To find α in rpm per second and in rps per second, from Eq. (*B*) of Table 17-2,

$$\left[\alpha = \frac{\omega - \omega_0}{t}\right] \qquad \alpha = \frac{2{,}700 - 300}{20} = 120\frac{\text{rpm}}{\text{sec}}$$

or

$$\alpha = \left(120\frac{\text{rpm}}{\text{sec}}\right)\left(\frac{1}{60}\right) = 2\frac{\text{rps}}{\text{sec}} \qquad \textit{Ans.}$$

Solving Eq. (*B*), Table 17-2, in units only, gives

$$\frac{\text{rpm}}{\text{sec}} = \frac{\text{rpm} - \text{rpm}}{\text{sec}} = \frac{\text{rpm}}{\text{sec}}$$

Before we can solve for θ, ω_0 and ω must be converted from rpm to rps. That is, $\omega_0 = 300$ rpm $= 5$ rps, and $\omega = 2{,}700$ rpm $= 45$ rps. Also, $\alpha = 2$ rps per sec and $t = 20$ sec. Then, from Eq. (17-2), we obtain

$$\left[\theta = \omega_0 t + \frac{1}{2}\alpha t^2\right] \qquad \theta = (5)(20) + \left(\frac{1}{2}\right)(2)(20)^2 \qquad \text{or} \qquad \theta = 500 \text{ rev} \qquad \textit{Ans.}$$

Solution of Eq. (17-2), in units only, gives

$$\text{rev} = \left(\frac{\text{rev}}{\text{sec}}\right)(\text{sec}) + \left(\frac{\text{rev}}{\text{sec}^2}\right)(\text{sec}^2) = \text{rev}$$

17-10. The angular speed ω of a motor increases uniformly from rest to 2,400 rpm in 10 sec. It continues at that speed for the next 480 rev, after which the speed decreases uniformly, until the motor stops in 160 additional revolutions. Draw complete motion diagrams. Determine the total number of revolutions θ of the motor, and the elapsed time t.

Solution: The statement indicates three consecutive time periods to be considered,

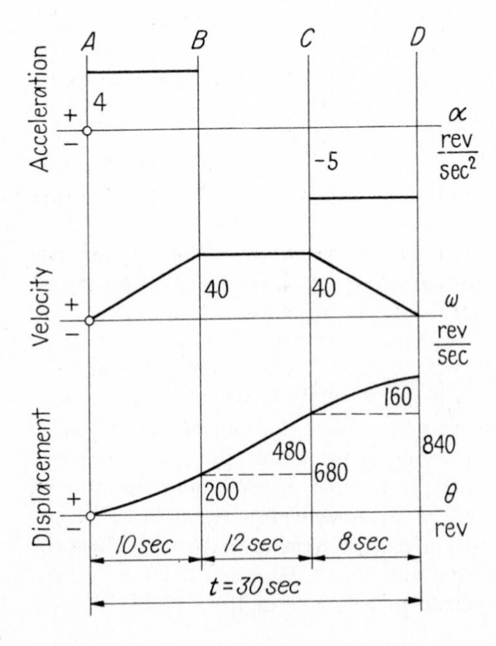

FIG. 17-3 Completed motion diagrams.

the first being 10 sec. During the second period, the motor turns 480 rev ($\theta_{BC} = 480$), and during the third it turns 160 rev ($\theta_{CD} = 160$). Let the four vertical axes in Fig. 17-3 be labeled A, B, C, and D. Also, from the statement, $\omega_A = 0$, $\omega_B = 2{,}400$ rpm $= 40$ rps, and $\omega_D = 0$.

The *uniformly increasing speed* from A to B indicates *uniform acceleration*. The *uniform speed* from B to C indicates *zero acceleration*, and the *uniformly decreasing speed* from C to D indicates *uniform deceleration*. With this information and by use of the two laws of motion diagrams outlined in Art. 14-5, the diagrams are easily drawn. By the second law, $\theta_B = (\frac{1}{2})(10)(40) = 200$ rev, and $\theta_{BC} = 40t_{BC} = 480$, from which $t_{BC} = 12$ sec. Then $\theta_C = \theta_B + \theta_{BC} = 680$ rev, and $\theta_D = \theta_C + \theta_{CD} = 840$ rev. Also, by the second law, $\theta_{CD} = (\frac{1}{2})(40)(t_{CD}) = 160$ rev and $t_{CD} = 8$ sec, from which $t = 10 + 12 + 8 = 30$ sec.

PROBLEMS

17-11. The flywheel of an engine attains its normal running speed of 12 rps after turning 18 rev from rest. Compute the required uniform angular acceleration α in rps per sec and the elapsed time t. *Ans.* $\alpha = 4$ rps per sec; $t = 3$ sec

17-12. Calculate the constant angular acceleration α in rps per sec which must be given a flywheel in order to increase its angular velocity ω uniformly from 8 to 32 rps in 4 sec. How many revolutions θ does it turn in the 4 sec?

17-13. In Prob. 17-12, let the angular velocity increase uniformly from 120 to 1,200 rpm in 6 sec, and solve. *Ans.* $\alpha = 3$ rps per sec; $\theta = 66.1$ rev

17-14. The flywheel of a steam engine starts from rest ($\omega_0 = 0$) and is accelerated 2 rad per sec^2 for 10 sec. What is its final angular velocity ω in rad per sec, in rad per min, and in rpm? *Ans.* $\omega = 20$ rad per sec $= 1{,}200$ rad per min $= 191$ rpm

17-15. A water turbine is brought to its normal operating speed of 180 rpm in 6 minutes. What was the angular acceleration if the angular velocity increased uniformly? How many revolutions did the turbine make in coming to normal speed? How many revolutions did the turbine make in the first 3 minutes?

Ans. $\alpha = 0.0524$ rad per sec^2; $\theta_6 = 540$ rev; $\theta_3 = 135$ rev

17-16. If friction brings a flywheel to rest from its rotational speed of 1,800 rpm in 3 minutes, how many turns will it make in coming to rest? How long will it take to complete the first half of these turns?

17-17. If the angular speed of the rotor of an electric motor can be changed from 100 to 1,100 rpm in 5 sec, what is the angular acceleration? What is its displacement θ in the 5-sec interval? Assume a uniform angular acceleration.

17-18. A pulley with an initial angular velocity ω_0 of 40 rad per sec is accelerated 10 rad per sec^2 for 6 sec. What is its final angular velocity ω in rad per sec, rps, and rpm? What is its displacement θ in the 6 sec in radians and in revolutions?

Ans. $\omega = 100$ rad per sec $= 15.9$ rps $= 956$ rpm; $\theta = 420$ rad $= 66.9$ rev

17-19. A flywheel rotating at 900 rpm is brought to rest in 60 sec by a constant frictional resistance. Compute the angular deceleration α in rpm per sec and in rps per sec, and the number of revolutions θ it turns before stopping.

Ans. $\alpha = -15$ rpm per sec $= -0.25$ rps per sec; $\theta = 450$ rev

17-20. Solve Prob. 17-9 by motion diagrams.

17-21. Solve Prob. 17-18 by motion diagrams.

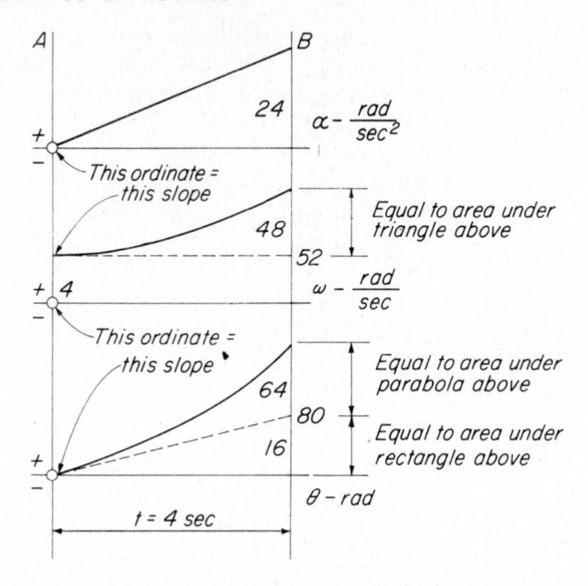

FIG. 17-4 Variable acceleration example solved by motion diagrams.

17-22. Solve Prob. 17-11 by motion diagrams.

17-23. A flywheel rotating at 1,800 rpm is slowed to 1,200 rpm in 12 minutes by air and journal friction. It was then brought to rest by the application of a brake which caused an angular deceleration of 300 rpm per min. Sketch the complete motion diagrams and find the total time and number of revolutions made in coming to rest from the 1,800 rpm initial speed. *Ans.* $t = 16$ minutes; $\theta = 20{,}400$ rev

17-24. A motor starting from rest is accelerated 2 rad per sec² for 20 sec. By application of a brake, its angular velocity is then decreased uniformly until the motor stops in 5 sec. Using motion diagrams, determine (*a*) the angular velocity ω_{20} and the displacement θ_{20} at the end of 20 sec, (*b*) the deceleration α during the last 5 sec, and (*c*) the total displacement θ_{25} during the entire 25 sec.

Ans. (*a*) $\omega_{20} = 40$ rad per sec; $\theta_{20} = 400$ rad;
(*b*) $\alpha = -8$ rad per sec²; (*c*) $\theta_{25} = 500$ rad

17-25. A pulley has an initial angular velocity of 12 rps which is increased at a uniform rate to 36 rps in 8 sec. The pulley is then decelerated 2 rps per sec until it stops. Using motion diagrams, determine (*a*) the angular acceleration during the first 8 sec, (*b*) the total elapsed time *t*, and (*c*) the total number of revolutions θ.

17-7 Rotation. Variable Acceleration

Problems with variable angular accelerations may be solved with elementary calculus, as is illustrated in Appendix D, but many such problems are also readily solved by motion diagrams, as developed in Arts. 14-5 and 14-6, and as illustrated below.

ILLUSTRATIVE PROBLEM

17-26. A flywheel rotating with an angular velocity ω_0 of 4 rad per sec is given an acceleration α, which increases at a uniform rate from 0 to 24 rad per sec^2 during 4 sec. Using motion diagrams, determine its final velocity ω and total displacement θ at the end of 4 sec. (Figure 17-4 shows the completed diagrams.)

Solution: From the given data, we may draw the complete acceleration diagram. The initial velocity $\omega_0 = 4$ rad per sec. In accordance with the first law, the slope of the velocity curve is positive and increasing, as are the ordinates in the acceleration diagram. According to the second law, the *increase* in velocity equals the area under the acceleration curve, which is $(\frac{1}{2})(4)(24) = 48$, giving a total velocity ω_B of 52 rad per sec. The slope of the displacement curve is positive and increasing, as are the ordinates in the velocity diagram. The total displacement equals the area under the velocity curve, which is $(4)(4) + (\frac{1}{3})(4)(48) = 80$ rad.

PROBLEMS

17-27. Solve Prob. 17-26 when $\omega_0 = 12$ rad per sec and α increases from 0 to 32 rad per sec^2 in 8 sec. *Ans.* $\omega = 140$ rad per sec; $\theta = 437$ rad

17-28. The diagram in Fig. 17-5 indicates the angular acceleration of a rotating body during two consecutive time periods of 4 and 8 sec. The motion starts from rest. Complete the velocity and displacement diagrams, and determine the final velocity ω and the total displacement θ at the end of the 12 sec.

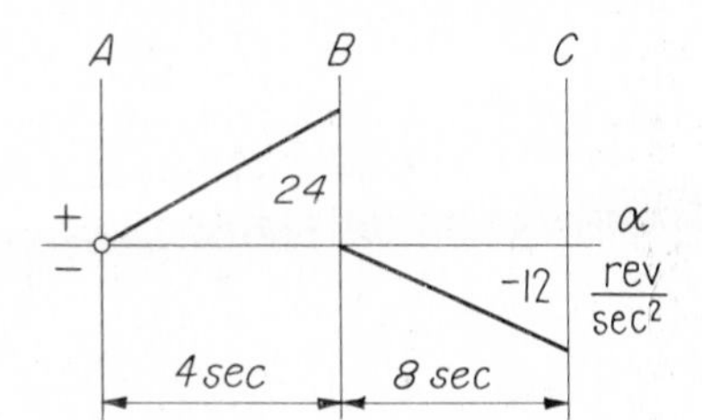

FIG. 17-5 Prob. 17-28.

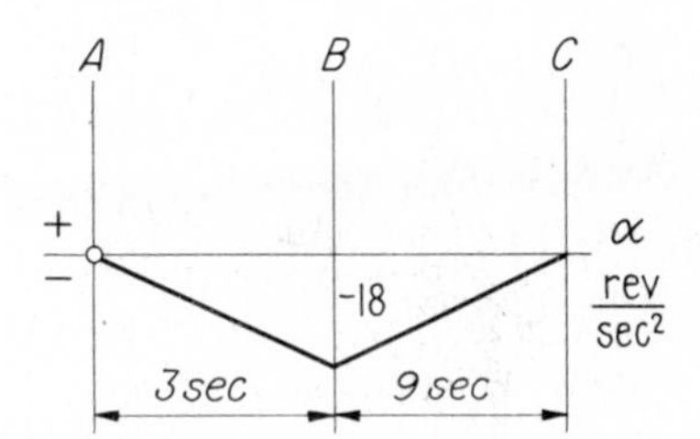

FIG. 17-6 Prob. 17-29.

17-29. A rotating flywheel is decelerated, as indicated by the diagram in Fig. 17-6, and comes to a stop at the end of 12 sec. Using motion diagrams, determine its initial angular velocity ω_θ and the total displacement θ.

Ans. $\omega_0 = 108$ rps; $\theta = 540$ rev

17-8 Relation between Linear and Angular Motion

The two pulleys C and D shown in Fig. 17-7 are keyed together. When they are rotated counterclockwise through an angle θ equal to 1 rad, particle A on the outer rim of the large diameter moves to position A'. Its *angular displacement* is 1 rad, and its *linear displacement* along the arc AA' is s_1, which equals r_1 when θ is 1 rad.

Particle B moves to position B', where its linear displacement s_2 equals r_2.

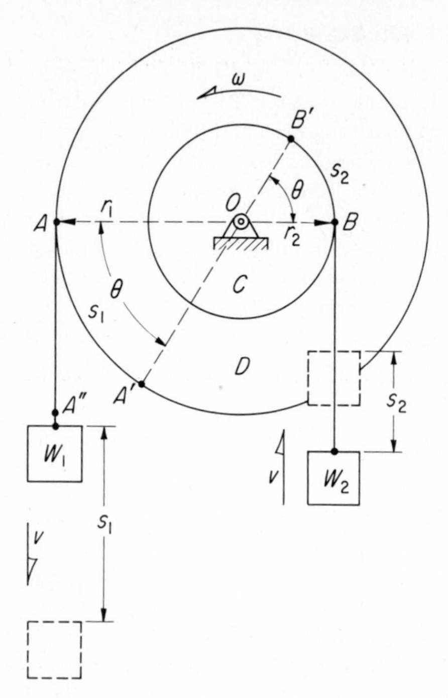

FIG. 17-7 Relation between linear and angular motion.

Clearly, then, the linear displacement of any particle is proportional to its distance r from the center of rotation. Hence, *the linear displacement s equals the radius r for every radian of angular displacement,* and, if θ is expressed in radians,

$$\boldsymbol{s = r\theta} \tag{17-4}$$

Also, if weights W_1 and W_2 are attached to cords wound, respectively, around the large and the small diameters, the linear displacement of W_1 is seen to be s_1 $(= r_1)$ for every radian of angular displacement of the pulley and that of W_2 to be s_2 $(= r_2)$, whence $s = r\theta$. Because θ is without unit, s and r will have like units, usually feet.

The relation between linear and angular velocity may be derived from Eq. (17-4). Since velocity is the rate of change of position (displacement) with respect to time, $v = s/t$ and $\omega = \theta/t$. Rewriting these, we obtain $s = vt$, and $\theta = \omega t$. Hence,

$$s = r\theta = vt = r\omega t \tag{a}$$

from which

$$\boldsymbol{v = r\omega} \tag{17-5}$$

The relation between linear and angular acceleration is now derived from Eq. (17-5). Since acceleration is the rate of change of velocity with respect

Table 17-3 RELATION BETWEEN LINEAR AND ANGULAR MOTION

	Linear			Angular		
Relation	Symbol	Unit	Quantity	Symbol	Unit	Relation
$a = r\alpha$	a	$\frac{\text{ft}}{\text{sec}^2}$	Acceleration	α	$\frac{\text{rad}}{\text{sec}^2}$	$\alpha = \frac{a}{r}$
$v = r\omega$	v	$\frac{\text{ft}}{\text{sec}}$	Velocity	ω	$\frac{\text{rad}}{\text{sec}}$	$\omega = \frac{v}{r}$
$s = r\theta$	s	ft	Displacement	θ	rad	$\theta = \frac{s}{r}$

to time, $a = v/t$ and $\alpha = \omega/t$. Rewriting these, we obtain $v = at$ and $\omega = \alpha t$. Hence

$$v = r\omega = at = r\alpha t \qquad (b)$$

from which

$$\mathbf{a = r\alpha} \qquad \mathbf{(17\text{-}6)}$$

For easy reference, Eqs. (17-4) to (17-6), together with their proper symbols and basic units, are listed in Table 17-3. *The conversions indicated in Table 17-3 cannot be made unless the quantities are in the basic units as listed.*

It is important to remember, in the solution of problems, that r must be in feet and that the radian is an abstract number without unit. For convenience, its unit may be considered as unity or 1. Hence, the unit of α is $1/\text{sec}^2$, that of ω is $1/\text{sec}$, and that of θ is 1. As an example, $v = r\omega$, or, in units only,

$$\frac{\text{ft}}{\text{sec}} = (\text{ft})\left(\frac{1}{\text{sec}}\right) = \frac{\text{ft}}{\text{sec}}$$

ILLUSTRATIVE PROBLEM

17-30. Block B in Fig. 17-8a is attached to a cord which is wound around pulley A, whose radius is 2 ft. The initial angular velocity ω_0 of the pulley is 4 rad per sec, and the weight of block B accelerates it 3 rad per sec^2 for 6 sec. Calculate (a) the final angular velocity ω and the angular displacement θ of the pulley, and (b) the linear acceleration a, velocity v, and displacement s of block B at the end of 6 sec. (c) Solve parts (a) and (b) independently by motion diagrams.

Solution: The angular velocity ω and the displacement θ are found by Eqs. (17-1) and (17-2) of Table 17-2. We have given, $\alpha = 3$ rad per sec^2, $\omega_0 = 4$ rad per sec, and $t = 6$ sec. Hence,

$[\omega = \omega_0 + \alpha t]$ $\quad \omega = 4 + (3)(6) = 4 + 18 \quad$ or $\quad \omega = 22$ rad per sec $\qquad$ *Ans.*

$\left[\theta = \omega_0 t - \frac{1}{2}\alpha t^2\right]$ $\quad \theta = (4)(6) + \left(\frac{1}{2}\right)(3)(6)^2 = 24 + 54 \quad$ or $\quad \theta = 78$ rad $\qquad$ *Ans.*

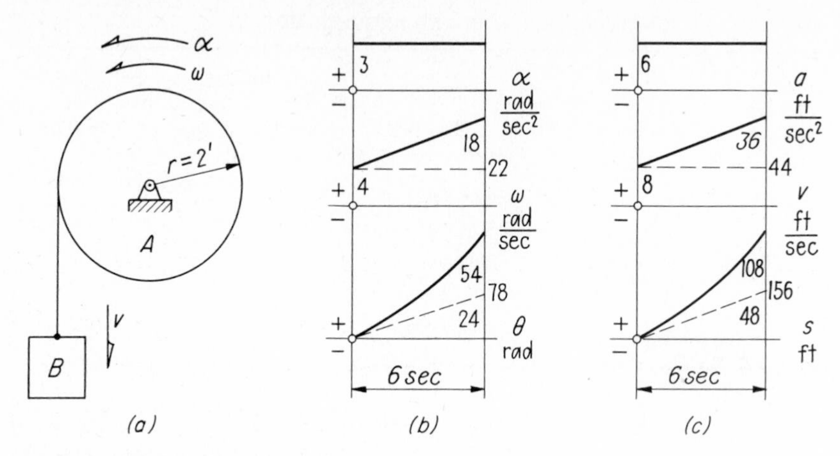

FIG. 17-8 Relation between linear and angular motion. (*a*) Space diagram. (*b*) Angular motion. (*c*) Linear motion.

The linear values of a, v, and s are found by Eqs. (17-6), (17-5), and (17-4), respectively. When $r = 2$ ft, we obtain

$[a = r\alpha]$	$a = (2)(3) = 6$ ft per sec^2	*Ans.*
$[v = r\omega]$	$v = (2)(22) = 44$ ft per sec	*Ans.*
$[s = r\theta]$	$s = (2)(78) = 156$ ft	*Ans.*

Solution by Motion Diagrams: The *angular motion diagrams* shown in Fig. 17-8*b* may be started directly from the given data: $\omega_0 = 4$ rad per sec, $\alpha = 3$ rad per sec^2, $t = 6$ sec. With this information, these diagrams are readily completed. Before starting the *linear motion diagrams* shown in Fig. 17-8*c*, the given values of ω_0 and α must be converted into linear quantities. That is, $v_0 = r\omega_0$ or $v_0 = (2)(4) = 8$ ft per sec, and $a = r\alpha = (2)(3) = 6$ ft per sec^2. The linear motion diagrams are then easily completed.

PROBLEMS

17-31. Solve Prob. 17-30 when $r = 1.5$ ft, $\omega_0 = 10$ rad per sec, $\alpha = 2$ rad per sec^2, and $t = 8$ sec. *Ans.* (*a*) $\omega = 26$ rad per sec; $\theta = 144$ rad; (*b*) $a = 3$ ft per sec^2; $v = 39$ ft per sec; $s = 216$ ft

17-32. Let block B in Fig. 17-8*a* move downward from rest with a linear acceleration of 2 ft per sec^2. Calculate the linear velocity v and displacement s of the block, and the angular velocity ω and displacement θ of the pulley, at the end of 5 sec.

17-33. Block B in Fig. 17-9 is raised by the clockwise rotation of pulley A, whose radius r is 3 in. and whose angular velocity ω_0 at a given instant is 6 rps. The pulley stops after B has been raised 14.14 ft with uniformly decreasing velocity. Compute (*a*) the number of revolutions θ made by the pulley, (*b*) the angular and linear decelerations (α and a) of pulley A and block B, respectively, and (*c*) the elapsed time t.

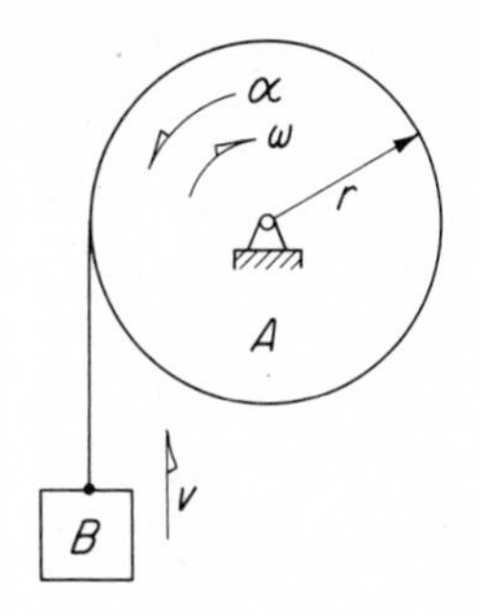

FIG. 17-9 Probs. 17-33 and 17-37.

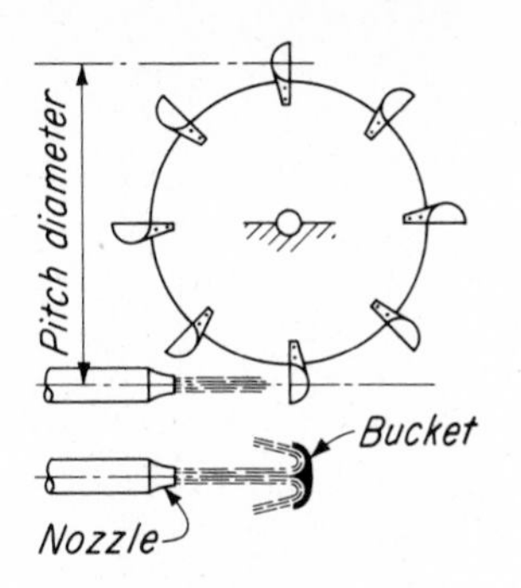

FIG. 17-10 Impulse turbine.

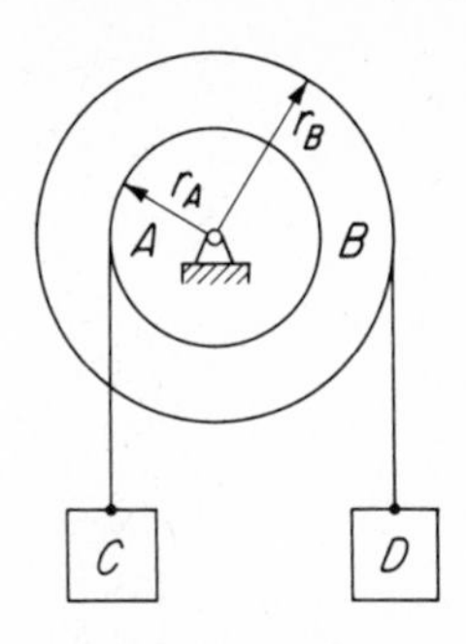

FIG. 17-11 Prob. 17-34.

17-34. In Fig. 17-11, pulleys A and B are keyed together. The initial counterclockwise angular velocity ω_0 of 12 rad per sec decreases at a uniform rate to 0 in 4 sec. Compute (*a*) the angular deceleration α, (*b*) the total displacement θ of the pulleys (in rev), and (*c*) the linear displacements s_C and s_D of blocks C and D, when $r_A = 2$ ft and $r_B = 4$ ft.

Ans. (*a*) $\alpha = 3$ rad per sec^2; (*b*) $\theta = 24$ rad; (*c*) $s_C = 48$ ft; $s_D = 96$ ft

17-35. A schematic diagram of an impulse water turbine is shown in Fig. 17-10. For maximum efficiency the peripheral speed of the buckets at the pitch diameter should be about 0.4 the velocity of the jet of water. If the jet velocity is 90 ft per sec, what is the best operating speed of the turbine in rpm? The pitch diameter is 3 ft.

Ans. $\omega = 229$ rpm

17-36. A 12-in. diam pulley is attached to the shaft of the turbine of Prob. 17-35. Find the linear velocity of a belt running over this pulley. If the belt is to drive a generator at 120 rpm, what diameter pulley should be used on the generator?

17-37. At a certain instant, disk A in Fig. 17-9 rotates clockwise with an angular speed of 360 rpm. Its radius r is 6 in. The weight of block B causes the disk to stop after B has moved 10 ft upward. Calculate the linear and angular decelerations a and α, and the elapsed time t.

Ans. $a = -17.75$ ft per sec^2; $\alpha = -35.5$ rad per sec^2; $t = 1.058$ sec

17-9 The Coriolis Acceleration

The analysis of the motion of a body moving on the surface of a rotating disk or sphere is relatively complex. Nevertheless, such motions are important. The earth is essentially a rotating sphere, and, of course, motion of bodies on its surface are common occurances. Fortunately the effect of the earth's rotation can be neglected in most engineering problems. However, the effect of this rotation on the motions of rockets, aircraft, satellites, and the ocean and atmospheric currents may require a numerical evaluation.

Consider first the *radial* motion of a particle on the surface of a rotating disk. Assume a constant radial velocity v_r of the particle and a constant angular

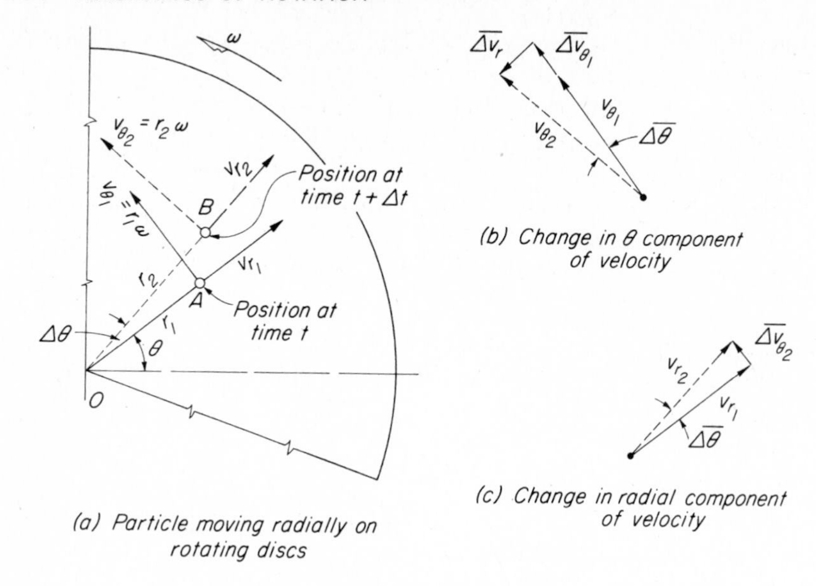

FIG. 17-12 Coriolis acceleration.

velocity ω of the disc (see Fig. 17-12*a*). This particle will have a tangential component of velocity ωr due to the disc rotation. At time t the particle will be at A with tangential and radial components of velocity $v_{\theta_1} = r_1\omega$ and v_{r_1} respectively. After a very short time interval Δt, the particle will be at B with velocity components $v_{\theta_2} = r_2\omega$ and $v_{r_2} = v_{r_1}$.

The velocity vectors v_{θ_1} and v_{θ_2} are shown in Fig. 17-12*b*. Two separate velocity changes are evident. The first, Δv_r, is caused by the change in *direction* of the tangential velocity vector. When Δv_r is divided by the time interval Δt we have the acceleration a_{n_1}.

$$a_{n_1} = \frac{\Delta v_r}{\Delta t} = \frac{(v_\theta)(\Delta\theta)}{\Delta t} = \frac{(r\omega)(\omega\ \Delta t)}{\Delta t} = r\omega^2 \tag{a}$$

Since $\omega = \dfrac{v}{r}$, the acceleration can be written

$$\boldsymbol{a}_{n_1} = \boldsymbol{r\omega^2} = \boldsymbol{r}\left(\frac{\boldsymbol{v}}{\boldsymbol{r}}\right)^2 = \frac{\boldsymbol{v}^2}{\boldsymbol{r}} \tag{16-1}$$

The second velocity change indicated in Fig. 17-12*b* is Δv_{θ_1}. This change in velocity arises because the tangential component of velocity increases as the particle moves radially outward on the disc, that is, $v = r\omega$ increases as r increases. Note that this velocity change is associated with the *relative motion* between the rotating disk and the radial movement of the particle. If either of the motions stop, Δv_{θ_1} will reduce to zero.

During the time interval Δt the *direction* of v_r changes by an amount Δ_θ. The resulting velocity change Δv_{θ_2} is shown in Fig. 17-12*c*. Note again that this velocity change is associated with the *relative motion* between the rotating disk and the radial movement of the particle. The total change in velocity due to the relative motion is in the tangential direction and is equal to $\Delta v_{\theta_2} + \Delta v_{\theta_1}$. The acceleration due to the relative motion is

$$\left[a = \frac{\Delta v}{\Delta t}\right] \qquad a_C = \frac{\Delta v_{\theta_2} + \Delta v_{\theta_1}}{\Delta t} = \frac{\Delta v_{\theta_2}}{\Delta t} + \frac{\Delta v_{\theta_1}}{\Delta t}$$

Since $\Delta v_{\theta_2} = v_r(\Delta\theta)$ and $\Delta v_{\theta_1} = v_{\theta_2} - v_{\theta_1}$, we have

$$a_C = \frac{v_r\,\Delta\theta}{\Delta t} + \frac{v_{\theta_2} - v_{\theta_1}}{\Delta t} = \frac{v_r\,\Delta\theta}{\Delta t} + \frac{r_2\omega - r_1\omega}{\Delta t}$$

Also, $\Delta\theta/\Delta t = \omega$ and $\Delta r/\Delta t = v_r$. Therefore

$$a_C = \frac{v_r\,\Delta\theta}{\Delta t} + \frac{(r_2 - r_1)\,\omega}{\Delta t} = v_r\omega + v_r\omega$$

or

$$\boldsymbol{a_C = 2v_r\omega} \tag{17-7}$$

This acceleration associated with the relative motion between the rotating disk and the radially translating body is called the **Coriolis acceleration.** Note that the acceleration is in the tangential direction and would disappear if either ω or v_r were zero. The *normal* acceleration given by Eq. (16-1) is directed along a line towards the axis of rotation (i.e., the inward radial direction) and does not require a radial velocity.

Motion on a curved surface results in an additional acceleration due to the continual change in the direction of the velocity vector. A body moving at high speed over an extensive portion of the earth's surface is a practical application of such a motion. Suppose that a body were moving towards the pole on a spherical surface as indicated in Fig. 17-13. In moving from A to B, the velocity vector rotates through an angle $\Delta\beta$. The resulting change in velocity Δv (see Fig. 17-13*b*) is $\Delta v = v(\Delta\beta)$ and the resulting acceleration is

$$a = \frac{\Delta v}{\Delta t} = \frac{v(\Delta\beta)}{\Delta t}$$

Since $\Delta\beta = \dfrac{\Delta s}{R}$ (see Fig. 17-9*a*) and $\Delta s/\Delta t = v$, we have

$$\boldsymbol{a = \frac{v\,\Delta s}{\Delta t(R)} = \frac{v^2}{R}} \tag{17-8}$$

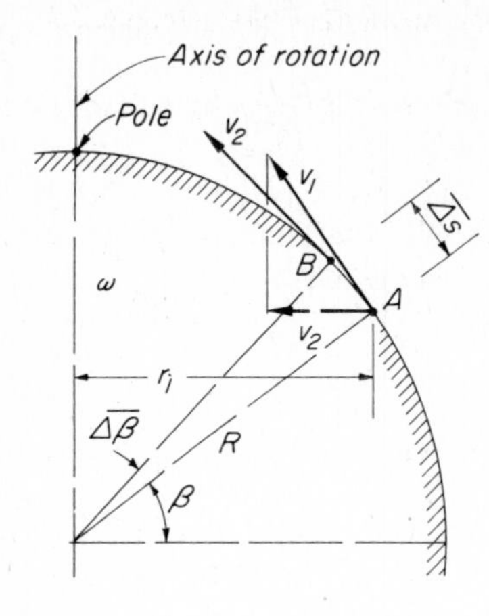

(a) Section through sphere

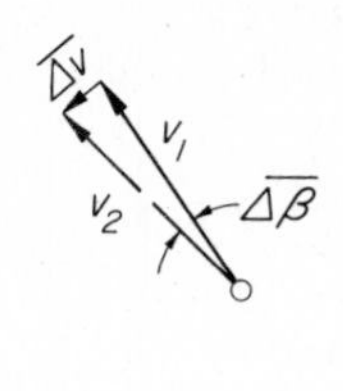

(b) Velocity vectors and change in velocity

FIG. 17-13 Normal acceleration due to motion on surface of a sphere.

Equation (17-8) gives the acceleration of a body moving with constant speed on a curved surface. Note that the acceleration is directed towards the *center of curvature* of the surface (i.e., in the R direction). If the sphere is rotating two additional accelerations would occur. These are the *normal acceleration* given by Eq. (16-1), directed towards the axis of rotation (i.e., in the r_1 direction) and the tangential *Coriolis acceleration* perpendicular to the cutting plane used in Fig. 17-13*a*.

ILLUSTRATIVE PROBLEMS

17-38. A slider moves radially outward with a velocity of 30 ft per sec along a rod which is rotating about its end with an angular velocity of 6 rad per sec. See

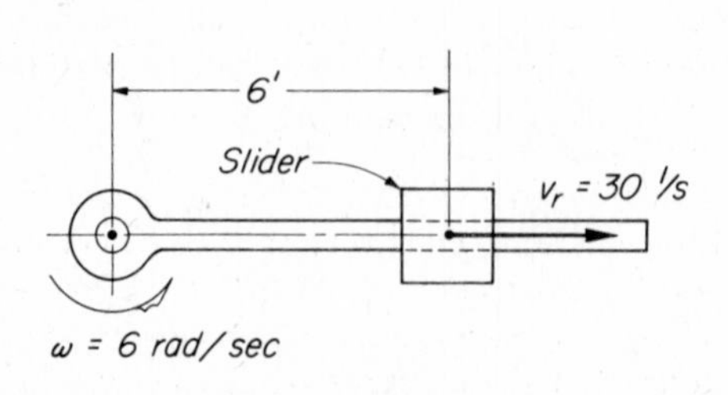

(a) Slider on rotating rod

(b) Computed acceleration

FIG. 17-14 Coriolis acceleration.

Fig. 17-14*a*. Calculate the acceleration components of the slider as it passes a point 6 ft from the axis of rotation.

Solution: The normal component of the acceleration is given by Eq. (16-1), with $v = r\omega$ or 36 ft per sec.

$$\left[a_n = \frac{v^2}{r}\right] \qquad a_n = \frac{36(36)}{6} = 216 \text{ ft per sec}^2 \qquad \textit{Ans.}$$

This acceleration component is directed towards the center of rotation.

The Coriolis acceleration is given by Eq. (17-6).

$$[a_C = 2v_r\omega] \qquad a_C = 2(30)(6) = 360 \text{ ft per sec}^2 \qquad \textit{Ans.}$$

This acceleration is normal to the rod in the direction of rotation (see Fig. 17-14*b*).

17-39. A hollow sphere has an inside diameter of 10 ft. It rotates about its geometric axis at a rate of 12 rad per sec. A particle moves along a meridian (north-south line) on its inside surface with a velocity of 16 ft per sec. Calculate the acceleration of the particle as it passes a point 30° above the equator.

Solution: The normal acceleration due to the rotation of the sphere is towards the axis of rotation and is given by

$$[a_n = r\omega^2] \qquad a_{n_1} = 5 \cos 30° \ (12)^2 = 4.33 \ (144) = 624 \text{ ft per sec}^2 \qquad \textit{Ans.}$$

The Coriolis acceleration due to the relative motion between the particle and the sphere is parallel to the lines of equal latitude and is given by

$$[a_C = 2\omega v_r] \qquad a_C = 2(12)(16) \sin 30° = 192 \text{ ft per sec}^2 \qquad \textit{Ans.}$$

The normal acceleration due to the fact that the particle is moving on a curved surface is directed towards the center of curvature and is given by

$$\left[a_{n_2} = \frac{v_r^2}{R}\right] \qquad a_{n_2} = \frac{16(16)}{5} = 51.2 \text{ ft per sec}^2 \qquad \textit{Ans.}$$

PROBLEMS

17-40. A northern-hemisphere bowler at latitude 40°N has been accustomed to bowling on an east-west lane. How shall he shift his aiming point to compensate for the Coriolis acceleration if he bowls (*a*) north on a north-south alley, and (*b*) south on a north-south alley? Assume velocity of ball is 20 ft per sec and length of alley is 80 ft.

17-41. A hydroplane travels due south at 60 mph. Calculate the lateral drift during 1 mile of travel near latitude 50°N. *Ans.* $s = 17.6$ ft E

17-42. Solve Prob. 17-41 if the course is Southeast.

17-43. A man walks radially outward on a merry-go-round platform that is turning at 0.4 rpm. Calculate his acceleration components when he is 20 ft from the center of rotation. *Ans.* $a_n = 0.035$ ft per sec^2; $a_C = 0.335$ ft per sec^2

17-44. A river flows northerly in the northern hemisphere. The surface velocity is high compared to that along the bottom. Will the flow have a clockwise or counterclockwise twist when viewed looking north?

Ans. If not free to "slip" the water particles would be subjected to a Coriolis acceleration to the left. They are free to slip, however, and the top particles would drift to the right, thus causing a clockwise rotation of the flow.

SUMMARY

(By article number)

17-1. In **rotational motion,** all particles of a body move in circular paths about a fixed axis, usually centroidal, and always perpendicular to the plane of motion.

17-2. Angular displacement is the distance moved by any particle of a rotating body. It is usually measured in radians or in revolutions and is denoted by the symbol θ.

17-3. Angular velocity is the rate of angular displacement with respect to time. It is usually measured in radians per second or revolutions per second (rps) and is denoted by the symbol ω.

17-4. Angular acceleration is the rate of change of angular velocity with respect to time. It is usually measured in radians per second per second or revolutions per second per second (rps per sec) and is denoted by the symbol α.

17-5. Conversions of units of angular motion from one form to another can readily be made by remembering that 1 min = 60 sec, or 1 sec = $\frac{1}{60}$ min, and that 1 rev = 2π rad or 1 rad = $1/2\pi$ rev.

17-6. The three basic equations of rotation with uniform acceleration, together with the corresponding basic *equations of rectilinear motion,* are as follows:

Rectilinear motion		*Rotational motion*	
$v = v_0 + at$	(14-1)	$\omega = \omega_0 + \alpha t$	(17-1)
$s = v_0 t + \frac{1}{2}at^2$	(14-2)	$\theta = \omega_0 t + \frac{1}{2}\alpha t^2$	(17-2)
$v^2 = v_0^2 + 2as$	(14-3)	$\omega^2 = \omega_0^2 + 2\alpha\theta$	(17-3)

17-7. Problems involving **rotation with variable acceleration** are solved by calculus, or by motion diagrams when the motion is not too complex.

17-8. The relation between linear and angular motion is expressed by the following equations:

$$s = r\theta \quad (17\text{-}4) \qquad v = r\omega \quad (17\text{-}5) \qquad a = r\alpha \quad (17\text{-}6)$$

17-9. Due to the relative motion between them, a body with a radial component of velocity will undergo an acceleration if moving on a rotating surface. The resulting acceleration is called the Coriolis acceleration and is given by

$$a_C = 2v_r\omega \qquad (17\text{-}7)$$

If the body is free to "slip" in the tangential direction, it will experience a Coriolis drift opposite in direction to the acceleration.

REVIEW PROBLEMS

17-45. The heavy flywheel of a large engine is accelerated $\frac{1}{2}$ rad per sec^2 from rest. What is its *speed in rpm* after 20 sec? (*b*) In what time will it reach a speed of 180 rpm? (*c*) How many revolutions does it turn before reaching that speed?

Ans. (*a*) $\omega = 95.5$ rpm; (*b*) $t = 37.7$ sec; (*c*) $\theta = 56.5$ rev

17-46. An electric motor reaches a speed of 2,400 rpm from rest in 10 sec. (*a*) Compute the average (constant) acceleration α in rpm per second, in rps per second, and in radians per second per second. (*b*) *How many revolutions are turned* in the first 5 sec and in the last 5 sec?

17-47. The angular speed of the flywheel of a steam engine reduces uniformly from 900 to 300 rpm in 20 sec. Compute (*a*) *the angular deceleration* α in rpm per second and in rps per second, and (*b*) *the number of revolutions* θ turned by the flywheel during this time.

Ans. (*a*) $\alpha = -30$ rpm per sec $= -0.5$ rps per sec; (*b*) $\theta = 200$ rev

17-48. The speed of a steam-turbine rotor is increased uniformly from rest to 6,000 rpm in 40 sec. (*a*) *Compute the angular acceleration* α in rps per second. (*b*) *How many revolutions does the rotor turn* in the first 20 sec and in the last 20 sec?

Ans. (*a*) $\alpha = 2.5$ rps per sec; (*b*) 500 rev; 1,500 rev

17-49. Solve Prob. 17-39 by *motion diagrams.*

17-50. Solve Prob. 17-18 by *motion diagrams.*

17-51. The angular speed of a flywheel is increased at a uniform rate from rest to 2,400 rpm in 10 sec and is then further increased uniformly to 6,000 rpm in 30 sec more. A constant braking force is then applied, bringing the flywheel to a stop with constant deceleration in an additional 20 sec. Using *motion diagrams,* determine (*a*) the *constant angular acceleration* α in each of the three time periods and (*b*) the total number of revolutions θ turned by the wheel.

Ans. (*a*) $\alpha = 4$ rps per sec, 2 rps per sec, -5 rps per sec; (*b*) $\theta = 3{,}300$ rev

17-52. The cable drum on a mine hoist is 4 ft in diameter (see Fig. 17-8*a*). As the bucket is lowered down the shaft its velocity increases uniformly from 5 to 15 ft per sec in 20 sec. What was the angular acceleration of the drum, in rad per sec^2, and the total number of turns made by the drum in the 20-sec period?

17-53. At a particular instant a mark on a rotating disk is 20 in. directly above the horizontal axis of rotation. The angular velocity at that instant is 3 rad per sec in a clockwise direction. The angular deceleration is 2 rad per sec^2. Locate the point 2 sec later and find its absolute (true) acceleration.

Ans. $x = +18.2$ in.; $y = -8.3$ in.; $a = 3.73$ ft per sec^2

17-54. Solve Prob. 17-38 if the initial angular velocity is 20 rpm.

17-55. The angular acceleration of a pulley increases at a uniform rate from 0 to 36 rad per sec^2 in 3 sec. If its initial velocity ω_0 is 10 rad per sec, determine, by *motion diagrams,* its final velocity ω at the end of 3 sec, and the total displacement in radians. *Ans.* $\omega_3 = 64$ rad per sec; $\theta = 84$ rad

17-56. The acceleration of a motor starting from rest increases uniformly from 0 to 12 rps per sec during 6 sec. Then it decreases at a uniform rate from 12 rps per sec to 0 in 3 sec, after which the motor is brought to a stop by a deceleration which varies uniformly from 0 to -9 rps per sec. Draw complete *motion diagrams,* and determine the time t elapsed from start to stop and the total number of revolutions θ. *Ans.* $t = 21$ sec; $\theta = 648$ rev

17-57. A long cable of small diameter is wound onto a large spool whose inner and outer diameters are 3 and 6 ft, respectively. The linear velocity v of the cable as it enters the spool remains constant at 9 ft per sec during the entire winding. What must be *the angular velocity* ω of the spool (in rps) when the winding begins and when the spool is full?

17-58. An endless belt passes over two pulleys A and B. The diameter of A is 3 ft and of B is 1 ft. If a splice in the belt moves with a linear velocity of 30 ft per sec, calculate (*a*) *the angular velocity* ω of each pulley in rad per sec and in rps, and (*b*) *the displacement* θ of each pulley in rad and in rev at the end of 4 sec.

Ans. (*a*) $\omega_A = 20$ rad per sec $= 3.182$ rps; $\omega_B = 60$ rad per sec $= 9.55$ rps; (*b*) $\theta_A = 80$ rad $= 12.7$ rev; $\theta_B = 240$ rad $= 38.2$ rev

17-59. A ballast sand bag is released from a balloon at latitude 45°N. The bag reaches its terminal velocity of 240 mph and then falls at this constant speed. Calculate the Coriolus acceleration, and state the direction of the drift.

Ans. $a_C = 0.0363$ ft per sec^2 E

REVIEW QUESTIONS

17-1. Define rotational motion and give examples of rotating bodies.

17-2. How is angular displacement measured, and what is its unit?

17-3. Do all particles of a rotating body at any instant have the same angular acceleration and velocity? Explain.

17-4. Give the symbols and the customary engineering units of angular acceleration, velocity, and displacement.

17-5. Write the three basic equations of uniformly accelerated angular motion.

17-6. Give the three equations that express the relation between linear and angular (*a*) acceleration, (*b*) velocity, and (*c*) displacement.

17-7. A tube is attached to a rotating disk with the tube axis perpendicular to the disk surface. Will a particle moving up or down in a tube undergo a Coriolis acceleration?

17-8. Will an artillery shell fired north undergo a Coriolis acceleration as it crosses the equator? the Arctic circle?

17-9. Would the Coriolis effect be more pronounced for a bowler in Alaska than in Florida?

CHAPTER 18

Kinetics of Rotation

18-1 Introduction

The forces acting on a particle (or a body small enough compared to the radius of its path that it might be treated as a particle) traveling in a circular path were studied in Chap. 16. The present chapter will deal with the kinetics of a larger body rotating about a fixed axis perpendicular to the plane of motion. All forces acting on such rotating bodies are presumed to lie in the plane of the motion. The axis of rotation passes through the center of gravity of a large majority of rotating bodies, such as pulleys, flywheels, rotors of electric motors and generators, and many others. If the center of rotation does not coincide with the center of gravity, the members supporting the rotating body may be subjected to reversed and repeated forces. The vibrations set up by these forces, or the stresses caused by them, may present a formidable problem to the design engineer.

18-2 Accelerating Moment

Linear motion of a body may be produced by a force and will take place in the direction of the force. *Rotation is revolving motion* and can be produced only by a *revolving effort* such as a *turning moment* or *couple.*

In Fig. 18-1*a* is shown a circular disk free to rotate about a frictionless shaft C, passing through its center of gravity. The weight W of the disk is balanced by the reaction R of the shaft. Let the two forces F, forming a turning moment, or a couple, be applied at opposite points on the rim of the disk.

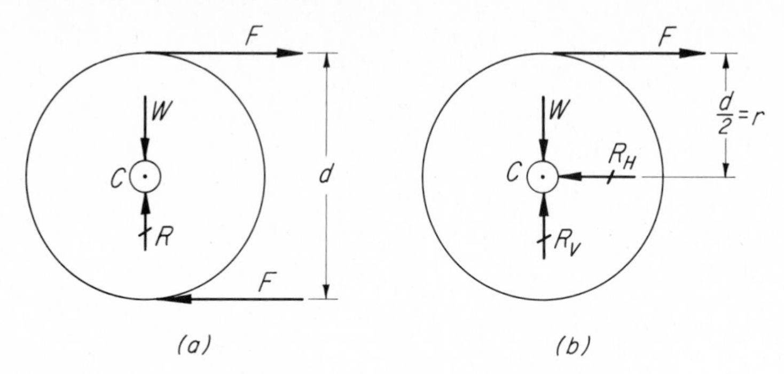

FIG. 18-1 Simple rotating bodies.
(*a*) Couple $F \cdot d$ is accelerating moment. (*b*) $F \cdot r$ is accelerating moment.

Force equilibrium exists ($\Sigma H = 0$, and $\Sigma V = 0$), but the moment Fd of the couple is *unbalanced* and is, therefore, an **accelerating moment** which will produce *accelerated rotational motion* of the disk.

When a single force F is applied, as in Fig. 18-1*b*, the force F and the horizontal reaction R_H of the shaft will form a couple, whose unbalanced clockwise moment ($M_C = \frac{1}{2}Fd = Fr$) will accelerate the disk clockwise. Several external forces may, of course, combine to cause the disk to rotate. Their *resultant moment* M_C about the axis of rotation then is the *accelerating moment,* which is usually called the **torque.**

18-3 Inertia Moment

In linear motion, *the reaction to an accelerating force is an inertia force* (Art. 15-3). Similarly, in rotation, **the reaction to an accelerating moment is an inertia moment.**

Let the body in Fig. 18-2 be accelerated counterclockwise about its axis of rotation C by an accelerating moment M_C. Consider now a single particle of the body as it is accelerated in its circular path about the center C. At any instant, the linear (tangential) acceleration of the particle a_t is *tangent* to its path. If the weight of the particle is w, its *tangential inertia force* is wa_t/g or $wr\alpha/g$, since $a_t = r\alpha$, and is directed opposite to the acceleration a_t. *The moment of the inertia force of the particle with respect to the center of rotation is* $(wr\alpha/g) \cdot r = wr^2\alpha/g$. *The inertia moment of all the particles comprising the body then is* $\Sigma wr^2\alpha/g$, since the angular acceleration α is the same for all particles of the body. Because inertia manifests itself only when a body is accelerated, *the inertia moment is a dynamic reaction.*

However, as was shown in Art. 16-2, a particle moving in a curved path is given an acceleration a_n, toward the center of rotation, equal to v^2/r or $r\omega^2$, since $v = r\omega$. Consequently, the particle is also acted upon by a centrifugal inertia force, directed away from the center, the value of which is wv^2/gr

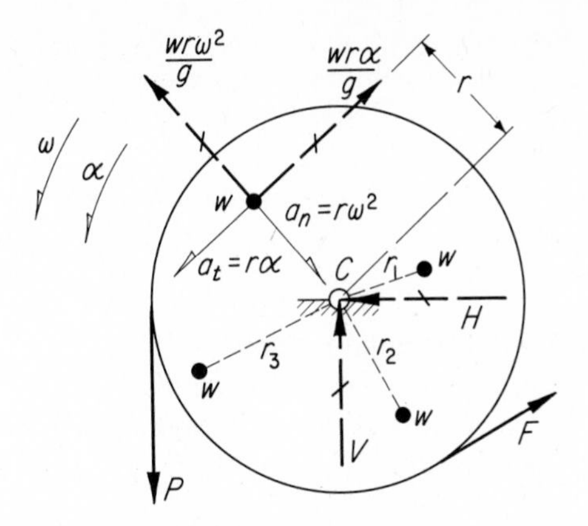

FIG. 18-2 ΣM_C is the accelerating moment and $\Sigma wr^2\alpha/g$ is the reacting inertia moment.

or $wr\omega^2/g$, since $v = r\omega$. Because the action line of this centrifugal inertia force passes through the center of rotation *C*, its moment about that center is zero.

18-4 Dynamic Equilibrium

When several forces combine to rotate a body, such as forces *P* and *F* in Fig. 18-2, their resultant moment ΣM_C about the center of rotation *C* is the *accelerating moment,* usually referred to as the **torque.** Since, **for dynamic equilibrium, action and reaction must be equal and opposite,** the torque, an externally applied action, must be equaled by the sum total of the internal reacting inertia moments of all particles comprising the body. This is in accordance with d'Alembert's principle, explained in Art. 15-5. Let all particles of the body in Fig. 18-2 be of equal weight *w*, and let their different radii be designated r_1, r_2, r_3, etc. Since these particles have the same angular acceleration α, we have

$$\text{Torque} = \sum M_C = \frac{wr_1{}^2\alpha}{g} + \frac{wr_2{}^2\alpha}{g} + \frac{wr_3{}^2\alpha}{g} \cdots = \sum \frac{wr^2\alpha}{g} \qquad (a)$$

The term $\Sigma wr^2/g$ is called the **moment of inertia** of the body and is denoted by the symbol *I*. Hence,

$$\textbf{Torque} = \Sigma M_C = I_C\alpha \qquad \textbf{(18-1)}$$

where ΣM_C = torque or accelerating moment, lb-ft
I_C = moment of inertia of the body, ft-lb-sec², with respect to the axis of rotation
α = angular acceleration, rad per sec²

In rotation, $I\alpha$ represents the internal *inertia-moment reaction* to an external *accelerating-moment action.* In linear motion, the corresponding relation is that of the internal inertia-force reaction to an external accelerating-force action (Art. 15-3). *The moment of inertia I of a body is a function only of its weight, size, and shape and is a measure of its resistance to changes in its motion.*

18-5 Moments of Inertia of Solid Bodies

In Art. 18-4, the term $\Sigma wr^2/g$ was defined as the *moment of inertia I* of a body with respect to its centroidal axis of rotation. When the weight of each particle is infinitesimally small, or dw, and when a variable radius ρ replaces r, this term becomes $I = \int \rho^2\, dw/g$. From this general mathematical expression we may derive the expression for the moment of inertia of many solids of revolution, such as a cylinder, a cone, or a sphere. The derivations of some of these are shown in Appendix C. For convenient reference, a number of the expressions commonly used are listed in Table 18-1.

Irregular-shaped bodies may generally be divided into *component* bodies of regular geometric shapes. Such a body may be called a *composite* body. The moment of inertia of each component body, with respect to the axis of rotation of the composite body, is then determined by the following principle:

The moment of inertia of a body with respect to any noncentroidal axis is equal to the moment of inertia of that body with respect to its own parallel centroidal axis plus the product of the mass (W/g) of the body and the square of the distance between the two axes. When this principle is stated in symbols, we have

$$I_A = I_C + \frac{Wd^2}{g} \tag{18-2}$$

where I_A = moment of inertia with respect to any axis of rotation parallel to a centroidal axis, ft-lb-sec²
I_C = moment of inertia with respect to the centroidal axis, ft-lb-sec²
W = weight of the body, lb
d = perpendicular distance between the parallel axes, ft
g = 32.2 ft per sec²

Equation (18-2) is called the **transfer formula,** since it "transfers" the moment of inertia of a body from its centroidal axis to any other parallel axis, as is shown in its complete derivation in Appendix C. *The moment of inertia of a composite body then is the sum of the moments of inertia of its component bodies,* all with respect to a common axis. Close study of this equation reveals that the moment of inertia of a body is always *least* with respect to a centroidal axis of rotation.

Table 18-1 MOMENTS OF INERTIA OF SOLID BODIES
(Units: r and l in ft; W in lb; g = 32.2 ft per sec²)

Body	Volume	Axis	Moment of inertia
Cross section of body			
(A) 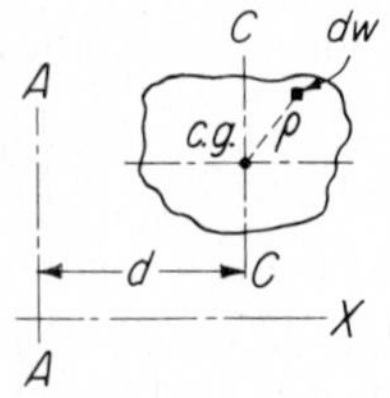		Geometric (CC)	$I_C = \int \frac{\rho^2\, dw}{g}$
(B)		Any axis (AA)	$I_A = I_C + \frac{Wd^2}{g}$
Cylinder (C) 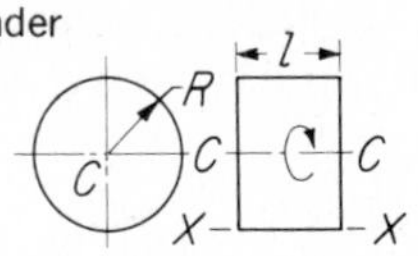	$\pi R^2 l$	Geometric (CC)	$I_C = \frac{1}{2} \cdot \frac{WR^2}{g}$
		Axis XX	$I_X = \frac{3}{2} \cdot \frac{WR^2}{g}$
Hollow cylinder (D) 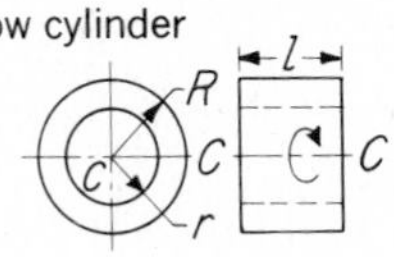	$\pi(R^2 - r^2)l$	Geometric (CC)	$I_C = \frac{1}{2} \cdot \frac{W(R^2 + r^2)}{g}$
Cone (E) 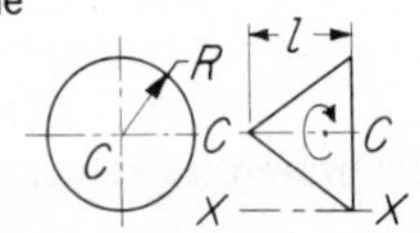	$\frac{\pi R^2 l}{3}$	Through apex and cg (CC)	$I_C = \frac{3}{10} \cdot \frac{WR^2}{g}$
		Axis XX	$I_X = \frac{13}{10} \cdot \frac{WR^2}{g}$
Sphere (F) 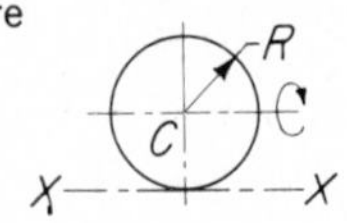	$\frac{4}{3} \cdot \pi R^3$	Any diameter	$I_C = \frac{2}{5} \cdot \frac{WR^2}{g}$
		Axis XX	$I_X = \frac{7}{5} \cdot \frac{WR^2}{g}$
Slender rod (G) 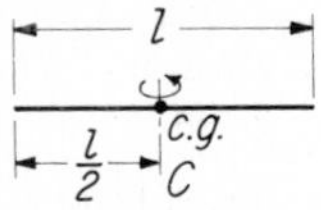	Al (A = cross-sectional area of uniform slender rod)	Normal through cg	$I_C = \frac{1}{12} \cdot \frac{Wl^2}{g}$
(H) 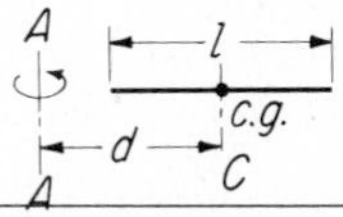		Normal through longitudinal axis	$I_A = \frac{1}{12} \cdot \frac{Wl^2}{g} + \frac{Wd^2}{g}$
(I) 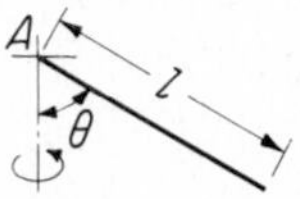		Through one end	$I_A = \frac{1}{3} \cdot \frac{Wl^2 \sin^2 \theta}{g}$ (When $\theta = 90°$, $\sin^2 \theta = 1$)

We find it interesting here to note the complete similarity of this transfer formula and the transfer formula for the moment of inertia of an area, given in Eq. (11-2) of Art. 11-4, which is $I_X = I_C + Ad^2$. Note that area A in Eq. (11-2) has merely been replaced by mass (W/g) in Eq. (18-2).

18-6 Units of Moment of Inertia

Since the inertia of a body manifests itself only when the body is accelerated, the quantity *moment of inertia I* can appear only in equations also containing acceleration α, as in Eq. (18-1) of Art. 18-4. Consequently, by itself, the term "moment of inertia" is merely a quantitative mathematical expression of the amount of resistance a given body will offer to a change in its state of motion. Likewise, the *unit of moment of inertia* (ft-lb-sec²) is of purely mathematical derivation and has as yet no separate name.

In engineering practice, the pound is the basic unit of force. Hence, when w is in pounds and r in feet, and $g = 32.2$ ft per sec², we obtain

Units of moment of inertia:

$$\left[\sum \frac{wr^2}{g} = I\right] \qquad \frac{(\text{lb})(\text{ft})^2}{\dfrac{\text{ft}}{\text{sec}^2}} = (\text{lb})(\text{ft})^2\left(\frac{\text{sec}^2}{\text{ft}}\right) = \text{ft-lb-sec}^2$$

18-7 Moment of Inertia by Experiment

By experiment, we may readily obtain the centroidal moment of inertia of a rotating body, as is illustrated in the following example.

ILLUSTRATIVE PROBLEM

18-1. A weight w of 3 lb, attached to a weightless cord, was found to rotate the flywheel, shown in Fig. 18-3, about its centroidal axis at constant speed, thus barely overcoming the bearing friction moment FM. When an additional weight W of 13.1 lb was attached to the cord, both weights were noted to descend 9 ft from rest in 6 sec. From these data, calculate the centroidal moment of inertia I_C of the flywheel whose radius r is 2 ft.

Solution: In Art. 18-4, we learned that torque $= I\alpha$. To obtain I, we need then to determine the effective torque which accelerates the flywheel, and its angular acceleration α. From Fig. 18-3, we note the following facts:

Total torque on flywheel, ΣM_C	$T \cdot r$
Torque required to overcome friction F.	$3 \cdot r$
Effective torque causing acceleration	$(T - 3) \cdot r$

The linear distance s traveled by bodies w and W from rest with constant acceleration is $s = \frac{1}{2}at^2$. Since $s = 9$ ft, $t = 6$ sec, and $\alpha = a/r$, we solve for a and α as follows:

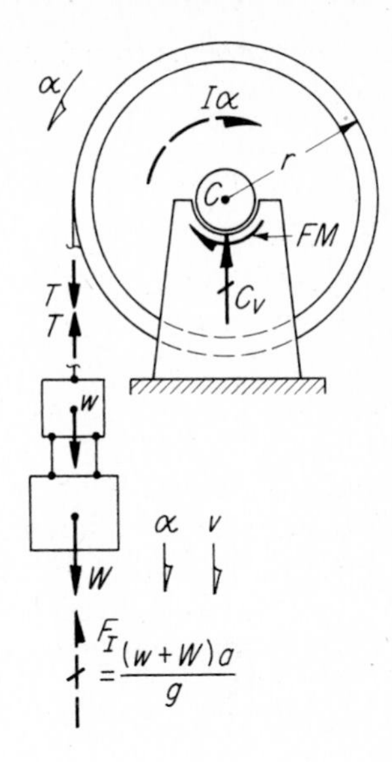

FIG. 18-3 Moment of inertia by experiment.

$$\left[a = \frac{2s}{t^2}\right] \qquad a = \frac{(2)(9)}{(6)^2} = \frac{18}{36} = 0.5 \text{ ft per sec}^2$$

$$\left[\alpha = \frac{a}{r}\right] \qquad \alpha = \frac{0.5}{2} = 0.25 \text{ rad per sec}^2$$

From Fig. 18-3, we find that $\left[T = (w + W) - \frac{(w + W)a}{g}\right]$

or
$$T = (3 + 13.1) - \frac{(3 + 13.1)(0.5)}{32.2} = 15.85 \text{ lb}$$

Since $r = 2$ ft, the effective torque is

[Eff. torque $= (T - 3)r$] $\quad (15.85 - 3)(2) = 25.70$ lb-ft

Now, $\alpha = 0.25$ rad per sec^2. Solving for I gives

[Eff. torque $= I\alpha$] $\quad 25.70 = 0.25I \quad$ or $\quad I = 102.8$ ft-lb-sec^2 $\quad$ *Ans.*

PROBLEMS

18-2. Solve Prob. 18-1 when $w = 4$ lb, $W = 60.4$ lb, bodies w and W descend 10 ft in 5 sec, and $r = 4$ ft.

18-3. The moment of inertia of the irregular-shaped cam, shown in Fig. 18-4, with respect to axis AA is determined experimentally by causing the cam to rotate in a horizontal plane about the vertical axis AA. The following data have been obtained: $w = 4$ lb; $W = 24$ lb; bodies w and W descended 10 ft in 4 sec; $r = 0.125$ ft (of the shaft). Calculate the moment of inertia of this cam about axis AA.

Ans. $I_A = 0.287$ ft-lb-sec^2

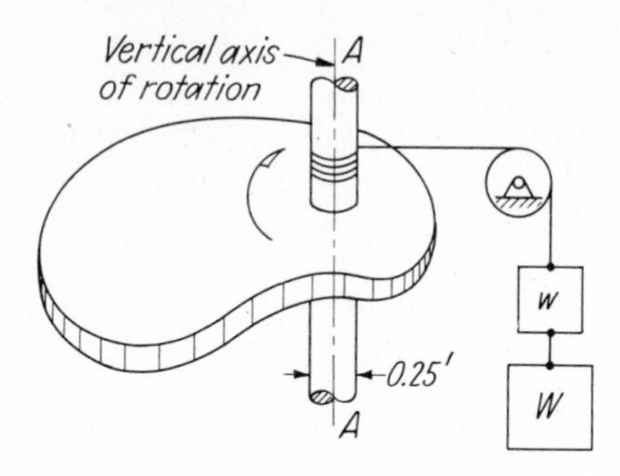

FIG. 18-4 Prob. 18-3.

18-8 Radius of Gyration

If the entire weight W of a body is assumed to be concentrated at a single point a distance k from the axis of rotation, no such summation of the elemental particles dw as that indicated in Art. 18-5, and by Eq. (*a*) of Appendix C, is necessary, and this equation then takes the form

$$I = \frac{Wk^2}{g}$$

from which

$$\mathbf{k} = \sqrt{\frac{\mathbf{Ig}}{\mathbf{W}}} \tag{18-3}$$

in which **k is the radius of gyration.** *The radius of gyration is defined as that distance from the axis of rotation at which the entire weight of the body may be assumed to be concentrated, without changing its moment of inertia.* Actually, k^2 is the *mean of the squares* of all the distances (r) from the axis of rotation to each of the equal-sized particles of the body.

The main usefulness of the radius of gyration is one of comparison in the design of rotating machine parts. That is, to obtain high moment of inertia per unit of weight, the bulk of the material must be placed away from the axis of rotation, as it is in flywheels where perhaps 80 percent or more of the weight is in the rim. When W and k of a body are known, its moment of inertia is readily computed by means of Eq. (18-3).

18-9 Moments of Inertia of Simple Bodies

The moment of inertia of a simple body may now readily be computed from the appropriate equation given in Table 18-1, as illustrated in the following examples. In engineering practice, approximations are often made with negligible error. For example, flywheel spokes and governor arms are often considered as slender rods of uniform cross section.

ILLUSTRATIVE PROBLEMS

18-4. The outside diameter of the cast-iron flywheel rim shown in Fig. 18-5 is 48 in. Its rim, 2 in. thick and 9 in. wide, is in the form of a hollow cylinder [(*D*) in Table 18-1]. Compute the moment of inertia I_C of the rim about its geometric axis of rotation, and its radius of gyration k. The density δ of cast iron is 450 pcf.

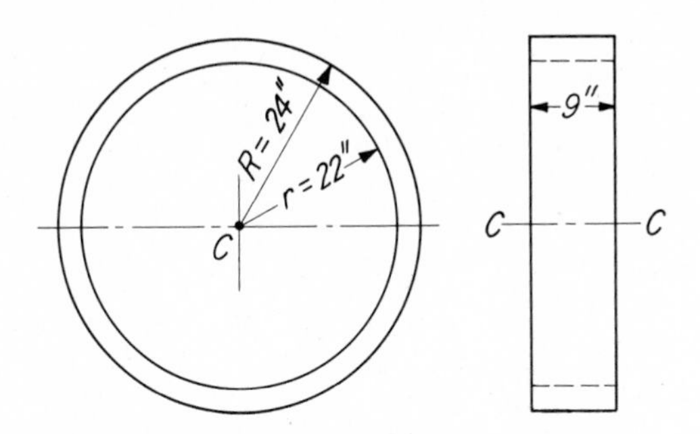

FIG. 18-5 Moment of inertia of flywheel rim.

Solution: From Eq. (*D*) of Table 18-1, the volume V of a hollow cylinder is $\pi(R^2 - r^2)l$ and its moment of inertia I_C is $W(R^2 + r^2)/2g$. The weight W equals (density) (volume), or δV. The radii R and r and the width l must be converted into feet. Four quantities are given: $R = 24/12 = 2$ ft; $r = 22/12 = 1.833$ ft; $l = 9/12 = 0.75$ ft; $\delta = 450$ pcf. Hence,

$$[W = \delta V] \qquad W = (450)(3.1416)[(2)^2 - (1.833)^2](0.75) = 679 \text{ lb}$$

Then, from Eq. (*D*),

$$I_C = \frac{W(R^2 + r^2)}{2g} = \frac{(679)[(2)^2 + (1.833)^2]}{(2)(32.2)} \quad \text{or} \quad I_C = 77.6 \text{ ft-lb-sec}^2 \qquad \textit{Ans.}$$

From Eq. (18-3) of Art. 18-8, the radius of gyration k is

$$k = \sqrt{\frac{Ig}{W}} = \sqrt{\frac{(77.6)(32.2)}{679}} \quad \text{or} \quad k = 1.92 \text{ ft} = 23.04 \text{ in.} \qquad \textit{Ans.}$$

Note that k is slightly over 23 in. as should be expected since the square root k of the *mean* of the squares of the distances r always exceeds the *average* of the distances, which here is 23 in.

18-5. The spherical cast-iron flyball of a governor is 4 in. in diameter and rotates in a circular path whose radius r is 8 in., as indicated in Fig. 18-6. Compute its moment of inertia I_A about the axis of rotation A, and its radius of gyration k. Cast iron weighs 450 pcf.

Solution: Since the axis of rotation does not pass through the center of gravity of the body, Eq. (*B*) in Table 18-1 must be used. In this equation, I_C is the centroidal moment of inertia of the sphere, given by Eq. (*F*) in Table 18-1.

If we analyze the physical conditions of this problem, we discover that in one revolution the ball turns once about a vertical centroidal axis, parallel to the vertical

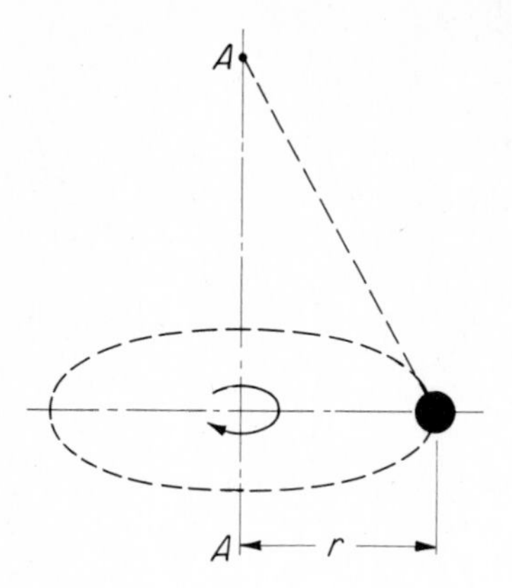

FIG. 18-6 Rotating flyball.

axis AA. This fact accounts for I_C in Eq. (B). It also turns once about axis A, which accounts for Wd^2/g. Of these two effects, the former I_C is negligible, as we shall see in the following solution.

The volume of a sphere is $\frac{4}{3}\pi R^3$ (Table 18-1). Also, $R = 2$ in. $= \frac{1}{6}$ ft; $d = 8$ in. $= \frac{2}{3}$ ft; $\delta = 450$ pcf; $W = \delta V$. Then,

$$[W = \delta V] \qquad W = (450)\left(\frac{4}{3}\right)(3.1416)\left(\frac{1}{6}\right)^3 = 8.75 \text{ lb}$$

Equations (B) and (F) of Table 18-1 give the moment of inertia:

$$\left[I_A = \frac{2}{5}\cdot\frac{WR^2}{g} + \frac{Wd^2}{g}\right] \qquad I_A = \left(\frac{2}{5}\right)\left(\frac{8.75}{32.2}\right)\left(\frac{1}{6}\right)^2 + \left(\frac{8.75}{32.2}\right)\left(\frac{2}{3}\right)^2$$

or

$$I_A = 0.00302 + 0.121 = 0.124 \text{ ft-lb-sec}^2 \qquad \textit{Ans.}$$

Note that I_C is comparatively negligible. The radius of gyration k is

$$k = \sqrt{\frac{Ig}{W}} = \sqrt{\frac{(0.124)(32.2)}{8.75}} = 0.676 \text{ ft or } 8.10 \text{ in.} \qquad \textit{Ans.}$$

If I_C were neglected, k would equal d.

PROBLEMS

In the following problems, use these densities in pcf: cast iron, 450; steel, 490; concrete, 150; wood, 40.

18-6. In Prob. 18-4, change r to 21 in. and the width of the rim to 12 in., and solve for I_C and k.

18-7. In Prob. 18-5, change the diameter to 8 in. and the radius of rotation d to 5 in., and solve for I_A and k. *Ans.* $I_A = 0.473$ ft-lb-sec^2; $k = 0.467$ ft

18-8. A circular grindstone has a diameter of 3 ft and a uniform thickness of 4 in. The density δ of the stone is 147 pcf. Compute the moment of inertia I_C of the stone with respect to its geometric axis, and its radius of gyration k.

18-9. A uniform slender rod 6 ft long rotates about an axis perpendicular to its geometric axis. Determine the radius of gyration of the rod about its axis of rotation if (*a*) this axis is 1 ft from the *c.g.* of the rod and (*b*) the axis is 3 ft from the *c.g.* of the rod. *Ans.* (*a*) $k = 2$ ft; (*b*) $k = 3.46$ ft

18-10. Derive the equation giving the radius of gyration k of a rod of length l rotating about an axis passing through the end of the rod and perpendicular to the geometric axis of the rod. *Ans.* $k = l/\sqrt{3}$

18-11. What is the radius of gyration of a solid cylinder 4 ft in diameter about an axis on the surface of the cylinder and parallel to its centroidal axis?

18-12. Compute the moment of inertia I_C and radius of gyration k of a right circular cone with a base radius of $\frac{1}{2}$ ft and an altitude l of 3 ft, with respect to an axis through its apex and center of gravity. The cone is of steel.

Ans. $I_C = 0.896$ lb-ft-sec^2; $k = 0.274$ ft

18-13. Using Eq. (18-3) of Art. 18-8, show that for a solid cylinder $k = \sqrt{0.5R^2}$, that for a cone $k = \sqrt{0.3R^2}$, and that for a sphere $k = \sqrt{0.4R^2}$. Then check the values of k as obtained in Probs. 18-11, 18-12, and 18-60.

18-14. Calculate the moment of inertia I_C of a flywheel spoke, considered as a uniform slender steel rod, 6 ft long and having a cross-sectional area of 6 sq in., about a normal axis through its center of gravity. Compute also its radius of gyration k.

18-15. The slender rod supporting the flyball of a governor rotates at an angle of 30° with a vertical axis of rotation A which passes through the upper end of the rod. The rod weighs 2 lb and is 15 in. long. Compute its moment of inertia I_A and its radius of gyration k. *Ans.* $I_A = 0.00809$ ft-lb-sec^2; $k = 0.361$ ft $= 4.33$ in.

18-16. The rim of a large cast-iron flywheel is 2 ft wide, and its inner and outer *diameters* are, respectively, 10 and 11 ft. Compute the moment of inertia I_C of the rim about its geometric axis, and the radius of gyration k.

18-17. The hub of the cast-iron flywheel in Prob. 18-16 is in the form of a hollow cylinder [Eq. (*D*) in Table 18-1]. Its inner and outer diameters are, respectively, 12 and 24 in., and the hub is 18 in. long. Calculate the moment of inertia I_C of the hub about its geometric axis, and its radius of gyration k.

18-18. Each spoke of the cast-iron flywheel in Prob. 18-16 has a cross-sectional area of 28.8 sq in. and a length of 4 ft. The spoke may be regarded as a uniform slender rod. The distance d is 3 ft from the center of gravity of the spoke to the axis of rotation A of the flywheel [Eq. (*H*), Table 18-1]. Determine the moment of inertia I_A of one spoke, and its radius of gyration k.

Ans. $I_A = 115.6$ ft-lb-sec^2; $k = 3.21$ ft

18-19. The radius of gyration k of a certain flywheel has been found to be 0.84 ft, and its weight W is 600 lb. Compute its moment of inertia I_C.

18-20. A large wooden pulley weighs 322 lb, and its radius of gyration k is 1.5 ft. Compute its moment of inertia I_C.

18-10 Moments of Inertia of Composite Bodies

In engineering practice, rotating bodies are seldom of simple geometric forms, but many of these bodies can usually be divided into several parts, each having or resembling closely one of the simple geometric shapes listed in Table 18-1. Such bodies are called *composite bodies*. **The moment of inertia of a composite body about any axis equals the sum of the moments of inertia of its several**

parts about the same axis. Similarly, if parts of a solid are removed, the moment of inertia of the remainder equals the moment of inertia of the solid *minus* the moment of inertia of the parts removed, all with respect to the same axis.

ILLUSTRATIVE PROBLEM

18-21. Calculate the moment of inertia, with respect to its normal axis of rotation, of the castiron flywheel, whose cross section is shown in Fig. 18-7. The flywheel has six similar spokes.

Solution: This flywheel can be divided into three simple parts: (1) the rim and (2) the hub, both of which are considered to be hollow cylinders, and (3) the spokes, which are regarded as uniform slender rods. First to be considered is the weight of each part. Equation (*D*) of Table 18-1 applies to the rim and the hub, and Eq. (*H*) applies to the spokes. In Eq. (*H*), $d = 15$ in., as may be seen in Fig. 18-7. Since all spokes are alike, they may be considered as a single spoke having six times the weight of one. The slight taper is disregarded.

The weight of each part is as follows:

Rim: $[W = \delta\pi(R^2 - r^2)l]$ $R = 30$ in., $r = 24$ in., $l = 12$ in.

$$W = (450)(\pi)\left[\left(\frac{30}{12}\right)^2 - \left(\frac{24}{12}\right)^2\right]\left(\frac{12}{12}\right) = 3{,}181 \text{ lb}$$

Hub: $[W = \delta\pi(R^2 - r^2)l]$ $R = 6$ in., $r = 3$ in., $l = 9$ in.

$$W = (450)(\pi)\left[\left(\frac{6}{12}\right)^2 - \left(\frac{3}{12}\right)^2\right]\left(\frac{9}{12}\right) = 199 \text{ lb}$$

Spokes: $[W = 6\delta Al]$ $A = 14.4$ sq in., $l = 18$ in.

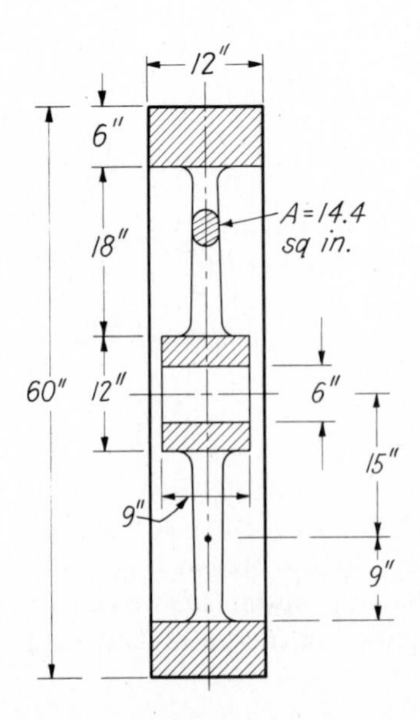

FIG. 18-7 Flywheel.

$$W = (6)(450)\left(\frac{14.4}{144}\right)\left(\frac{18}{12}\right) = 405 \text{ lb}$$

The weight of the entire flywheel then is

$$W = 3{,}181 + 199 + 405 = 3{,}785 \text{ lb}$$

The moment of inertia of each part is

Rim:

$$I = \frac{W(R^2 + r^2)}{2g} = \left[\frac{3{,}181}{(2)(32.2)}\right]\left[\left(\frac{30}{12}\right)^2 + \left(\frac{24}{12}\right)^2\right] = 507 \text{ ft-lb-sec}^2$$

Hub:

$$I = \frac{W(R^2 + r^2)}{2g} = \left[\frac{199}{(2)(32.2)}\right]\left[\left(\frac{6}{12}\right)^2 + \left(\frac{3}{12}\right)^2\right] = 0.965 \text{ ft-lb-sec}^2$$

Spokes:

$$I = \left[\frac{Wl^2}{12g} + \frac{Wd^2}{g}\right] = \left[\left(\frac{405}{(12)(32.2)}\right)\left(\frac{18}{12}\right)^2 + \left(\frac{405}{32.2}\right)\left(\frac{15}{12}\right)^2\right] = 22 \text{ ft-lb-sec}^2$$

The moment of inertia of the entire flywheel then is

$$I = 507 + 0.965 + 22 \qquad \text{or} \qquad I = 530 \text{ ft-lb-sec}^2 \qquad \textit{Ans.}$$

The radius of gyration k of the entire flywheel is

$$\left[k = \sqrt{\frac{Ig}{W}}\right] \qquad k = \sqrt{\frac{(530)(32.2)}{3785}} \qquad \text{or} \qquad k = 2.126 \text{ ft} = 25.5 \text{ in.} \qquad \textit{Ans.}$$

Note that virtually all (95.6 percent) of the inertia is in the rim, that the inertia of the hub is negligible, and that the total inertia of the spokes is relatively small. These facts are also indicated by the value of the radius of gyration (25.5 in.). That is, if the entire weight of the flywheel (3,785 lb) is assumed to be concentrated at a point in the rim only 4½ in. from the outer surface, it will have the same moment of inertia about the axis of rotation as has the actual distributed weight.

A simple *approximate method of obtaining the total moment of inertia of the spokes and hub* is to disregard the hub and consider the six separate spokes as three spokes, 48 in. long, extending through the hub and rotating about their centers of gravity which then coincide with the axis of rotation. Their total weight must be recomputed, and Eq. (*G*) of Table 18-1 now applies. Since $W = 3\delta Al$, and $A = 0.1$ sq ft,

$$\left[I = \frac{Wl^2}{12g}\right] \qquad I = \frac{(3)(450)(0.1)(4)(4)^2}{(12)(32.2)} = 22.4 \text{ ft-lb-sec}^2$$

a result virtually the same as that obtained previously for the hub and spokes.

PROBLEMS

18-22. In Prob. 18-21 and Fig. 18-7, let the outside diameter of the flywheel be 72 in. Then, omitting the hub and using the approximate method of obtaining the total moment of inertia of the spokes, solve for I_C and k. The hub diameter is 12 in., and the rim thickness is unchanged.

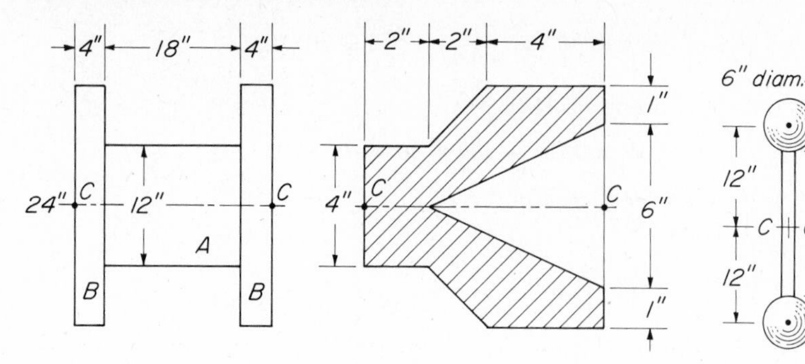

FIG. 18-8 Probs. 18-23 and 18-24.

FIG. 18-9 Prob. 18-25.

FIG. 18-10 Prob. 18-26.

18-23. The cable spool shown in Fig. 18-8 is of wood (40 pcf) and revolves about its geometric axis *CC*. Calculate its moment of inertia I_C and its radius of gyration *k*. *Ans.* $I_C = 1.158$ ft-lb-sec^2; $k = 6.98$ in.

18-24. In Prob. 18-23, assume that the spool is wound full with a 6-in. layer of cable weighing 200 pcf in place, and solve for I_C and *k*

18-25. The steel machine part shown in Fig. 18-9 has the form of a solid of revolution formed by rotating the section shown about axis *CC*. Calculate I_C.
Ans. $I_C = 0.1077$ ft-lb-sec^2

18-26. Each of the two steel spheres shown in Fig. 18-10 weighs 32.2 lb. They are connected by a round steel rod weighing 15 lb. Calculate the moment of inertia and the radius of gyration *k* of this assembly about the axis of rotation *CC*.
Ans. $I_C = 2.14$ ft-lb-sec^2; $k = 0.931$ ft $= 11.2$ in.

18-27. The rotating element of a helicopter is jet-propelled and consists of four jet motors weighing 40 lb each mounted on the outboard ends of four 12-ft blades weighing 180 lb each. Find the moment of inertia of the assembly about the axis of rotation. Treat the blades as slender rods and assume the moment of inertia of the jet motors about their own centroidal axis is negligible compared with their moment of inertia about the axis of rotation. *Ans.* $I = 447.5$ ft-lb-sec^2

18-11 Friction Moment

In Fig. 18-11, the pull *P* causes drum *A* to rotate about the center of its centroidal axle *C*, which rests in two like and opposite bearing blocks *B* (usually considered as one single bearing having the frictional resistance of both). Here, $N = W + P$, and the friction force *F* (not shown) at the surface of contact is $F = fN$. The *moment* of the friction force *F* about the center of the shaft is $F \cdot r$. It is called the **friction moment** and is denoted by *FM*. In lubricated bearings, the coefficient of friction *f* may vary from less than 0.01 to more than 0.1, but is often negligible.

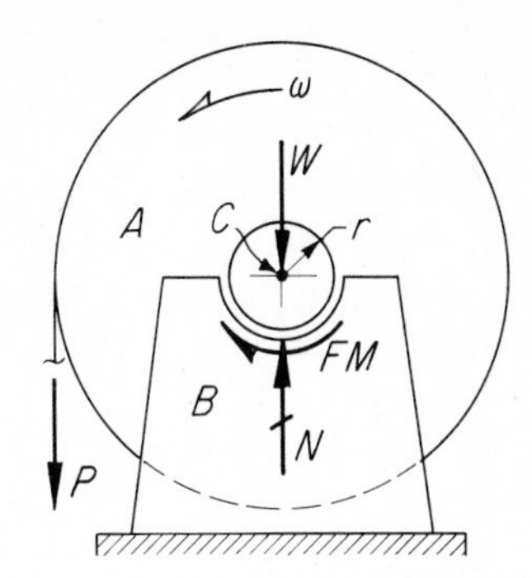

FIG. 18-11 *FM* is friction moment.

18-12 Dynamic Equilibrium of Rotating Bodies

The force required to produce a given rotational motion, or the rotational motion produced by a given force, may now be determined. As in rectilinear motion, use of d'Alembert's principle—regarding the inertia of a body as a force—places all force systems in *dynamic equilibrium*. Therefore, in any one direction, $\Sigma F = 0$, and about any axis, usually the axis of rotation, $\Sigma M = 0$. This method of solution is referred to as **the force method.**

We must remember here that, in the equation $\Sigma M_C = I\alpha$ (Art. 18-4), both sides bear the same units. *The torque* ΣM_C *is an accelerating* (*or decelerating*) *moment,* usually in pound-feet, and *the term* $I\alpha$ *is a resisting inertia moment,* also in pound-feet, because

$$I\alpha = (\text{ft-lb-sec}^2)(\text{rad/sec}^2) = \text{ft-lb} \qquad \text{or} \qquad \text{lb-ft}$$

since a radian has no unit. Hence, *the quantity* $I\alpha$*, being a moment, is always included directly in a summation of moments.*

Signs of Inertia Forces and Moments. In rectilinear motion, the inertia-force reaction F_I is always directed opposite to the linear acceleration. Similarly, in rotation, *the inertia-moment reaction* $I\alpha$ *is always directed opposite to the angular acceleration.* When forces and moments on a free-body diagram are directed by these or other means, as they usually are, no sign convention for summations of forces or moments is necessary.

ILLUSTRATIVE PROBLEMS

18-28. The pulley shown in Fig. 18-12*a* is subjected to belt tensions T_1 and T_2. The moment of inertia of the pulley I_C about its axis of rotation is 40 ft-lb-sec². The average bearing-friction moment *FM* is 20 lb-ft. Calculate the angular acceleration α of the pulley when $T_1 = 190$ lb, $T_2 = 100$ lb, and $R = 2$ ft.

Solution: Since T_1 exceeds T_2, the pulley will rotate and accelerate clockwise. The inertia-moment reaction $I\alpha$ and the friction moment *FM* will therefore be directed counterclockwise. We find α from a moment equation, using the axis of rotation C as the moment center. That is,

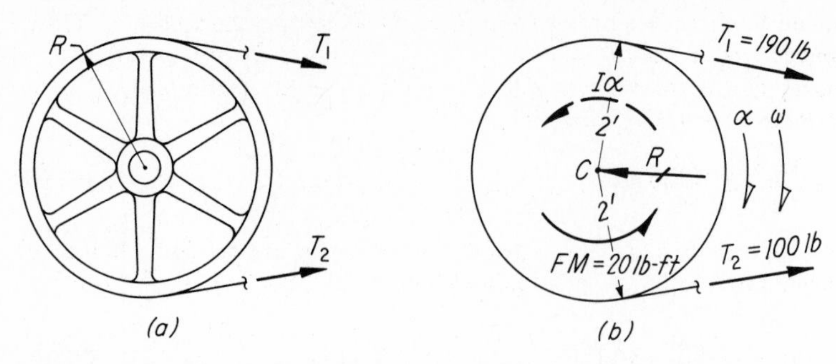

FIG. 18-12 Dynamic equilibrium of a rotating body. (*a*) Rotating pulley. (*b*) Free-body diagram.

$[\Sigma M_C = 0] \qquad I\alpha + FM + RT_2 = RT_1$

or $\quad 40\alpha + 20 + 2(100) = 2(190) \quad$ and $\quad \alpha = 4$ rad per sec^2 $\quad$ *Ans.*

18-29. Block B in Fig. 18-13*a* is attached to a weightless cord which is wrapped around disk A, causing it to rotate about its centroidal axis C. Block B weighs 64.4 lb. The moment of inertia I_C of the disk is 12 ft-lb-sec^2, and its radius r is 2 ft. A constant bearing-friction moment of 5 lb-ft resists rotation. Determine the angular acceleration α of disk A, the linear acceleration a of block B, and the tension T in the cord.

Solution: The unbalanced moment rT acting on the disk is clockwise, causing clockwise acceleration and motion, indicated by the motion arrows α and ω shown on the free-body diagram in Fig. 18-13*b*. The inertia moment $I\alpha$ is directed *opposite*

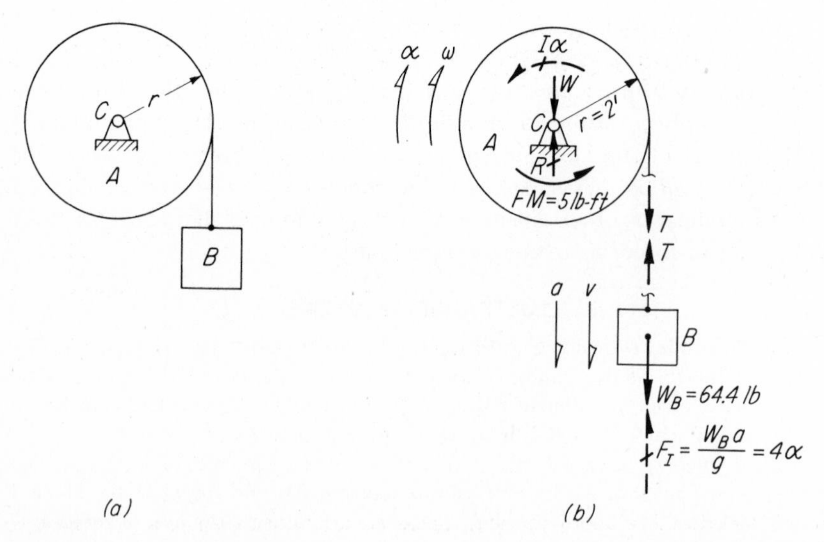

FIG. 18-13 Dynamic equilibrium. (*a*) Space diagram. (*b*) Free-body diagram.

to α, and the friction moment FM is directed *opposite* to the motion ω. These terms $I\alpha$ and FM are *moments*, indicated directly on the body by curved arrows. Block B moves downward with accelerated motion and its inertia force F_I is directed *opposite* to a. Since $a = r\alpha$, we have

$$\left[F_I = \frac{W_B a}{g} = \frac{W_B r\alpha}{g}\right] \qquad F_I = \frac{(64.4)(2)\alpha}{32.2} = 4\alpha$$

Two equations, a moment equation about the center C and a force equation involving the free body B, will now determine α. That is,

$$[\Sigma M_C = 0] \qquad rT = FM + I\alpha \qquad \text{or} \qquad 2T = 5 + 12\alpha \qquad (a)$$
$$[\Sigma V = 0] \qquad T + F_I = W_B \qquad \text{or} \qquad T = 64.4 - 4\alpha \qquad (b)$$

Multiplying Eq. (*b*) by 2 and subtracting it from Eq. (*a*) give

$$\begin{aligned} 2T &= 5 + 12\alpha \\ 2T &= 128.8 - 8\alpha \\ \hline 0 &= -123.8 + 20\alpha \end{aligned} \qquad \text{and} \qquad \alpha = 6.19 \text{ rad per sec}^2 \qquad \textit{Ans.}$$

To obtain the acceleration a, we have

$$[a = r\alpha] \qquad a = (2)(6.19) \qquad \text{or} \qquad a = 12.38 \text{ ft per sec}^2 \qquad \textit{Ans.}$$

Substituting this value of α in Eq. (*a*) to obtain T gives

$$2T = 5 + (12)(6.19) = 79.28 \qquad \text{and} \qquad T = 39.64 \text{ lb} \qquad \textit{Ans.}$$

18-30. In Fig. 18-14*a*, as block A is descending at the rate of 24 ft per sec, a braking force P is suddenly applied to the brake lever DEF. The weight of A is 161 lb. Pulley B weighs 322 lb, and its radius of gyration k is 2.5 ft. The coefficient of friction f

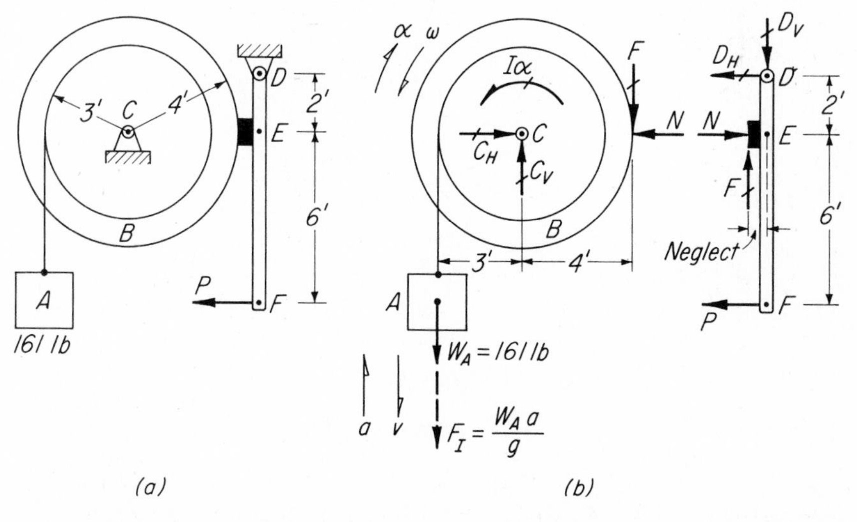

FIG. 18-14 Rotating pulley stopped by brake. (*a*) Space diagram. (*b*) Free-body diagram.

is 0.5, between the brake block and the pulley. (Neglect thickness of brake block and lever, and bearing friction.) Calculate the constant force P required to stop the system in 4 sec.

Solution: The motion of block A is downward, indicated by v in Fig. 18-14*b*, and that of pulley B is counterclockwise, indicated by ω. Because the system is *decelerated*, a is opposed in direction to v and α is opposed to ω. Also, the inertia force F_I is opposed to a, and the inertia moment $I\alpha$ is opposed to α. When the brake lever with its block is removed, N is the normal force exerted by the brake block against the pulley, and F is the friction force, directed against the motion ω. On the brake lever, N and F are oppositely directed. Since v_0, v, and t are known, we may solve for a and α. A moment equation about C will then give F, from which N and finally P are obtained. Since $v_0 = 24$ ft per sec, $v = 0$, and $t = 4$ sec, we obtain

$$\left[a = \frac{v - v_0}{t}\right] \qquad a = \frac{0 - 24}{4} \qquad \text{or} \qquad a = -6 \text{ ft per sec}^2 \text{ (opposite to initial } v)$$

$$\left[\alpha = \frac{a}{r}\right] \qquad \alpha = \frac{6}{3} = 2 \text{ rad per sec}^2$$

We may now evaluate $I\alpha$ and F_I.

$$\left[F_I = \frac{W_A a}{g}\right] \qquad F_I = \frac{(161)(6)}{32.2} = 30 \text{ lb}$$

The moment of inertia I of the pulley is $W_B k^2/g$. Hence,

$$I\alpha = \frac{W_B k^2 \alpha}{g} = \frac{(322)(2.5)^2(2)}{32.2} = 125 \text{ lb-ft}$$

The friction force F may now be found by taking moments about C. That is, for the system as a whole,

$$[\Sigma M_C = 0] \qquad 4F = I\alpha + 3W_A + 3F_I$$

or $\quad 4F = 125 + (3)(161) + (3)(30) = 698 \quad$ and $\quad F = 174.5$ lb

Since $F = fN$, and $f = 0.5$,

$$\left[N = \frac{F}{f}\right] \qquad N = \frac{174.5}{0.5} = 349 \text{ lb}$$

Using the free body of the lever and taking moments about D to obtain P, neglecting thickness of lever and brake block, gives

$[\Sigma M_D = 0] \qquad 8P = 2N = (2)(349) = 698 \quad$ and $\quad P = 87.3$ lb $\quad$ *Ans.*

PROBLEMS

18-31. Solve Prob. 18-28 when $I_C = 10$ ft-lb-sec^2, $FM = 10$ lb-ft, $T_1 = 100$ lb, $T_2 = 40$ lb, and $R = 1.5$ ft.

18-32. The flywheel shown in Fig. 18-7 is placed between a centrifugal pump and an electric motor. Calculate the rpm 30 sec after a power failure to the pump motor if the average decelerating torque on the flywheel is 265 lb-ft and the unit was rotating at 720 rpm at the instant of the power outage. The moment of inertia of the pump and motor is negligible compared to that of the flywheel.

18-33. How long will it take to bring a flywheel having a moment of inertia of 200 ft-lb-sec² from rest to a speed of 600 rpm if the applied torque is 100 lb-ft?
Ans. $t = 2$ min 5.8 sec

18-34. A cast-iron cylinder 2 ft in diameter and 1 ft long is mounted on a shaft through its geometric axis. A 322-lb weight is suspended by a wire wrapped around the cylinder. Compute the linear acceleration of the falling weight and the tension in the wire. Neglect journal friction and weight of wire, and assume cast iron weighs 450 pcf.

18-35. The cylinder in Fig. 18-15 is made to rotate about its centroidal axis C by the constant pull P in the cord. The cord is wound around the cylinder, whose moment of inertia I_C is 20 ft-lb-sec². Determine (*a*) the force P required to accelerate the cylinder 3 rad per sec² and (*b*) the angular acceleration when P is 100 lb. Neglect bearing friction. *Ans.* (*a*) $P = 30$ lb; (*b*) $\alpha = 10$ rad per sec²

18-36. In Prob. 18-35, change the radius of the cylinder to 2.5 ft, let I_C equal 70 ft-lb-sec², and solve.

18-37. The weight B shown in Fig. 18-13*a* weighs 200 lb and has an upward velocity of 12 ft per sec. The cylinder weighs 322 lb and the radius r is 2 ft. How far will the block move before coming to an instantaneous rest? Neglect all friction.
Ans. $s = 4.03$ ft

18-38. In Prob. 18-29, let $W_B = 96.6$ lb and let $FM = 10$ lb-ft; then solve.

18-39. Solve Prob. 18-30 when $v_0 = 36$ ft per sec and $W_A = 64.4$ lb.
Ans. $P = 54.3$ lb

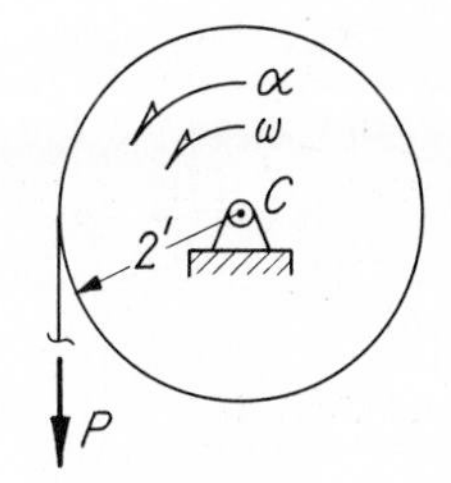

FIG. 18-15 Probs. 18-35 and 18-36.

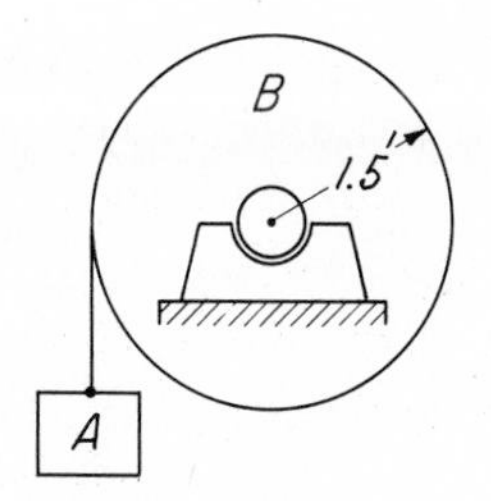

FIG. 18-16 Probs. 18-40, 18-43 to 18-45.

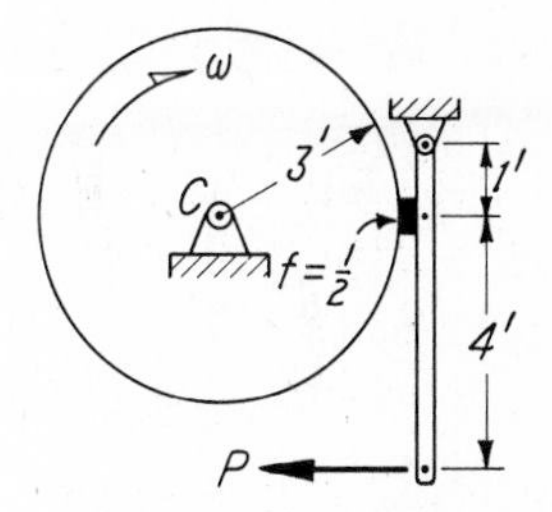

FIG. 18-17 Probs. 18-41 and 18-42.

18-40. The drum B in Fig. 18-16 weighs 590 lb. A weight A of 10 lb will keep the drum rotating at uniform speed, thus just overcoming the bearing friction. If the shaft is 6 in. in diameter ($\frac{1}{2}$ ft), calculate the coefficient of friction f between shaft and bearing.

18-41. The circular drum shown in Fig. 18-17 rotates clockwise at the rate of 180 rpm and is stopped by application of the force P to the brake lever. (Neglect the thickness of brake block and lever.) If the moment of inertia I_C of the drum is 40 ft-lb-sec², what force P will stop it in 6 sec? *Ans.* $P = 16.76$ lb

18-42. In Prob. 18-41, change the speed to 300 rpm and I_C to 30 ft-lb-sec² and solve.

18-43. To determine experimentally the moment of inertia I_C of the pulley B shown in Fig. 18-16, block A of known weight is attached to a weightless cord which is wrapped around the pulley, and its motion is then observed. The bearing-friction moment *FM* is known to be 5.75 lb-ft. Calculate I_C, if *A* weighs 48.3 lb and is observed to drop 12 ft in 4 sec after being released from rest. *Ans.* $I_C = 63.33$ ft-lb-sec²

18-44. Solve Prob. 18-43 when $W_A = 64.4$ lb, $FM = 14.8$ lb-ft, and *A* drops 36 ft from rest in 12 sec.

18-45. Calculate the weight W_A of block *A* in Fig. 18-16 required to give pulley *B* an angular acceleration of 4 rad per sec², if I_C is 14 ft-lb-sec² and the bearing-friction moment *FM* is 5 lb-ft. *Ans.* $W_A = 50$ lb

18-46. Solve Prob. 18-45 if $I_C = 20$ ft-lb-sec² and $FM = 10$ lb-ft.

18-13 Rotation about Noncentroidal Axis

If the axis of rotation passes through the center of gravity of the body, the algebraic sum of the tangential inertia forces acting on the particles of a rotating body undergoing an acceleration is zero (the particles on each side of the axis of rotation are accelerating in opposite directions, see Fig. 18-18*a*). If the axis of rotation does not coincide with the gravity axis of the body, the algebraic sum of the tangential inertia forces acting on the individual particles is not zero, but is $W\bar{r}\alpha/g$ as will now be shown.

The tangential inertia force acting on *each particle* of weight w of the rod shown in Fig. 18-18*b* is wa_T/g or $wr\alpha/g$ and the resultant inertia force *acting*

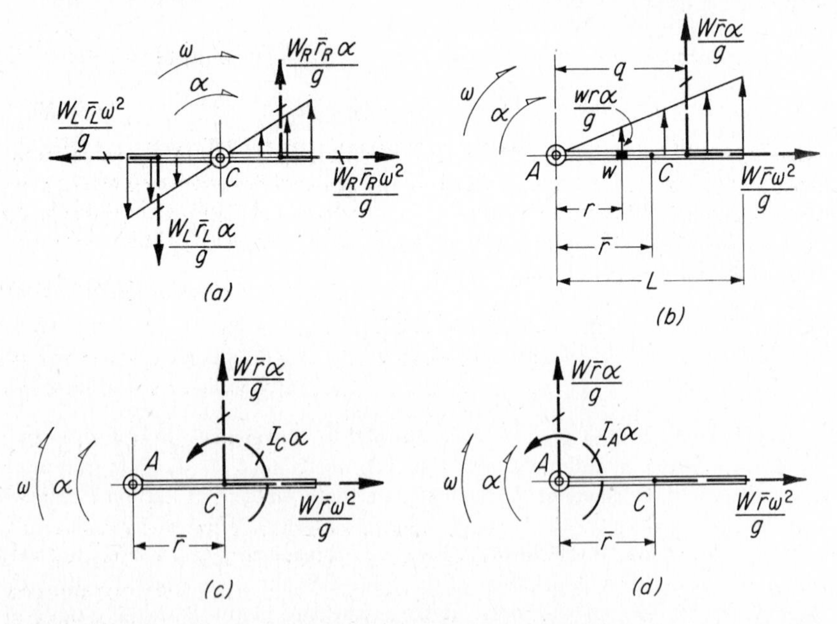

FIG. 18-18 Inertia forces acting on a rotating rod. Note that the above sketches are not complete free-body diagrams—only the inertia forces and moments have been shown. (*a*) Axis of rotation through gravity axis. (*b*) Axis of rotation not through gravity axis.

on the rod is the sum of these or $\Sigma wr\alpha/g$. Since α and g are constants, we may write this summation as $\Sigma wr(\alpha/g)$. But from Eq. (10-1), $\Sigma wr = W\bar{r}$; therefore

$$F_I = \frac{W\bar{r}\alpha}{g} \tag{18-4}$$

where F_I = resultant tangential inertia force acting on the rotating body, generally in pounds
W = total weight of rotating body, generally in pounds
$\bar{r}$ = distance between the axis of rotation and the centroidal axis, generally in feet

The normal inertia force has already been shown to be

$$F_I = Wa_n/g = Wv^2/g\bar{r}$$

[see Eq. (16-2)]. Since $v = \bar{r}\omega$ [see Eq. (17-5)], we have

$$F_I = \frac{W(\bar{r}\omega)^2}{g\bar{r}} \quad \text{or} \quad F_I = \frac{W\bar{r}\omega^2}{g} \tag{18-5}$$

where F_I = resultant normal inertia force acting on the rotating body, generally in pounds
ω = angular velocity of body, generally in radians per second

ILLUSTRATIVE PROBLEM

18-47. An 80-lb boy sits 6 ft from the center of a rotating horizontal disk at an amusement park. If the coefficient of friction f is 0.1 and the angular acceleration of the disk is 0.5 rad per sec², how long until the boy starts to slide? Also determine the direction of his initial motion with respect to a radial line.

Solution: The tangential inertia force is

$$\left[F_I = \frac{W\bar{r}\alpha}{g}\right] \qquad \text{Tangential } F_I = \frac{(80)(6)(0.5)}{32.2} = 7.45 \text{ lb}$$

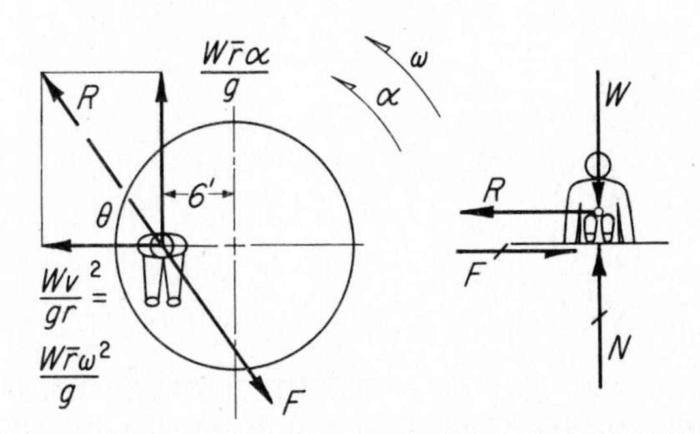

FIG. 18-19 Boy sitting on a rotating disk.

This tangential inertia force is opposite in direction to the tangential acceleration and would be directed as shown in Fig. 18-19.

The centrifugal inertia force is opposite in direction to the normal acceleration and is given by

$$\left[F_I = \frac{W\bar{r}\omega^2}{g}\right] \qquad \text{Normal } F_I = \frac{80(6)\omega^2}{32.2} = 14.9\omega^2$$

but from Eq. (17-1), we have

$$[\omega = \omega_0 + \alpha t] \qquad \omega = 0 + 0.5t = 0.5t$$

Substituting this equation for ω into the above equation for the normal inertia force, we have

$$\text{Normal } F_I = 14.9(0.5t)^2 = 3.73t^2$$

From Fig. 18-19 we see that $R = \sqrt{(W\bar{r}\alpha/g)^2 + (W\bar{r}\omega^2/g)^2}$

or $R = \sqrt{(7.45)^2 + (3.73t^2)^2}$

The resultant force R tends to slide the boy on the disk and this tendency to slide is opposed by the friction force F shown in the sketch. When sliding is impending, $F = fN = 0.1(80) = 8$ lb, and $F = R$. Hence,

$$[F = R] \qquad 8 = \sqrt{(7.45)^2 + (3.73t^2)^2}$$

Squaring both sides, we obtain

$$64 = (7.45)^2 + (3.73t^2)^2$$

from which

$$t = 0.883 \text{ sec} \qquad \textit{Ans.}$$

Substituting this time in the above equation for the normal inertia force, we have

$$\text{Normal } F_I = 3.73t^2 = (3.73)(0.833)^2 = 2.91 \text{ lb}$$

From Fig. 18-19 we see that

$$\tan\theta = \frac{W\bar{r}\alpha/g}{W\bar{r}\omega^2/g} = \frac{7.45}{2.91} = 2.56 \qquad \text{and therefore} \qquad \theta = 68°40' \qquad \textit{Ans.}$$

PROBLEMS

18-48. Solve Prob. 18-47 if $\alpha = 0.3$ rad per sec^2 and $f = 0.15$.

18-49. At what speed N in rpm will an object placed 4 ft from the center of a horizontal rotating disk start to slide if $f = 0.2$? The angular acceleration is very small and may be disregarded. *Ans.* $N = 12.1$ rpm

18-50. The angular velocity of the assembly shown in Fig. 18-20 is 180 rpm. The rod CD weighs 32.2 lb. What is the tension in the string AB?

Ans. $T = 738$ lb

18-51. The axis of rotation of a 2-ft diameter circular disk weighing 322 lb is $\frac{1}{8}$ inch from the geometric axis (and the center of gravity) and is horizontal. A support bearing is located on each side of the disk, and under static conditions each bearing

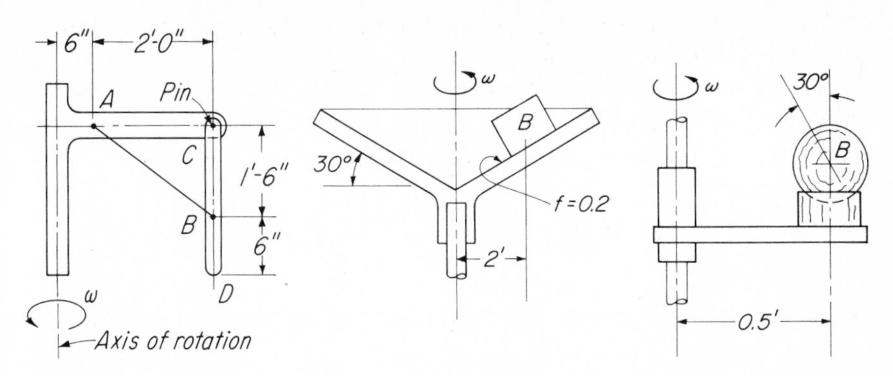

FIG. 18-20 Probs. 18-50, 18-71, and 18-72.

FIG. 18-21 Probs. 18-52 and 18-53.

FIG. 18-22 Prob. 18-54.

exerts a 161-lb force on the axle. What will be the maximum and minimum values of the bearing reactions if the angular velocity of the disk is 1,000 rpm?

18-52. The block B shown in Fig. 18-21 weighs 32.2 lb. Calculate the angular velocity ω of the cone at which the motion of the block B will be impending up the incline.

18-53. Solve Prob. 18-52 for the minimum angular velocity ω of the cone to prevent motion down the incline. *Ans.* $\omega = 22.3$ rpm

18-54. Calculate the angular velocity ω at which the ball B shown in Fig. 18-22 will roll from the socket.

18-14 Location of the Tangential Inertia Force

The previous article evaluated the tangential inertia force acting on a rotating body as $F_I = W\bar{r}\alpha/g$. The location of this resultant tangential inertia force will now be determined.

It has already been shown that the inertia moment is $I\alpha$. If the tangential inertia force is a distance q from the axis of rotation (see Fig. 18-18*b*), then its moment is $F_I q$ or $(W\bar{r}\alpha/g)q$. Equating this moment to $I\alpha$, we have

$$\frac{W\bar{r}\alpha}{g}q = I\alpha$$

Since the moment of inertia I about the axis of rotation is $I = Wk^2/g$ (see Art. 18-8), we have

$$\frac{W\bar{r}\alpha}{g}q = \frac{Wk^2}{g}\alpha$$

and therefore

$$q = \frac{k^2}{\bar{r}} \tag{18-6}$$

Obviously, for the case shown in Fig. 18-18*b*, $q = \frac{2}{3}L$. The radius of gyration of a slender rod about an axis through one end of the rod is $k = L/\sqrt{3}$ (see Prob. 18-10). Then $q = k^2/\bar{r} = (L/\sqrt{3})^2/L/2 = \frac{2}{3}L$, which agrees with the above statement.

The location of the resultant tangential inertia force is sometimes called the **center of percussion** because a blow at this point will not cause a force reaction parallel to the blow at the axis of rotation. Striking a baseball far from the center of percussion of the bat may cause a stinging sensation in the hands.

Sometimes in the analysis of a dynamics problem it is convenient to place the resultant inertia force at the center of gravity (see Fig. 18-18*c*) or at the center of rotation (see Fig. 18-18*d*). A corrective moment must be added to the free-body diagram to compensate for the misplacement. If the inertia force is moved to the centroidal axis, the corrective moment is $I_C\alpha$; if it is moved to the center of rotation *A*, the corrective moment is $I_A\alpha$.

That the systems shown in Fig. 18-18*c* and *d* are equivalent may be shown by use of the transfer formula in Eq. (18-2). Taking moments about *A* in Fig. 18-18*c*, we have

$$M_A = (W\bar{r}\alpha/g)\bar{r} + I_C\alpha = [(W\bar{r}^2/g) + I_C]\alpha$$

From the transfer formula, $I_A = I_C + (W/g)d^2$ or $I_A = (W\bar{r}^2/g) + I_C$. Therefore, $M_A = I_A\alpha$ as is shown in Fig. 18-18*d*.

ILLUSTRATIVE PROBLEMS

18-55. A uniform slender rod weighing 322 lb is supported as shown in Fig. 18-23*a*. Determine the reaction at end *A* immediately after the cord is cut. $L = 9$ ft.

Solution: The free-body diagram is shown in Fig. 18-23*b*. The tangential inertia force is

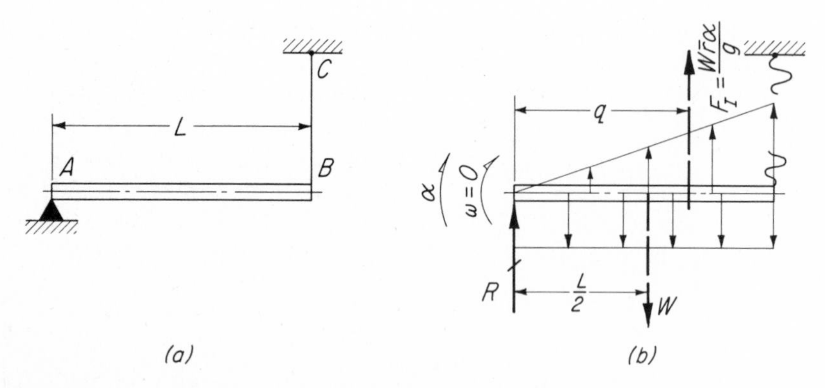

FIG. 18-23 An example of dynamic equilibrium for rotation about a noncentroidal axis. (*a*) Rod before cord *BC* was cut. (*b*) Free-body diagram of rod immediately after cord was cut.

$\left[F_I = \frac{Wr\alpha}{g}\right] \qquad F_I = \frac{322(4.5)\alpha}{32.2} = 45\alpha$

The radius of gyration of a slender rod about one end is $k = L/\sqrt{3}$ (see Prob. 18-10) and therefore q is

$\left[q = \frac{k^2}{\bar{r}}\right] \qquad q = \frac{(L/\sqrt{3})^2}{L/2} = \frac{81/3}{4.5} = \frac{27}{4.5} = 6 \text{ ft}$

Taking moments about A in order to find the angular acceleration α,

$[\Sigma M_A = 0] \qquad W\left(\frac{L}{2}\right) - F_I q = 0 \quad \text{or} \quad 322(4.5) - (45\alpha)6 = 0$

and $\qquad \alpha = 5.37 \text{ rad per sec}^2$

Substituting this value of α into the above expression for the tangential inertia force, we have $F_I = 45\alpha = 45(5.37) = 241.5$ lb. To find the reaction R_A, we sum up the forces in the vertical direction.

$[\Sigma V = 0] \qquad R_A + 241.5 - 322 = 0 \quad \text{or} \quad R_A = 80.5 \text{ lb} \qquad$ *Ans.*

18-56. A slender rod of uniform cross section rotates about a vertical axis passing through its upper end, as shown in Fig. 18-24*a*. Its constant angular velocity is 6 rad per sec. The weight W of the rod is 32.2 lb and its length L is 6 ft. Find the angle of inclination θ of the rod with the vertical.

Solution: The free-body diagram of the rod is shown in Fig. 18-24*b*. The distance $\bar{r}$ is $3 \sin \theta$. The centrifugal inertia force then is

$\left[F_I = \frac{W\bar{r}\omega^2}{g}\right] \qquad F_I = \frac{(32.2)(3 \sin \theta)(6)^2}{32.2} = 108 \sin \theta$

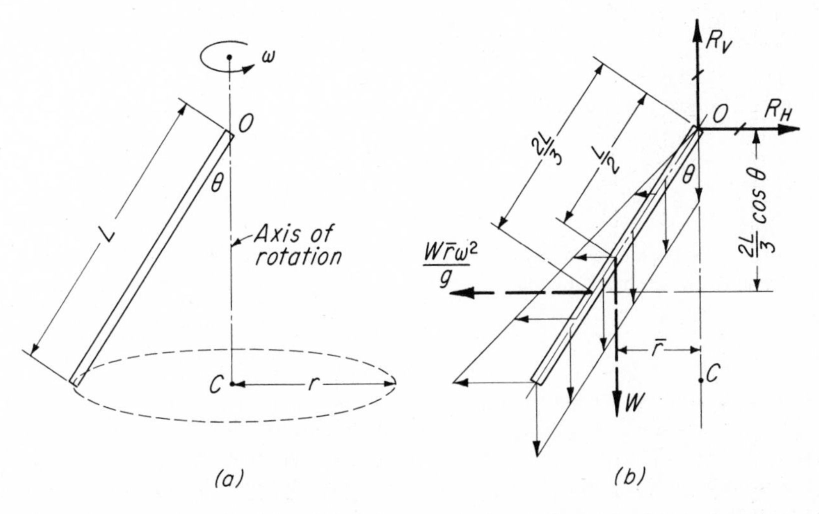

FIG. 18-24 Inclined rod rotating about a vertical axis through one end of the rod. (*a*) Rotating rod. (*b*) Free-body diagram.

Taking moments about the top of the rod, we obtain

$$[\Sigma M_O = 0] \qquad \left(\frac{W\bar{r}\omega^2}{g}\right)\left(\frac{2L}{3}\cos\theta\right) = W(3\sin\theta)$$

Therefore, $(108\sin\theta)(4\cos\theta) = 32.2(3\sin\theta)$ or $432\cos\theta = 96.6$

and $\cos\theta = 0.2236$ from which $\theta = 77°05'$ *Ans.*

PROBLEMS

18-57. Solve Prob. 18-51 if $W = 40$ lb and $L = 3$ ft.

18-58. Solve Prob. 18-51 if the left-hand knife-edge is 3 ft from the left end and 6 ft from the string. *Ans.* $R = 241.5$ lb

18-59. A slender rod 4.5 ft long and weighing 64.4 lb is attached to a support at one end by a smooth pin. The rod is free to rotate about this pin in a vertical plane. The rod is released from such a position that its angular velocity as it reaches the horizontal is 3 rad per sec. Find the vertical and horizontal reactions at the pin as the rod passes through the horizontal position.

Ans. $V = 16.1$ lb; $H = 40.5$ lb

18-60. Solve Prob. 18-56 if $L = 9$ ft, $W = 100$ lb, and $\omega = 3$ rad per sec.

18-61. Find the vertical and horizontal components of the reaction at the support in Prob. 18-56.

18-62. Find the angular velocity necessary to make the rod shown in Fig. 18-24*a* stand at 30° to the vertical axis of rotation. $L = 6$ ft.

Ans. $\omega = 3.05$ rad per sec

SUMMARY

(By article number)

18-1. The *resultant moment,* about the center of rotation, of all external forces tending to rotate a body is the **accelerating moment,** also called the **torque.** It is denoted by ΣM_C.

18-3. The inertia moment $I\alpha$ is the moment, about the axis of rotation, of the internal inertia forces of all the particles of an accelerated rotating body. An inertia moment is *the reaction to an accelerating moment.* For any given body, *the inertia moment is the product of the moment of inertia of the body and its acceleration.* (See Art. 18-6 below.)

18-4. For **dynamic equilibrium,** the external accelerating moment, or torque ΣM_C, must equal the internal inertia moment $I_C\alpha$, or,

$$\textbf{Torque} = \Sigma M_C = I_C\alpha \tag{18-1}$$

where I_C is the moment of inertia of the body with respect to its center of rotation and α is its angular acceleration.

18-6. The moment of inertia of a body is a property that is a function of its size, shape, and weight. The general expression for *the moment of inertia of any solid of revolution* about a *centroidal axis of rotation* is

$$I_C = \int \frac{\rho^2\, dw}{g}$$

The moment of inertia of a solid with respect to *any axis of rotation* A parallel to its centroidal axis C is

$$I_A = I_C + \frac{Wd^2}{g} \qquad (18\text{-}2)$$

where d is the perpendicular distance between the parallel axes. The equations giving the moments of inertia of several bodies of common geometric shapes are given in Table 18-1.

18-8. The radius of gyration of a body is that distance k from its axis of rotation to the point at which the entire weight could be assumed to be concentrated without changing its moment of inertia. Hence

$$I = \frac{Wk^2}{g} \qquad \text{from which} \qquad k = \sqrt{\frac{Ig}{W}} \qquad (18\text{-}3)$$

18-10. The moment of inertia of a composite body with respect to any axis equals the sum of the moments of inertia of its several parts with respect to the same axis.

18-12. The inertia-moment reaction $I\alpha$ is always directed opposite to the angular acceleration and is *always* included directly in a summation of moments.

18-13 and 18-14. If the axis of rotation of a body does *not* pass through the center of gravity of a body, the **resultant tangential inertia force** is given by the equation

$$F_I = \frac{W\bar{r}\alpha}{g} \qquad (18\text{-}4)$$

and passes through a point a distance from the axis of rotation equal to

$$q = \frac{k^2}{\bar{r}} \qquad (18\text{-}6)$$

The **resultant normal inertia** force is given by the equation

$$F_I = \frac{W\bar{r}\omega^2}{g} \qquad (18\text{-}5)$$

and acts radially outward. The line of action of this force passes through the center of rotation and through the center of gravity of the body.

REVIEW PROBLEMS

18-63. A solid, circular wooden pulley has a diameter of $1\frac{1}{2}$ ft and a uniform thickness of $\frac{1}{2}$ ft. *Compute its moment of inertia* I_C and its radius of gyration k with respect to its geometric axis. The density of wood is 40 pcf.

Ans. $I_C = 0.309$ ft-lb-sec^2; $k = 0.53$ ft

18-64. The spherical cast-iron flyball of a governor is 6 in. ($\frac{1}{2}$ ft) in diameter and revolves about its centroidal axis, as well as about the vertical axis of rotation A of the entire governor. *Compute its moment of inertia* I_C *and its radius of gyration* k about its centroidal axis. *Ans.* $I_C = 0.0229$ ft-lb-sec^2; $k = 0.158$ ft $= 1.9$ in.

18-65. The disk shown in Fig. 18-25 is made of steel (490 pcf) and has a uniform thickness of 1 in. Six equally spaced holes, each 6 in. in diameter, have been cut in the disk. *Calculate the moment of inertia* I_C of the disk with respect to a normal axis through its center of gravity.

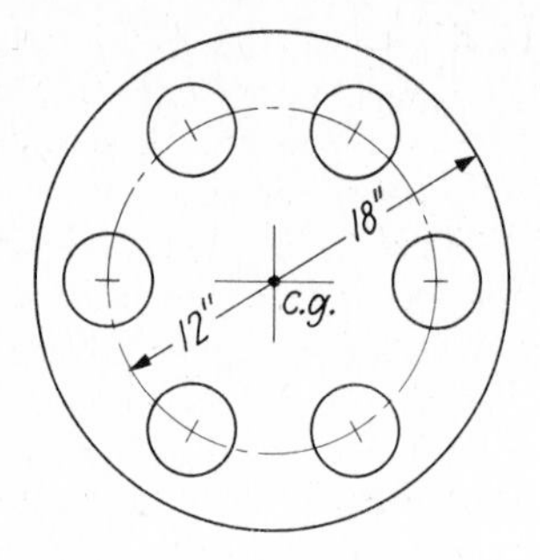

FIG. 18-25 Prob. 18-65.

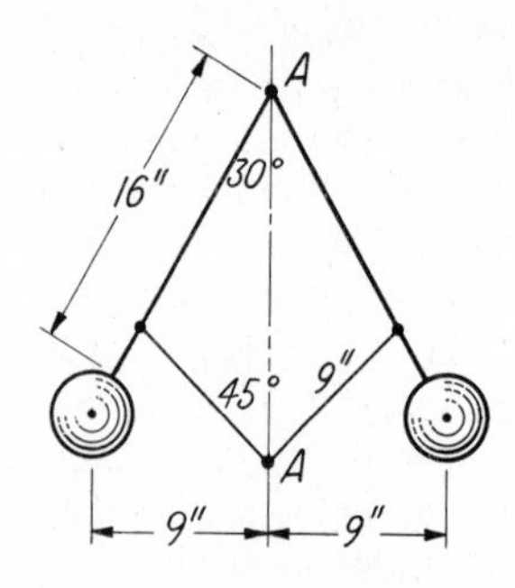

FIG. 18-26 Prob. 18-67.

18-66. Compute the moment of inertia of the rotating element of the impulse turbine shown in Fig. 17-10. Each bucket weighs 4 lb and its centroid is on the pitch circle (the pitch diameter is 3 ft 0 in.). The steel disk is $\frac{3}{8}$ in. thick and 2 ft 8 in. in diameter (steel weighs 490 pcf). The I of each bucket about its own centroidal axis is negligible compared with the I about the axis of rotation. Disregard the hub.

18-67. In the governor assembly shown in Fig. 18-26, each of the two 4-in. diam cast-iron flyballs weighs 8.74 lb. Each upper arm weighs 2 lb and each lower arm 1 lb. *Calculate the moment of inertia I_A and the radius of gyration k of this assembly* about the axis of rotation A.

Ans. $I_A = 0.338$ ft-lb-sec^2; $k = 0.681$ ft $= 8.16$ in.

18-68. *Calculate the constant tension T* in the cord in Fig. 18-27 required to give body D an angular velocity ω of 24 rad per sec in 8 sec from rest, if I_C is 30 ft-lb-sec^2, $R = 3$ ft, and the bearing-friction moment is 60 lb-ft. *Ans.* $T = 50$ lb

18-69. The moment of inertia I_C of the disk D in Fig. 18-27 is 7.75 ft-lb-sec^2, and its radius is 1.5 ft. A constant bearing friction moment FM of 11.7 lb-ft resists rotation. Block B weighs 32.2 lb. The disk D has an initial clockwise angular velocity. Calculate the angular acceleration α of the disk. *Ans.* $\alpha = 6$ rad per sec^2

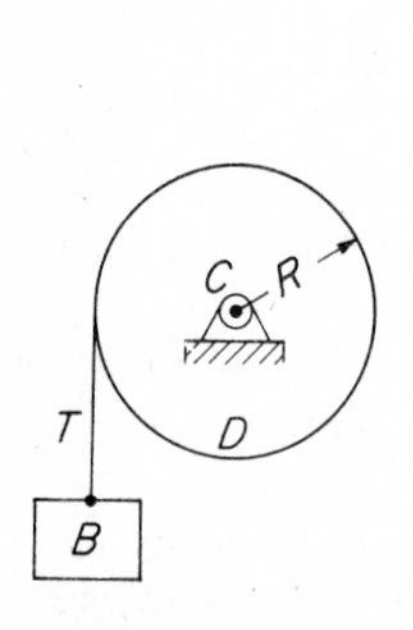

FIG. 18-27 Probs. 18-68, 18-69, and 18-74.

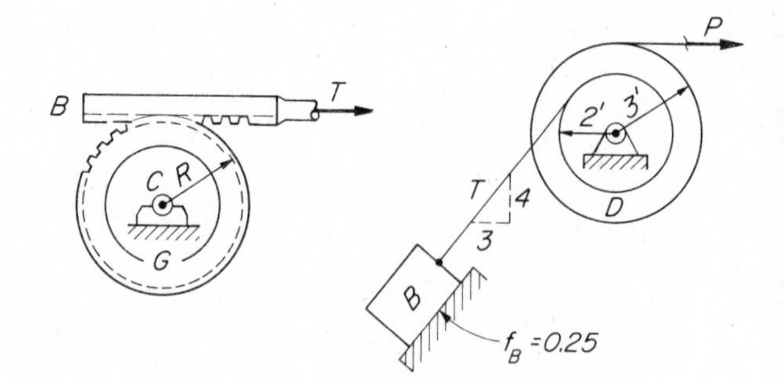

FIG. 18-28 Prob. 18-70.

FIG. 18-29 Probs. 18-73 and 18-75.

18-70. The gear G shown in Fig. 18-28 has a centroidal moment of inertia $I_C = 15$ ft-lb-sec², a pitch radius of 9 in., and a bearing-friction moment FM of 20 lb-ft. The rack R accelerates to the right with $a = 20$ ft per sec². Calculate the required tensile force T. The mass of the rack is negligible compared to that of the gear.

18-76. The bar CD shown in Fig. 18-20 weighs 16.1 lb. If the breaking strength of the cord is 10 lb, compute the angular velocity at which the cord should break.

18-72. The angular velocity of the element shown in Fig. 18-20 is 2 rad per sec. The bar CD weighs 16.1 lb. Find the horizontal and vertical components of the pin reactions at C.

18-73. The block B in Fig. 18-29 weighs 100 lb. The moment of inertia of the rotating part D is 60 ft-lb-sec². Calculate the required pull P to give part D a clockwise angular acceleration α of 4 rad per sec². *Ans.* $P = 160$ lb

18-74. Block B in Fig. 18-27 weighs 100 lb. The moment of inertia I_C of disk D is 20 ft-lb-sec², and its radius is 2 ft. A constant bearing-friction moment FM of 40 lb-ft resists motion. *Determine the angular acceleration* α of disk D, *the linear acceleration* a of block B, and *the tension* T *in the cord.*

18-75. The pull P in Fig. 18-29 is 40 lb. Calculate the angular acceleration of the part D if its centroidal moment of inertia is 50 ft-lb-sec² and the weight of block B is 200 lb. The system started from rest. *Ans.* $\alpha = 5.59$ rad per sec²

18-76. Each of the two balls shown in Fig. 18-30 weighs 32.2 lb and the distance between their centers of gravity is 24 in. If the axis of rotation GE is ½ in. from the center of gravity of the two balls, find the maximum tensile and compressive forces in member CD if $N = 120$ rpm. *Ans.* $F_{CD} = 18.4$ lb T; $F_{CD} = 18.4$ lb C

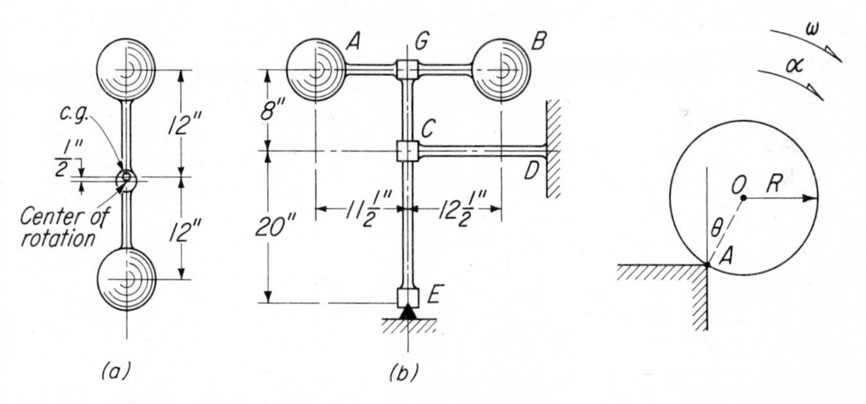

FIG. 18-30 Prob. 18-76. (*a*) Plan view. (*b*) Elevation view.

FIG. 18-31 Probs. 18-77 and 18-78.

18-77. The cylinder shown in Fig. 18-31 is rolling off the end of a loading platform. Find its angular acceleration when $\theta = 30°$, if $R = 2$ ft and the cylinder weighs 644 lb. Assume slipping does not occur at A. *Ans.* $\alpha = 5.36$ rad per sec²

18-78. Determine the vertical and horizontal components of the reaction at A in Prob. 18-77 if $\theta = 45°$ and $\omega = 2$ rad per sec.

REVIEW QUESTIONS

18-1. What must be applied to a rotating body in order to increase its angular velocity?

18-2. What is meant by a torque?

18-3. When an accelerating moment changes the speed of a rotating body, what reaction manifests itself? Is this reaction dependent on the acceleration of the body? What symbols express it?

18-4. Explain what is meant by dynamic equilibrium. Discuss also d'Alembert's principle.

18-5. Of what is the moment of inertia of a solid body a measure? By what methods may we determine it? Is it dependent on the acceleration of the body? What are the common engineering units of moment of inertia?

18-6. Explain the meaning of radius of gyration. Discuss briefly its usefulness.

18-7. What is meant by friction moment?

18-8. If ΣM_C is the resultant moment or torque that changes the angular speed of a rotating body, write the equation expressing dynamic equilibrium for the body.

18-9. What is the main function of a flywheel? Why is most of its weight placed in the rim?

18-10. Is a heavy rim desirable in a pulley used merely for transmitting power? Explain.

CHAPTER 19

Plane Motion

19-1 Plane Motion Defined

When the motion of a body is such that *all particles move in parallel planes*, the body is said to have **plane motion.** The *reference plane* in which the forces are shown is usually one passing through the center of gravity of the body.

Plane motion is usually a combination of rotation and translation, as in the case of a **rolling body,** such as a wheel, a cylinder, or a sphere, which will be discussed in the present chapter. *When no slipping occurs on the supporting surface, such bodies are said to be* **free rolling.** More complex examples of plane motion are the connecting rods of a stationary engine and the side rods connecting the driving wheels of a locomotive. Only *free-rolling bodies* are considered in this chapter.

19-2 Instantaneous Center

A body in plane motion may, at any instant, be considered as being in rotation about some axis, referred to as the *instantaneous axis*. The intersection of the instantaneous axis and the reference plane is called the **instantaneous center.**

Consider now the free-rolling body shown in Fig. 19-1. Let a diameter be drawn through the point of contact A and its center of gravity C. A little study will show that, at any instant, the particle at A is momentarily at rest, but that *all other particles on this diameter have velocity components parallel to the incline whose magnitudes are proportional to their normal distances from* A. Since the particle at A also has zero velocity *normal* to the incline, we

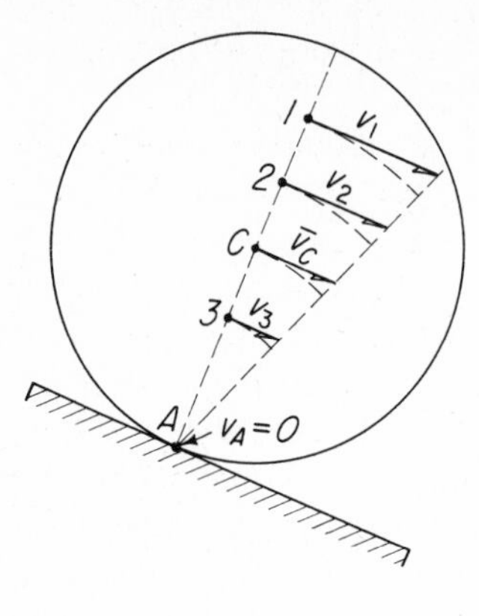

FIG. 19-1 Point A is the instantaneous center of a free-rolling body.

may conclude that the body at any instant is rotating about point A, which, then, is the *instantaneous center.* That is, **the instantaneous center of a free-rolling body is the point of contact between the body and the supporting surface.**

19-3 Kinematics of Plane Motion

Since plane motion is a combination of translation and rotation, *the kinematic relations of plane motion are combinations of those of translation and rotation.* There were outlined in Chaps. 14 and 17. The *linear* acceleration, velocity, and displacement ($\overline{a}$, $\overline{v}$, $\overline{s}$) of a body in plane motion are those of its center of gravity. The relations $\overline{a} = \overline{r}\alpha$, $\overline{v} = \overline{r}\omega$, and $\overline{s} = \overline{r}\theta$ (see Art. 17-8) hold for *free-rolling* (nonslipping) bodies.

Further study of Fig. 19-1 will show that the linear accelerations, *parallel to the incline,* of the particles of the body vary in the same manner as the velocities, being zero for the particle at the point of contact A, and maximum for the particle farthest from the contacting surface. We may then state that, *in the direction of motion,* **the linear acceleration and velocity components of a particle of a free-rolling body, lying on a diameter passing through its instantaneous center, are proportional to its distance from the instantaneous center.**

19-4 Kinetics of Plane Motion

In plane motion, as in translation and rotation, *forces and moments will be in dynamic equilibrium when the inertia of the body is considered.* The concept of a moving instantaneous center of rotation makes the kinetic treatment of a free-rolling body identical to that of noncentroidal rotation about a fixed axis. If the free-rolling disk shown in Fig. 19-2a is accelerated to the right by the force P applied to its axle, the rotation will take place about the instantaneous center A, and the resultant inertia force will oppose the acceleration. It has already been shown that resultant inertia force acts through the center of percussion (see Art. 18-14). Recall that the center of percussion lies a distance $q = k^2/\bar{r}$ from the axis of rotation.

The three optional methods of solving noncentroidal rotation problems are equally applicable here. We may place the resultant inertia force at the center of gravity and add the corrective moment $I_C\alpha$ to the free-body diagram (Fig. 19-2*b*). Doing so gives a very useful and enlightening concept. We see that the inertia force $W\bar{r}\alpha/g$ is $W\bar{a}/g$ where $\bar{a}$ is the linear acceleration of the center of gravity of the body and that the inertia moment is $I_C\alpha$, which is what it would be for pure centroidal rotation. We may think of plane motion, then, as a combination of translation and rotation. The translation is that of the center of gravity and the rotation is centroidal.

Although the concept of a moving coordinate system is a little difficult to grasp, its use does simplify the arithmetic in many instances. The use of the moving point of contact as a reference axis is illustrated in Fig. 19-2*c*. If we take moments about the instantaneous center A, the entire rotational inertia effect is contained in the one term $I_A\alpha$.

The relationship discussed in Arts. 19-3 and 19-4 assumed no slipping on the supporting surface. Just as the inertia forces oppose the acceleration, so does the friction force F oppose the direction of motion $\bar{v}$. To have force equilibrium, $P = F_I + F$, and $N = W$. *When slipping occurs*, and the coefficient of friction f is known, the friction force F may be evaluated. That is, $F = fN$. We may then find α from a moment equation, using C as the moment center in Fig. 19-2*b*, and $\bar{a}$ from a summation of forces, if P is known. *If slipping does not occur*, F will be unknown, and $\bar{a}$ and α may then be found ($\bar{a} = \bar{r}\alpha$) from a summation of moments about A.

Rolling resistance on hard, plane surfaces is usually negligible and therefore has been disregarded in the following examples and problems. In Fig. 19-2, the force F *is not the rolling resistance but is the static frictional resistance*

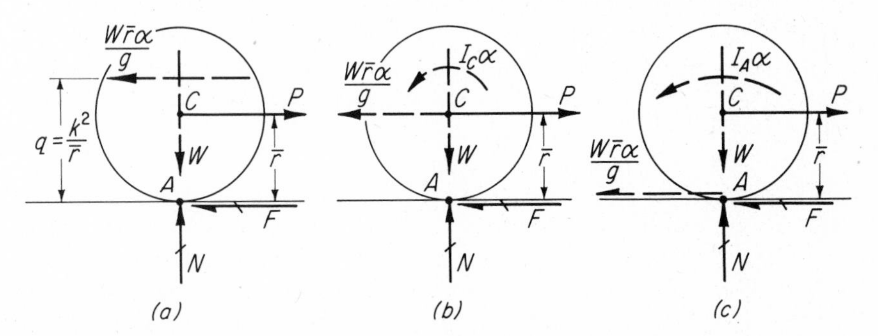

FIG. 19-2 Equivalent free-body diagrams for a free-rolling body. (*a*) Inertia force at the center of percussion. (*b*) Plane motion treated as a combination of translation and rotation. (*c*) Plane motion treated as rotation about the instantaneous center.

which prevents slipping. When F is insufficient to prevent slipping, and **when slipping occurs,** the instantaneous center is no longer the point of contact A, but lies between it and the center of gravity C. Then, the relation $\bar{a} = r\alpha$ no longer holds. Problems involving slipping are not dealt with in this text.

ILLUSTRATIVE PROBLEMS

19-1. A solid cylinder, 4 ft in diameter and weighing 1,288 lb, rolls freely (without slipping) down an incline, as shown in Fig. 19-3*a*. Its centroidal moment of inertia I_C is 80 ft-lb-sec². Calculate (*a*) the linear acceleration $\bar{a}$ of its center of gravity and (*b*) the minimum coefficient of static friction f required to prevent slipping.

Solution: The forces acting on this body should be resolved into a system lying parallel and normal to the incline. The velocity $\bar{v}$ and acceleration $\bar{a}$ are, of course, both directed down the incline; the friction force F and the inertia force F_I are consequently directed up the incline, as indicated on the free-body diagram in Fig. 19-3*b*. We see also that the angular acceleration α will be counterclockwise and that the inertia moment $I\alpha$ therefore will be clockwise. Because F is unknown, we must determine α from a moment equation about the point of contact A. Then, forces F_I, F, and N are computed, and f is found from the relation $f = F/N$.

The components of W are

$$W_T = \frac{5}{13}(1{,}288) = 495 \text{ lb} \qquad \text{and} \qquad W_N = \frac{12}{13}(1{,}288) = 1{,}189 \text{ lb}$$

To determine α, we now take moments about A.

$$[\Sigma M_A = 0] \qquad 2W_T = (2)\frac{W\bar{r}\alpha}{g} + I_C\alpha$$

$$\text{or} \qquad (2)(495) = 2\left(\frac{1{,}288(2)\alpha}{32.2}\right) + 80\alpha \qquad \text{and} \qquad \alpha = 4.125 \text{ rad per sec}^2$$

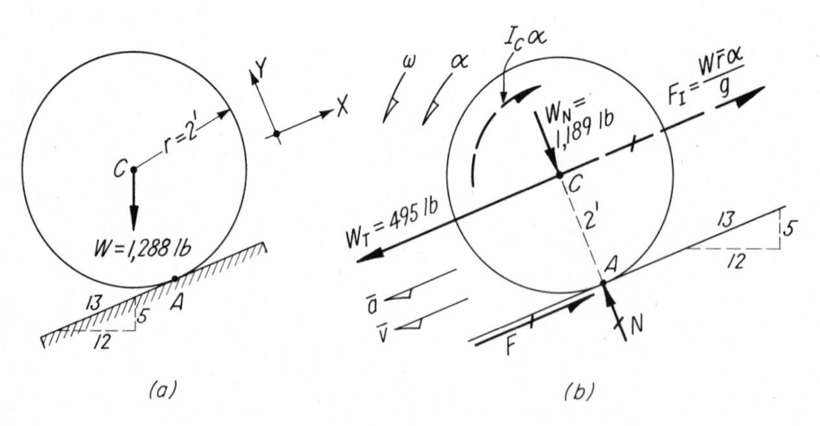

FIG. 19-3 Plane motion of a free-rolling body. (*a*) Space diagram. (*b*) Free-body diagram.

From $\bar{a} = \bar{r}\alpha$,

$$\bar{a} = 2(4.125) = 8.25 \text{ ft per sec}^2 \qquad \textit{Ans.}$$

The inertia force F_I then is

$$\left[F_I = \frac{W\bar{r}\alpha}{g}\right] \qquad F_I = \frac{1{,}288(2)4.125}{32.2} = 330 \text{ lb.}$$

To obtain the friction force F, we have

$$[\Sigma F_X = 0] \qquad F + 330 = 495 \quad \text{and} \quad F = 165 \text{ lb}$$

and to obtain N, we have

$$[\Sigma F_Y = 0] \qquad N = W_N = 1{,}189 \text{ lb}$$

The minimum coefficient of friction f required to prevent slipping then is

$$\left[f = \frac{F}{N}\right] \qquad f = \frac{165}{1{,}189} \quad \text{or} \quad f_{\min} = 0.138 \qquad \textit{Ans.}$$

Alternate Solution: The inertia force may be placed at the center of percussion as shown in Fig. 19-4*a*. The line of action of the resultant inertia force passes through this center of percussion which lies a distance q from the instantaneous center of rotation. Since $q = k^2/r$, we must determine k from $\sqrt{I_A/(W/g)}$. From the transfer formula [Eq. (18-2)] we see that $I_A = I_C + Wd^2/g$. Therefore, $I_A = 80 + 1{,}288(2)^2/32.2 = 240$ ft-lb-sec². Then k is

$$\left[k = \sqrt{\frac{I_A}{W/g}}\right] \qquad k = \sqrt{\frac{240(32.2)}{1{,}288}} = 2.45$$

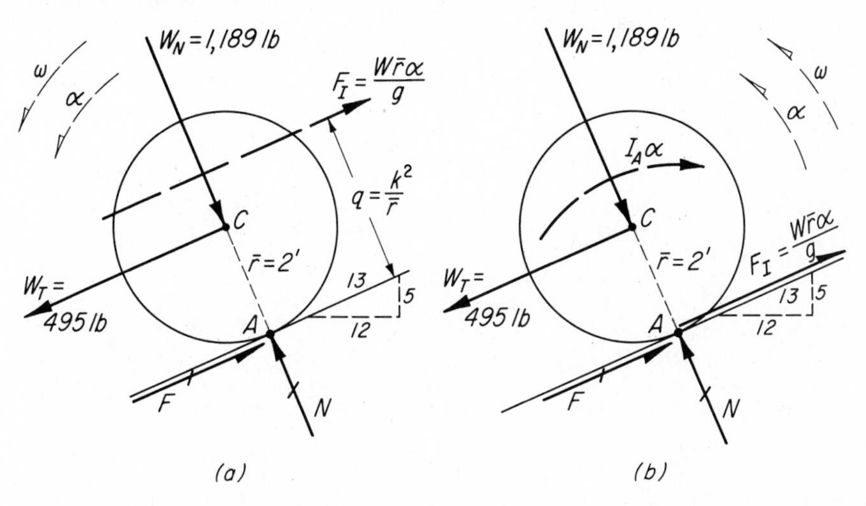

FIG. 19-4 Alternate free-body diagrams for a free-rolling body. (*a*) Inertia force at center of percussion. (*b*) Inertia force at instantaneous center of rotation.

The distance q is then

$$\left[q = \frac{k^2}{\bar{r}}\right] \qquad q = \frac{(2.45)^2}{2} = 3 \text{ ft}$$

To determine α, we now take moments about A in Fig. 19-4*a*.

$$[\Sigma M_A = 0] \qquad 3(F_I) = (W_T) \quad \text{or} \quad 3\left[\frac{(1{,}288)(2)\alpha}{32.2}\right] = 2(495)$$

and

$$\alpha = 4.125 \text{ rad per sec}^2$$

This value checks that found by the first solution. The rest of the solution follows the first solution.

Another Alternate Solution: The inertia force may be placed at the instantaneous center A to take care of the linear acceleration of this reference axis and the moment $I_A\alpha$ added to take care of the rotational effects of inertia about the instantaneous center of rotation. The free-body diagram is shown in Fig. 19-4*b*.

To determine α, we now take moments about A.

$$[\Sigma M_A = 0] \qquad 2W_T = I_A\alpha \quad \text{or} \quad 2(495) = 240\alpha$$

from which $\alpha = 4.125$ rad per sec^2. This value checks that found by the first solution. The rest of the solution would be identical to that shown in the first solution.

19-2. A section of reinforced-concrete pipe, 6 ft in outside diameter and weighing 6,440 lb, is rolled along a level surface by means of a rope wound around it and exerting a pull P, as shown in Fig. 19-5*a*. The *centroidal moment of inertia* I_C of this pipe is 3,300 ft-lb-sec^2. What acceleration $\bar{a}$ will its center of gravity be given when the constant force P is 100 lb? What time t will it take to roll a distance s of 50 ft from rest? The coefficient of friction is of such value that slipping will not occur.

Solution: From the free-body diagram in Fig. 19-5*b*, we see that the body will move to the right. Hence the angular acceleration α is clockwise and the linear

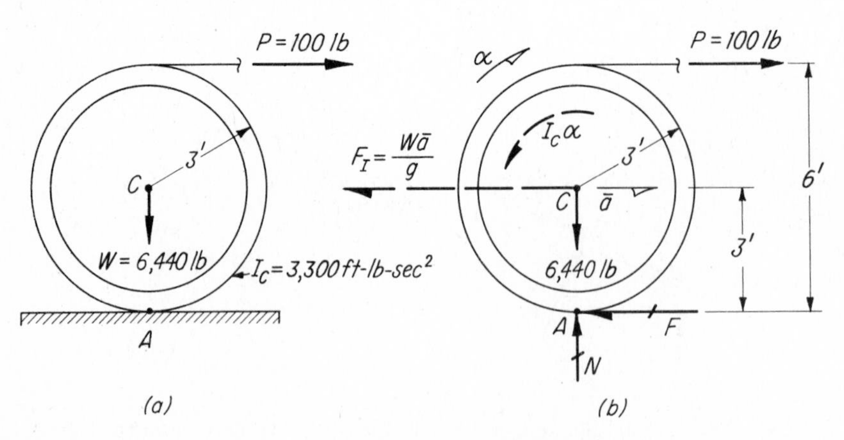

FIG. 19-5 Plane motion of a free-rolling body. (*a*) Space diagram. (*b*) Free-body diagram.

acceleration $\bar{a}$ is to the right. The friction force is unknown, since slipping is presumed not to occur. Hence, a summation of moments about point A must be made which, then, will give $\bar{a}$ and α, since $\bar{a} = r\alpha$. That is,

$$[\Sigma M_A = 0] \qquad 6P = (3)\frac{W\bar{a}}{g} + I_C\alpha$$

But $\alpha = \bar{a}/r = \bar{a}/3$. Therefore, since $P = 100$ and $I_C = 3{,}300$,

$$(6)(100) = (3)\left(\frac{6{,}440\bar{a}}{32.2}\right) + (3{,}300)\left(\frac{\bar{a}}{3}\right) \qquad \text{and} \qquad \bar{a} = 0.353 \text{ ft per sec}^2 \qquad \textit{Ans.}$$

The time t required to roll a distance of 50 ft, when $\bar{a}$ is constant, then is

$$[s = \tfrac{1}{2}at^2] \qquad 50 = (\tfrac{1}{2})(0.353)t^2 \qquad \text{and} \qquad t = 16.8 \text{ sec} \qquad \textit{Ans.}$$

19-3. In Fig. 19-6*a*, block B is attached to a weightless cord which passes over a weightless pulley and winds around a groove 4 ft in diameter in the center of the wheel. Determine the acceleration $\bar{a}$ of the center of gravity of the wheel and the tension T in the cord. $W_D = 644$ lb and $W_B = 96.6$ lb.

Solution: From the free-body diagram in Fig. 19-6*b*, we see that motion of block B will be downward and that the wheel will roll to the right. Hence, a_B is directed downward, $\bar{a}_D$ is directed to the right, and α is clockwise. As is important here to recognize, a_B and $\bar{a}_D$ are not equal but are proportional to their normal distances from A, as is indicated in the free-body diagram and discussed in Art. 19-3. (Note here, incidentally, that when the horizontal cord is in line with the center of gravity of the wheel, $\bar{a}_D$ will equal a_B.)

To find $\bar{a}_D$, using the wheel as a free body, we have

$$[\Sigma M_A = 0] \qquad 5T = 3F_{ID} + I_C\alpha \qquad (a)$$

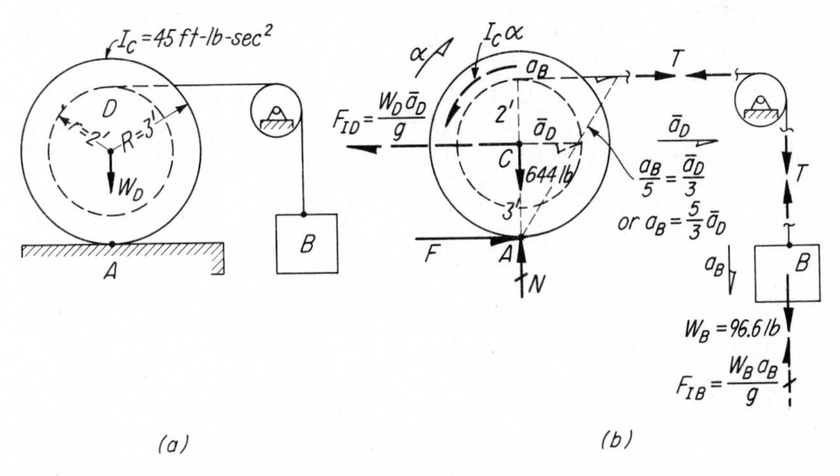

FIG. 19-6 A system of bodies in plane motion. (*a*) Space diagram. (*b*) Free-body diagram.

With block B as a free body, we have

$$[\Sigma F_Y = 0] \qquad T = W_B - F_{IB} \tag{b}$$

Substituting Eq. (b) in Eq. (a), we obtain

$$5(W_B - F_{IB}) = 3F_{ID} + I_C\alpha \tag{c}$$

Equation (c) might be obtained directly by leaving the cord uncut. But

$$F_{IB} = \frac{W_B a_B}{g} \qquad F_{ID} = \frac{W_D \bar{a}_D}{g} \qquad a_B = \frac{5}{3}\bar{a}_D \qquad \text{and} \qquad \alpha = \frac{\bar{a}_D}{r}$$

Hence,

$$(5)\left[96.6 - \left(\frac{96.6}{32.2}\right)\left(\frac{5\bar{a}_D}{3}\right)\right] = (3)\left(\frac{644\bar{a}_D}{32.2}\right) + 45\frac{\bar{a}_D}{3}$$

or $\quad 483 - 25\bar{a}_D = 60\bar{a}_D + 15\bar{a}_D \quad$ and $\quad \bar{a}_D = 4.83$ ft per sec² $\quad$ *Ans.*

We may now find T from Eq. (b). Since $a_B = 5/3\bar{a}_D$,

$$[t = W_B - F_{IB}] \; T = 96.6 - \left(\frac{96.6}{32.2}\right)\left(\frac{(5)(4.83)}{3}\right) \qquad \text{or} \qquad T = 72.5 \text{ lb} \qquad \textit{Ans.}$$

PROBLEMS

19-4. Solve Prob. 19-1 when $W = 2{,}576$ lb and $r = 2$ ft. $I_C = 160$ ft-lb-sec².

19-5. In Prob. 19-2 let $W = 9{,}660$ lb and $I_C = 2{,}300$ ft-lb-sec², and solve.

19-6. If, in Fig. 19-5a, the weight of the concrete pipe is 9,660 lb and its centroidal moment of inertia I_C is 2,300 ft-lb-sec², what constant force P will roll it a distance of 90 ft from rest in 30 sec?

19-7. A solid wooden spherical body, 4 ft in diameter and weighing 1,288 lb, rolls without slipping down an incline but is restrained by the force P of 200 lb, as shown in Fig. 19-7. (a) Determine the acceleration $\bar{a}$ of its center of gravity. (b) Will it skid when $f = 0.3$? Show proof.

Ans. (a) $\bar{a} = 7.24$ ft per sec²; (b) no; $F = 116$ lb; $F_{max} = 341$ lb

19-8. The wooden sphere shown in Fig. 19-7 rolls down the incline. What minimum restraining force P must be applied to prevent slipping, if $f = 0.15$?

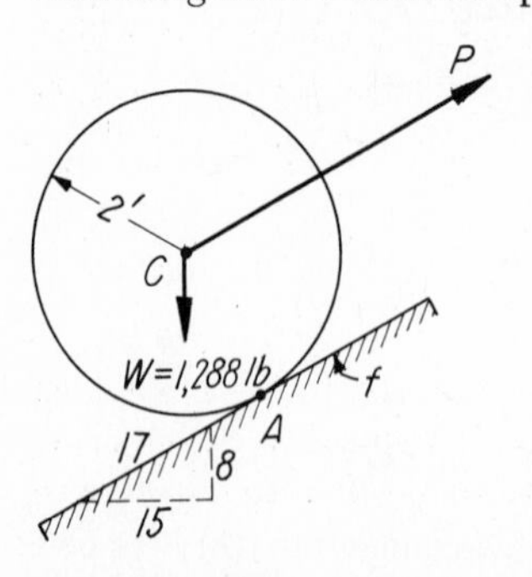

FIG. 19-7 Probs. 19-7 and 19-8.

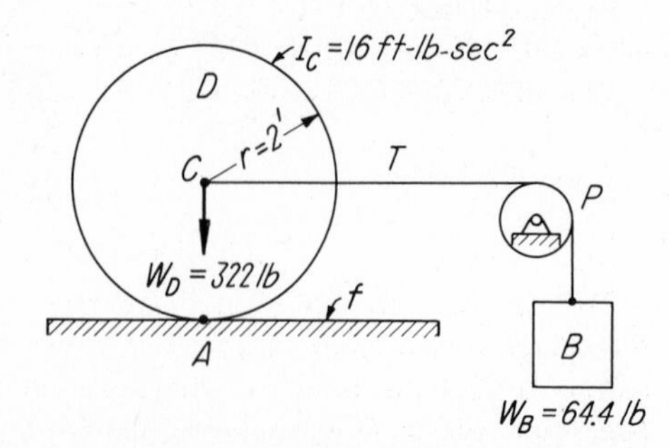

FIG. 19-8 Probs. 19-9 and 19-10.

19-9. In Fig. 19-8, block B is attached to a weightless cord which passes over the weightless pulley P and fastens to an axle passing through the centroid of disk D. Determine (*a*) the acceleration $\bar{a}_D$ of the center of gravity of the disk, (*b*) the tension T in the cord, and (*c*) the friction force F between the disk and its supporting surface. (*d*) If the coefficient of friction f is 0.1, will the disk slip? Show proof.

Ans. (*a*) $\bar{a}_D = 4.025$ ft per sec^2; (*b*) $T = 56.35$ lb; (*c*) $F = 16.1$ lb; (*d*) no

19-10. Solve Prob. 19-9 when $W_D = 644$ lb, $r = 3$ ft, $W_B = 96.6$ lb, and $I_C = 80$ ft-lb-sec^2.

19-11. Solve Prob. 19-3 when $W_D = 966$ lb, $R = 5$ ft, $r = 3$ ft, $I_C = 100$ ft-lb-sec^2, and $W_B = 161$ lb.

19-12. A sphere, a cylinder, and a pipe section are released simultaneously on an inclined plane. All three bodies have the same external diameters and weigh the same. Assume each of them rolls without slipping. Which will reach the bottom first? Which last? Show all work.

19-13. The 644-lb cylinder shown in Fig. 19-9 is wrapped with rope as shown and released from rest. Find the angular acceleration and the distance traveled by the center of gravity in 3 sec.

Ans. $\alpha = 14.3$ rad per sec^2; $s = 96.7$ ft

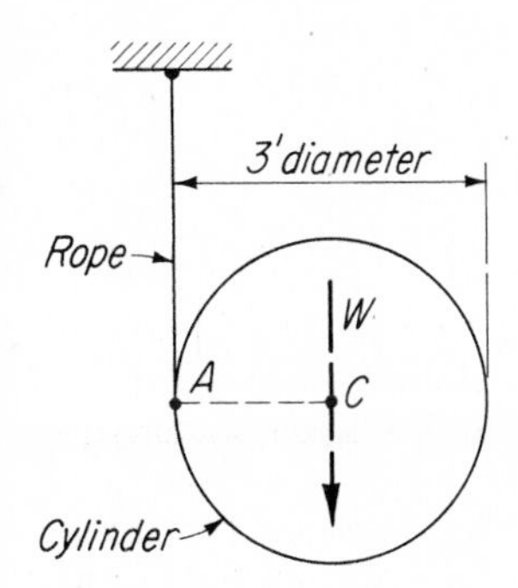

FIG. 19-9 Probs. 19-13

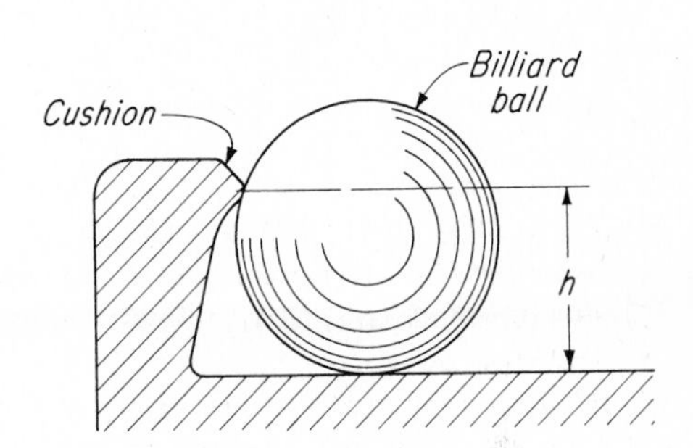

FIG. 19-10 Prob. 19-15. Billiard ball striking cushion.

19-14. Solve Prob. 19-13 if $W = 322$ lb and $t = 4$ sec.

19-15. Find the height h shown in Fig. 19-10 so that there will be no friction force between the ball and the table during the impact against the cushion. This height will minimize the wear on the felt covering of the table. Assume the ball is $2\frac{1}{4}$ in. in diameter. (The idea of Fig. 19-2*a* is suggested.)

Ans. $h = 1.575$ in.

SUMMARY

(By article number)

19-1. In **plane motion,** all particles of a body move in parallel planes, of which *the reference plane passes through its center of gravity*. In rolling bodies, the plane motion is a combination of rotation and translation. *When no slipping occurs, such bodies are said to be* **free rolling,** and $\bar{a} = \bar{r}\alpha$.

19-2. The instantaneous center is the point about which a body in plane motion may be considered to be rotating at any instant. *For free-rolling bodies, the instantaneous center is the point of contact between the body and the contacting surface.*

19-3. Kinematics of plane motion. The linear motion of a body in plane motion is that of its center of gravity. Its angular motion is generally considered with respect to its centroidal axis.

19-4. The kinetics of plane motion may be treated in three alternate ways. The inertia force may be placed at the center of percussion; the inertia force may be placed at the center of gravity and a moment $I_C\alpha$ (C is the centroidal axis) added to the free-body diagram; or the inertia force may be placed at the instantaneous center of rotation and a moment $I_A\alpha$ (A is the instantaneous center) added to the free-body diagram. The first and last of these concepts are based upon the use of the instantaneous center as the reference axis. The second concept is based upon the use of the centroidal axis as the reference axis, and in this case plane motion may be thought of as a combination of translation and rotation.

REVIEW PROBLEMS

19-16. A pair of cast-iron freight-car wheels (Fig. 19-11), together with their connecting axle, weighs 966 lb. The diameter of each wheel is 2.75 ft (33 in.). The centroidal moment of inertia I_C of the entire assembly is 45 ft-lb-sec². The wheels are moved on horizontal rails by a force P of 200 lb applied at the mid-point of the axle. *Calculate the acceleration $\bar{a}$ of the center of gravity of this assembly.*

Ans. $\bar{a} = 3.72$ ft per sec²

19-17. Solve Prob. 19-16 when $W = 1{,}288$ lb, $I_C = 60$ ft-lb-sec², and $P = 300$ lb. Will the wheels skid on the rails, if $f = 0.3$? Show proof.

19-18. The system shown in Fig. 19-12 is released from rest. Calculate the angular acceleration α of part D if $W_B = 128.8$ lb, $W_D = 322$ lb, $r = 2$ ft, and the moment of inertia of disk E is negligible.

19-19. Solve Prob. 19-18 if the centroidal moment of inertia of disk E is 5 lb-ft-sec² and its radius is 0.5 ft.

Ans. $\alpha = 0.265$ rad per sec²

19-20. The rolling part D of Fig. 19-12 weighs 200 lb. Calculate the required weight W_B to limit the downhill angular acceleration of part D to 0.4 rad per sec². The moment of inertia of E is negligible and $r = 1.5$ ft.

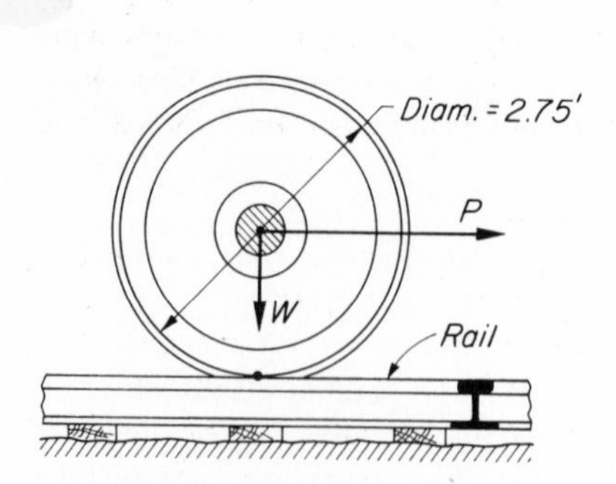

FIG. 19-11 Probs. 19-16 and 19-17.

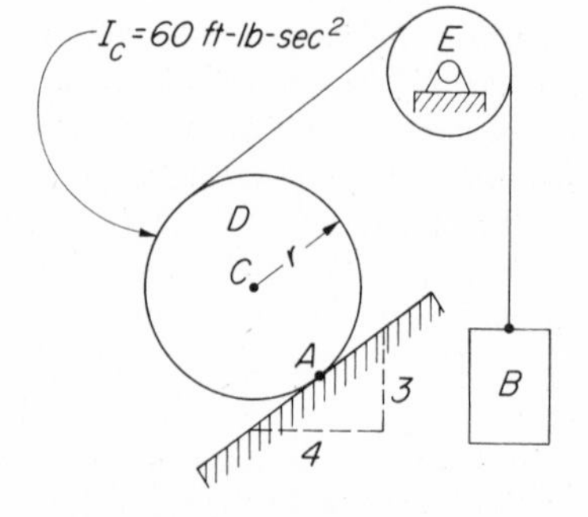

FIG. 19-12 Probs. 19-18 to 19-20.

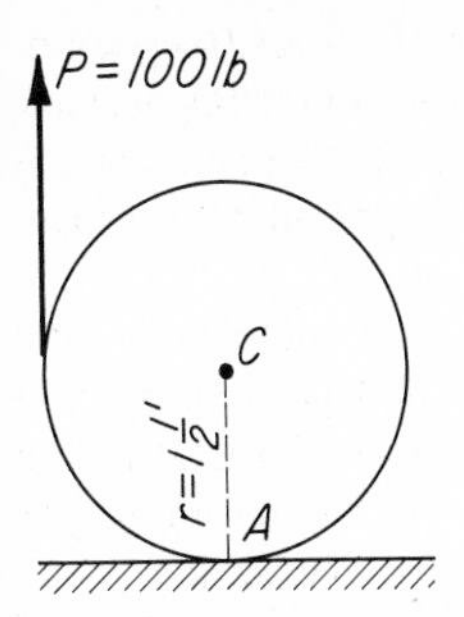

FIG. 19-13 Probs. 19-22 and 21-35.

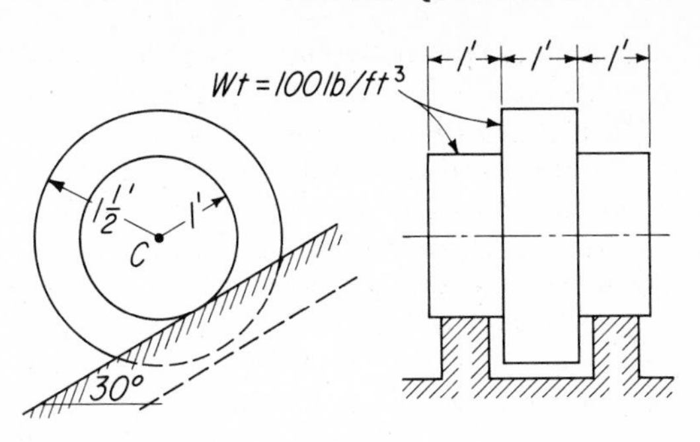

FIG. 19-14 Prob. 19-23.

19-21. A 2-ft diam solid cylinder weighing 500 lb rolls without slipping down a plane inclined at 30° to the horizontal. Determine the minimum coefficient of friction to prevent slipping. *Ans.* $f = 0.192$

19-22. The cylinder shown in Fig. 19-13 weighs 966 lb. Find the coefficient of friction f necessary to prevent slipping. *Ans.* $f = 0.077$

19-23. The concentric solid cylinders shown in Fig. 19-14 roll without slipping down the 30° incline. Find the angular acceleration α and the velocity of the center of gravity v after the assembly has traveled 30 ft from rest.

Ans. $\alpha = 8.79$ rad per sec²; $v = 23$ ft per sec

REVIEW QUESTIONS

19-1. Define plane motion.

19-2. Mention at least three examples of plane motion.

19-3. When is a body said to be free rolling?

19-4. What is meant by the instantaneous axis?

19-5. Define instantaneous center.

19-6. What particular point is the instantaneous center of a free-rolling body?

19-7. Consider a number of particles of a free-rolling homogeneous sphere along a diameter through its instantaneous center and its center of gravity. What is the particular variation of the acceleration and velocity components of these particles, parallel to the supporting surface?

19-8. What do the symbols $\bar{a}$, $\bar{v}$, and $\bar{s}$ represent in plane motion?

19-9. Give the mathematical relation between the angular acceleration of a free-rolling body and the linear acceleration of its center of gravity.

19-10. Does the relation mentioned in the preceding question hold when the body slips?

19-11. In the problems in this chapter, does the friction force F represent rolling or static frictional resistance?

19-12. If one thinks of plane motion as a combination of translation and centroidal rotation: (*a*) What is the expression for the translational inertia force? (*b*) Define its line of action. (*c*) What is the expression for the rotational inertia effect?

19-13. If one thinks of plane motion as rotation about the instantaneous center of rotation, define the line of action of the resultant tangential inertia force.

19-14. If one thinks of plane motion as rotation about the instantaneous center of rotation and chooses to put the tangential inertia force at the instantaneous center of rotation, what is the expression for the corrective moment that must be added to the free-body diagram?

CHAPTER 20

Work, Energy, and Power

20-1 Introduction

The conditions of motion of a body, or the forces producing such motion, were determined in Chaps. 15 and 18 by a method using *force systems in dynamic equilibrium,* referred to as the *force method.* The quantities involved were force, mass (W/g), and acceleration.

A second method, called *the work-energy method* and based on the law of conservation of energy, will now be developed. In this method, problems are solved by using certain known relations between force and displacement on one side, and mass (W/g) and velocity on the other.

A third method, using the concepts of impulse (force and time) and momentum (mass and velocity), is developed in the next chapter. Of these three methods, the force method is most applicable to the solution of problems involving acceleration. Other problems, involving velocity and displacement, are most easily solved by the work-energy method. The method of impulse and momentum is useful in solving problems involving time and velocity. A solution by one method is often readily checked by one or both of the others.

20-2 Work

When a body is displaced by force in the direction of the force, *work* is done. Work may be done *by* or *against* forces. **Positive work is done by forces acting in the direction of the initial displacement,** and **negative work is done against forces opposed to the direction of the initial displacement.** *Only external forces* *can do work.* Internal (inertia) forces can do no work. The weight of a body is an external force, caused by gravitational attraction.

In evaluating work done, three types of external forces are usually considered: applied, gravitational, and frictional. Work may be done *by* (+) or *against* (−) applied forces, since they may act either with or against the direction of motion. The gravitational force W is always downward acting. Hence, when a body moves to a lower level, positive work is done *by* W; when it moves to a higher level, negative work is done *against* W. *Friction forces always do negative work,* since they always act against the direction of motion. As must be clearly understood, *an inertia force can do no work,* since it is merely an internal reaction to an external force that *does* work.

20-3 Work Done by Constant Forces

The measure of *the work done by or against a constant force is the product of the force and the linear displacement, in the direction of the force, of its point of application.* If the symbol U denotes work, F force, and s displacement,

$$\textbf{Work} = \textbf{(force)(displacement)} \qquad \text{or} \qquad U = \boldsymbol{F} \cdot \boldsymbol{s} \tag{20-1}$$

Examples of work done by constant forces are shown in Figs. 20-1 and 20-2. In Fig. 20-1, the work done *by* the constant applied force F in pushing block A a distance s *against* frictional resistance is Fs. The work done by the constant pull P, in Fig. 20-2, in lifting the block a vertical distance h against the gravitational pull W, is Ph or Wh.

In Fig. 20-3, a block is pushed up an incline a distance s by a horizontal force P. Here, all the positive work is done *by* the component of P parallel to the incline ($P \cdot \cos\theta$), whose point of application is displaced a distance s also parallel to the incline. The work done is $U = (P \cdot \cos\theta)(s)$. No work is done by the normal component $P \cdot \sin\theta$, since there is no displacement normal to the incline of its point of application.

Note especially here that *the equation $U = Fs$ can be used only when the force F is constant, or when F is the average value of a variable force.* Negative

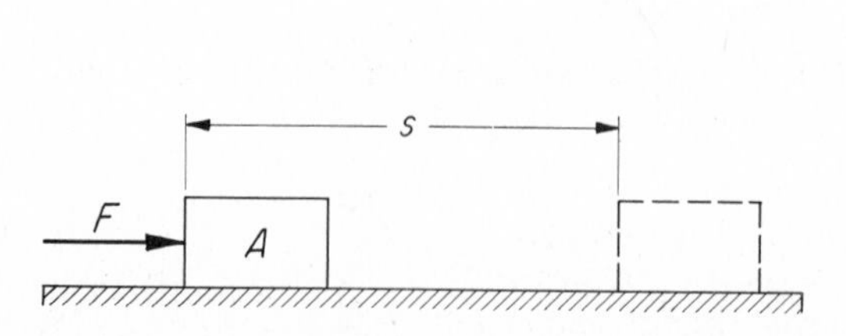

FIG. 20-1 Work done is $F \cdot s$.

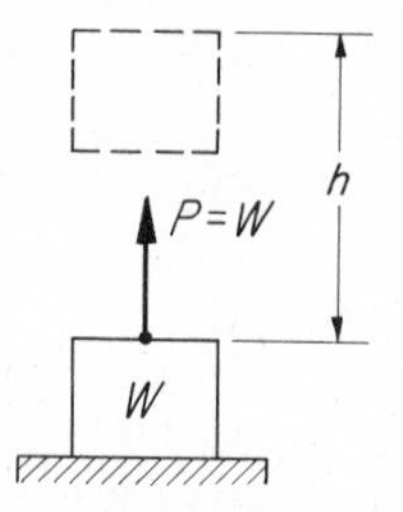

FIG. 20-2 Work done is $P \cdot h = W \cdot h$.

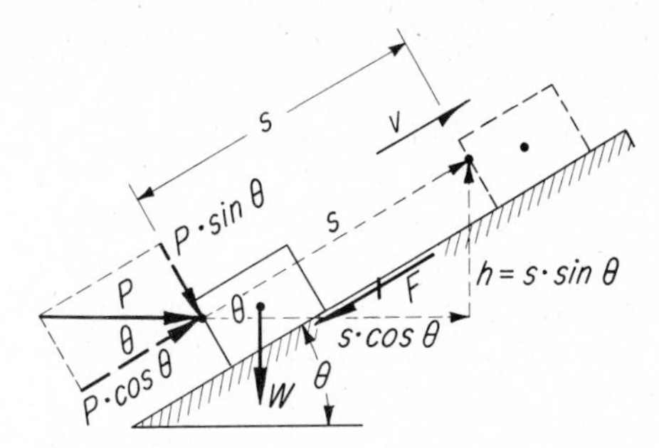

FIG. 20-3 Work done is $P \cdot \cos \theta \cdot s$.

work is done against the friction force F and the gravitational force W, since they oppose the displacement. The work done in displacing a body often must be evaluated in terms of the resistance to be overcome, as is illustrated in the example below.

Units. In engineering practice, force is commonly measured in pounds and displacement in feet. The common *unit of work*, then, is the foot-pound (ft-lb). Other units are the inch-pound (in.-lb) and the foot-kip (ft-kip).

ILLUSTRATIVE PROBLEM

20-1. Calculate the work done in moving the block in Fig. 20-3 a distance s up the incline, when $W = 100$ lb, $s = 6$ ft, $\theta = 30°$, and $F = 30$ lb. Find also the constant value of P required to do this work.

Solution: Since P is unknown, the work done is found by calculating the total resistance to be overcome in moving the block. Of the two forces W and F resisting motion, W has its point of application (the *c.g.* of the block) displaced a vertical distance h, and F has its point of application (the surface of contact between block and incline) displaced a distance s. Since $W = 100$ lb, $h = s \cdot \sin 30°$, $F = 30$ lb, and $s = 6$ ft, the total work done is

$$[U = Wh + Fs] \qquad U = (100)(6)(0.5) + (30)(6) \qquad \text{and} \qquad U = 480 \text{ ft-lb} \qquad \textit{Ans.}$$

But the work done is also equal to $(P \cdot \cos \theta)(s)$. Hence, since $\cos 30° = 0.866$ and $s = 6$ ft,

$$[U = (P \cdot \cos \theta)(s)] \qquad 480 = P(0.866)(6) \qquad \text{and} \qquad P = 92.4 \text{ lb} \qquad \textit{Ans.}$$

PROBLEMS

20-2. Calculate the amount of work U done in sliding a 100-lb block a distance of 12 ft along a level floor against a frictional resistance of 26 lb.

Ans. $U = 312$ ft-lb

20-3. A rocket and its fuel weighs 4,600 lb at lift-off. What work was done in lifting the assembly 200 ft above the launch pad if 6 lb of fuel was consumed during each foot of ascent? *Ans.* $U = 800{,}000$ ft-lb

20-4. Calculate the work done in pushing a 100-lb weight 20 ft (measured along the incline) up a 30° incline if the coefficient of friction between the surfaces is 0.2.

20-5. A 1,000-lb weight is pushed slowly up a 20-ft incline (length measured along the incline) by a horizontal force P. Calculate the work done if the coefficient of friction $f = 0.25$ and the incline make an angle of $17\frac{1}{2}°$ with the horizontal.

Ans. $U = 11{,}700$ ft-lb

20-6. Each turn of the crank of a hand-driven hoist winds up 1 ft of rope. The crank arm is 18 in. long. What load can be lifted vertically by a 40-lb force on the crank if all the work done on the crank goes into lifting the load?

20-7. A locomotive pulls a 1,000-ton train up a 1 percent grade. (*a*) How much work (in ft-tons) does it do in pulling the train a distance of 1 mile (5,280 ft) against an average train resistance (rolling friction, wind resistance, etc.) of 10 lb per ton? (*b*) Determine also the average drawbar pull P, exerted by the locomotive on the train, parallel to the track. *Ans.* (*a*) $U = 79{,}200$ ft-tons; (*b*) $P = 15$ tons

20-8. An airplane weighing 4,000 lb rises 125 ft above the ground while traveling a horizontal distance of 5,000 ft after take-off. (*a*) What is the total amount of work done by the motor if air resistance in the direction of flight averages 200 lb? (*b*) What average pull P must be exerted by the propeller in the direction of flight, if the ascent is made at a uniform speed?

20-9. A jackscrew is used to raise a 6,000-lb load P as indicated in Fig. 20-4. Assume that the constant horizontal force F remains at right angles to the horizontal bar during turning. (*a*) If screw friction increases the work to be done by 15 percent, what total amount of work is done in raising the load 6 in.? (*b*) What force F must be applied if this work is done in 12 complete turns of the bar?

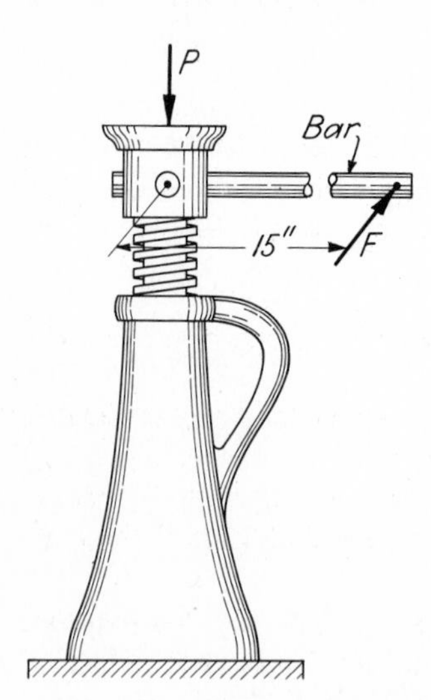

FIG. 20-4 Prob. 20-9.

20-4 Work Done by Variable Forces

The work done by a variable force is a product of the *average* value of the force and the displacement. That is,

$$\textbf{Work} = \textbf{(average force)(displacement)}$$

or

$$U = F_{avg} \cdot s \tag{20-1a}$$

The simplest variable force is one that increases or decreases at a uniform rate. Its average value is then one-half the sum of its minimum and maximum values. Such variation occurs in many instances. For example, the force required to pull up a vertically suspended rope, hand over hand, decreases uniformly from a maximum equal to the full weight of the rope to zero when all the rope is up. Also the force required to deform a spring is proportional to the deformation of the spring, within the elastic limit of the material.

Coil Springs. The coil spring furnishes perhaps the simplest example of a variable force. The *undeformed length* of a coil spring is called its **free length.** The force required to deform a spring a unit distance, 1 in. or 1 ft, from its free length is called the **spring modulus** and is denoted by the symbol k. Hence, if a force of 5 lb is required to stretch a given spring 1 in. from its free length, 10 lb to stretch it 2 in., and 15 lb to stretch it 3 in., its *spring modulus* is said to be 5 lb per in., and it is referred to as a 5-lb spring. The force required to deform a spring s in. then is

$$F = k \cdot s \tag{20-2}$$

Since the force required to deform a spring is proportional to the deformation, *the work done in deforming a spring from its free length is the product of the average force F_{avg} and the deformation s.* But $F_{avg} = \frac{1}{2}F$. Hence, $U = \frac{1}{2}(ks)s$, or

Work of deforming coil spring from free length:

$$U = \frac{1}{2} ks^2 \tag{20-3}$$

where k = spring modulus, lb per in. (or lb per ft)
s = deformation of spring from its free length, in. (or ft)

A little study will show that *the work done in deforming a spring is always negative,* since the spring decreases the velocity of the body or force deforming it. Likewise, *the work done by a deformed spring in releasing itself is always positive,* since the spring increases the velocity of the body upon which it acts.

In other problems involving work done, *successive displacements may be variable while the force remains constant,* as, for example, if a number of bricks of equal weight are lifted one at a time to successively higher levels, as in the building of a chimney. That is, the force remains constant, but the successive displacements increase uniformly.

When the variation of the force is difficult to determine, *the work done in displacing a body may be measured as the product of its weight and the total vertical displacement of its center of gravity, plus the work of overcoming frictional resistances.* (See Probs. 20-11 and 20-12.)

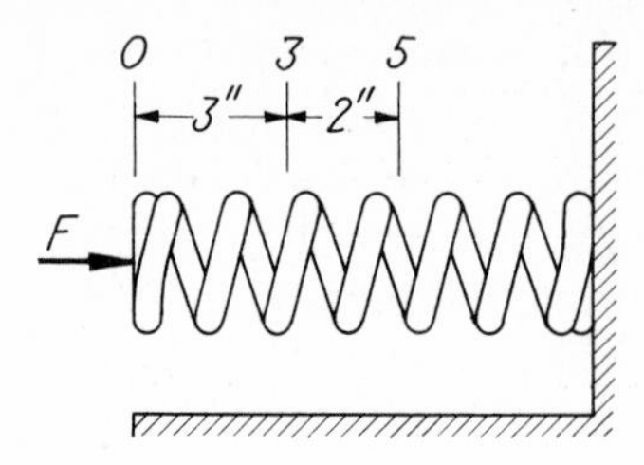

FIG. 20-5 Work done in compressing spring.

ILLUSTRATIVE PROBLEMS

20-10. The coil spring in a spring bumper (Fig. 20-5) has a modulus k of 200 lb per in. How much work must be done to compress it (*a*) the first 3 in. from its free length and (*b*) two additional inches from 3 to 5 in.?

Solution: When the spring is compressed *from its free length,* Eq. (20-3) applies. Therefore,

$$\left[U = \frac{1}{2}ks^2\right] \qquad U = \left(\frac{1}{2}\right)(200)(3)^2$$

or

$$U = 900 \text{ in.-lb} \qquad \textit{Ans.}$$

When the spring is not compressed from its free length, as in part (*b*), Eq. (20-3) is not applicable. The forces F_3 and F_5 required to compress the spring 3 and 5 in., respectively, must be determined. The work done is then the product of the average of F_3 and F_5, and the distance which the spring is compressed. Hence, when $s =$ 3 in.,

$$[F = ks] \qquad F_3 = (200)(3) = 600 \text{ lb}$$
$$[F = ks] \qquad F_5 = (200)(5) = 1{,}000 \text{ lb}$$

The average force then is

$$\left[F_{\text{avg}} = \frac{1}{2}(F_3 + F_5)\right] \qquad F_{\text{avg}} = \frac{1}{2}(600 + 1{,}000) = 800 \text{ lb}$$

and the work done in compressing the spring from 3 to 5 in. is

$$[U = F_{avg} \cdot s] \qquad U = (800)(2) \qquad \text{or} \qquad U = 1{,}600 \text{ in.-lb} \qquad \textit{Ans.}$$

As a check, the sum of the answers to (*a*) and (*b*) must equal the work done in compressing the spring a total of 5 in. Therefore,

$$\left[U = \frac{1}{2}ks^2\right] \qquad U = \left(\frac{1}{2}\right)(200)(5)^2 = 2{,}500 \text{ in.-lb} \qquad \textit{Check.}$$

which equals 900 in.-lb + 1,600 in.-lb.

20-11. A man wishes to move the solid block of wood shown in Fig. 20-6 a distance of 2 ft to the right with the least amount of work. If the block weighs 100 lb and

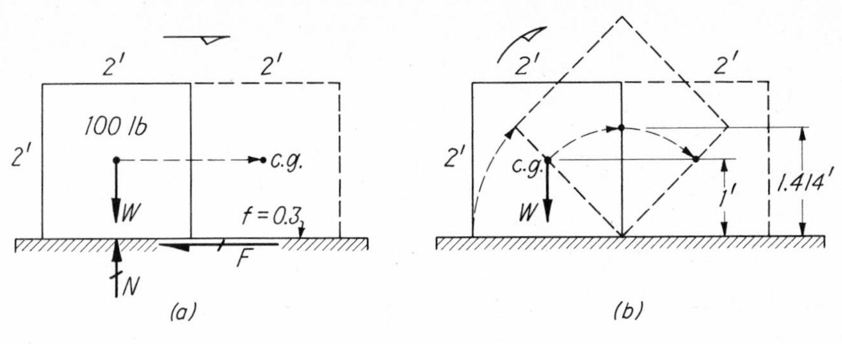

FIG. 20-6 Least work to move block. (*a*) Sliding block. (*b*) Tipping block.

the coefficient of friction f is 0.3, should he slide it, as indicated in (*a*), or should he tip it, as indicated in (*b*)?

Solution: The friction force F is $fN = (0.3)(100)$, or 30 lb, and the work done in sliding the block then is $Fs = (30)(2)$, or 60 ft-lb. If the block is tipped, the friction becomes negligible. Hence all of the work is done *against* gravity in raising its center of gravity vertically a distance of 0.414 ft. Thereafter gravity will complete the tipping. The work done in tipping the block then is

$[U = Wh]$ $\qquad U = (100)(0.414)$

or $\qquad U = 41.4$ ft-lb $\qquad$ *Ans.*

which is the least work with which the block can be moved.

20-12. The circular water tank shown in Fig. 20-7 is filled by pumping from a reservoir whose water surface remains at a constant level. Neglecting pipe friction, how much work must be done to fill the tank? Water weighs 62.5 pcf.

Solution: Since the intake pipe enters the tank at a point 2 ft above its bottom, all of the water in the lower 2 ft must be raised 20 ft. The water in the upper 8 ft must be raised a distance equal to the *average* between 20 and 28 ft which is 24 ft. The total work done is obtained by multiplying the weight of each volume of water by the vertical distance it must be raised.

$$\text{Area of tank: } A = \pi r^2 = (3.1416)(4)^2 = 50.3 \text{ sq ft}$$

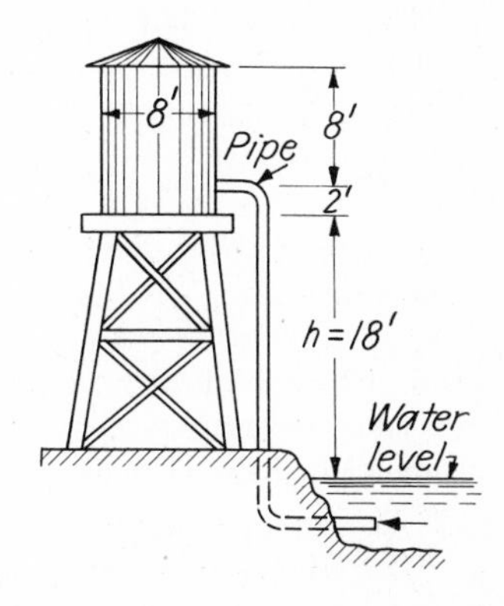

FIG. 20-7 Work done to fill water tank.

In the lower 2 ft of the tank,

$$\begin{aligned}\text{Volume of water: } V &= (50.3)(2) = 100.6 \text{ cu ft}\\ \text{Weight of water: } W &= (62.5)(100.6) = 6{,}290 \text{ lb}\\ \text{Work to be done: } U &= (6{,}290)(20) = 125{,}800 \text{ ft-lb}\end{aligned}$$

In the upper 8 ft of the tank,

$$\begin{aligned}\text{Volume of water: } V &= (50.3)(8) = 402.4 \text{ cu ft}\\ \text{Weight of water: } W &= (62.5)(402.4) = 25{,}150 \text{ lb}\\ \text{Work to be done: } U &= (25{,}150)(24) = 603{,}600 \text{ ft-lb}\end{aligned}$$

The total work to be done then is

$$U = 125{,}800 + 603{,}600 \qquad \text{or} \qquad U = 729{,}400 \text{ ft-lb} \qquad \textit{Ans.}$$

PROBLEMS

20-13. In Prob. 20-10, determine the work done in compressing the spring (*a*) the first 2 in. and (*b*) four additional inches from 2 to 6 in.

20-14. The coil spring on a screen door requires a force of 20 lb to stretch it 4 in. (*a*) What is the modulus k of this spring? (*b*) How much will a force of 30 lb stretch it? *Ans.* (*a*) $k = 5$ lb per in; (*b*) $s = 6$ in.

20-15. In compressing a coil spring, whose modulus is 10 lb per in., 80 in.-lb of work was done. How many inches was the spring compressed? *Ans.* $s = 4$ in.

20-16. Three springs support a 7,500-lb load. The two outer springs have spring modulus of 1,000 lb per in. and the center spring has a modulus of 500 lb per in. If the springs are located so that they deform equal amounts under the 7,500-lb load, how much work is done in deforming the middle spring?

Ans. $U = 2{,}250$ in.-lb

20-17. If an additional 80 in.-lb of work is done on the spring of Prob. 20-15, how many additional inches will the spring be compressed? *Ans.* $s = 1.66$ in.

20-18. Solve Prob. 20-11 if the block is 3 by 3 ft and the man wishes to move it 3 ft.

20-19. The excavation for a building is 80 by 120 ft by 16 ft deep. Find the work done in bringing the excavated material to the original ground surface if it weighed 120 pcf in place.

20-20. Water is drawn 60 ft vertically out of a well by means of a bucket fastened to the lower end of a rope the top of which winds around a drum. The rope weighs $\frac{1}{2}$ lb per ft. Neglecting all friction, how much work is done (*a*) in winding up the 60 ft of rope alone and (*b*) in pulling up a bucket of water weighing 20 lb the full 60 ft? *Ans.* (*a*) $U = 900$ ft-lb; (*b*) $U = 2{,}100$ ft-lb

20-21. Solve Prob. 20-20 when the rope weighs $\frac{1}{3}$ lb per ft and the filled bucket weighs 25 lb.

20-22. The car shown in Fig. 20-8 travels an actual distance of 1,000 ft along the road surface, from rest at A to a stop at B, against an average road resistance (rolling friction, wind resistance, etc.) F of 40 lb. If the car weighs 3,000 lb and if B is 60 ft higher than A, what total amount of work is done by the motor?

Ans. $U = 220{,}000$ ft-lb

20-23. Solve Prob. 20-22 when the car weighs 4,000 lb and F is 50 lb.

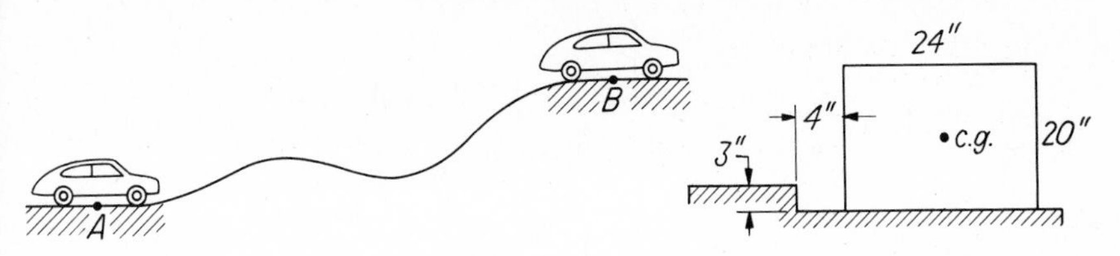

FIG. 20-8 Probs. 20-22 and 20-23.

FIG. 20-9 Prob. 20-24.

20-24. The solid concrete block shown in Fig. 20-9 weighs 200 lb. How much work must be done to tip it up until its left end rests on the higher level? (Assume that friction is negligible and that the block does not slip while being tipped.)

Ans. $U = 1{,}200$ in.-lb

20-25. Solve Prob. 20-12 (*a*) when the tank is 8 ft high and its inside diameter is 6 ft, and h is 22 ft. How much work must be done to fill this tank if the pipe enters the tank (*b*) at the top and (*c*) at the bottom?

20-5 Potential and Kinetic Energy

Energy is the capacity to do work. It exists in many forms, such as chemical, heat, mechanical, and electrical. One of the fundamental laws of nature is that *energy can be neither created nor destroyed*, but can only be changed from one form to another. Man takes chemical energy from nature in the form of wood, coal, and oil, heat energy from the sun's rays, and mechanical energy from falling water—all of which he readily transforms into other forms to suit his convenience.

An interesting cycle of energy transformation is that of a power plant using coal to produce heat in the form of steam, with which to drive a steam engine which in turns drives a generator producing electricity. When this electricity is used for heating and lighting, heat is again produced. The cycle of energy transformations here is from heat energy to mechanical energy, to electrical energy, and back to heat energy. Only the study of *mechanical energy* lies within the scope of this book.

Mechanical energy exists in two forms: *potential* and *kinetic*. **Potential energy is possessed by a body by virtue of its position,** since it is capable of doing work in changing to a lower position. For example, when the block B in Fig. 20-10 is held in the position shown, it has potential energy PE equal to $W \cdot h$ with respect to any lower plane, because, as B drops the distance h after being released, the force W, which of course is the gravitational pull on the body, is capable of doing work in pulling body A *up* the incline. That is,

Potential energy:

$$\boldsymbol{PE = W \cdot h} \tag{20-4}$$

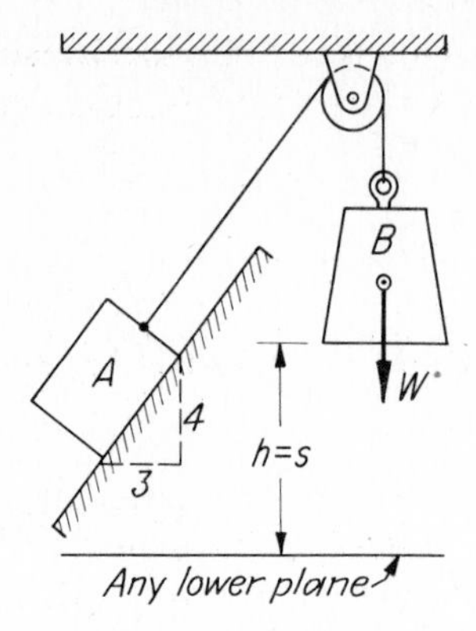

FIG. 20-10 Potential energy of a body at rest.

Kinetic energy is possessed by a body by virtue of its motion and is capable of doing work. This motion may be rectilinear, curvilinear, or rotational. **The kinetic energy possessed by a moving body is equal to the amount of work that must be done to bring it to rest.**

Energy is readily changed from the potential to the kinetic form, or vice versa. For instance, if the two bodies in Fig. 20-10 are held at rest, neither has kinetic energy *KE*, but both possess potential energy *PE* with respect to the lower plane. When released, *B* will move downward, if it is heavier than A, and will therefore lose *PE*, but will have gained *KE* by virtue of its motion. Body *A* will move up the incline. It will thereby gain *PE* and, being in motion, it will also have gained *KE*.

Similarly, the potential energy of water in a reservoir, with respect to a water turbine at some lower level, will have changed to kinetic energy by the time the water has reached the turbine through a pipe. Most of this kinetic energy is imparted to the turbine, which in turn may drive a generator. Thus we obtain hydroelectric power.

20-6 Kinetic Energy of Translation

The simple example of a body (Fig. 20-11), in motion at *A* and sliding from *A* to a gradual stop at *B*, will serve in deriving the mathematical expression for the kinetic energy of a body with linear motion. The constant friction force *F* decelerates the body. The reaction to this decelerating force is the inertia force whose value is Wa/g. The kinetic energy which the body possesses at *A* does work $[U = F \cdot s = (Wa/g) \cdot s]$ in moving it to *B* against the friction force *F*, until all of the energy has been dissipated and the body stops.

In Art. 14-1 we found that $s = (v^2 - v_0^2)/2a$, in which v_0 is the initial velocity, v is the final velocity, and a, the acceleration, is negative, when the body is decelerated. When the body stops, v is zero and $s = v_0^2/2a$. Now, since the kinetic energy of the body at *A* is equal to the work that must be

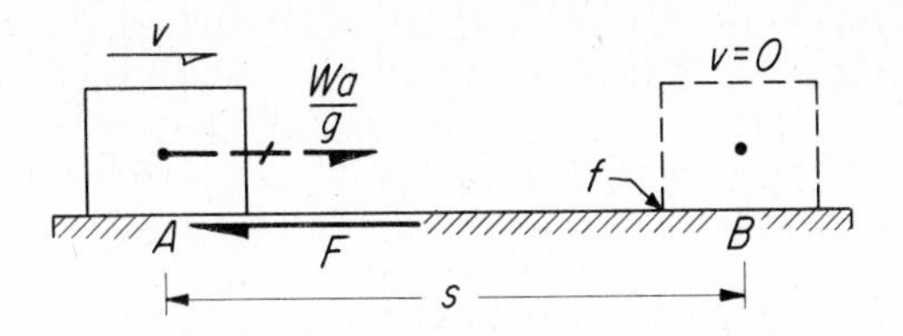

FIG. 20-11 Kinetic energy of block A does work in moving it to B.

done to stop it ($U = (Wa/g) \cdot s$), we have

$$KE = \frac{Wa}{g} \cdot s = \frac{Wa}{g} \cdot \frac{v_0^2}{2a} = \frac{Wv_0^2}{2g} \qquad (a)$$

As this expression indicates, *the kinetic energy of a body at any instant is proportional to its mass* (W/g) *and to the square of its instantaneous velocity* v, regardless of the direction of motion. Therefore,

Kinetic energy:
$$\mathbf{KE = \frac{Wv^2}{2g}} \qquad \mathbf{(20\text{-}5)}$$

Units. Since energy is measured in terms of work, *the unit of energy is the same as that of work*, usually the foot-pound. To obtain this unit, by use of Eq. (20-5), W must be in pounds and v in feet per second; g is constant at 32.2 feet per second per second. To illustrate,

Unit of kinetic energy:

$$\frac{Wv^2}{2g} = \frac{(\text{lb})(\text{ft}^2/\text{sec}^2)}{\text{ft}/\text{sec}^2} = \text{ft-lb}$$

ILLUSTRATIVE PROBLEM

20-26. Calculate the kinetic energy of an automobile weighing 3,220 lb and traveling at 30 mph. (Disregard the energy of rotation of the wheels.)

Solution: Converting 30 mph to ft per sec we have $v = (30)(^{88}/_{60}) = 44$ ft per sec. Then,

$$\left[KE = \frac{Wv^2}{2g}\right] \qquad KE = \frac{(3{,}220)(44)^4}{(2)(32.2)} = 96{,}800 \text{ ft-lb} \qquad \textit{Ans.}$$

PROBLEMS

20-27. Solve Prob. 20-26 when the car travels at 60 mph. (Note that *the kinetic energy of a body in motion varies as the square of its velocity.*)

20-28. The kinetic energy of a certain body moving with a velocity of 1 ft per sec is 100 ft-lb. Its kinetic energy varies as the square of its velocity. Calculate its KE when its velocity is 2, 3, and 4 ft per sec.

Ans. 400 ft-lb; 900 ft-lb; 1,600 ft-lb

20-29. A certain body moving with a velocity of 2 ft per sec possesses kinetic energy in the amount of 200 ft-lb. Calculate its *KE* at 3, 4, and 5 ft per sec.

Ans. 450 ft-lb; 800 ft-lb; 1,250 ft-lb

20-30. Compare the kinetic energies of a 100-lb body with a velocity of 10 ft per sec with that of a 10-lb body with a velocity of 100 ft per sec.

Ans. The 10-lb body would have 10 times as much *KE* as the 100-lb body.

20-31. The block shown in Fig. 20-11 weighs 322 lb, and the coefficient of sliding friction is 0.2. How much kinetic energy must it possess at *A* to slide along the level to a stop at *B*, if *s* is 10 ft?

20-32. At what velocity must a car weighing 3,220 lb travel in order to possess kinetic energy in the amount of 54,500 ft-lb?

20-33. A 12,000-ton boat strikes a pier with a velocity of 1 mph. How much kinetic energy must be dissipated to bring the boat to a stop?

Ans. $U = 803{,}000$ ft-lb

20-34. How much work must be done to change the velocity of a 3,220-lb car from 30 to 60 mph? Is this the same amount of work required to bring the car's velocity to 30 mph from rest? Neglect frictional resistance.

20-35. An automobile weighing 3,220 lb is approaching a traffic stop on a level street. The brakes are applied with a constant force for a distance of 200 ft before the car stops. What total frictional force *F* must be produced between the wheels and the pavement if, when the brakes are applied, the velocity of the car is (*a*) 15 mph, (*b*) 30 mph, (*c*) 60 mph?

Ans. (*a*) $F = 121$ lb; (*b*) $F = 484$ lb; (*c*) $F = 1{,}936$ lb

20-36. In what minimum distance *s* can a 3,220-lb car be stopped on a level road by a pressure on the brake pedal which produces a constant frictional resistance *F* of 1,000 lb between the tires and the pavement, if, when the brakes are applied, the car travels at (*a*) 15 mph, (*b*) 30 mph, (*c*) 60 mph?

20-37. A machine-gun bullet weighing 2 oz leaves the muzzle of the gun barrel with a velocity of 2,000 ft per sec. (*a*) Calculate its kinetic energy. (*b*) At what velocity would a 3,220-lb car have to travel to possess as much kinetic energy?

Ans. (*a*) $KE = 7{,}760$ ft-lb; (*b*) $v = 12.47$ ft per sec $= 8.5$ mph

20-7 The Work-Energy Equation

Since the kinetic energy of a body depends upon its velocity, the **initial kinetic energy** *IKE* possessed by a body will be increased if positive work is done on it tending to increase its velocity, and will be decreased if negative work is done against it tending to decrease its velocity. The **final kinetic energy** *FKE* will then be the algebraic sum of these quantities. When this principle is stated in the form of an equation, we have

$$\begin{bmatrix}\textbf{Initial}\\\textbf{kinetic}\\\textbf{energy}\end{bmatrix} + \begin{bmatrix}\textbf{positive work}\\\textbf{done by forces}\\\textbf{tending to } \textbf{\textit{increase}}\\\textbf{the velocity}\end{bmatrix} - \begin{bmatrix}\textbf{negative work}\\\textbf{done by forces}\\\textbf{tending to } \textbf{\textit{decrease}}\\\textbf{the velocity}\end{bmatrix} = \begin{bmatrix}\textbf{final}\\\textbf{kinetic}\\\textbf{energy}\end{bmatrix}$$

or, in abbreviated form,

The work-energy equation:

$$IKE + \text{pos work} - \text{neg work} = FKE \qquad (20\text{-}6)$$

Units. The quantities in the work-energy equation are those of force, mass, displacement, and velocity. Without exception, in the problems that follow, forces should be expressed in pounds (lb), displacements in feet (ft), and velocities in feet per second (ft per sec).

Signs. Kinetic energy is always a positive quantity. Positive work is done by forces acting in the direction of motion; negative work is done by forces opposing the motion.

The quantity (pos work − neg work) is called the *net work*. Consequently, *the net work done upon a body is equal to the change in its kinetic energy*, or, rewriting Eq. (20-6),

Net work:

$$\text{Pos work} - \text{neg work} = FKE - IKE \qquad (20\text{-}7)$$

20-8 The Work-Energy Method. Translation. Constant Forces

The work-energy equation provides a second method of solving motion problems, which is especially useful when the quantities involved are force, velocity, and displacement. In rectilinear translation, all particles of the body have, at any instant, the same velocity and direction of motion. Where the path of motion is slightly curved, such as that of a car traversing a hilly winding road, as in Fig. 20-8, the resulting rotation of the body is comparatively slight and the energy of rotation is therefore negligible. Such bodies are usually considered as undergoing rectilinear translation. The rotational energy of the wheels of a moving car is also comparatively negligible. An application of the work-energy method is shown below.

ILLUSTRATIVE PROBLEMS

20-38. At a mine, a train of small dump cars is pulled along a level track from A to B (Fig. 20-12), and then up a 5-percent grade. At point C, 100 ft up the grade, the last car becomes uncoupled while its velocity is 15 mph. If the car weighs 4,000 lb and the track resistance is 1 percent of its weight, how far up the grade will

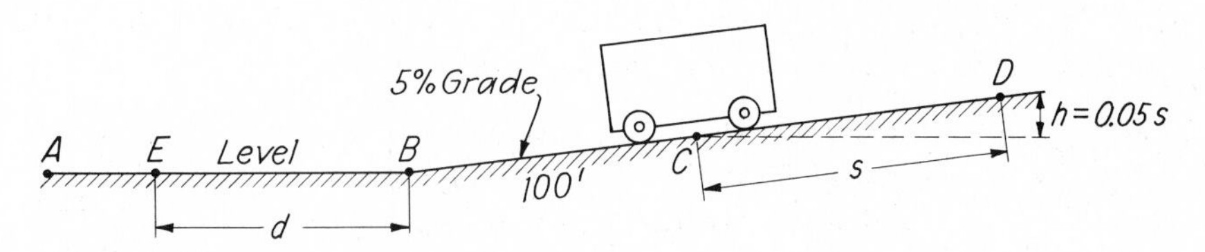

FIG. 20-12 Application of the work-energy method. Translation.

the car continue before stopping at D? How far beyond B will the car stop after coasting back down the grade to E?

Solution: As the train moves up the grade from B to C, positive work is done on the car by the pulling force in the drawbar connecting it to the car ahead. From the moment the car becomes uncoupled at C, its kinetic energy is gradually dissipated in overcoming friction and gravity as the car continues up the incline. That is, its kinetic energy IKE at C must do negative work $(F \cdot s)$ *against* friction and *against* the force of gravity $(W \cdot h)$ as long as the car continues *up* the grade. When the car stops at D, its final kinetic energy FKE is zero. Now, $KE = Wv^2/2g$, $W = 4{,}000$ lb, $v_C = 15$ mph $= (15)(88/60) = 22$ ft per sec, $F = (0.01)(4{,}000) = 40$ lb, and $h = 0.05s$. Therefore, by the work-energy equation,

$$IKE + \text{pos work} - \text{neg work} = FKE$$

$$\left[\frac{Wv_C^2}{2g} + 0 - Fs - Wh = 0\right] \qquad \frac{(4{,}000)(22)^2}{(2)(32.2)} + 0 - 40s - (4{,}000)(0.05s) = 0$$

or $\qquad 30{,}000 + 0 - 40s - 200s = 0 \qquad$ and $\qquad s = 125$ ft $\qquad$ *Ans.*

Immediately after stopping at D, the car begins to coast *down* the grade. Starting from rest, its IKE at D is zero. The force of gravity now does positive work $(U = W \cdot h)$ while the car travels back from D to B. Negative work is done against friction $[U = Fs = F(DB + d)]$ as long as the car is in motion from D to E. Its FKE at E is zero. The distance $BD = 225$ ft and its vertical projection $h = (0.05)(225) = 11.25$ ft. Then,

$$IKE + \text{pos work} - \text{neg work} = FKE$$

$$[0 + Wh - Fs = 0] \quad 0 + (4{,}000)(11.25) - (40)(225 + d) = 0$$

or $\qquad 0 + 45{,}000 - 9{,}000 - 40d = 0 \qquad$ and $\qquad d = 900$ ft $\qquad$ *Ans.*

20-39. In Fig. 20-13, body A weighs 90 lb and B weighs 100 lb. As body A drops, B is pulled up the incline against a constant frictional resistance F of 20 lb. Calculate the velocity v of this system after A has dropped 10 ft from rest. Consider weight of cord and sheave to be negligible.

Solution: The manner in which these two bodies are connected indicates that both will move with the same velocity. Connected bodies form a *system*. When we apply

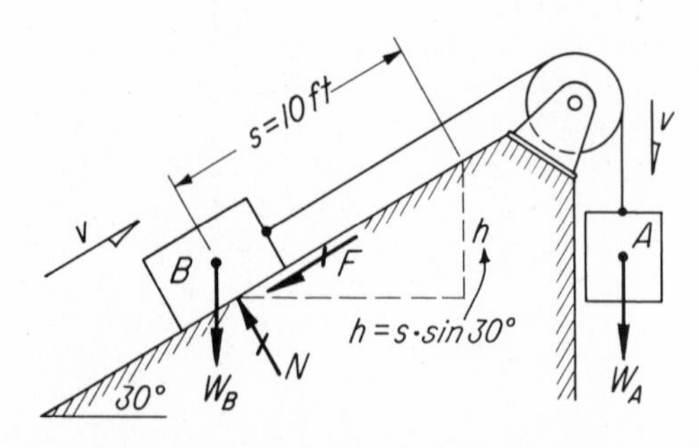

FIG. 20-13 The work-energy method.

the work-energy equation to *any* system, we must remember that *the initial and final kinetic energies (IKE and FKE) are of the entire system of bodies.* That is, *the total kinetic energy of any system of connected bodies is equal to the sum of the kinetic energies of the separate bodies,* even though the several bodies may move with different velocities and in different directions.

Because the system starts from rest, its initial kinetic energy is zero. To determine the work done which changes the kinetic energy of the system, we must consider the motion of each body and the work done by or against it. Each body moves s ft. As body A drops to the lower position, the gravitational force W_A does positive work in the amount of $W_A \cdot s$. As body B moves up the incline, negative work is done against the gravitational force W_B: $(U = -W_B \cdot s \cdot \sin 30°)$ and against the friction force F: $(U = -F \cdot s)$. The final kinetic energy is $(W_1 + W_2)v^2/2g$. Restating the work-energy equation and substituting in it, we have

$$IKE + \text{pos work} - \text{neg work} = FKE$$

or

$$0 + W_A \cdot s - W_B \cdot s \cdot \sin 30° - F \cdot s = \frac{(W_A + W_B)v^2}{2g}$$

in which $W_A = 90$ lb, $W_B = 100$ lb, $s = 10$ ft, $F = 20$ lb, and v is to be solved for. We then get

$$0 + (90)(10) - (100)(10)(0.05) - (20)(10) = \frac{(90 + 100)v^2}{(2)(32.2)}$$

and

$$v = 8.24 \text{ ft per sec} \qquad \textit{Ans.}$$

PROBLEMS

20-40. Solve Prob. 20-38 when BD is a 3-percent grade.

20-41. The car in Fig. 20-12 weighs 3,220 lb and moves against an average frictional resistance F of 30 lb. (*a*) If the distance AB is 400 ft, what initial velocity v_A must the car have at A in order to *coast* from A to C? (*b*) What distance d will it coast back beyond B on its return trip from C?

Ans. (*a*) $v_A = 25$ ft per sec; (*b*) $d = 437$ ft

20-42. The car in Fig. 20-14 weighs 3,400 lb. It coasts from rest at A down the hill to B and on up to C, which is 8 ft lower than A. With what velocity v_C will it reach C, if the actual road distance A to C is 170 ft and the average road resistance is 60 lb? (Disregard the slight rotation of the car.)

Ans. $v_C = 18$ ft per sec $= 12.2$ mph

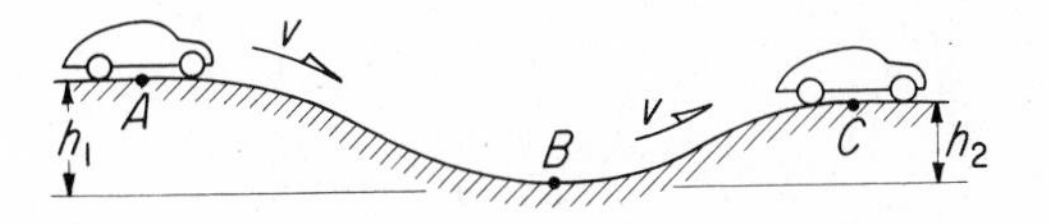

FIG. 20-14 Probs. 20-42 and 20-43.

20-43. Solve Prob. 20-42, if the car leaves A with a velocity of 15 mph and coasts from there to C.

20-44. A 100-lb block is pushed up a 30° incline by a 200-lb horizontal force. In what distance will the velocity be changed from 5 to 20 ft per sec? Assume $f = 0.4$.

Ans. $s = 12$ ft

20-45. A bobsled and its riders weigh 1,000 lb. If the wind and sliding resistance are constant at 50 lb, what will be the velocity of the sled after traveling 500 ft from rest on a 10° incline? If the hill levels out at this point, how far will the sled coast on the level?

20-46. A package chute is to deliver packages from the second floor of a building to a loading platform 15 ft below. How far out on the platform will the packages slide if the chute is 25 ft long and the coefficient of friction is 0.25?

Ans. $s = 40$ ft

20-47. A 3,200-lb car is traveling 30 mph when the brakes are applied. If the skid marks show that the car was brought to rest in 50 ft, what was the coefficient of friction between the tires and the road, if the road is level?

20-48. A bullet weighing 1 oz strikes the trunk of a tree with a velocity of 2,500 ft per sec. If the average resistance to penetration of the bullet is 7,000 lb, how deeply will in penetrate the wood? *Ans.* $d = 10.4$ in.

20-49. Solve Prob. 20-48 when the bullet weighs 0.8 oz, $v = 2{,}000$ ft per sec, and $F = 4{,}000$ lb.

20-50. A 1,200-lb rocket is fired vertically upward. The rocket motor develops a constant thrust of 2,200 lb. What was the velocity of the rocket when it was 1 mile above its starting point? Neglect air friction and the decrease in weight of the rocket because of fuel expenditure. *Ans.* $v = 363$ mph

20-51. The blocks A and B shown in Fig. 20-15 each weigh 130 lb. Calculate the velocity of block B after it has moved 26 ft from rest. Assume the coefficient of friction f for all surfaces is 0.2 and that the pulley C is weightless and frictionless.

Ans. $V_B = 15.2$ ft per sec

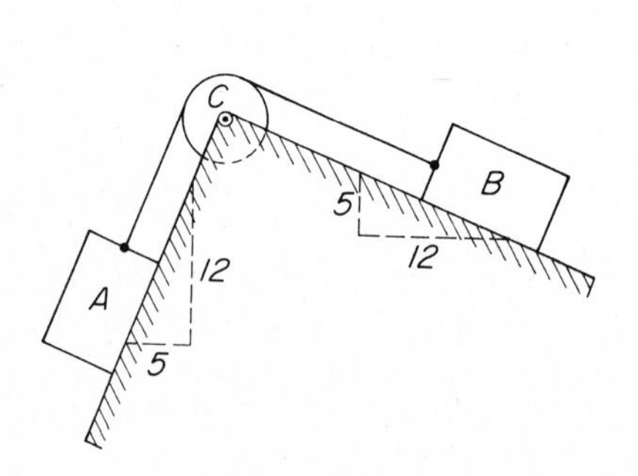

FIG. 20-15 Prob. 20-51.

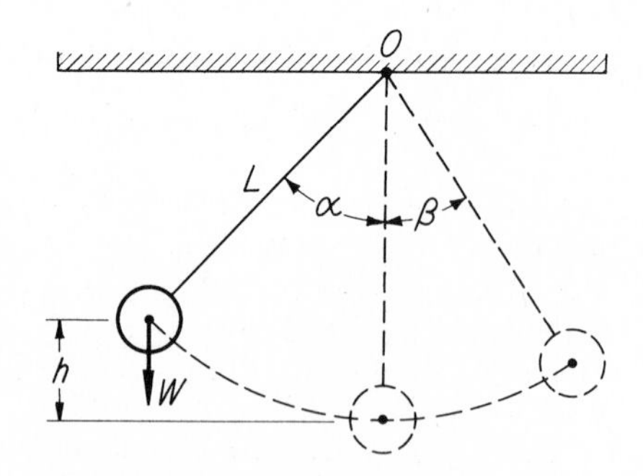

FIG. 20-16 Probs. 20-52 and 20-53.

20-52. The steel ball shown in Fig. 20-16 weighs 10 lb and is attached to a light cord of such length that the distance L from O to the center of gravity of the ball is 6 ft. Air resistance and weight of cord are negligible. If the ball is released when α is 30°, what will be its velocity (*a*) when the cord is vertical ($\alpha = 0$), and (*b*) when β is 15°? *Ans.* (*a*) $v = 7.18$ ft per sec; (*b*) $v = 6.22$ ft per sec

20-53. Solve Prob. 20-52 when (*a*) $\alpha = 45°$ and (*b*) $\beta = 30°$.

20-54. A locomotive pulls a train of 40 cars, each weighing 60 tons, with a constant velocity of 45 mph along a level track. The train resistance is 10 lb per ton. (*a*) What constant drawbar pull P must the locomotive exert? (*b*) If this pull is maintained constant, what will be the velocity of the train after it has traveled a distance of 2,000 ft up a 1-percent grade?

Ans. (*a*) $P = 24{,}000$ lb; (*b*) $v = 55.3$ ft per sec $= 37.8$ mph

20-55. Solve Prob. 20-54 when the train resistance is 8 lb per ton and the grade is 1.5 percent.

20-56. Solve Prob. 20-39 when W_A is 60 lb, W_B is 80 lb, and F is 15 lb.

20-57. How many foot-pounds of energy must be provided to catapult a 16,000-lb airplane into the air at 90 mph? If the catapult run is 60 ft in length, what average force must be provided? Neglect friction. Assume motion to be level.

20-58. Calculate the velocity v of the system shown in Fig. 20-17 after each body has moved 6 ft from rest. Body A weighs 80 lb and B weighs 60 lb. Find also the tension T in the cord. Consider weight of cord and sheave to be negligible. [HINT: After finding v, find T by cutting cord and applying Eq. (20-6) to block B alone.]

Ans. $v = 7.44$ ft per sec; $T = 68.6$ lb

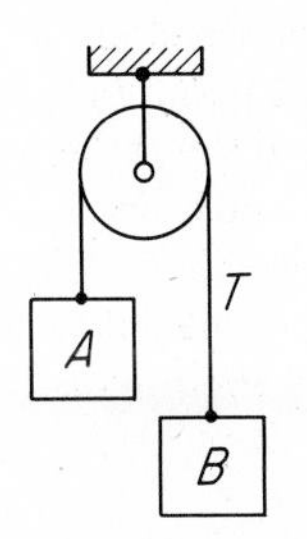

FIG. 20-17 Probs. 20-58 and 20-59.

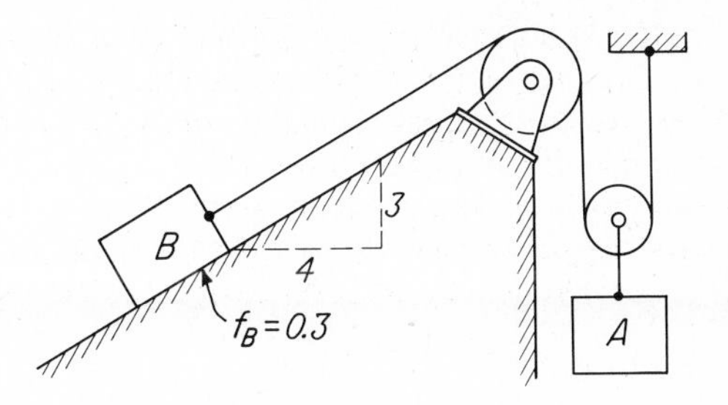

FIG. 20-18 Probs. 20-60 and 20-61.

20-59. Solve Prob. 20-58 when A weighs 80 lb, B weighs 100 lb, and the system has an initial velocity of 6 ft per sec, with B moving downward.

20-60. In Fig. 20-18, A weighs 90 lb and B weighs 50 lb. The coefficient of friction f_B is 0.3. Calculate the velocity v_A of body A after it has moved 5 ft from rest. Consider weight of cord and sheave to be negligible. (HINT: The velocity and displacement of B are double those of A.)

Ans. $v_A = 2.58$ ft per sec ↓

20-61. Solve Prob. 20-60 when A weighs 60 lb and B weighs 100 lb.

20-9 The Work-Energy Method. Translation. Variable Forces

In coil springs we have perhaps the best example of a simple, variable force; that is, a force varying uniformly from zero to a maximum, as in compressing a coil spring from its free length, or from a maximum to zero, when the spring is fully released from a compressed position. As is shown by Eq. (20-3) of Art. 20-4, the work done in deforming a coil spring from its free length, or the work done by the spring in releasing itself, is $\frac{1}{2}ks^2$, where k is the modulus of the spring and s is the deformation. The assumption is that a coil spring, when released, will give up all of the energy expended in compressing it.

Units. *A common source of trouble* in the use of this term $\frac{1}{2}ks^2$ in the work-energy equation (IKE + pos work − neg work = FKE) is that of units. The first and last terms involve kinetic energy ($Wv^2/2g$), for which the unit is the foot-pound. For this reason, the spring modulus must be given in pounds per foot and all distances in the work-energy equation must be given in feet. Strict adherence to these requirements will eliminate any difficulty concerning units.

ILLUSTRATIVE PROBLEMS

20-62. The weight W shown in Fig. 20-19 falls freely a distance h when it strikes a coil spring whose modulus k is 480 lb per ft. If W is 20 lb, what distance h must the weight fall from rest to compress the spring a distance s of $\frac{1}{2}$ ft?

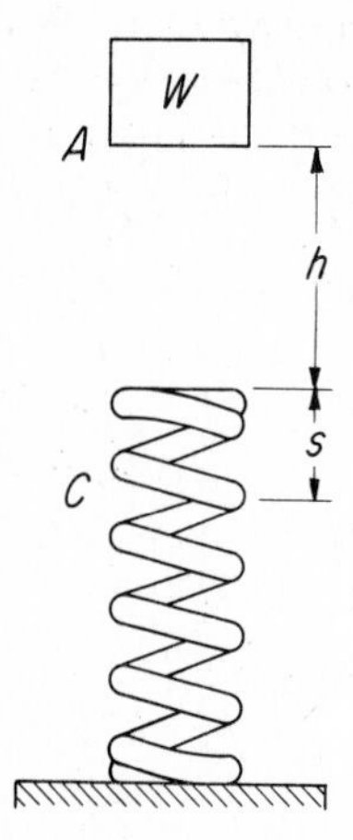

FIG. 20-19 Spring compressed by falling weight.

Solution: Since the weight starts from rest at A and again comes to rest at the point of maximum compression C, its initial and final kinetic energies are zero. We need, therefore, only determine the positive and negative work done. As the body falls from A to C, the force of gravity W does positive work equal to $W(h + s)$. The spring will exert an increasing upward-acting force as it resists being compressed; this force does negative work equal to $\frac{1}{2}ks^2$. We then have

$$IKE + \text{pos work} - \text{neg work} = FKE$$

or, since $W = 20$ lb, $s = \frac{1}{2}$ ft, and $k = 480$ lb per ft,

$$\left[0 + W(h + s) - \frac{1}{2}ks^2 = 0\right] \qquad 20\left(h + \frac{1}{2}\right) = \frac{1}{2}(480)\left(\frac{1}{2}\right)^2$$

and $$h = 2.5 \text{ ft} \qquad Ans.$$

20-63. At a mine, the end of a sidetrack is to be provided with a spring bumper. The spring must be capable of stopping a 10,000-lb ore car which has a velocity

v of 4 ft per sec down the incline at a point 120 ft up the incline from B, and then coasts from there to the bumper, as indicated in Fig. 20-20. The track resistance F remains constant at 60 lb. What modulus k must this spring have in order to stop the car after being compressed $1\frac{1}{2}$ ft? What distance d will the car roll back along the level track after rebounding from the point of maximum compression?

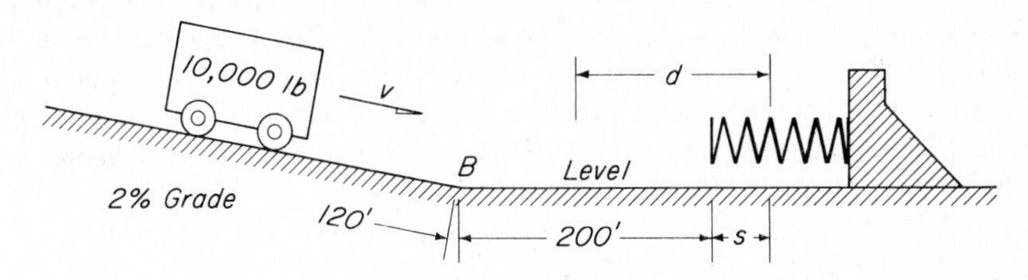

FIG. 20-20 The work-energy method. Variable forces.

Solution: The car has initial kinetic energy by virtue of its velocity of 4 ft per sec down the incline. Positive work ($U = W \cdot h$) is done by the gravity force W acting through a vertical distance h equal to 2 percent of 120 ft, or 2.4 ft. Negative work is done against the track resistance $[U = -F(120 + 200 + s)]$, and in compressing the spring ($U = -\frac{1}{2}ks^2$). At the point of maximum compression the velocity of the car, and hence its final kinetic energy, is zero. We then have

$$IKE + \text{pos work} - \text{neg work} = FKE$$

or

$$\frac{Wv_0^2}{2g} + Wh - F(120 + 200 + s) - \frac{1}{2}ks^2 = 0$$

in which $W = 10{,}000$ lb, $v_0 = 4$ ft per sec, $g = 32.2$ ft per sec², $F = 60$ lb, and $s = 1.5$ ft. Therefore,

$$\frac{(10{,}000)(4)^2}{(2)(32.2)} + (10{,}000)(2.4) - (60)(321.5) - \frac{1}{2}k(1.5)^2 = 0$$

$$2{,}490 + 24{,}000 - 19{,}290 - 1.125k = 0 \quad \text{and} \quad k = 6{,}400 \text{ lb per ft} \qquad \textit{Ans.}$$

At the point of maximum compression the velocity of the car, and hence its initial kinetic energy, is zero. The spring does positive work ($U = \frac{1}{2}ks^2$) in releasing itself, thereby giving the car a velocity to the left. Friction does negative work ($U = -Fd$) in bringing the car to a stop after it has traveled a distance d. Then, since $IKE = 0$, $k = 6{,}400$ lb per ft, $s = 1.5$ ft, and $F = 60$ lb,

$$IKE + \text{pos work} - \text{neg work} = FKE$$

$$\left[0 + \frac{1}{2}ks^2 - Fd = 0\right] \quad 0 + \left(\frac{1}{2}\right)(6{,}400)(1.5)^2 - 60d = 0 \quad \text{or} \quad d = 120 \text{ ft} \quad \textit{Ans.}$$

PROBLEMS

20-64. Solve Prob. 20-62 when the spring modulus is 240 lb per ft.

20-65. The weight W in Fig. 20-19 falls freely from rest at A a distance h, then strikes 100 lb per ft coil spring which compresses 1 ft. Calculate W if h is 4 ft.

Ans. $W = 10$ lb

20-66. If, in Prob. 20-63, a 4,000-lb per ft spring is used, what distance s will it compress? (NOTE: Disregard the negligible amount of track resistance while the spring compresses. If this is not done, a quadratic equation must be solved.)

20-67. The force required to stretch a given coil spring is 80 lb per ft. If the spring hangs vertically, calculate the maximum stretch s produced from its free length when a 30-lb weight, attached to the lower end of the spring, is suddenly released. What distance s will the 30-lb weight stretch the spring, if it is lowered *very* slowly?

Ans. $s = 0.75$ ft $= 9$ in.; $s = 0.375$ ft $= 4.5$ in.

20-68. Solve Prob. 20-67 when the spring modulus k is 120 lb per ft and W is 40 lb.

20-69. The spring gun shown in Fig. 20-22 can be aimed at any angle from the horizontal to the vertical. If the 180 lb per ft spring is compressed a distance s of $\frac{2}{3}$ ft (8 in.) while in the horizontal position, what maximum velocity v will it give a 2-lb ball when released? Disregard negligible air resistance and friction.

Ans. $v = 35.9$ ft per sec

20-70. A 200-lb weight slides 30 ft down a 30° incline before striking an 8,000 lb per ft spring. What was the coefficient of kinetic friction between the block and the incline if the spring deflection was 0.7 ft?

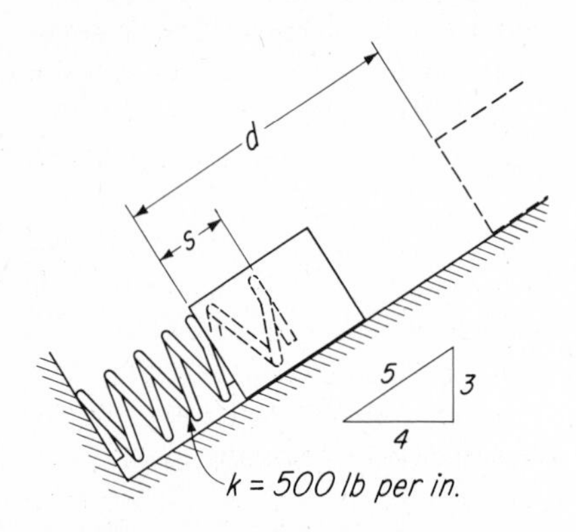

FIG. 20-21 Probs. 20-71 and 20-72.

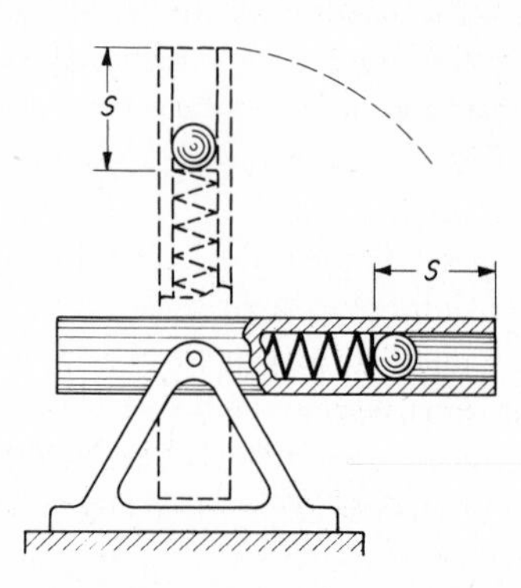

FIG. 20-22 Spring gun. Probs. 20-69, 20-74, and 20-75.

20-71. The 100-lb block is pushed against the spring shown in Fig. 20-21 until it is compressed 9 in. The block is then released. How far up the incline will the block slide if the coefficient of friction is 0.3?

Ans. $d = 20.1$ ft

20-72. How far should the spring shown in Fig. 20-21 be compressed if it is to project a 60-lb block up the incline (assume $f = 0.2$) so that the velocity v will be 8.03 ft per sec when the distance d is 2.5 ft?

20-73. The 100-lb block shown in Fig. 20-21 is pushed against the spring until its deflection s is 1.5 ft. What will be the velocity of the block as it passes the point where d is 10 ft? Assume $f = 0.5$.

Ans. $v = 60.9$ ft per sec

20-74. What distance s should the 180 lb per ft spring in the horizontal spring gun in Fig. 20-21 be compressed in order to impart a velocity v of 50 ft per sec to a 2-lb ball? Assume that all the energy of the spring is imparted to the ball.

20-75. If the spring gun in Fig. 20-22 is aimed vertically, to what height h above its compressed position will a 240 lb per ft spring project a 3-lb ball when the compression s is $\frac{1}{2}$ ft? What will be the velocity v of the ball after it has risen 6 ft?
Ans. $h = 10$ ft; $v = 16.07$ ft per sec

20-76. What distance s will the 240 lb per ft spring in Fig. 20-19 be compressed by a 10-lb weight which falls freely from rest a distance h of 2.5 ft? (NOTE: A quadratic equation must be solved.)
Ans. $h = \frac{1}{2}$ ft

20-77. In punching a hole through a steel plate, the force exerted by the punch may be assumed to vary from a maximum at the beginning of the operation to zero when the punch has just passed through the metal. The cylindrical area of material sheared is the product of the circumference of the hole and the thickness of the plate. If the initial maximum resistance of the metal is 40,000 psi, calculate the work done in punching a 1-in. diam hole in a plate $\frac{1}{2}$ in. thick.
Ans. $U = 15{,}700$ in.-lb

20-78. Solve Prob. 20-77 when the diameter is $\frac{3}{4}$ in. and the thickness is $\frac{1}{4}$ in.

20-10 Kinetic Energy of Rotation

In Art. 20-6, we found that the kinetic energy of a translating body was $Wv^2/2g$, a function of the mass (W/g) of the body and the square of its linear velocity. We shall now show that the kinetic energy of a rotating body is a function of its moment of inertia I and the square of its angular velocity.

In Fig. 20-23 is shown a body rotating about an axis through the center C with an angular velocity ω. Let this body be composed of a large number of very small particles, each having a weight of w. We may now consider the motion of one of these particles, such as w_1. At any instant, this particle has a linear velocity of v which equals $r\omega$ in rotational units.

The linear kinetic energy of this particle is $w_1v_1{}^2/2g$. Replacing v_1 with $r_1\omega_1$, we find its **rotational kinetic energy** to be $w_1r_1{}^2\omega_1{}^2/2g$. Now, the angular velocity ω is the same for *all* particles of a rotating body. The rotational kinetic energy of *all* particles (w_1, w_2, w_3, etc.) comprising the entire body then is

$$\frac{\omega^2}{2}\left(\frac{w_1r_1{}^2}{g} + \frac{w_2r_2{}^2}{g} + \frac{w_3r_3{}^2}{g} \cdots \text{etc.}\right) = \frac{\omega^2}{2}\left(\sum\frac{wr^2}{g}\right)$$

But in Art. 18-4 we found that $\sum\frac{wr^2}{g}$ is the moment of inertia I of a body with respect to its axis of rotation. Therefore,

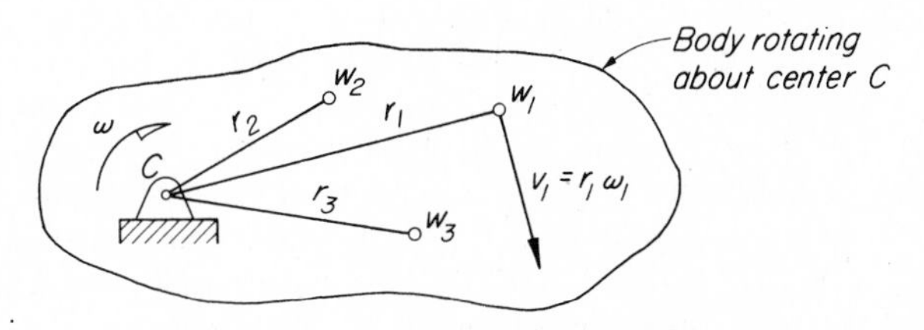

FIG. 20-23 Rotational kinetic energy.

Kinetic energy of a rotating body:

$$KE = \frac{1}{2} I\omega^2 \qquad (20\text{-}8)$$

where KE = rotational kinetic energy, ft-lb
I = moment of inertia, ft-lb-sec^2
ω = angular velocity, rad per sec (1/sec)

Units. *The unit of rotational kinetic energy is the foot-pound,* and results will be obtained in foot-pounds when the units of I and ω are as shown above. We must remember here that a radian is a quantity without unit, but we may think of it as having a unit of 1 or unity. To illustrate, then,

$$KE = \frac{1}{2} I\omega^2 = (\text{ft-lb-sec}^2)\left(\frac{1^2}{\text{sec}^2}\right) = \text{ft-lb}$$

ILLUSTRATIVE PROBLEM

20-79. Calculate the rotational kinetic energy of a steam-engine flywheel whose moment of inertia I is 530 ft-lb-sec^2, when its angular speed is 300 rpm.

Solution: We must convert the angular speed of 300 rpm to rad per sec. 1 rev = 2π rad, and 1 rpm = $2\pi/60 = \pi/30$ rad per sec. (See also Table 17-1, Art. 17-5.) Then, $\omega = (300)(\pi/30) = 31.4$ rad per sec, and

$[KE = \frac{1}{2}I\omega^2]$ $\quad KE = (\frac{1}{2})(530)(31.4)^2$ $\quad$ or $\quad KE = 261{,}000$ ft-lb $\quad$ *Ans.*

PROBLEMS

20-80. The flywheel on a steam engine weighs 3,220 lb and has a radius of gyration k of 2 ft. Compute its rotational kinetic energy when its angular velocity ω is 450 rpm. ($I = Wk^2/g$). *Ans.* KE = 444,000 ft-lb

20-81. Solve Prob. 20-80 when ω is 225 rpm. (Note that the KE varies as the square of the velocity.)

20-82. How much energy must be supplied to increase the rotational speed of a flywheel having a moment of inertia I_C of 800 ft-lb-sec^2 from 300 to 600 rpm? *Ans.* U = 1,180,000 ft-lb

20-83. If the rotational kinetic energy stored in a 6-ft diam and 1-ft-thick solid rotating disk weighing 10,000 lb is used to lift a 100-lb weight vertically, how high can it be lifted if the disk is brought to rest from a rotational speed of 300 rpm?

20-84. The motor of a small airplane is started by using the kinetic energy stored, by hand cranking, in the high-speed flywheel of a lightweight *inertia starter.* Calculate the kinetic energy possessed by a 4-lb flywheel, whose radius of gyration is 3 in., when it rotates at 18,000 rpm. (Compare this KE with that of the 3,220-lb steam-engine flywheel in Prob. 20-80.) *Ans.* KE = 13,820 ft-lb

20-85. Solve Prob. 20-84 when W = 5 lb, k = 4 in., and ω = 15,000 rpm.

20-86. A large punching machine uses the kinetic energy stored in a solid cast-iron (450 pcf) flywheel, 4 ft in diameter and ½ ft thick, as a smooth source of energy for doing the work of punching holes in steel plates. Calculate its kinetic energy when it rotates at 600 rpm. *Ans.* $KE = 346{,}000$ ft-lb

20-87. The flyball governor shown in Fig. 16-4 rotates at such a speed that $\theta = 30°$. Each ball is of cast iron (450 pcf) and is 6 in. in diameter. If $L = 18$ in., calculate the kinetic energy stored in the rotating balls (the weight of the arms is negligible). *Ans.* $KE = 13.35$ ft-lb

20-88. The arms L supporting the flyball governor shown in Fig. 16-4 each weigh 4 lb and can be considered to be slender rods rotating about an axis through the upper ends of the rods. If their length is 15 in., $\theta = 30°$, and $N = 100$ rpm, calculate the rotational KE stored in the rods.

20-11 The Work-Energy Method. Rotation and Translation. Constant Forces

Problems involving a single rotating body are usually most easily solved by the force method outlined in Art. 18-12. Others, especially those involving *systems of connected bodies whose motions are related,* and in which some rotate while others undergo translation, are readily solved by the force method as well as by the work-energy method, as illustrated in the following problems.

The work done by a couple or moment is the product of the moment and the angle, measured in radians, through which the moment turns. The work done by the force F shown in Fig. 20-24 in moving to its new position is the product of the force and the distance through which the force moves or, in symbols, the work done is $F \cdot s$. If the angle θ is measured in radians, $s = r\theta$, and therefore the work done by the force F is $Fr\theta$. But Fr is the moment of the force about the axis of rotation. Therefore the work done by the moment M is

$$U = M\theta \tag{20-9}$$

where M = moment, lb-ft
θ = angle through which moment turns, rad

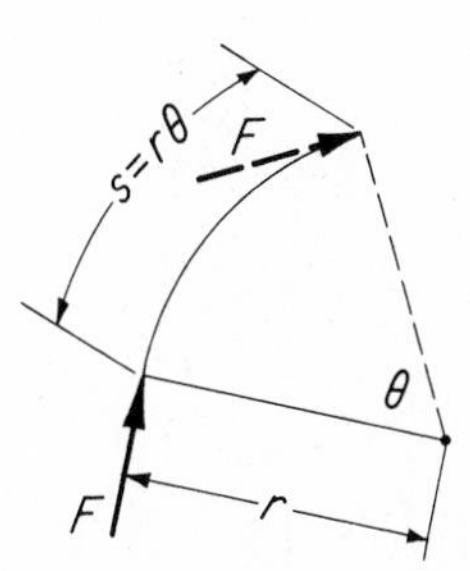

FIG. 20-24 Work done by a moment.

It is important to remember that **the kinetic energy of a system of connected bodies is the sum of the kinetic energies of the separate bodies.**

Units. The units of quantities in the work-energy equation (W, v, s, I, ω) and of other related quantities (a, α, θ) must be as follows: all forces, in pounds (lb); moments, in pound-feet (lb-ft); I, in foot-pounds-second² (ft-lb-sec²); a, in feet per second per second (ft per sec²); v, in feet per second (ft per sec); s, in feet (ft); α, in radians per second per second (rad per sec²); ω, in radians per second (rad per sec); θ, in radians (rad).

ILLUSTRATIVE PROBLEMS

20-89. Disk A in Fig. 20-25*a* is mounted in bearings whose frictional resistance is negligible. Block B weighs 128.8 lb and is attached to a (weightless) cord wrapped around the disk. Calculate the velocity v_B of block B after it has dropped a distance s of 10 ft from rest. Find also the tension T in the cord.

Solution: Since the system starts from rest ($v_B = 0$ and $\omega_A = 0$), its initial kinetic energy *IKE* is zero. The velocities of bodies A and B are obtained from their final kinetic energies *FKE* after B has dropped 10 ft. To obtain these, we must first determine the positive and negative work done on the system. Force W_B (Fig. 20-25*b*) increases the velocity of the system and hence does positive work in the amount of $W_B \cdot s$. Since no forces act to decrease the velocity, the negative work is zero. We then have

$$IKE + \text{pos work} - \text{neg work} = FKE$$

or

$$0 + W_B \cdot s - 0 = \frac{W_B v_B^2}{2g} + \frac{1}{2} I \omega_A^2$$

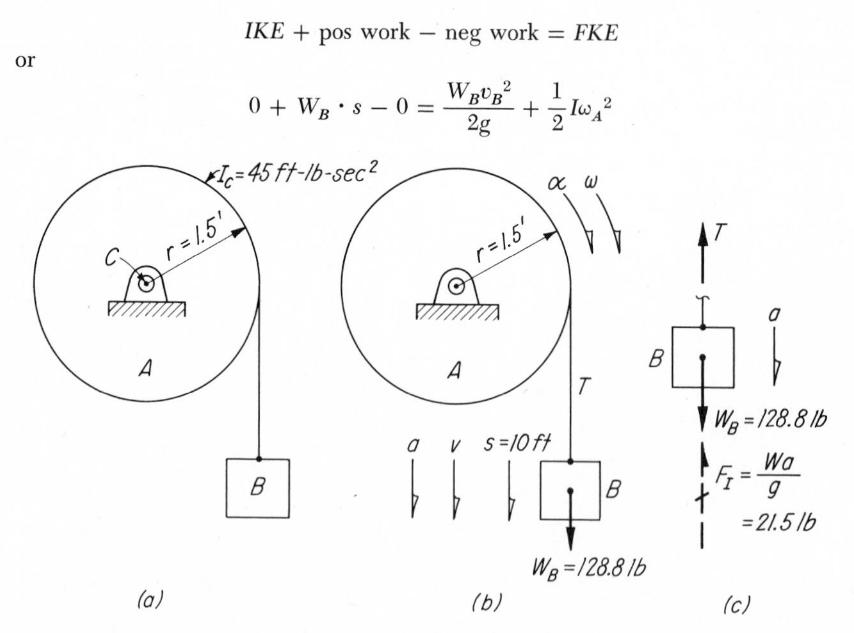

FIG. 20-25 The work-energy method. (*a*) Space diagram. (*b*) Rotation. (*c*) Translation.

Since $\omega = v/r$ and $r = 1.5$ ft, we get

$$(128.8)(10) = \frac{(128.8)v_B^2}{(2)(32.2)} + \frac{1}{2}(45)\frac{v_B^2}{(1.5)^2}$$

from which

$$1{,}288 = 2v_B^2 + 10v_B^2 \qquad \text{and} \qquad v_B = 10.37 \text{ ft per sec} \qquad \textit{Ans.}$$

Now we may obtain the tension T in the cord by noting that all of the kinetic energy possessed by disk A was given it by force T acting through a distance s of 10 ft. This energy ($\frac{1}{2}I\omega^2$) was shown above to equal $10v_B^2$. Therefore,

$$10T = 10v_B^2 = (10)(10.37)^2 = 1{,}073 \qquad \text{and} \qquad T = 107.3 \text{ lb} \qquad \textit{Ans.}$$

A simple check on T is obtained by the force method by computing the acceleration a and the inertia force F_I for block B, which is shown isolated as a free body in Fig. 20-25*c*. From $v^2 = v_O^2 + 2as$, $a = v^2/2s$ when $v_O = 0$. Also, $F_I = Wa/g$. That is,

$$a_B = \frac{(10.37)^2}{(2)(10)} = 5.37 \text{ ft per sec}^2 \qquad \text{and} \qquad F_{IB} = \frac{(128.8)(5.37)}{32.2} = 21.5 \text{ lb}$$

Then, from Fig. 20-24*c*, we see that

$$[T = W - F_I] \qquad T = 128.8 - 21.5 \qquad \text{or} \qquad T = 107.3 \text{ lb} \qquad \textit{Check.}$$

20-90. Block B in Fig. 20-26 is attached to a (weightless) cord which is wrapped around disk A, causing it to rotate about its centroidal axis. A constant bearing-friction moment FM of 5 lb-ft resists rotation. Determine the linear acceleration a of block B, the angular acceleration α of disk A, and the tension T in the cord. (The following

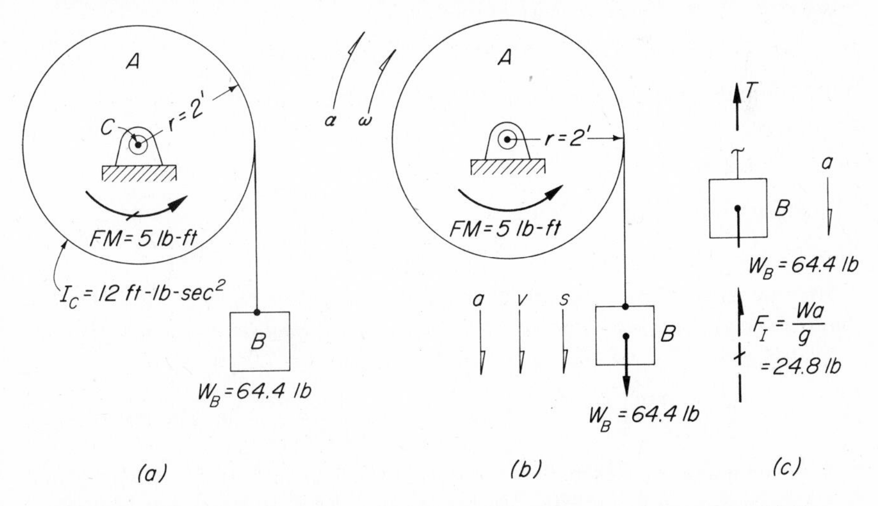

FIG. 20-26 The work-energy method. (*a*) Space diagram. (*b*) Rotation. (*c*) Translation.

is the *work-energy solution* of Prob. 18.29, the *force solution* of which is shown in the illustrative problems of Art. 18-12.)

Solution: The quantities used in the work-energy equation are force, distance, and velocity. Since, in the statement of this problem, no distance is given, we may assume one, because acceleration is independent of velocity. For convenience, let the angular distance θ_A through which disk A rotates be 1 rad. Then, the linear distance s_B traveled by the block is $s_B = r\theta = (2)(1) = 2$ ft, which is the radius of the disk. We may also assume that the system starts from rest. Hence, $IKE = 0$.

Positive work tending to increase the velocity of the system is done by W_B acting through distance s_B. The friction moment FM of 5 lb-ft does negative work equal to

$[U = M\theta]$ $$U = (5)(1) = 5 \text{ ft-lb}$$

The final kinetic energy FKE of the system is the sum of the kinetic energies of bodies A and B. Summing up these quantities, we get

$$IKE + \text{pos work} - \text{neg work} = FKE$$

or

$$0 + W_B \cdot s_B - FM \cdot \theta = \frac{W_B v^2}{2g} + \frac{1}{2} I\omega^2$$

Since $\theta = 1$ rad, $s = 2$ ft, $v = r\omega$, and $r = 2$ ft, we have

$$0 + (64.4)(2) - (5)(1) = \frac{(64.4)(2\omega)^2}{(2)(32.2)} + \frac{1}{2}(12)\omega$$

Solving for the angular velocity ω of the wheel, we get

$$128.8 - 5 = 4\omega^2 + 6\omega^2 \qquad \text{and} \qquad \omega = 3.52 \text{ rad per sec}$$

Since $r = 2$ ft, the linear velocity of the block is

$[v = r\omega]$ $$v = (2)(3.52) = 7.04 \text{ ft per sec}$$

To obtain the linear acceleration a, when $v_0 = 0$, $v = 7.04$ ft per sec, and $s_B = 2$ ft, we have

$[v^2 = v_0^2 + 2as]$ $\quad (7.04)^2 = 0 + 2a(2) \quad$ or $\quad a = 12.38$ ft per sec^2 *Ans.*

The angular acceleration α then is

$[a = r\alpha]$ $\quad 12.38 = 2\alpha \quad$ and $\quad \alpha = 6.19$ rad per sec^2 *Ans.*

We may obtain T by noting that *the positive work done by force W on block B minus the negative work done by the tension T must equal its final kinetic energy.* That is, since $W = 64.4$ lb, $s_B = 2$ ft, $v_0 = 0$, and $v = 7.04$ ft per sec,

$\left[0 + Ws - Ts = \frac{Wv^2}{2g}\right]$ $\quad 0 + (64.4)(2) - 2T = \frac{(64.4)(7.04)^2}{(2)(32.2)} \quad$ or $T = 39.6$ lb *Ans.*

To obtain a simple check on the tension T, by the force method, block B is isolated as a free body, as in Fig. 20-26c, and the inertia force F_I is computed. That is,

$\left[F_I = \frac{Wa}{g}\right]$ $$F_I = \frac{(64.4)(12.38)}{32.2} = 24.8 \text{ lb}$$

Solving for T, we get

$[T = W_B - F_I]$ $T = 64.4 - 24.8 = 39.6$ lb *Check.*

PROBLEMS

20-91. A pipefitter maintains a 40-lb pull on a 12-in. pipe wrench as he screws a fitting onto a pipe. How much work has been done when the wrench has turned through an angle of 90°? *Ans.* $U = 62.8$ ft-lb

20-92. What constant torque must be maintained on a 4-ft diam flywheel if it is to be brought to 600 rpm from rest in 200 turns? The flywheel weighs 1,610 lb, $I_C = 300$ ft-lb-sec², and friction moment = 40 lb-ft.

20-93. Solve Prob. 20-89 when $W_B = 161$ lb, $s = 15$ ft, $I = 60$ ft-lb-sec², and $r = 2$ ft.

20-94. Calculate the weight W_B of block B in Fig. 20-25*a* required to give it a velocity v_B of 15 ft per sec after it has dropped 10 ft from rest. *Ans.* $W_B = 346$ lb

20-95. Solve Prob. 20-90 when $W_B = 96.6$ lb, $FM = 6$ lb-ft, $I = 36$ ft-lb-sec², and $r = 3$ ft.

20-96. What must be the weight W_B of block B in Fig. 20-26*a* to give it a velocity of 10 ft per sec after it has dropped 5 ft from rest? *Ans.* $W_B = 47.2$ lb

20-97. The weight B in Fig. 20-26*a* weighs 64.4 lb and drops 20 ft from rest before striking the ground. How many turns from its initial at rest position will the disk A make before the 5 lb-ft friction moment brings it to rest?

20-98. A 4-ft diam solid disk 1 ft thick and weighing 3,220 lb is mounted on a 6-in. diameter short horizontal axle. If the coefficient of friction between the axle and the bearing is 0.05, how many turns will the disk make in coming to rest from a rotational speed of 300 rpm? (See Art. 18-11 for method of computing the friction moment.) *Ans.* $\theta = 390$ turns

20-99. The ore bucket A in Fig. 20-27 is attached to a rope which winds around hoisting drum B. The bucket is allowed to drop a distance of 30 ft from rest before the constant force N is applied to the brake block. Bearing friction is negligible.

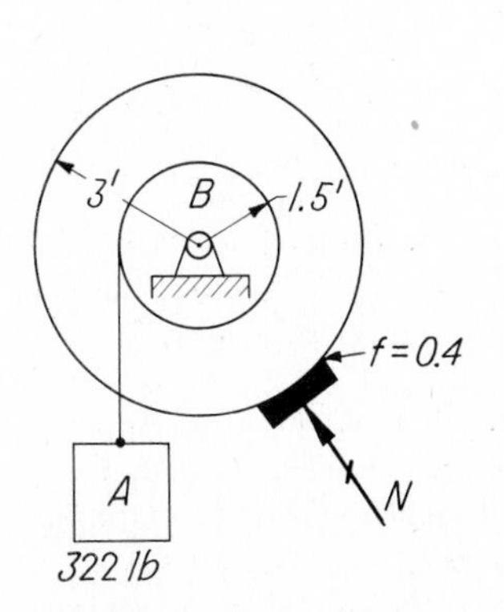

FIG. 20-27 Probs. 20-99 and 20-100.

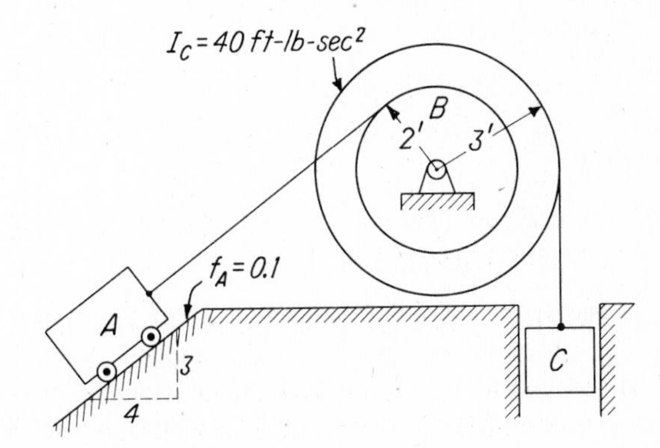

FIG. 20-28 Probs. 20-101 and 20-102.

In what additional distance s can a constant braking force N of 800 lb stop it? (HINT: Note that the initial kinetic energy of the system when N is applied is 30 W_A.)

Ans. $s = 30.3$ ft

20-100. In Fig. 20-27, the ore bucket is allowed to drop a distance of 50 ft from rest before the constant force N is applied to the brake block. Calculate the required force N if the bucket is to be stopped after dropping an additional 60 ft.

20-101. To conserve power and to reduce wear on braking equipment, the counterweight A in Fig. 20-28 is attached to the hoisting-drum assembly B which operates mine cage C. (*a*) If all frictional resistances are neglected, what should be the weight W_A of the counterweight A in order to balance the 2,000-lb weight of the empty cage C when at rest? (*b*) If A weighs 5,000 lb and the coefficient of rolling friction f is 0.1, what should be the total weight W_C of the loaded cage in order to gain a velocity of 15 ft per sec after it had dropped 30 ft from rest? Bearing and cage frictions are negligible.

Ans. (*a*) $W_A = 5{,}000$ lb; (*b*) $W_C = 2{,}880$ lb

20-102. Solve Prob. 20-101 when cage C gains a velocity of 20 ft per sec after dropping 50 ft from rest.

20-103. The block B shown in Fig. 20-29 weighs 200 lb and is released from rest from the position shown. A flexible rope is attached to the stiff but light member AB at C. Calculate the angular velocity of disk D when member AB is vertical. The centroidal moment of inertia of disk D is 20 ft-lb-sec^2 and $r = 2$ ft. Assume the pulley is very small, neglect all friction, and neglect the rotational KE of block B.

Ans. $\omega = 3.24$ rad per sec

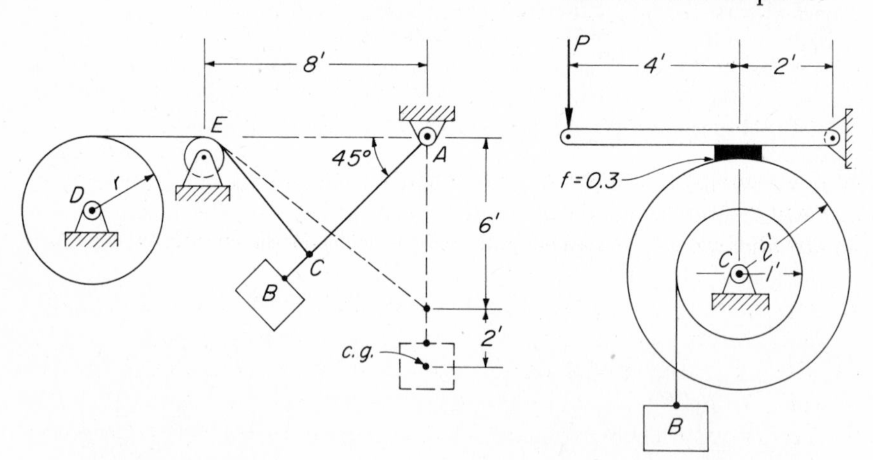

FIG. 20-29 Prob. 20-103.

FIG. 20-30 Prob. 20-104.

20-104. The block B shown in Fig. 20-30 weighs 322 lb and is descending with $V_B = 20$ ft per sec when the brake is applied. What constant force P must be applied if the block B is brought to rest after traveling an additional 24 ft? The centroidal moment of inertia of the rotating assembly is 40 ft-lb-sec^2.

20-105. The engine of a small airplane is started by using the kinetic energy stored, by hand cranking, in the high-speed flywheel of a lightweight inertia starter, views of which are shown in Fig. 20-31. Assume that an average force F of 30 lb is applied

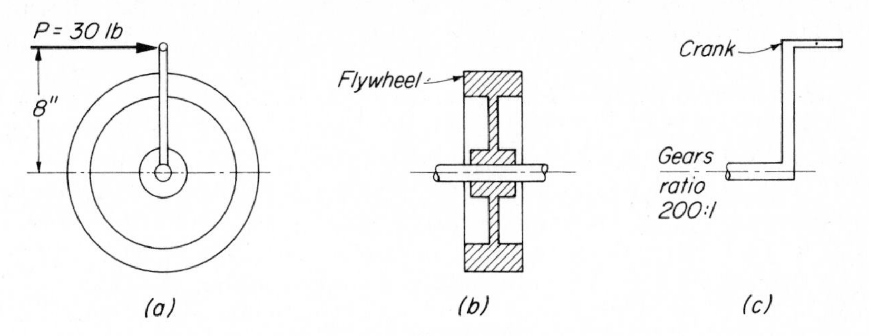

FIG. 20-31 Prob. 20-105. Inertia starter.

perpendicularly to the crank and that frictional resistance is negligible. The flywheel weighs 4 lb and has a radius of gyration of 3 in. (= $\frac{1}{4}$ ft). (*a*) Calculate the average angular acceleration α of the flywheel. (*b*) How many seconds are required to give the flywheel a velocity of 18,000 rpm from rest? (*c*) How many revolutions will the energy in the flywheel turn the engine from rest against a resisting torque of 150 lb-ft? (NOTE: Because of the gear ratio, the acceleration of the flywheel is 200 times that of the crank.) *Ans.* (*a*) α = 12.9 rad per sec^2; (*b*) t = 146 sec; (*c*) 14.7 rev

20-12 The Work-Energy Method. Plane Motion. Constant Forces

In this article, plane motion is restricted to that of a rolling body, either alone or connected by a flexible cord to a translating body. *In the application of the work-energy method to plane motion, we must remember that*

1. A free-rolling body may possess energy of translation ($W\bar{v}^2/2g$) as well as energy of rotation ($\frac{1}{2}I_c\omega^2$).
2. The kinetic energy of a system (*IKE* or *FKE*) is the sum total of the kinetic energies of all bodies in the system.
3. Only *external* forces do work (the weight of a body W is an *external* force; inertia forces and inertia moments are *internal* effects).
4. Where no slipping of a rolling body occurs, the friction force F does no work.
5. Forces must be expressed in pounds (lb), distances in feet (ft), and moments of inertia, in foot-pound-second^2 (ft-lb-sec^2).

ILLUSTRATIVE PROBLEMS

20-106. The solid cylinder shown in Fig. 20-32*a* weighs 1,288 lb, its radius r is 2 ft, and its moment of inertia I_C is 80 ft-lb-sec^2. It rolls without slipping down the incline. Assume rolling friction to be negligible. Calculate the acceleration $\bar{a}$ of its center of gravity, (NOTE: This problem is the same as the first part of Prob. 19.1, solved here by the work-energy method.)

Solution: The acceleration of the cylinder is independent of any initial velocity

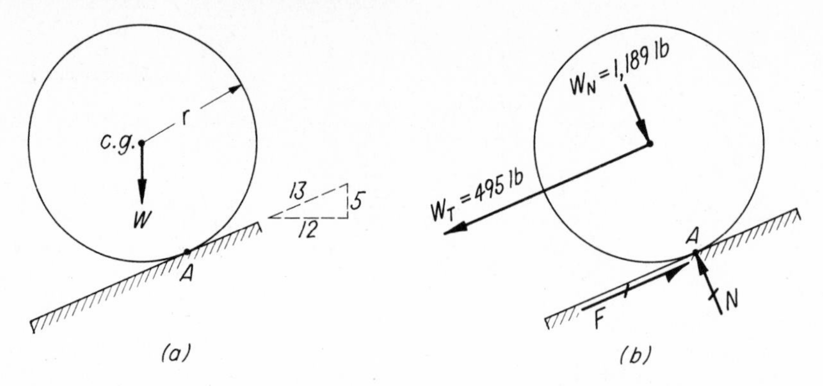

FIG. 20-32 The work-energy method. (*a*) Space diagram. (*b*) Plane motion.

the cylinder may have. Hence, we will assume it ($\bar{v}_0$) to be zero. From the diagram in Fig. 20-32*b*, we see that the component of the weight W_T acting parallel to the incline does positive work. Since no slipping occurs, the friction force F does no work. The negative work therefore is zero. The final kinetic energy of the cylinder is that of translation ($W\bar{v}^2/2g$) plus that of rotation ($\frac{1}{2}I_C\omega^2$). Computing the components of W, we get

$$W_T = \frac{5}{13}(1{,}288) = 495 \text{ lb} \qquad \text{and} \qquad W_N = \frac{12}{13}(1{,}288) = 1{,}189 \text{ lb}$$

When we state these quantities in symbols in the work-energy equation, we obtain

$$[IKE + \text{pos work} - \text{neg work} = FKE] \qquad 0 + W_T \cdot s - 0 = \frac{W\bar{v}^2}{2g} + \frac{1}{2}I_C\omega^2 \qquad (a)$$

The acceleration is also independent of any displacement s. Let us therefore assume it to be equal to the radius, which is 2 ft. We may now express $\bar{v}^2$ and ω^2 in terms of $\bar{a}$. If the body starts from rest, $\bar{v}_0 = 0$, and

$$[v^2 = v_0{}^2 + 2as] \qquad \bar{v}^2 = 0 + (2\bar{a})(2) = 4\bar{a} \qquad (b)$$

$$\left[\omega = \frac{v}{r}\right] \qquad \omega^2 = \frac{v^2}{r^2} = \frac{4\bar{a}}{4} = \bar{a} \qquad (c)$$

When these values are substituted in Eq. (*a*), we get

$$(495)(2) = \frac{(1{,}288)(4\bar{a})}{(2)(32.2)} + \frac{1}{2}(80\bar{a})$$

or $\quad 990 = 80\bar{a} + 40\bar{a} = 120\bar{a} \quad$ and $\quad \bar{a} = 8.25$ ft per sec^2 *Ans.*

20-107. Calculate the angular velocity ω of disk D in Fig. 20-33 and the linear velocity v_B of block B after it has dropped a distance of 10 ft from rest. Compute also the acceleration a_B of block B, and the tension T in the cord. Consider that the cord and the small pulley are weightless, and that the disk does not slip. $W_B = 161$ lb. The weight of disk D is 966 lb and its centroidal moment of inertia I_C is 100 ft-lb-sec^2.

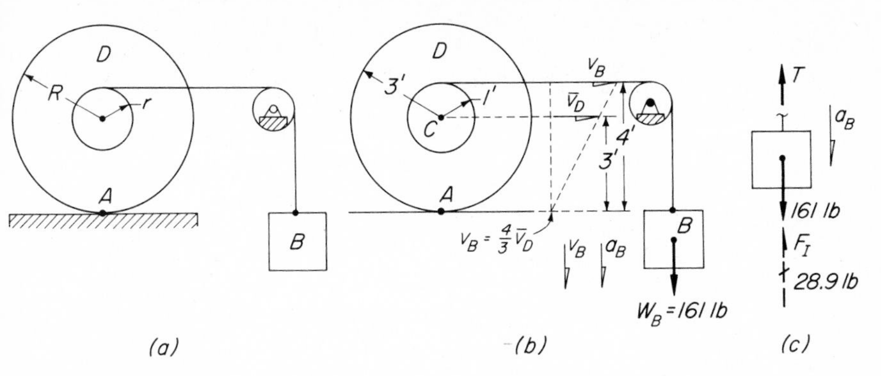

FIG. 20-33 The work-energy method. (*a*) Space diagram. (*b*) Plane motion. (*c*) Translation.

Solution: Since the system starts from rest, its initial kinetic energy is zero. Positive work equal to $W_B \cdot s$ is done by force W_B acting through the distance $s = 10$ ft. Since no slipping occurs, no negative work is done against friction. The final kinetic energy of the system equals the sum of the energy of translation $(W_D\bar{v}_D{}^2/2g)$ and of rotation $(\frac{1}{2}I_C\omega^2)$ of disk, D, plus the energy of translation of block B $(W_Bv_B{}^2/2g)$.

We shall find it simplest here to express $\bar{v}_D$ and v_B in terms of ω. When ω is found, the quantities $\bar{v}_D$, v_B, a_B, and T are easily determined. The work-energy equation is

$$IKE + \text{pos work} - \text{neg work} = FKE$$

Substituting in this equation the appropriate symbols for this problem, we get

$$0 + W_B \cdot s_B - 0 = \frac{W_D\bar{v}_D{}^2}{2g} + \frac{1}{2}I_C\omega^2 + \frac{W_Bv_B{}^2}{2g} \qquad (d)$$

But $\bar{v}_D = r\omega = 3\omega$. From Fig. 20-33*b*, we see that $v_B = \frac{4}{3}\bar{v}_D$, or $v_B = \frac{4}{3}(3\omega) = 4\omega$. Also, $W_B = 161$ lb, $s_B = 10$ ft, $W_D = 966$ lb, and $I_C = 100$ ft-lb-sec². Substituting these values in Eq. (*d*), we have

$$(161)(10) = \frac{(966)(3\omega)^2}{(2)(32.2)} + \frac{1}{2}(100\omega^2) + \frac{(161)(4\omega)^2}{(2)(32.2)}$$

or $\quad 1{,}610 = 135\omega^2 + 50\omega^2 + 40\omega^2 \qquad$ and $\qquad \omega = 2.68$ rad per sec $\qquad$ *Ans.*

Now we may obtain v_B and a_B. Since $v_B = 4\omega$,

$$v_B = 4\omega = 4(2.68) \qquad \text{or} \qquad v_B = 10.72 \text{ ft per sec} \qquad \textit{Ans.}$$

$$[v^2 = v_0{}^2 + 2as] \qquad (10.72)^2 = 0 + (2a_B)(10) \quad \text{and} \quad a_B = 5.75 \text{ ft per sec}^2 \qquad \textit{Ans.}$$

The tension T in the cable may be obtained by noting the fact that the final kinetic energy of block B is the result of positive work ($U = W_B \cdot s$) done by force W_B less the negative work ($U = -T \cdot s$) done by T. Since $IKE = 0$, $W_B = 161$ lb, $s = 10$ ft, and $v_B = 10.72$ ft per sec, we have

$$[0 + W_B \cdot s - Ts = FKE] \qquad 0 + (161)(10) - 10T = \frac{(161)(10.72)^2}{(2)(32.2)}$$

or $\quad 1{,}610 - 10T = 288 \quad$ and $\quad T = 132.2$ lb $\quad$ *Ans.*

A simple check on T by the force method may be obtained by isolating block B as a free body and computing F_I and T. That is, from Fig. 20-33c,

$$\left[F_I = \frac{Wa}{g}\right] \qquad F_{IB} = \frac{(161)(5.75)}{32.2} = 28.8 \text{ lb}$$

$$[T = W_B - F_{IB}] \qquad T = 161 - 28.8 \quad \text{and} \quad T = 132.2 \text{ lb} \qquad \textit{Check.}$$

PROBLEMS

20-108. Solve Prob. 20-106 when $W = 2{,}576$ lb and $r = 3$ ft.

20-109. Solve Prob. 20-107 when $W_D = 1{,}288$ lb, $r = 1$ ft, $R = 2$ ft, $I_C = 70$ ft-lb-sec^2, $W_B = 161$ lb, and $s_B = 10$ ft.

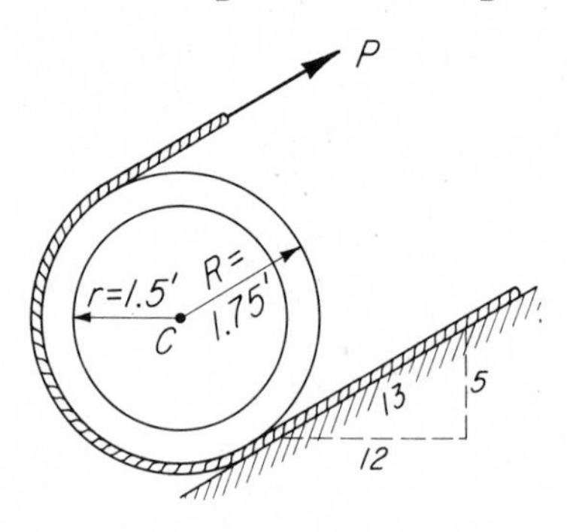

FIG. 20-34 Prob. 20-110.

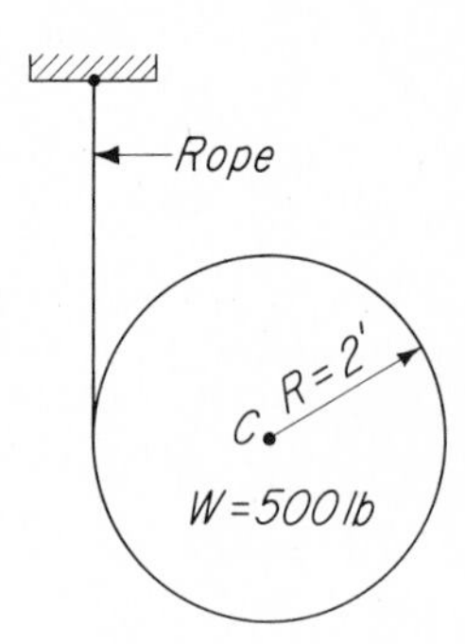

FIG. 20-35 Probs. 20-113 and 21-36.

20-110. What constant pull P must be applied to the rope to roll the 644-lb pipe in Fig. 20-34 up the incline 26 ft in 4 sec from rest? Ans. $P = 184.7$ lb

20-111. A 1-ft diam steel (490 lb per ft^3) sphere is released from rest on an incline making an angle of 20° with the horizon. What will be its velocity after it has rolled 50 ft down the incline? *Ans.* $v = 28.1$ ft per sec

20-112. Solve Prob. 20-111 if the sphere has an initial velocity of 10 ft per sec.

20-113. The rope shown in Fig. 20-35 is wrapped around the 500-lb solid cylinder. One end of the rope is securely attached to an overhead support. The cylinder is released from rest. What will be its angular velocity after it has dropped 10 ft? *Ans.* $v = 20.7$ ft per sec

20-114. The rotating body shown in Fig. 20-36 weighs 644 lb and rolls freely down a grooved incline. Rolling resistance may be neglected. Its centroidal moment of inertia is 80 ft-lb-sec^2. Also, $r = 2$ ft and $R = 3$ ft. Compute its angular velocity ω, the linear velocity $\bar{v}$, and the acceleration $\bar{a}$ of its center of gravity, after it has rolled 10 ft down the incline from rest.

Ans. $\omega = 6.15$ rad per sec; $\bar{v} = 12.3$ ft per sec; $\bar{a} = 7.58$ ft per sec^2

20-115. Solve Prob. 20-114 when the body has an initial velocity $\bar{v}_0$ of 4 ft per sec.

20-116. Solve Prob. 20-114 when $W = 322$ lb, $r = 1$ ft, $R = 2$ ft, $I_C = 30$ ft-lb-sec^2.

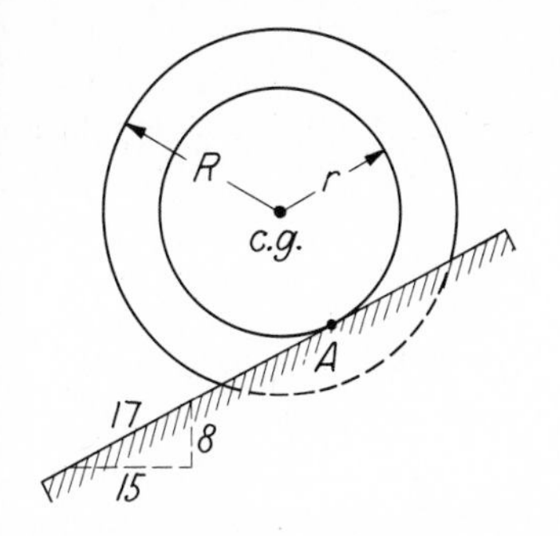

FIG. 20-36 Probs. 20-114 to 20-116.

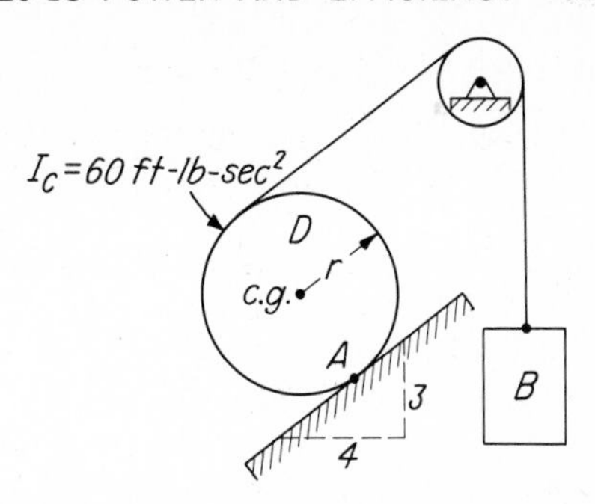

FIG. 20-37 Probs. 20-117 and 20-118.

20-117. Determine the direction of motion of block B in Fig. 20-37. Then solve for the angular velocity ω and the linear velocity $\bar{v}_D$ of body D, and the velocity v_B and acceleration a_B of block B, after B has dropped 10 ft. Find also the tension T in the cord. $W_B = 128.8$ lb, $W_D = 322$ lb, and $r = 2$ ft.

Ans. $\omega = 1.98$ rad per sec;
$\bar{v}_D = 3.96$ ft per sec; $\bar{v}_B = 7.92$ ft per sec; $\bar{a}_D = 3.13$ ft per sec²; $T = 116.3$ lb

20-118. Calculate the weight W_B of the block B shown in Fig. 20-37 so that the angular velocity ω of disk D will be 4 rad per sec after the disk has moved 15 ft *down* the incline from rest. The disk weighs 500 lb, its centroidal moment of inertia is 70 ft-lb-sec², and $r = 3$ ft.

20-13 Power and Efficiency

Power is the time rate of doing work, that is, the amount of work done in a unit of time. Thus,

$$\text{Power} = \frac{\text{work}}{\text{time}} = \frac{U}{t} \tag{20-10}$$

Units of Work and Power. When mechanical work is measured in foot-pounds (ft-lb) and time in seconds (sec), the unit of power is the foot-pound per second (ft-lb per sec). A larger and more convenient *unit of mechanical power is the* **horsepower** (hp). One horsepower does work at the rate of 550 ft-lb per sec (or 33,000 ft-lb per min). The corresponding *unit of electrical power is the* **kilowatt** (kw), which does work at the rate of 737 ft-lb per sec. *Approximately,* then, 1 hp = ¾ kw, and 1 kw = ⁴⁄₃ hp.

Larger and more convenient *units of work* (and energy) are the horse-power-hour (hp-hr) and the kilowatt-hour (kw-hr); that is, the expenditure of energy at the rate of one horsepower or one kilowatt for one hour. Consumption of electrical energy is measured by meters in kilowatt-hours, and costs, on the average, from a fraction of a cent to a few cents per kilowatt-hour.

When power is multiplied by time, we again obtain *work done,* or *energy.* Thus, a 10-hp motor, working at its rated capacity for 4 hr, has done work equal to

$$(10 \text{ hp})\left(550 \frac{\text{ft-lb/sec}}{\text{hp}}\right)(4 \text{ hr})\left(3{,}600 \frac{\text{sec}}{\text{hr}}\right) = 79{,}200{,}000 \text{ ft-lb}$$

Conversion Factors for Energy and Power. The conversion factors given in Table 20-1 are frequently needed in the solution of problems involving power.

Efficiency. In all machines by which work is done or in which energy is transformed from one form to another, such as steam engines, internal-combustion engines, water turbines, and electrical generators, losses occur due to friction and other factors. That is, the energy put out by such a machine is less than the energy put into it by the amount of the loss. *The ratio of output to input* of a machine is called its *efficiency* and is denoted by e. That is,

Efficiency:

$$e = \frac{\textbf{output}}{\textbf{input}} \qquad \textbf{(20-11)}$$

The *over-all efficiency* of a number of machines working in series is the product of their separate efficiencies. For example, if in a hydroelectric power plant the efficiency of the water turbine is 93 percent and of the generator is 95 percent, the over-all efficiency of the plant is $(0.93)(0.95) = 0.884$ or 88.4 percent.

Some confusion in the application of the efficiency factor may arise if efficiency is thought of merely as a mathematical ratio. A decided help is to recognize whether, because of losses, the desired result should be more or

Table 20-1 ENERGY AND POWER CONVERSION FACTORS

Given↓	Multiply	by	to obtain
Energy	Foot-pounds	$5.05 \cdot 10^{-7}$ $3.77 \cdot 10^{-7}$	Horsepower-hours Kilowatt-hours
	Horsepower-hours	$1.98 \cdot 10^{6}$ 0.746	Foot-pounds Kilowatt-hours
	Kilowatt-hours	$2.66 \cdot 10^{6}$ 1.34	Foot-pounds Horsepower-hours
Power	Foot-pounds per second	$1.82 \cdot 10^{-3}$ $1.36 \cdot 10^{-3}$	Horsepower Kilowatts
	Horsepower	550 0.746 or ¾	Foot-pounds per second Kilowatts
	Kilowatts	738 1.34 or ⁴⁄₃	Foot-pounds per second Horsepower

less than the given quantity. Having so recognized, we can easily decide whether to divide (giving more) or to multiply (giving less) by the efficiency factor e, which of course is always less than unity.

20-14 Power Equations

When the rate at which work is done is constant,

$$\text{Power} = \frac{\text{work}}{\text{time}} = \frac{Fs}{t} \tag{a}$$

But s/t is velocity v. Hence, we may write Eq. (*a*) thus: $P = Fv$. To obtain horsepower, where v is the instantaneous velocity, we merely divide by 550; thus,

Horsepower:

$$\mathbf{hp} = \frac{Fv}{550} \tag{20-12}$$

where F = force doing work, lb
v = velocity with which the point of application of F moves, ft per sec

Water Power. The kinetic energy ($Wv^2/2g$) gained by water as it passes through a pipeline to a lower level may be converted into mechanical energy by running it through a water turbine. When friction losses in the pipe are disregarded, the kinetic energy ($Wv^2/2g$) delivered to the turbine equals the potential energy (Wh) lost. *The vertical drop h is called the* **head.** That is,

Energy of water:

$$\frac{Wv^2}{2g} = Wh \tag{b}$$

Pipeline and penstock frictional and other losses have the effect of reducing the actual head h to an **effective head** h_e. The difference is called the **friction head** h_f.

The flow of water is generally expressed as a quantity Q in cubic feet per second (cu ft per sec), and the weight of water is w, in pounds per cubic foot (lb per cu ft). The product of these gives flow, in pounds per second (lb per sec). When the flow is multiplied by the *head h* in feet, we have power in foot-pounds per second (ft-lb per sec). Dividing this total by 550 gives horsepower. That is, for 100 percent efficiency,

(Water) horsepower:

$$\mathbf{hp} = \frac{Qwh}{550} \tag{20-13}$$

where Q = quantity of water, cu ft per sec
w = weight of water, pcf
h = head, ft (vertical drop of water)

ILLUSTRATIVE PROBLEMS

20-119. A locomotive exerts a constant drawbar pull P of 10,000 lb while pulling a train along the level from A to B. The velocity increases from 30 mph at A to 45 mph at B. What horsepower does the locomotive deliver to the drawbar at A and at B?

Solution: Equation (20-12) gives the instantaneous horsepower delivered by a force F whose point of application is being displaced at the rate of v ft per sec. When $v_A = (30)(88/60) = 44$ ft per sec, and $v_B = (45)(88/60) = 66$ ft per sec,

$$\left[\text{hp} = \frac{Fv}{550}\right] \quad \text{At } A\text{:hp} = \frac{(10{,}000)(44)}{550} \quad \text{or} \quad 800 \text{ hp} \quad \textit{Ans.}$$

$$\left[\text{hp} = \frac{Fv}{500}\right] \quad \text{At } B\text{:hp} = \frac{(10{,}000)(66)}{550} \quad \text{or} \quad 1{,}200 \text{ hp} \quad \textit{Ans.}$$

20-120. On a construction job, the vertical cable of a materials hoist operates with a maximum tension T of 2,200 lb at a maximum velocity v of 4 ft per sec. If the hoist is 70 percent efficient, what horsepower must the motor supply? If the motor is 90 percent efficient, how many kilowatts (kw) will it draw from the power line? If electrical energy cost ¾ cent per kilowatt-hour (kw-hr), what is the cost of operating the hoist per 8-hr day?

Solution: The power delivered by the cable is Tv, which is the *output* of the hoist. Because of losses in the hoist ($e = 70$ percent $= 0.7$), *the input to the hoist exceeds its output.* From Eq. (20-11), input = output/e. We see also that the output of the motor equals the input to the hoist. The required motor horsepower then is

$$\left[\text{hp} = \frac{Fv}{550 \cdot e}\right] \quad \text{Motor hp} = \frac{(2200)(4)}{(550)(0.7)} \quad \text{or} \quad 22.9 \text{ hp} \quad \textit{Ans.}$$

Similarly, to obtain the input to the motor (kw drawn from power line), we divide its output by its efficiency (0.9). To obtain this input in kilowatts, we multiply horsepower by 0.746 (from Table 20-1). The kilowatts drawn from the power line then is

$$\left[\text{kw} = \frac{\text{hp} \cdot 0.746}{e}\right] \quad \text{kw} = \frac{(22.9)(0.746)}{0.9} \quad \text{or} \quad 18.97 \text{ kw} \quad \textit{Ans.}$$

At ¾ cent per kilowatt-hour, the cost per 8-hr day is

$$\text{Cost} = (18.97)(8)\frac{3}{4} = \$1.14 \text{ per day} \quad \textit{Ans.}$$

20-121. A small stream flows with an average velocity of 6 ft per sec at a point where its cross-sectional area is 110 sq ft. A water turbine can be located at a point downstream which is 100 ft vertically lower. Losses in the pipeline are 10 percent, thus reducing the effective head h_e to 90 ft. What hydroelectric power (in hp and kw) can be obtained from this stream if the following efficiencies can be expected: turbine 92 percent, generator 95 percent? Water weighs 62.5 pcf.

Solution: Equation (20-13) gives the horsepower output for 100 percent efficiency. This output multiplied by the over-all efficiency will give the net power output. The over-all efficiency of this plant is $(0.92)(0.95) = 0.874$, or 87.4 percent. The quantity of water is the product of the cross-sectional area and the stream velocity. Therefore,

$[Q = Av]$ $\qquad Q = (110)(6) = 660$ cu ft per sec

and the *net horsepower output* is

$$\left[\text{hp} = \frac{Qwh}{550} \cdot e\right] \qquad \frac{(660)(62.5)(90)(0.874)}{550} \quad \text{or} \quad 5{,}900 \text{ hp} \qquad \textit{Ans.}$$

Since 1 hp = 0.746 kw, the electrical power output is

$$(5{,}900)(0.746) \quad \text{or} \quad 4{,}400 \text{ kw} \qquad \textit{Ans.}$$

PROBLEMS

20-122. Solve Prob. 20-119 when $P = 15{,}000$ lb, $v_A = 15$ mph, and $v_B = 60$ mph.

20-123. Solve Prob. 20-120 when $T = 3{,}300$ lb, $v = 3$ ft per sec, and hoist and motor are, respectively, 60 and 90 percent efficient.

Ans. 30 hp, 24.9 kw; $1.49 per day

20-124. Solve Prob. 20-121 when $A = 140$ sq ft, $v = 5$ ft per sec, and $h_e = 60$ ft.

20-125. A jet engine can develop a 5,500-lb thrust at 480 mph. What horsepower is being delivered? *Ans.* hp = 7,050

20-126. A 4-cu-yd bucket is used in placing concrete during construction of a dam. The empty bucket weighs 3,800 lb and the concrete weighs 150 pcf. What is the maximum speed at which a 100-hp motor can lift (*a*) the empty bucket, (*b*) the loaded bucket? *Ans.* $v = 14.45$ ft per sec; $v = 2.75$ ft per sec

20-127. A locomotive delivers 4,000 hp to the drawbar in pulling a train along a level track with a velocity of 30 mph. Compute the drawbar pull P.

Ans. $P = 50{,}000$ lb

20-128. A locomotive can deliver a maximum of 6,000 hp to the drawbar. If train resistance is 10 lb per ton, what maximum number N of 60-ton freight cars can it pull up a 2-percent grade at a speed of 30 mph? How many cars can it pull on the level?

20-129. An electric locomotive uses its motor as a generator to assist in braking the train on a long downhill grade. How many kilowatts can be generated if the train weighs 4,000 tons, the train resistance is 15 lb per ton, and the train travels 30 mph down a 2-percent grade? Assume the generator has an efficiency of 70 percent.

Ans. $P = 4{,}170$ kw

20-130. A pile-driver hammer weighing 1,100 lb is raised vertically a distance of 10 ft in 5 sec at approximately constant velocity. Assume the initial acceleration and the guide friction to be negligible. If the hoisting equipment is 80 percent efficient, what horsepower must the engine deliver to the hoist? *Ans.* 5 hp

20-131. An automobile weighing 3,600 lb is driven up a 5-percent grade at a constant velocity of 45 mph. If the resistance to motion is 260 lb, what horsepower must the motor deliver to the rear wheels? (Consider W_T to equal 5 percent of W.)

20-132. What horsepower must be developed by a locomotive if it is to pull 50 cars weighing 60 tons each up a $1\frac{1}{2}$-percent grade at 15 mph? The train resistance may be assumed at 10 lb per ton.

20-133. A continuous-belt bucket elevator raises grain a vertical distance of 70 ft before discharging it at the top of a bin. The efficiency of the elevator is 75 percent and that of the connected motor is 90 percent. If electrical energy costs 2 cents per kilowatt-hour, what is the cost per ton? *Ans.* 0.156 cent

20-134. At a small pumping plant water is pumped from a constant-level reservoir to a tank 100 ft above, at the rate of $\frac{1}{2}$ cu ft per sec. The motor is 90 percent efficient, the pump 80 percent. The friction head h_f is 15 ft (caused by the pipe friction), giving a total head of 115 ft to be overcome. (*a*) What horsepower must the motor deliver to the pump? (*b*) If the capacity of the tank is 1,200 cu ft, how long will it take to fill? (*c*) If electrical energy costs $1\frac{1}{2}$ cents per kilowatt-hour, what will be the cost of filling the tank? *Ans.* (*a*) 8.17 hp; (*b*) 40 min; (*c*) 6.77 cents

20-135. A 6-in. diam jet of water enters a water turbine with a velocity v of 80.25 ft per sec under an effective head h_e of 100 ft ($Q = A_v$). What power output (in kw) can be expected from an electric generator connected to the turbine, if the turbine is 80 percent efficient, the generator 93 percent?

20-136. Assume that the power plant in Prob. 20-121 operates at an average capacity of 30 percent during each 24 hr. [Average output = (4,400)(0.3) = 1,320 kw.] Losses in transmission lines from plant to consumers average 10 percent. If 60 percent of the average output is sold at $\frac{1}{2}$ cent per kilowatt-hour for industrial use, and 40 percent at 2 cents per kilowatt-hour for residential use, what is the daily income of the plant? *Ans.* $313.60

SUMMARY

(By article number)

20-2. Work is done when forces displace material bodies. It is positive when the force acts in the direction of the motion. *Only external forces do work.* Internal (inertia) forces can do no work.

20-3. The work done by a constant force is the product of the force F and the displacement s of its point of application in the direction of the force.

$$\textbf{Work} = \textbf{(force)(displacement)} \qquad \text{or} \qquad U = F \cdot s \tag{20-1}$$

20-4. The work done by a variable force is the product of the *average force* and the displacement of its point of application in the direction of the force.

20-5. Energy is the capacity to do work. By virtue of its position in space, **potential energy** *is possessed by a body with respect to any lower plane,* since it is capable of doing work while moving downward. The work so done ($W \cdot h$) is the product of its weight W and the vertical distance h through which it moves.

Kinetic energy *is possessed by a body by virtue of its motion.* It is equal, at any instant, to the amount of work that must be done to bring the body to rest.

20-6. The kinetic energy of translation of a body at any instant is a function of its mass (W/g) and the square of its velocity. That is,

Kinetic energy of translation:
$$KE = \frac{Wv^2}{2g} \tag{20-5}$$

in which v is the velocity of the center of gravity of the body.

20-7. The work-energy equation is

$$\begin{bmatrix}\text{Initial}\\ \text{kinetic}\\ \text{energy}\end{bmatrix} + \begin{bmatrix}\text{positive work done}\\ \text{by forces tending to}\\ \textit{increase}\text{ the velocity}\end{bmatrix} - \begin{bmatrix}\text{negative work done}\\ \text{by forces tending to}\\ \textit{decrease}\text{ the velocity}\end{bmatrix} \begin{bmatrix}\text{final}\\ \text{kinetic}\\ \text{energy}\end{bmatrix}$$

or

$$IKE + \text{pos work} - \text{neg work} = FKE \tag{20-6}$$

This equation holds for all conditions of motion—translational, rotational, or plane motion. It is especially useful in the solution of problems in which the quantities involved are force, displacement, and velocity. With the aid of simple kinematic relationships, it may also be used to solve problems involving force and acceleration.

20-10. The kinetic energy of rotation of a body at any instant is a function of its centroidal moment of inertia I and the square of its angular velocity. That is,

Kinetic energy of rotation:
$$KE = \frac{1}{2}I\omega^2 \tag{20-8}$$

20-12. The work-energy method, based on the work-energy equation, is especially useful in the solution of problems involving *force, displacement, and velocity,* linear as well as angular.

20-13. Power *is the time rate of doing work,* or the rate at which energy is expended. **The efficiency** *of a machine is the ratio of its output to its input.* Since all machines are less than 100 percent efficient, output is always less than input.

Efficiency:
$$e = \frac{\text{output}}{\text{input}} \tag{20-11}$$

20-14. When the point of application of a force of F lb is displaced at the rate of v ft per sec, the power involved is Fv, in foot-pounds per second. One horsepower (hp) does 550 ft-lb of work per second. Therefore,

Horsepower:
$$\text{hp} = \frac{Fv}{550} \tag{20-12}$$

When *power* is derived from falling water and losses are disregarded,

(Water) horsepower:
$$\text{hp} = \frac{Qwh}{550} \tag{20-13}$$

REVIEW PROBLEMS

20-137. A truck pulls a 5-ton trailer up a 4-percent grade a distance of 1,000 ft, then along the level for 600 ft more, and finally up a 2 percent grade a distance of 400 ft. If the average road resistance (rolling and wind resistance) is 20 lb per ton, *what total amount of work is done* by the truck in pulling the trailer? (NOTE: For grades of 10 percent or less, the slope distance may be taken as equal to the horizontal distance.) *Ans.* $U = 680{,}000$ ft-lb

20-138. A warehouse truck is pushed up a ramp by a force P acting parallel to the ramp. The ramp is 20 ft long and has a slope of 10 percent. If the loaded truck weighs 300 lb and the total frictional resistance F is 20 lb, *how much work is done* in pushing the truck the full length up the ramp? What force P is required?

Ans. $U = 1{,}000$ ft-lb; $P = 50$ lb

20-139. A *valve spring* on a motor has a modulus of 160 lb per in. and is compressed ½ in. when inserted in place. It operates between a compression of ½ and ¾ in. *How much work is done* each time it is compressed between these limits?

20-140. A solid brick column, 2 by 2 ft square, is to be built to a height of 16 ft above the ground level. The material weighs 120 pcf. If all material is to be raised from the ground level, *how much work* must be done? *Ans.* $U = 61{,}440$ ft-lb

20-141. A belt passes over a 3-ft diam pulley which is turning at 180 rpm. If the belt tension on one side of the pulley is 300 lb and on the other is 100 lb, what horsepower is being transmitted?

20-142. An airplane weighing 22,000 lb is flying horizontally at 240 mph. What is the drag force if the engine is developing 1,600 hp? *Ans.* $F = 2{,}500$ lb

20-143. What total power must be developed by the engine of the plane of Prob. 20-142 if the rate of climb is 400 ft per mile and the speed of 240 mph is maintained?

Ans. 2,665 hp

20-144. The velocity of the water in a 4-ft diam pipeline is 8 ft per sec. The pipeline is 1 mile long. How many foot-pounds of energy must be dissipated if a valve at the end of the line is closed quickly? *Ans.* $U = 4{,}125{,}000$ ft-lb

20-145. Ore is hoisted out of a vertical mine shaft 100 ft deep by means of a bucket attached to a cable weighing 3 lb per ft. (*a*) If the bucket weighs 80 lb and carries 320 lb of ore each trip, *how much work is done* per trip? (*b*) What percentage of this work is useful in raising the ore? *Ans.* (*a*) $U = 50{,}000$ ft-lb; (*b*) 64 percent

20-146. What horsepower motor is needed to drive a pump that is to lift 1,000,000 gal of water each day from a lake whose water surface is at elev 4,800 ft to a reservoir whose water surface is at elev 5,200 ft? Assume the pump efficiency is 70 percent. (1 gal of water weighs 8.33 lb.)

20-147. A 4-ft diam cylinder starts up a 20° incline with an angular velocity of 120 rpm. How far will it go before starting back?

20-148. A 1-in. diam hole is to be punched through a ½-in.-thick plate. The initial maximum resistance of the metal is 40,000 psi. The energy for driving the punch is received from a solid cast-iron flywheel 2 ft in diameter and 4 in. thick. What will be the final speed of rotation if the initial speed is 360 rpm? (Cast iron weighs 450 pcf.)

20-149. A locomotive weighing 128.8 tons travels on a level track at 15 mph. Then the brakes are applied with constant pressure, stopping it in a distance of 100 ft. If all of the *kinetic energy* is assumed to be dissipated at the rails, what total friction force F must be developed between the wheels and the rails?

Ans. $F = 9.68$ tons

20-150. The coefficient of static friction between the rails and the wheels of a railway train is 0.25. If all the *kinetic energy* of the train is assumed to be dissipated at the rails, in what minimum distance s can a train traveling on a level track be stopped if, at the instant the brakes are applied, its speed is (*a*) 15 mph? (*b*) 30 mph? (*c*) 60 mph? *Ans.* (*a*) $s_{15} = 30$ ft; (*b*) $s_{30} = 120$ ft; (*c*) $s_{60} = 480$ ft

20-151. A railroad car is brought to a stop by the frictional resistance between

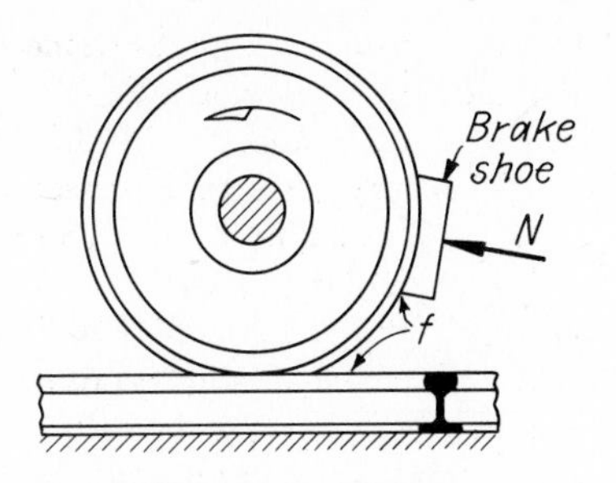

FIG. 20-38 Prob. 20-151.

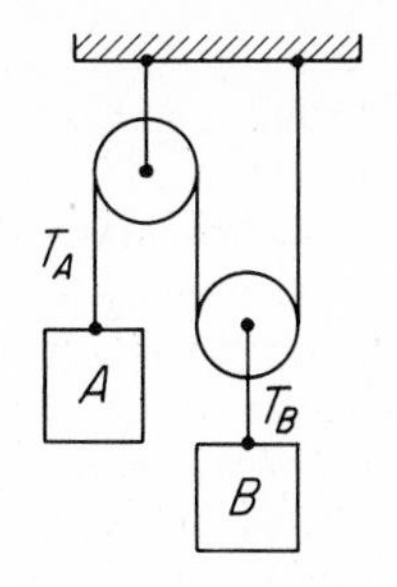

FIG. 20-39 Probs. 20-153 and 20-154.

the brake shoes and each of its eight wheels. If the total kinetic energy to be dissipated at each brake shoe is 380,000 ft-lb and if f is 0.25, what constant force N, in Fig. 20-38, will stop the car in 500 ft? Assume that the wheels do not skid on the rails. (NOTE: A point on the rim of the wheel travels the same distance relative to the brake shoe as the entire car travels relative to the rails.) *Ans.* $N = 3{,}040$ lb

20-152. In what major respect will Prob. 20-39 change when A weighs 28.8 lb, B weighs 100 lb, and F is 15 lb? Using *the work-energy method,* solve for v after the system has moved 10 ft from rest.

20-153. Body A in Fig. 20-39 weighs 60 lb and B weighs 100 lb. By *the work-energy method* compute the velocity v_A of body A after it has moved 5 ft from rest. Determine also tensions T_A and T_B. Consider cords and sheaves to be weightless and frictionless. (HINT: Using v_A as found, solve for a_A, F_I of A, and T_A. Then $T_B = 2T_A$.)

Ans. $v_A = 6.15$ ft per sec; $T_A = 52.94$ lb; $T_B = 105.88$ lb

20-154. If body A in Fig. 20-39 weighs 40 lb, what must be the weight W_B of B in order to give it a downward velocity v_B of 4 ft per sec after it has moved 6 ft from rest? Use *the work-energy method.* Cords and sheaves are assumed to be weightless.

20-155. In Fig. 20-10, let A weigh 20 lb, let B weigh 30 lb, and let the frictional resistance F between A and the incline be 4 lb. By *the work-energy method,* calculate the velocity v_B of B when it has dropped 10 ft after being released from rest. (HINT: The *FKE* is that possessed by both A and B.) *Ans.* $v_B = 11.35$ ft per sec

20-156. What distance s should the 240 lb per ft *coil spring* in Fig. 20-22 be compressed in order to project a 3-lb ball 20 ft up into the air when the gun is aimed vertically? Use *the work-energy method.*

20-157. The disk D shown in Fig. 20-40 has a centroidal moment of inertia of 20 ft-lb-sec^2. Block B weighs 32.2 lb and the weight of rod AB is negligible. If the disk has an angular velocity ω of 15 rad per sec and the spring is unstressed when in the position shown, calculate the angular velocity after the disk has rotated 90°. The velocity of B is the same as A when the rod is in the position shown.

20-158. The 20-lb block in Fig. 20-41 slides from rest down the incline and strikes a *coil spring.* The frictional resistance F is 4 lb. (a) Calculate the required modulus k of the spring, if the amount of compression s is limited to ½ ft. (b) What distance d up the incline will it rebound from the point of maximum compression? Use *the work-energy method.* *Ans.* (a) $k = 384$ lb per ft; (b) $s = 3.43$ ft

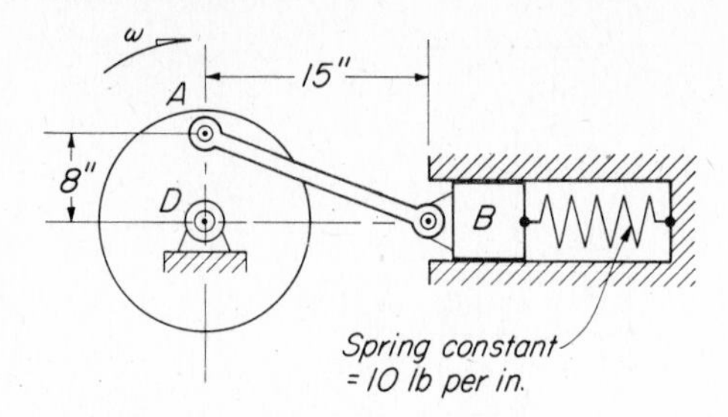

FIG. 20-40 Prob. 20-157.

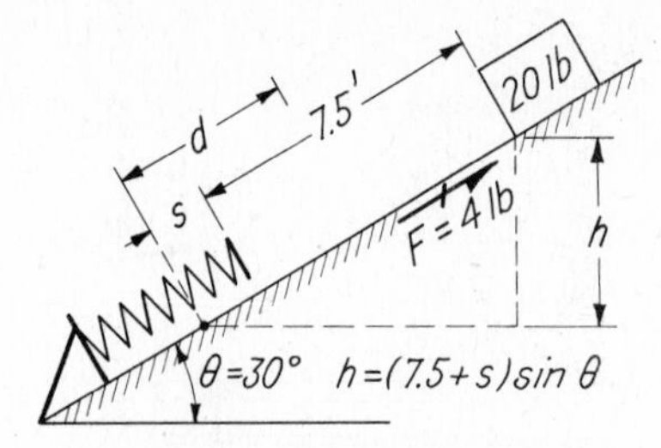

FIG. 20-41 Probs. 20-158 and 20-159.

20-159. Using *the work-energy method,* calculate the distance s the *spring* in Fig. 20-41 will be compressed, if its modulus k is 120 lb per ft. (A quadratic equation must be solved.)

20-160. Calculate the *rotational kinetic energy* of the governor assembly in Prob. 18-67 when it rotates at 300 rpm. *Ans.* KE = 167 ft-lb

20-161. When a 100,000-lb railroad car is traveling at 30 mph, its total kinetic energy is 3,049,600 ft-lb. The moment of inertia of each of its eight wheels (and rotating shafts) is 10 ft-lb-sec^2, and at 30 mph they turn at 300 rpm. Compute the total *rotational kinetic energy* of the eight wheels. Is it relatively negligible compared with the kinetic energy of the entire car?

Ans. KE = 39,400 ft-lb; yes, 1.29 percent

20-162. Using *the work-energy method,* calculate the required weight W_C of block C in Fig. 20-42 to give the system a velocity of 10 ft per sec from rest while C drops 20 ft. Find also the acceleration of the system, and the tensions T_A and T_C. Bearing friction is negligible. (Note the relatively negligible effect of the inertia of pulley B.) *Ans.* W_C = 197 lb; a = 2.5 ft per sec^2; T_A = 178.8 lb; T_C = 181.7 lb

20-163. Solve Prob. 20-162 when I_B = 40 ft-lb-sec^2. (Note the considerable effect on tension T_C of the increased inertia of the pulley.)

20-164. In Fig. 20-43 the moment of inertia of the rotating parts C is 30 ft-lb-sec^2. Block B weighs 322 lb. Bearing friction is negligible. Using *the work-energy method,* calculate the required weight W_A of block A to give it a velocity of 15 ft per sec after it has dropped 12 ft from rest. *Ans.* W_A = 405.5 lb

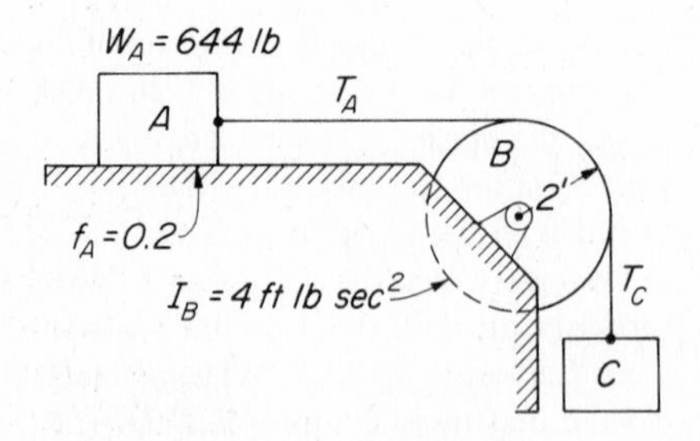

FIG. 20-42 Probs. 20-162 and 20-163.

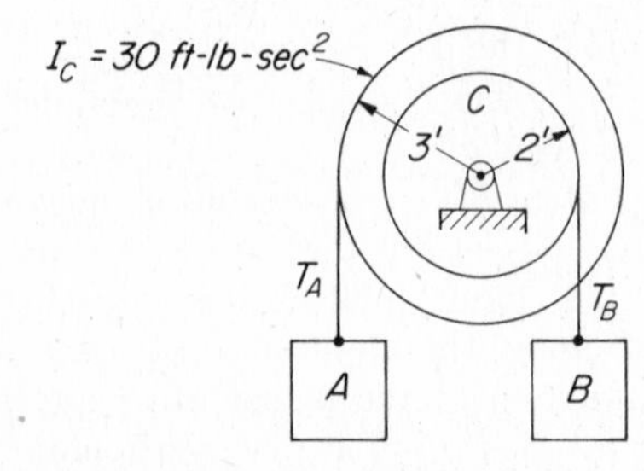

FIG. 20-43 Probs. 20-164 and 20-165.

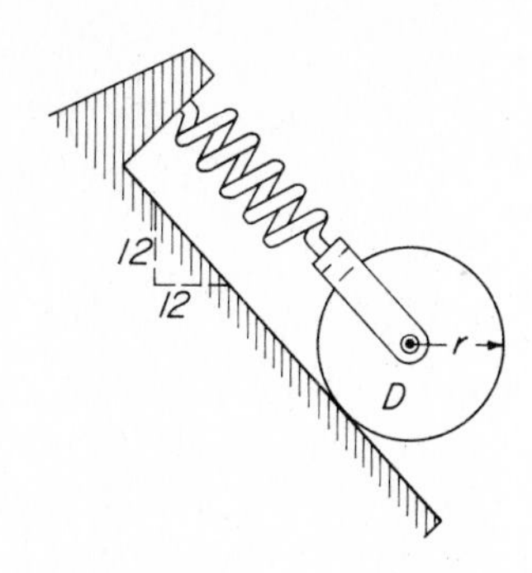

FIG. 20-44 Prob. 20-166.

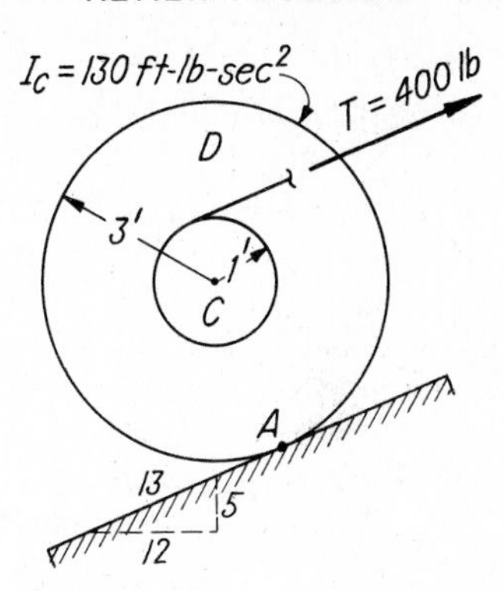

FIG. 20-45 Prob. 20-167.

20-165. By *the work-energy method,* calculate the velocities v_A and v_B of bodies A and B shown in Fig. 20-43 after B has moved 10 ft from rest. A weighs 161 lb, and B weighs 322 lb. Find also the acceleration of each of bodies A and B, and the tensions T_A and T_B. Bearing friction is negligible.

20-166. The disk D shown in Fig. 20-44 weighs 32.2 lb, has a radius $r = 0.5$ ft, and a centroidal moment of inertia $I_C = \frac{1}{8}$ ft-lb-sec^2. The spring constant is 5 lb per in. How far will disk roll if released from rest with no initial tension in the spring?

20-167. In Fig. 20-45 use *the work-energy method* to find ω and $\bar{v}$ after body D has rolled 6 ft from rest. Find also the acceleration $\bar{a}$ of its center of gravity. What minimum coefficient of friction f will prevent slipping? $W_D = 966$ lb.

Ans. $\omega = 2.2$ rad per sec; $\bar{v} = 6.6$ ft per sec; $\bar{a} = 3.63$ ft per sec^2; $f_{\min} = 0.0905$

20-168. Solve Prob. 19-2 by *the work-energy method.*

20-169. Solve Prob. 19-3 by *the work-energy method.*

20-170. A loaded mine cage weighing 3,220 lb is given a constant acceleration which brings it from rest to a velocity of 10 ft per sec in 5 sec. (*a*) Compute the tension T in the vertical cable. (*b*) If the efficiency e of the hoisting equipment is 0.75, what *horsepower* must the motor deliver? (*c*) If the motor is 90 percent efficient, how many *kilowatts* will it draw from the power line? Acceleration is upward.

Ans. (*a*) $T = 3{,}420$ lb; (*b*) 83 hp; (*c*) 69 kw

20-171. At a power plant, a motor-driven bucket elevator raises 1 ton of coal per minute a height h of 60 ft. The efficiency of the elevator is 70 percent and of the motor is 90 percent. (*a*) What *horsepower* must the motor deliver to the elevator? (*b*) If electrical energy costs $\frac{1}{2}$ cent per kilowatt-hour, what is the cost per 8-hr day of continuous operation of the elevator?

20-172. A locomotive pulls a 4,000-ton train up a 1-percent grade a distance of 3,000 ft from A to B. The train resistance is 10 lb per ton. If the train velocity decreases from 45 mph at A to 30 mph at B, what constant drawbar pull P does the locomotive exert, and what *horsepower* does it deliver at A and at B?

Ans. $P = 20{,}000$ lb; at A, 2,400 hp; at B, 1,600 hp

20-173. A *water turbine* is driven by a jet of water issuing from a 4-in. diameter nozzle under pressure of an effective head of 90 ft. If the over-all efficiency of the turbine and the connected generator is 80 percent, how many *kilowatts of power* can this plant produce?

Ans. 40.5 kw

REVIEW QUESTIONS

20-1. Define work, positive work, and negative work.

20-2. By what forces can work be done? Can work be done by an inertia force?

20-3. What is meant by spring modulus?

20-4. Give the equation expressing the force F required to compress a coil spring a distance s, if its modulus is k.

20-5. What equation expresses the work done in compressing a coil spring a distance s from its free length, if its modulus is k?

20-6. Define potential energy and kinetic energy; give the equation expressing the kinetic energy of translation.

20-7. Give the full word statement of the work-energy equation, also its abbreviated form.

20-8. What is meant by net work, and what is its relation to the change in the kinetic energy of a body?

20-9. Give the equation expressing the kinetic energy of a rotating body.

20-10. What equation expresses the work done by a moment turning through an angle θ?

20-11. The angle θ in the above expression must be in what units?

20-12. Complete the following statement: In the solution of problems, the work-energy method is especially useful when the quantities involved are ____________.

20-13. What are the most commonly used engineering basic units of work, potential energy, kinetic energy of translation, and kinetic energy of rotation?

20-14. Define power, and give its basic engineering unit. What are the commonly used larger units of mechanical and electrical power? How many basic units does each contain?

20-15. What is the appropriate numerical relationship between a horsepower and a kilowatt?

20-16. In connection with power, what is meant by efficiency? Is the over-all efficiency of a power plant the sum or the product of the separate efficiencies?

20-17. Give the equation for horsepower (*a*) in terms of F and v, and (*b*) in connection with hydroelectric power, in terms of Q, w, and h.

CHAPTER 21

Impulse, Momentum, and Impact

21-1 Introduction

In the study of the kinetics of rigid bodies, we may variously be concerned with (1) acceleration, (2) velocity and displacement, or (3) velocity and time.

In Chap. 15, problems involving force, mass, and *acceleration* were solved by *the force method.* In Chap. 20, similar problems involving force, mass, and *velocity and displacement* were solved by *the work-energy method.* Other similar problems involving force, mass, and *velocity and time* are readily solved by *the impulse-momentum method* to be developed in this chapter.

21-2 Impulse

A force acting on a body during an interval of time is said to give the body an **impulse,** the magnitude of which is the product of the force F and the time t during which it acts. When the force is constant,

$$\textbf{Impulse} = \textbf{(force)(time)} = \boldsymbol{F \cdot t} \qquad \textbf{(21-1)}$$

The unit of impulse is the pound-second, when the force is expressed in pounds (lb) and time in seconds (sec).

Only external forces give impulses to a body. When R is the *resultant* of a number of external forces, their combined or net impulse is $R \cdot t$. The *direction* of an impulse is the same as that of the force involved. When the motion of the body is restricted to a given path, the only impulses considered are those parallel to that path. Because it contains force, a vector quantity, impulse is also a vector quantity, having the direction and sign (+ or −) of the force producing it.

21-3 Momentum

Momentum is *quantity of motion.* It is the property of a body that determines the length of time required to bring it to rest. The momentum of a body is the product of its mass W/g and its instantaneous velocity v. That is,

$$\textbf{Momentum} = \textbf{(mass)(velocity)} = \frac{W}{g} \cdot v \qquad \textbf{(21-2)}$$

Since it contains velocity, a vector quantity, momentum is also a vector quantity.

The unit of momentum is the pound-second, when mass W/g is expressed in the customary engineering units, and velocity is in feet per second. That is, from Eq. (21-2),

$$\frac{\text{lb}}{\frac{\text{ft}}{\text{sec}^2}} \cdot \frac{\text{ft}}{\text{sec}} = \frac{(\text{lb})(\text{sec}^2)}{\text{ft}} \cdot \frac{\text{ft}}{\text{sec}} = \text{lb-sec}$$

21-4 The Linear Impulse-Momentum Equation

In Art. 20-7, we found that the *net work* done upon a body (pos work − neg work) is equal to the *change in its kinetic energy* (*FKE* − *IKE*). When R is the (constant) resultant of the *external* forces and s is the displacement, the net work is $R \cdot s$. Also, when a body starts from rest ($v_0 = 0$ and $IKE = 0$), the change in its kinetic energy is $Wv^2/2g$. For a body starting from rest, then,

$$[\text{Net work done on body}]\ R \cdot s = \frac{Wv^2}{2g}\ [\text{Change in its kinetic energy}] \qquad (a)$$

But $s = v_{\text{avg}} \cdot t$, and $v_{\text{avg}} = \frac{1}{2}v$ when a is constant and the initial velocity v_0 is zero. Hence, $s = \frac{1}{2}v \cdot t$, and

$$R \cdot \frac{1}{2} v \cdot t = \frac{1}{2} \cdot \frac{W}{g} \cdot v^2 \qquad (b)$$

When we divide both sides of Eq. (b) by $\frac{1}{2}v$, we get, for a body starting from rest,

$$[\text{Net impulse on body}]\ R \cdot t = \frac{W}{g} \cdot v\ [\text{Change in its momentum}] \qquad (c)$$

The *net impulse* on a body is (pos imp − neg imp) and, when the body has initial velocity, *the change in momentum* is (final mom − initial mom).

The complete similarity of the *work-energy relation,* expressed in Eq. (a), and the *impulse-momentum relation,* expressed in Eq. (c), is clearly apparent. Consequently, net impulse = positive impulse − negative impulse, and change in momentum = final momentum − initial momentum. From this reasoning we obtain the following word statement:

$$\begin{bmatrix}\text{Initial}\\ \text{linear}\\ \text{momen-}\\ \text{tum}\end{bmatrix} + \begin{bmatrix}\text{positive impulses}\\ \text{by forces tend-}\\ \text{ing to } \textit{increase}\\ \text{the velocity}\end{bmatrix} - \begin{bmatrix}\text{negative impulses}\\ \text{by forces tend-}\\ \text{ing to } \textit{decrease}\\ \text{the velocity}\end{bmatrix} = \begin{bmatrix}\text{final}\\ \text{linear}\\ \text{momen-}\\ \text{tum}\end{bmatrix}$$

or, putting this statement into the form of an equation, we get

The impulse-momentum equation:

$$\textbf{Initial mom + pos imp − neg imp = final mom} \qquad \text{(21-3)}$$

In symbols, these quantities are

$$\frac{Wv_0}{g} + F_{pos}t - F_{neg}t = \frac{Wv}{g} \qquad \text{(21-3S)}$$

21-5 The Impulse-Momentum Method

Problems are solved by this method in a manner similar to that of the work-energy method, the application of which was illustrated in Art. 20-8. The impulse-momentum method applies especially to problems involving *time and velocity*, as illustrated in the following example.

ILLUSTRATIVE PROBLEM

21-1. If the block shown in Fig. 21-1*a* weighs 26 lb and its initial velocity down the incline is 8 ft per sec, what will be its velocity 5 sec later?

Solution: First we determine the components of W, parallel and perpendicular to the incline, the normal force N, and the friction force F (Fig. 21-1*b*). Since the motion of the body is parallel to the incline, only impulses given by forces acting in that direction will change the momentum of the body. Since W_T acts *with* the direction of motion, its impulse ($W_T \cdot t$) is positive. Because F acts *against* the direction of motion, its impulse ($F \cdot t$) is negative. The initial and final momenta are respectively $W/g \cdot v_0$ and $W/g \cdot v$.

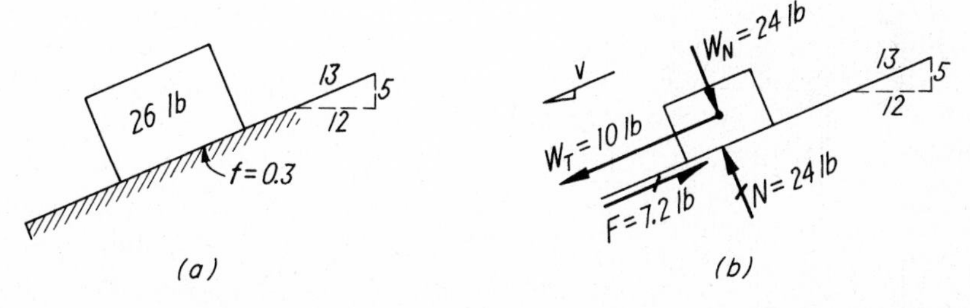

FIG. 21-1 The impulse-momentum method. (*a*) Space diagram. (*b*) Forces W_T and F produce impulses.

The components of W are

$$W_N = \frac{12}{13}(26) = 24 \text{ lb} \qquad \text{and} \qquad W_T = \frac{5}{13}(26) = 10 \text{ lb}$$

The impulse-momentum equation is

$$\text{Initial mom} + \text{pos imp} - \text{neg imp} = \text{final mom}$$

In symbols,
$$\frac{W}{g} \cdot v_0 + W_T \cdot t - F \cdot t = \frac{W}{g} \cdot v$$

or, when $W = 26$ lb, $v_0 = 8$ ft per sec, $W_T = 10$ lb, $t = 5$ sec, and $F = 7.2$ lb,

$$\left(\frac{26}{32.2}\right)(8) + (10)(5) - (7.2)(5) = \left(\frac{26}{32.2}\right)v$$

from which

$$6.46 + 50 - 36 = 0.807v \qquad \text{and} \qquad v = 25.4 \text{ ft per sec} \qquad \textit{Ans.}$$

PROBLEMS

21-2. A trailer is pulled along a level road by a car exerting a constant pull of 60 lb. The average road and wind resistance is 25 lb. What net impulse is given the trailer in 10 sec? *Ans.* 350 lb-sec

21-3. What is the momentum of an automobile weighing 3,220 lb when it travels with a speed of 30 mph? *Ans.* 4,400 lb-sec

21-4. Compute the momentum of a 64.4-lb projectile fired from a cannon with a muzzle velocity of 2,000 ft per sec.

21-5. A stalled automobile weighing 3,220 lb is pushed from rest along a level road by another car which exerts a constant force of 100 lb acting parallel to the direction of motion. If the average wind and road resistance is 20 lb, calculate the velocity given the automobile in 15 sec. *Ans.* $v = 12$ ft per sec

21-6. Solve Prob. 21.1 when $W = 130$ lb, $v = 5$ ft per sec, and $t = 10$ sec.

21-7. An automobile weighing 3,220 lb coasts down a 5 percent grade against an average road and wind resistance of 40 lb. In what time t will its velocity increase from 20 to 60 ft per sec? *Ans.* $t = 33.1$ sec

21-8. A 16,000-lb airplane lands on the deck of a carrier. If it is brought to rest in 2 sec from a speed of 90 mph, what force is exerted by the arresting gear? *Ans.* 33,000 lb

21-9. How long will it take a freely falling body to increase its velocity from 20 to 40 ft per sec? From 40 to 60 ft per sec? *Ans.* $t = 0.622$ sec

21-10. The exhaust gases from a rocket mounted on a static-test stand have a velocity of 6,000 ft per sec. If the fuel is consumed (expelled from the rocket) at the rate of 16.1 lb per sec, estimate the thrust being developed by the rocket motor. (HINT: Apply the impulse-momentum relationship to the rocket and its contents during a 1-sec interval of time.) *Ans.* $T = 3,000$ lb

21-11. A 10,000-ton ship is brought to rest from a speed of 1 mph by a constant pull maintained by a hawser wrapped several turns around a bollard. If the ship is brought to rest in 22 sec, what was the tension in the hawser? How many foot-pounds of energy were dissipated as heat by the rubbing of the hawser on the bollard? *Ans.* $T = 41,300$ lb; $U = 667,500$ ft-lb

21-12. A truck weighing 9,660 lb travels on a level street at 30 mph. (*a*) Compute the total frictional resisting force F which must be developed between its tires and the pavement to stop it in 5 sec. (*b*) What minimum coefficient of friction f is necessary to do this?

21-13. A machine gun fires 720 bullets per minute. Each bullet weighs 0.8 oz (1/20 lb) and leaves the muzzle of the gun with a velocity of 3,220 ft per sec. Calculate the *average reaction* R which must be exerted against the gun by its support. [Use Eq. (*c*).]
Ans. $R = 60$ lb

21-6 Conservation of Linear Momentum

According to Newton's third law of motion a force can act on one body only if an *equal and opposite* force acts on a second body. Let us consider that these two bodies constitute a system, and that no external force having a component in the direction of motion acts on either of these two bodies. Now if the two bodies collide or are pushed apart by a spring or by some other means, a mutual action and reaction exists between the two bodies. The forces exerted by each body on the other are equal and opposite, and since the time of interaction must necessarily be the same, the impulses are equal but opposite in direction. The two middle terms in Eq. (21-3) are then zero. Therefore, for the system, **initial momentum equals final momentum,** or in symbols

$$\frac{W_1 v_1}{g} + \frac{W_2 v_2}{g} = \frac{W_1 v_1'}{g} + \frac{W_2 v_2'}{g} \tag{21-4}$$

where v_1 = velocity of body one before the mutual reaction
v_2 = velocity of body two before the mutual reaction
v_1' = velocity of body one after the mutual reaction
v_2' = velocity of body two after the mutual reaction

In applying the equation, care should be used in assigning signs to the velocity terms. Choose a positive direction for velocity and then be consistent. If the direction of a velocity is unknown, it may be assumed positive, and the sign of the result will then agree with the previously chosen sign convention.

One should keep in mind that, although there is no loss of momentum in the interaction of two bodies, there is usually a loss of kinetic energy due to the generation of heat. This makes the momentum principle particularly useful in the solution of impact problems.

ILLUSTRATIVE PROBLEM

21-14. A motionless block of wood weighing 5 lb is suspended on a long string. A bullet weighing 2 oz is fired horizontally into the block, and strikes it with a velocity of 1,200 ft per sec. Calculate the velocity of the block and bullet after the bullet becomes fully embedded in the block.

Solution: Let W_1 be the weight of the wooden block and W_2 be the weight of the bullet. Then, since the initial velocity v_1 of the block was zero and the block and bullet move together after impact, we obtain, from Eq. (21-4),

$$\left[\frac{W_1 v_1}{g} + \frac{W_2 v_2}{g} = \frac{W_1 v_1'}{g} + \frac{W_2 v_2'}{g}\right] \qquad 0 + \frac{(2/16)(1{,}200)}{32.2} = 0 + \frac{[5 + (2/16)]v}{32.2}$$

or $$\frac{2{,}400}{16} = 5.125v \qquad \text{and} \qquad v = 29.3 \text{ ft per sec} \qquad \textit{Ans.}$$

PROBLEMS

21-15. A 3,220-lb car sliding with a velocity of 15 mph strikes the rear of a truck weighing 16,100 lb and moving in the same direction as the car with a velocity of 6 mph. The bumpers become locked during the collision. What is the velocity of the vehicles immediately after impact? *Ans.* $v = 7.5$ mph

21-16. Solve Prob. 21-15 if the truck had been moving toward the car with a velocity of 6 mph instead of away from it.
Ans. $v = 2.5$ mph (in direction of truck's original velocity)

21-17. A 32.2-lb body is moving to the right with a velocity of 6 ft per sec when it is struck by a 16.1-lb body moving to the left with a velocity of 15 ft per sec. The bodies become interlocked during impact and move as a unit thereafter. Find the velocity after impact. *Ans.* $v = 1$ ft per sec to the left

21-18. Solve the above problem if the velocity of the 16.1-lb body was 6 ft per sec. *Ans.* $v = 2$ ft per sec to the right

21-19. A 600-ton submarine fires 2 torpedoes weighing 2 tons each. If the torpedoes are fired from the bow tubes with an initial velocity of 15 mph, what is the momentary reduction in the speed of the submarine? *Ans.* Change in velocity is 0.1 mph

21-7 Direct Central Impact

When two colliding bodies are each directed normal to the surfaces at the point of contact, the impact is said to be **direct.** This term contrasts with a glancing or *oblique* impact. A **central** impact occurs when the force of impact between two bodies is along the line joining the centers of gravity of the bodies. A head-on collision between two cars would be classed as a *direct central impact.* A less gruesome example would be the direct central impact occurring between the head of a golf club and a well-hit golf ball. If the hit is neither direct nor central, the ball may not follow the desired course.

During the first part of the impact, the bodies **deform.** The second part of the impact process is one of **restitution.** If the bodies are at all elastic, they will tend to return to their original shape. The **coefficient of restitution** e is a measure of the ability of the bodies to regain their original shape. The general equation defining e is

$$e = \frac{\textbf{relative velocity after impact}}{\textbf{relative velocity before impact}} = \frac{v_2' - v_1'}{v_1 - v_2} \tag{21-5}$$

Since this equation will generally be used in conjunction with the equation expressing the law of conservation of linear momentum, the same sign convention should be assumed.

For perfectly elastic bodies the value of e is unity; for completely inelastic bodies it is zero. Most materials are neither completely elastic nor completely inelastic. The coefficient of restitution of common materials is given in most engineering handbooks.

ILLUSTRATIVE PROBLEMS

21-20. A golf ball was dropped from a height of 7 ft on a concrete slab. Find the coefficient of restitution if the rebound height was 5 ft.

Solution: Since the slab was stationary, $v_2 = 0$ and $v_2' = 0$. The velocity of the ball before impact may be found from Eq. (16-9).

$$[v^2 = v_0{}^2 + 2gh] \qquad v_1{}^2 = 0 + 2(32.2)(7) \qquad \text{or} \qquad v_1 = 21.2 \text{ ft per sec}$$

The velocity after impact is

$$[v^2 = v_0{}^2 + 2gh] \qquad 0^2 = (v_1')^2 + 2(32.2)(5) \qquad \text{or} \qquad v_1' = -17.9 \text{ ft per sec}$$

The coefficient of restitution is

$$\left[e = \frac{v_2' - v_1'}{v_1 - v_2}\right] \qquad e = \frac{0 - (-17.9)}{21.2 - 0} = 0.846 \qquad \textit{Ans.}$$

21-21. Direct central impact occurs between a 64.4-lb body moving to the right with a velocity of 20 ft per sec and a 32.2-lb body moving to the left with a velocity of 30 ft per sec. Find the velocity of each body after impact if the coefficient of restitution e is 0.8.

Solution: Signs are important. Velocities to the right will be taken as positive. From the conservation of linear momentum principle, Eq. (21-4),

$$\left[\frac{W_1 v_1}{g} + \frac{W_2 v_2}{g} = \frac{W_1 v_1'}{g} + \frac{W_2 v_2'}{g}\right]$$

$$\frac{64.4(+20)}{32.2} + \frac{32.2(-30)}{32.2} = \frac{64.4 v_1'}{32.2} + \frac{32.2 v_2'}{32.2} \qquad \text{or} \qquad 10 = 2v_1' + v_2' \qquad (a)$$

A second equation between these same unknowns may be obtained from Eq. (21-5).

$$\left[e = \frac{v_2' - v_1'}{v_1 - v_2}\right]$$

$$0.8 = \frac{v_2' - v_1'}{(+20) - (-30)} = \frac{v_2' - v_1'}{50} \qquad \text{or} \qquad 40 = v_2' - v_1' \qquad (b)$$

Solving Eq. (b) for v_2', $v_2' = 40 + v_1'$. Substitution of this value into Eq. (a) gives

$$10 = 2v_1' + 40 + v_1' = 3v_1' + 40$$

Therefore

$$3v_1' = -30 \qquad \text{and} \qquad v_1' = -10 \text{ ft per sec} \qquad \textit{Ans.}$$

(The minus sign indicates that the rebound velocity of body one is to the left.)

Substituting this value of v_1' into Eq. (*b*),

$$40 = v_2' - (-10) = v_2' + 10$$

Therefore

$$v_2' = +30 \text{ ft per sec (to the right)} \qquad \textit{Ans.}$$

Substituting these values into Eq. (21-4)

$$\left[\frac{W_1 v_1}{g} + \frac{W_2 v_2}{g} = \frac{W_1 v_1'}{g} + \frac{W_2 v_2'}{g}\right]$$

$$\frac{64.4(+20)}{32.2} + \frac{32.2(-30)}{32.2} = \frac{64.4(-10)}{32.2} + \frac{32.2(+30)}{32.2}$$

$$\text{or} \qquad +10 = +10 \qquad \textit{Check.}$$

The linear momentum of the system *before* impact was +10 lb-sec and the linear momentum of the system was +10 lb-sec after impact, as it should be from the principle of conservation of momentum.

PROBLEMS

21-22. How high would the golf ball of Prob. 21-20 rise on the second bounce?

21-23. Solve Prob. 21-21 if $e = 0.5$.

Ans. $v_1' = 5$ ft per sec to left; $v_2' = 20$ ft per sec to the right

21-24. A 100-lb body moving to the right with a velocity of 18 ft per sec strikes a 400-lb body moving to the right with a velocity of 4 ft per sec. What are the velocities of the two bodies after impact if the coefficient of restitution e is 0.785?

Ans. $v_1' = 2.0$ ft per sec to the left; $v_2' = 9.0$ ft per sec to the right

21-25. A loaded coal barge weighing 100 tons has a velocity of 10 mph to the right when it strikes an empty barge weighing 20 tons which is moving to the right with a velocity of 6 mph. Find the velocity of each barge after impact if the coefficient of restitution is 0.5.

21-8 The Angular Impulse-Momentum Equation

Recall that the net work done upon a rotating body (positive work minus negative work) is equal to the change in its *kinetic energy* (*FKE* − *IKE*). When the constant moment M is the resultant moment acting on the body, the net work is $M \cdot \theta$. Also, when the body starts from rest ($\omega_0 = 0$ and $IKE = 0$), the change in kinetic energy is $I\omega^2/2$. For a body starting from rest, then,

$$\left[\begin{array}{c}\text{Net work}\\ \text{on body}\end{array}\right] \qquad M \cdot \theta = \frac{I\omega^2}{2} \qquad \left[\begin{array}{c}\text{change in its}\\ \text{kinetic energy}\end{array}\right] \tag{a}$$

But $\theta = \omega_{avg} \cdot t$ and $\omega_{avg} = \frac{1}{2}\omega$ when the angular acceleration α is constant and the initial angular velocity ω_0 is zero. Hence, $\theta = \frac{1}{2}\omega t$ and

$$M \cdot \frac{1}{2}\omega t = \frac{I\omega^2}{2} \tag{b}$$

When we divide both sides of the equation by $\frac{1}{2}\omega$, we get, for a body starting from rest,

$$\begin{bmatrix}\text{Net angular}\\ \text{impulse}\end{bmatrix} \quad Mt = I\omega \quad \begin{bmatrix}\text{change in angu-}\\ \text{lar momentum}\end{bmatrix} \tag{c}$$

The *net angular impulse* on a body is the positive angular impulse minus the negative angular impulse, and when the body has initial angular velocity, the *change in its angular momentum* is its final angular momentum minus its initial angular momentum. We have, then, the following word statement:

$$\begin{bmatrix}\textbf{Initial}\\ \textbf{angular}\\ \textbf{momen-}\\ \textbf{tum}\end{bmatrix} + \begin{bmatrix}\textbf{positive angular}\\ \textbf{impulse by mo-}\\ \textbf{ments tending to}\\ \textbf{increase the ve-}\\ \textbf{locity}\end{bmatrix} - \begin{bmatrix}\textbf{negative angular}\\ \textbf{impulse by mo-}\\ \textbf{ments tending to}\\ \textbf{decrease the ve-}\\ \textbf{locity}\end{bmatrix} = \begin{bmatrix}\textbf{final}\\ \textbf{angular}\\ \textbf{momen-}\\ \textbf{tum}\end{bmatrix}$$

or, putting this statement into the form of an equation, we get the angular impulse-momentum equation:

$$\textbf{Initial mom + pos imp − neg imp = final mom} \tag{21-6}$$

In symbols, these quantities are

$$I\omega_0 + M_{pos}t - M_{neg}t = I\omega \tag{21-6S}$$

Note that equation (21-6) is identical in form with the corresponding linear impulse-momentum equation (21-3). However, the quantities involved and also the units of these quantities differ, as is clearly shown in the corresponding symbol equations (21-3S) and (21-6S).

ILLUSTRATIVE PROBLEM

21-26. Block B in Fig. 21-2 is attached to a weightless cord which is wrapped around disk A causing it to rotate about its centroidal axis. A constant bearing-friction moment FM of 5 lb-ft resists rotation. Determine the linear acceleration a of block B, the angular acceleration α of the disk A, and the tension T in the cord. (The following is the *impulse-momentum solution* of Prob. 18-29, the *force solution* of which is shown in Art. 18-12 and the *work-energy solution* of which is shown in Prob. 20.90, Art. 20-11. In this *impulse-momentum* solution we shall find it simplest to solve first for the angular velocity ω, next for the angular acceleration α, then for the linear acceleration a, and finally for the tension T.)

Solution: The quantities given in the impulse-momentum equation are velocity and time. Since in the statement of this problem no time was given, we may assume one, because acceleration is independent of time. We may also assume that the block starts from rest. Hence, the initial momentum is zero. Applying the linear impulse-momentum equation to the block B, we have

$$\textbf{Initial mom + pos imp − neg imp = final mom}$$

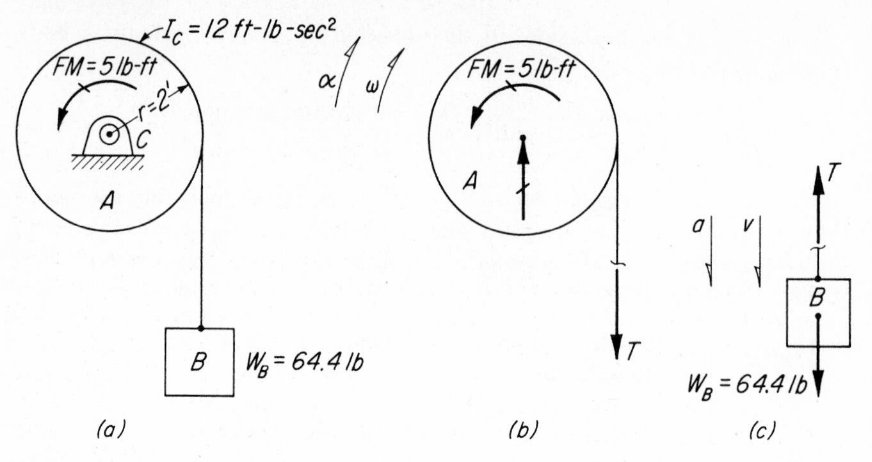

FIG. 21-2 The impulse-momentum method. Rotation and translation. (*a*) Space diagram. (*b*) Disk. (*c*) Block.

and in symbols, these quantities are

$$\frac{W}{g}v_0 + Wt - Tt = \frac{W}{g}v$$

or

$$\frac{64.4}{32.2}(0) + 64.4(t) - T(t) = \frac{64.4}{32.2}(v)$$

Let us now assume that $t = 1$ sec. Then, since $r = 2$ ft, $v_0 = 0$, and $v = r\omega = 2\omega$, we have

$$0 + 64.4(1) - T(1) = \frac{64.4}{32.2}(2\omega)$$

from which

$$T = 64.4 - 4\omega \qquad (a)$$

Next we apply the angular impulse-momentum equation (21-6) to disk A:

$$I\omega_0 + M_{pos}t - M_{neg}t = I\omega$$

Since $I_C = 12$ ft-lb-sec², $\omega_0 = 0$, $M_{pos} = Tr$, $M_{neg} = 5$ lb-ft, and $t = 1$ sec, we obtain

$$12(0) + (Tr)t - 5t = 12\omega \qquad \text{or} \qquad 2T - 5 = 12\omega \qquad (b)$$

Substituting Eq. (*a*) for T in Eq. (*b*), we have

$$2(64.4 - 4\omega) - 5 = 12\omega \qquad \text{or} \qquad 20\omega = 123.8$$

and

$$\omega = 6.19 \text{ rad per sec}$$

But $\omega = \alpha t$, or $\alpha = \omega/t$. Hence

$$\alpha = \frac{\omega}{t} = \frac{6.19}{1} = 6.19 \text{ rad per sec}^2 \qquad \textit{Ans.}$$

Since $a = r\alpha$ and $r = 2$ ft,

$$a = r_{\alpha} = 2(6.19) = 12.38 \text{ ft per sec}^2 \qquad \textit{Ans.}$$

We now obtain T from Eq. (*a*). That is,

$$T = 64.4 - 4\omega = 64.4 - 4(6.19) = 64.4 - 24.8$$

or

$$T = 39.6 \text{ lb} \qquad \textit{Ans.}$$

PROBLEMS

21-27. A rope is wrapped around a solid cylinder whose diameter is 4 ft, as shown in Fig. 18-14. The cylinder weighs 644 lb, and it rotates in a vertical plane about a horizontal axis coinciding with its geometric axis. How many seconds t will it take to bring the cylinder to an angular velocity of 300 rpm from rest, if a constant pull P of 206 lb is maintained on the rope and if the friction moment FM is constant at 12 lb-ft. Disregard weight of rope. *Ans.* $t = 3.14$ sec

21-28. If the force P in Prob. 21-27 is discontinued at the instant the cylinder reaches an angular velocity of 300 rpm, how many seconds t will it continue rotating before the friction moment brings it to rest? *Ans.* $t = 104.6$ sec

21-29. How long will it take the block B of Prob. 18-29 to reach a velocity of 20 ft per sec from rest?

21-30. Block B in Fig. 20-43 weighs 322 lb. Calculate the required weight W_A of block A to give block B a velocity of 15 ft per sec in 1.6 sec from rest.

21-9 Plane Motion

The impulse-momentum principles are applicable to the analysis of problems in plane motion. The linear impulse-motion equation (21-3) can be applied directly to a body undergoing plane motion. The velocity of the centroid should be used in computing the momentum of the body.

The angular impulse-momentum principle may be applied in two ways. The centroidal axis may be chosen as the reference axis or the instantaneous center may be chosen. The first of these concepts would be used if we think of plane motion as rotation about the centroidal axis combined with translation. The second concept will eliminate the friction force from the equations and may, in some cases, simplify the work. If the centroidal axis is chosen as the reference axis, the angular momentum about this axis is $I_C\omega$ and the angular impulse is Mt where M is the moment of the *external* acting forces about the centroidal axis C. If the instantaneous center A is chosen as the reference axis, the angular momentum about this axis is $I_A\omega$ and the angular impulse is Mt where M is the moment of the *external* acting forces about the instantaneous center A.

ILLUSTRATIVE PROBLEMS

21-31. The solid cylinder shown in Fig. 21-3 weighs 1,288 lb, its radius r is 2 ft and its moment of inertia I_C is 80 ft-lb-sec^2. It rolls without slipping down the incline. Assume rolling friction to be negligible. Calculate (*a*) the linear acceleration $\bar{a}$ of its center of gravity and (*b*) the minimum coefficient of static friction f required

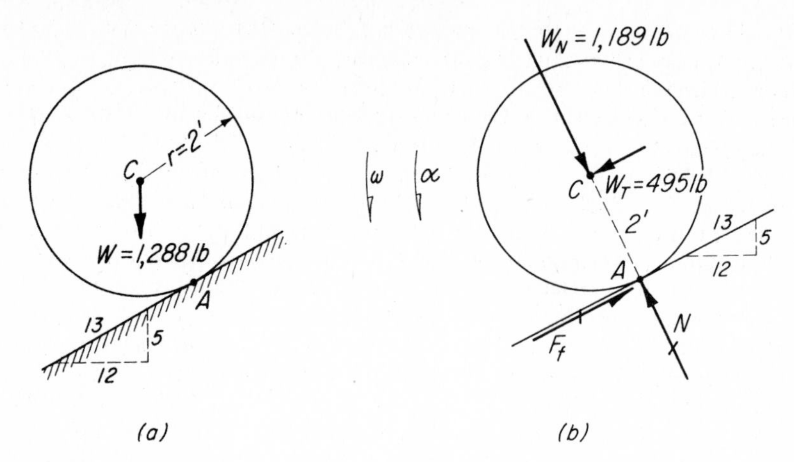

FIG. 21-3 Plane motion of a free-rolling body. (*a*) Space diagram. (*b*) Forces W_T and F produce impulses.

to prevent slipping. (NOTE: This problem is the same as Prob. 19.1, solved here by the impulse-momentum method.)

Solution: Since the acceleration is independent of both initial velocity and time, we shall assume that $\bar{v}_0 = 0$, $\omega_0 = 0$, and that $t = 1$ sec. Then, applying the *angular* impulse-momentum equation (21-6S) to the cylinder (which has centroidal rotation), we have

$$[I_C\omega_0 + M_{pos}t - M_{neg}t = I_C\omega] \qquad \text{or} \qquad 0 + F(2)t - 0 = 80\omega$$

and

$$F = 40\omega \tag{a}$$

Next, applying the *linear* impulse-momentum equation (21-3S) to the cylinder, when $\bar{v}_0 = 0$, $t = 1$ sec, and $\bar{v} = r\omega = 2\omega$, we have

$$\left[\frac{W\bar{v}_0}{g} + F_{pos}t - F_{neg}t = \frac{W\bar{v}}{g}\right] \qquad \text{or} \qquad 0 + 495(1) - F(1) = \frac{1{,}288}{32.2}(2\omega)$$

and

$$F = 495 - 80\omega \tag{b}$$

Now, from Eq. (*B*), Table 17-2, $\alpha = (\omega - \omega_0)/t = (\omega - 0)/1$, or $\alpha = \omega$. Also, $\bar{a} = r\alpha = r\omega = 2\omega$, since $r = 2$ ft. If now we equate (*a*) and (*b*), we obtain

$$40\omega = 495 - 80\omega \qquad \text{or} \qquad 120\omega = 495 \qquad \text{and} \qquad \omega = 4.125 \text{ rad per sec}$$

from which, since $\bar{a} = 2\omega$, we derive

$$\bar{a} = 2(4.125) = 8.25 \text{ ft per sec}^2 \qquad \textit{Ans.}$$

To determine F and f_{min}, we have, from Eq. (*a*), since $N = 1189$ lb,

$$F = 40\omega = 40(4.125) = 165 \text{ lb}$$

$$\left[f = \frac{F}{N}\right] \qquad f_{min} = \frac{165}{1{,}189} = 0.138 \qquad \textit{Ans.}$$

The above solution can readily be checked by an alternate solution, in which rotation about an axis through the instantaneous center is assumed.

21-32. Calculate the angular velocity ω of the disk D in Fig. 21-4 and the linear velocity v_B of the block B after it has dropped a distance of 10 ft from rest. Compute also the acceleration a_B of the block, and the tension T in the cord. Consider that the cord and the small pulley are weightless, and that the disk does not slip. $W_B = 161$ lb. The weight of disk D is 966 lb and its centroidal moment of inertia I_C is 100 ft-lb-sec². (NOTE: This problem is the same as Prob. 20.107, solved here by the impulse-momentum method.) Let $R = 3$ ft, $r = 1$ ft, and let $t = 1$ sec.

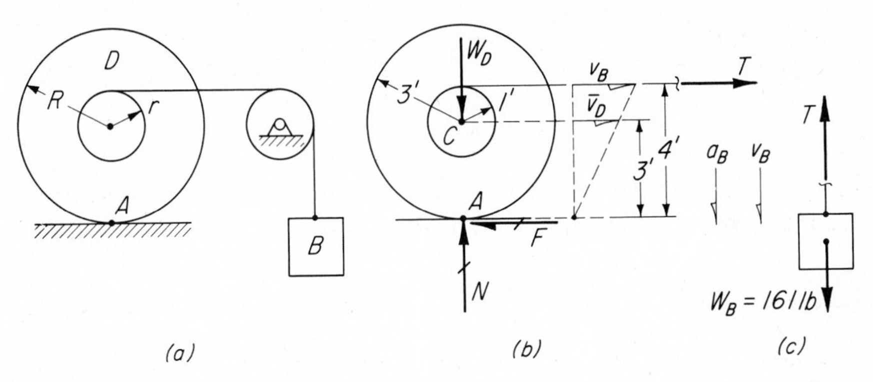

FIG. 21-4 The impulse-momentum method. (*a*) Space diagram. (*b*) Plane motion. (*c*) Translation.

Solution: The plane motion may be treated as rotation about the center of gravity combined with a linear translation of the center of gravity.

Applying the impulse-momentum equation to the disk

$$\left[\frac{W\bar{v}_0}{g} + F_{\text{pos}}t - F_{\text{neg}}t = \frac{W\bar{v}}{g}\right]$$

$$0 + Tt - Ft = \frac{966}{32.2}\bar{r}\omega \qquad \text{or} \qquad \frac{t}{\omega} = \frac{(30)(3)}{T - F} \qquad (a)$$

Applying the angular impulse-momentum equation to the disk, considering the rotation to be about the centroid,

$$[I_C\omega_0 + M_{\text{pos}}t - M_{\text{neg}}t = I_C\omega]$$

$$0 + T(1)t + F(3)t = 100\omega \qquad \text{or} \qquad \frac{t}{\omega} = \frac{100}{T + 3F} \qquad (b)$$

Equating (*a*) and (*b*),

$$\frac{(30)(3)}{T - F} = \frac{100}{T + 3F}$$

Cross-multiplying and solving for F gives

$$90T + 270F = 100T - 100F \qquad \text{or} \qquad F = \frac{T}{37}$$

Applying the linear impulse-momentum equation to the block B, since $v_0 = 0$ and $v_B = r\omega$,

$$\left[\frac{Wv_0}{g} + F_{\text{pos}}t - F_{\text{neg}}t = \frac{Wv}{g}\right] \qquad 0 + W_Bt - Tt = \frac{W_B(r\omega)}{g}$$

Since $r = 4$ ft,

$$0 + 161t - Tt = \frac{161}{32.2}(4\omega) \qquad \text{or} \qquad \frac{t}{\omega} = \frac{20}{161 - T} \tag{c}$$

Equating (a) and (c),

$$\frac{90}{T - F} = \frac{20}{161 - T}$$

Cross-multiplying gives

$$14{,}500 - 90T = 20T - 20F$$

Since $F = \frac{T}{37}$, this last equation may be written

$$14{,}500 = 20T + 90T - 20\left(\frac{T}{37}\right)$$

from which $\qquad 109.5T = 14{,}500 \qquad$ and $\qquad T = 132.2$ lb $\qquad$ *Ans.*

From Eq. (B), Table 17-2, $\alpha = (\omega - \omega_0)/t$. Since $\omega_0 = 0$, $\alpha = \omega/t$. Therefore, using Eq. (c) above, inverted,

$$\alpha = \frac{\omega}{t} = \frac{161 - T}{20} = \frac{161 - 132.2}{20} = 1.44 \text{ rad per sec}^2$$

From $a_B = r\alpha$,

$$a_B = 4(1.44) = 5.75 \text{ ft per sec}^2 \qquad \textit{Ans.}$$

From Eq. (H), Table 14-1,

$$\left[v = \sqrt{v_0^2 + 2as}\,\right] \quad v_B = \sqrt{0 + 2(5.75)(10)} \quad \text{or} \quad v_B = 10.72 \text{ ft per sec} \qquad \textit{Ans.}$$

Since $\omega = v/r$,

$$\omega = \frac{10.72}{4} = 2.68 \text{ rad per sec} \qquad \textit{Ans.}$$

Alternate Solution: The work will be shortened if we consider the motion of the disk as a rotation about the instantaneous axis A. From the transfer formula,

$$\left[I_A = I_C + \frac{W}{g}d^2\right] \qquad I_A = 100 + \frac{966}{32.2}(3)^2 = 370 \text{ ft-lb-sec}^2$$

Applying the linear impulse-momentum equation to the block B, since $v_0 = 0$, $v = r\omega$, and $r = 4$ ft,

$$\left[\frac{Wv_0}{g} + F_{\text{pos}}t - F_{\text{neg}}t = \frac{Wv}{g}\right]$$

$$0 + 161t - Tt = \frac{161}{32.2}(4\omega) \qquad \text{or} \qquad \frac{\omega}{t} = \frac{161 - T}{20} \tag{d}$$

Applying the angular impulse-momentum equation, considering the rotation to take place about the instantaneous center,

$$[I_A\omega_0 + M_{\text{pos}}t - M_{\text{neg}}t = I_A\omega]$$

$$0 + 4Tt - 0 = 370\omega \qquad \text{or} \qquad \frac{\omega}{t} = \frac{4T}{370} \tag{e}$$

Equating (d) and (e),

$$\frac{4T}{370} = \frac{161 - T}{20} \qquad \text{or} \qquad T = 132.2 \text{ lb} \qquad \textit{Ans.}$$

Since $\alpha = (\omega - \omega_0)/t$,

$$\alpha = \frac{\omega}{t} = \frac{4T}{370} = \frac{4(132.2)}{370} = 1.44 \text{ rad per sec}^2 \qquad \textit{Ans.}$$

Knowing $s = 10$ ft and $\alpha = 1.44$ rad per sec^2, we may find v_B and a_B from kinematics as in the first solution.

PROBLEMS

21-33. A cylinder 3 ft in diameter and weighing 240 lb starts up a 30° incline with an angular velocity of 360 rpm. How long before it comes to rest?
Ans. $t = 5.26$ sec

21-34. A solid steel sphere 1 ft in diameter is released from rest on a 20° incline. Find the angular velocity 10 sec after it is released. (Steel weighs 490 pcf.)

21-35. The cylinder shown in Fig. 19-13 weighs 966 lb. Find the angular velocity of the disk 10 sec from rest. (HINT: The solution will be expedited if the rotation is considered to be about the instantaneous center A.) *Ans.* $\omega = 141$ rpm

21-36. How long will it take for the velocity of the centroid of the cylinder shown in Fig. 20-33 to change from 10 to 25 ft per sec? *Ans.* $t = 0.7$ sec

21-37. The solid disk shown in Fig. 21-5 weighs 100 lb. Find the minimum coefficient of friction to prevent sliding. (NOTE: The solution will be simplified if t is assumed to be 1 sec.)

21-38. If the disk shown in Fig. 21-5 weighs 322 lb, how long will it take for its angular velocity to change from 200 to 280 rpm? *Ans.* $t = 0.325$ sec

21-39. The disk D shown in Fig. 21-6 weighs 322 lb and its moment of inertia about the instantaneous center is 100 ft-lb-sec^2. The block B weighs 161 lb. Find the angular velocity of the disk 20 sec after the system is released from rest.
Ans. $\omega = 32.5$ rad per sec

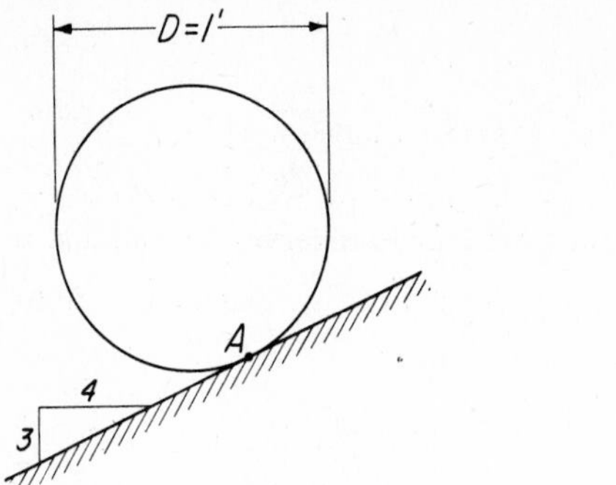

FIG. 21-5 Probs. 21-37 and 21-38.

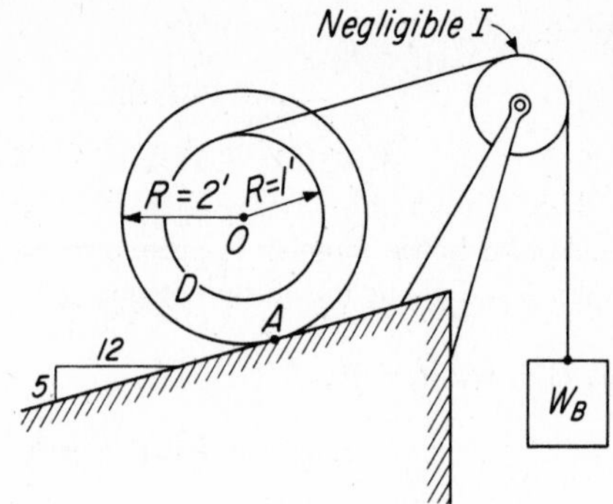

FIG. 21-6 Probs. 21-39 and 21-54.

SUMMARY

(By article number)

21-1. The principles of impulse and momentum are especially useful in the solution of motion problems involving force, mass, *velocity, and time.*

21-2. Impulse is the product of force and time. When the force is constant,

$$\textbf{Impulse} = \textbf{(force)(time)} = \mathbf{F \cdot t} \tag{21-1}$$

External forces acting *with* the direction of motion give *positive impulses* to a body, and those acting *against* the direction of motion give *negative impulses.*

21-3. The momentum of a body is the product of its mass and its instantaneous velocity. That is,

$$\textbf{Momentum} = \textbf{(force)(velocity)} = \frac{\mathbf{W}}{\mathbf{g}} \cdot \mathbf{v} \tag{21-2}$$

21-4. The *initial momentum* possessed by a body may be increased by positive impulses and decreased by negative impulses. The algebraic sum of these quantities gives its *final momentum.* From this principle we obtain **the linear impulse-momentum equation,** which is

$$\textbf{Initial mom + pos imp + neg imp = final mom} \tag{21-3}$$

or, in symbols,

$$\frac{Wv_0}{g} + F_{pos}t - F_{neg}t = \frac{Wv}{g} \tag{21-3S}$$

21-6. The algebraic sum of the momentum of two bodies before impact is equal to the algebraic sum of their momenta after impact. This principle is known as the law of **conservation of linear momentum,** and stated briefly, initial momentum equals final momentum. In symbols

$$\frac{W_1v_1}{g} + \frac{W_2v_2}{g} = \frac{W_1v_1'}{g} + \frac{W_2v_2'}{g} \tag{21-4}$$

21-7. Real bodies are not completely elastic. The ability of bodies to regain their original shape after an impact is measured by the **coefficient of restitution** e or specifically

$$e = \frac{\textbf{Relative velocity after impact}}{\textbf{Relative velocity before impact}} = \frac{v'_2 - v'_1}{v_1 - v_2} \qquad \textbf{(21-5)}$$

21-8. The impulse-momentum principle may be applied to rotation problems as well as to linear-translation problems. The angular impulse-momentum equation is

Initial ang mom + pos ang imp − neg ang imp = final ang mom (21-6)

or, in symbols,

$$I\omega_0 + M_{pos}t + M_{neg}t = I\omega \qquad \textbf{(21-6S)}$$

REVIEW PROBLEMS

21-40. A 100-lb block slides for 10 sec down a 45° incline whose coefficient of friction f is 0.3. Compute (*a*) the *positive impulse,* (*b*) the *negative impulse,* and (*c*) the *net impulse,* during this period of time.

Ans. (*a*) 707 lb-sec; (*b*) 212 lb-sec; (*c*) 495 lb-sec

21-41. What is the *momentum* of the block in Prob. 21.40 after sliding for 10 sec, (*a*) if it starts from rest and (*b*) if its initial velocity is 9 ft per sec?

21-42. A switch engine exerts a constant drawbar pull of 2,000 lb for a period of 5 sec, on a freight car weighing 70 tons. The track is level. If the track resistance is 10 lb per ton, what is the final velocity of the car (*a*) if it starts from rest and (*b*) if its initial velocity is 15 mph?

Ans. (*a*) v = 1.5 ft per sec; (*b*) v = 23.5 ft per sec

21-43. How long will the freight car in Prob. 21-42 continue in motion, if released by the switch engine with the velocity given in answers (*a*) and (*b*)?

Ans. (*a*) t = 9.3 sec; (*b*) t = 146 sec

21-44. A rifle weighing 12.5 lb fires a projectile weighing 0.05 lb with a muzzle velocity of 2,000 ft per sec. What is the velocity of the rifle immediately after the projectile leaves the gun? What is the kinetic energy of the rifle immediately after the projectile leaves the gun? By using the work-energy method, find the average force exerted by the rifle against the shoulder if the rifle is brought to rest in 0.1 ft.

21-45. A spring having a modulus of 40 lb per ft is placed between a 16.1-lb block and a 32.2-lb block on a smooth floor. Find the maximum velocity of each block if they are pushed together until the spring is compressed 0.5 ft and then suddenly released. (Obtain one equation from the conservation of momentum principle and one from the work-energy relationship. Disregard weight of spring.)

21-46. A motionless boat weighs 322 lb. What will be the velocity of the boat immediately after a 161-lb man leaps from a dock and lands on the boat with a horizontal velocity component of 9 ft per sec away from the dock?

Ans. v = 3 ft per sec

21-47. Solve Prob. 21-46 if the boat had an initial velocity toward the dock of 3 ft per sec.

21-48. What velocity will be imparted to a motionless boat weighing 322 lb if a 161-lb man leaps from it to a dock with a velocity of 9 ft per sec?

Ans. v = 4.5 ft per sec

21-49. A bullet weighing 0.05 lb embeds itself in a 9.95-lb wood block resting on a horizontal surface. The block is moved 4 ft along the surface by the impact. The coefficient of friction between the block and the horizontal surface is 0.3. Determine the velocity of the bullet before impact.

21-50. The head of a croquet mallet weighs 2 lb. What will be the velocity of the mallet head immediately after it strikes a 1-lb croquet ball if the initial velocity of the mallet was 30 ft per sec? Assume the coefficient of restitution e is 0.8.

Ans. $v = 12$ ft per sec

21-51. What force F must be applied to the brake handle shown in Fig. 21-7 to bring the disk A in 60 sec if its initial angular velocity is 900 rpm and the coefficient of friction between the brake shoe and the disk is 0.5. Neglect the journal friction.

Ans. $F = 25$ lb

21-52. Solve Prob. 21-51 if the initial angular velocity was in the opposite direction to that shown.

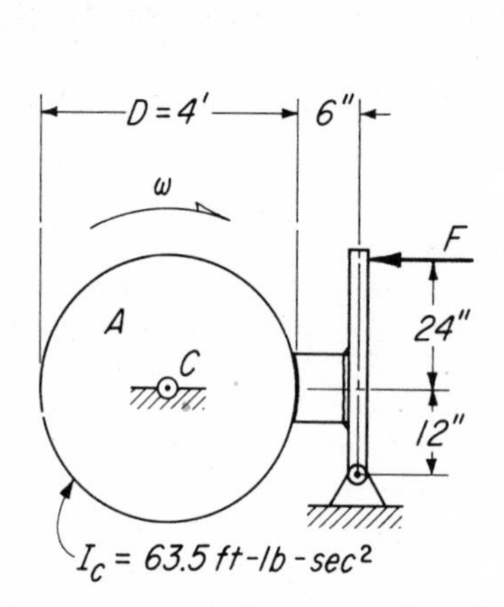

FIG. 21-7 Probs. 21-51 and 21-52.

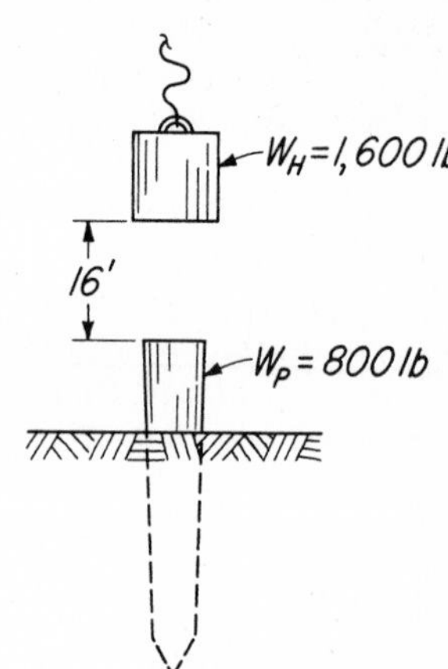

FIG. 21-8 Prob. 21-53.

21-53. The pile shown in Fig. 21-8 weighs 800 lb and is driven into the ground by a drop hammer weighing 1,600 lb. (*a*) What will be the velocity of the pile immediately after the blow, if the hammer drops 16 ft? Assume that hammer and pile move together as a unit after impact. (*b*) What will be the penetration per blow, if the soil resistance is constant at 16,000 lb?

21-54. How much energy was lost during the impact of the drop hammer and the top of the pile of Prob. 21-53?

21-55. The disk shown in Fig. 21-6 weighs 644 lb, and its moment of inertia about the instantaneous center of rotation is 150 ft-lb-sec^2. The block B weighs 500 lb. Find the tension in the cable.

Ans. $T = 339$ lb

21-56. An earth satellite has a moment of inertia of 1,200 ft-lb-sec^2 about its geometric axis. Calculate the required firing time of a pair of tangential control jets to remove an angular rotation of 1 rpm about the geometric axis. Each control jet expells 6 lb of gas per minute with a relative velocity of 66 ft per sec. The gas streams lie in a plane normal to the geometric axis and are 5 ft from it. Assume $g = 30$ ft per sec^2.

REVIEW QUESTIONS

21-1. What is meant by a linear impulse?

21-2. In what quantities is an impulse measured, and what is the unit of linear impulse?

21-3. What is meant by linear momentum?

21-4. In what quantities is momentum measured, and what is the unit of momentum?

21-5. What is meant by a positive impulse? a negative impulse? net impulse?

21-6. State the relation between net impulse and momentum.

21-7. Give the full word statement of the linear impulse-momentum equation.

21-8. Give the word statement of the conservation of momentum principle.

21-9. What name is given to the ratio of the relative velocity of two objects before impact to their relative velocity after impact?

21-10. What is the numerical value of e for impact between perfectly elastic bodies? Between inelastic bodies?

21-11. Give the word statement of the angular impulse-momentum equation.

APPENDIX A

Centroids of Areas by Integration

To locate the centroid of an area, as was stated in Art. 10-4, we need only equate the moment $A\bar{x}$ of the entire area, with respect to some reference axis, to the sum of the moments of the component area Σax. When each component area is infinitesimally small, or dA, the equations with respect to the Y and X axes become

$$A\bar{x} = \int x\, dA \qquad (a)$$

and

$$A\bar{y} = \int y\, dA \qquad (b)$$

Rules. In selecting the differential element dA, one of the following rules must be observed:

1. All points of the element must be the same distance from the moment axis.
2. The position of the centroid of the element must be known, so that the moment of the element with respect to the moment axis is the product of the element and the distance of its centroid from the axis. The locations of the centroids of a triangle, a parabola, and a sector of a circle are shown in Probs. A-1 to A-3.

ILLUSTRATIVE PROBLEMS

A-1. Locate the centroid of the tirangle shown in Fig. A-1 with respect to the Y axis.

Solution: Let the differential area dA be a vertical strip parallel to the side a.

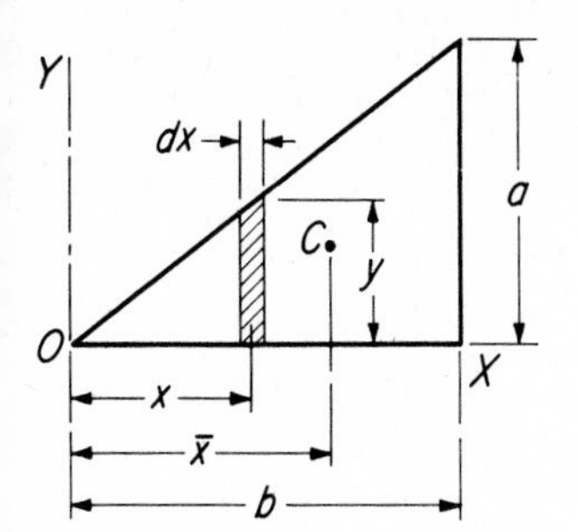

FIG. A-1 Centroid of a triangle.

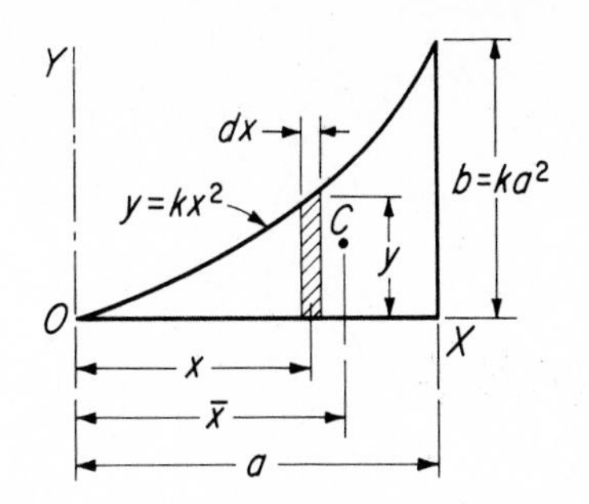

FIG. A-2 Centroid of a parabolic segment.

This area $dA = y\,dx$. From Eq. (*a*), we have

$$\left[A\bar{x} = \int x\,dA\right] \qquad \frac{1}{2}ba \cdot \bar{x} = \int_0^b xy\,dx \qquad (c)$$

But by similar triangles

$$\frac{y}{x} = \frac{a}{b} \quad \text{or} \quad y = \frac{ax}{b} \qquad (d)$$

When this value of y is substituted in Eq. (*c*), we obtain

$$\frac{1}{2}ba \cdot \bar{x} = \frac{a}{b}\int_0^b x^2\,dx \quad \text{and} \quad \bar{x} = \frac{2}{3}b \quad \textit{Ans.}$$

A-2. Locate the centroid with respect to the Y axis of the parabolic segment shown in Fig. A-2 and bounded by the X axis, the vertical side b, and the parabola $y = kx^2$.

Solution: As the differential area dA we select a vertical strip parallel to side b. This area $dA = y\,dx$. But $y = kx^2$ and when $x = a$, $b = ka^2$. Hence

$$\left[A = \int dA\right] \qquad A = \int_0^a y\,dx = \int_0^a kx^2\,dx = \frac{1}{3}ka^3 \qquad (e)$$

To find $\bar{x}$ from Eq. (*a*), we have

$$\left[A\bar{x} = \int x\,dA\right] \qquad \frac{1}{3}ka^3 \cdot \bar{x} = \int_0^a xy\,dx = k\int_0^a x^3\,dx \qquad (f)$$

from which

$$\frac{1}{3}ka^3 \cdot \bar{x} = \frac{1}{4}ka^4 \quad \text{and} \quad \bar{x} = \frac{3}{4}a \quad \textit{Ans.}$$

To find $\bar{y}$, we may use the same elemental strip. But our moment axis is now the X axis, and all points of the element are not at the same distance from the axis. Hence Rule 2 applies, and the moment of each elemental strip with respect to the X axis then is the product of its area $y\,dx$ and its centroidal distance $\frac{1}{2}y$, or $\frac{1}{2}y \cdot y\,dx$. Then, since $y = kx^2$ and $b = ka^2$,

$$\left[A\bar{y} = \int y\,dA\right] \qquad \frac{1}{3}ka^3 \cdot \bar{y} = \int_0^a \frac{1}{2}y \cdot y\,dx = \frac{1}{2}k^2\int_0^a x^4\,dx \qquad (g)$$

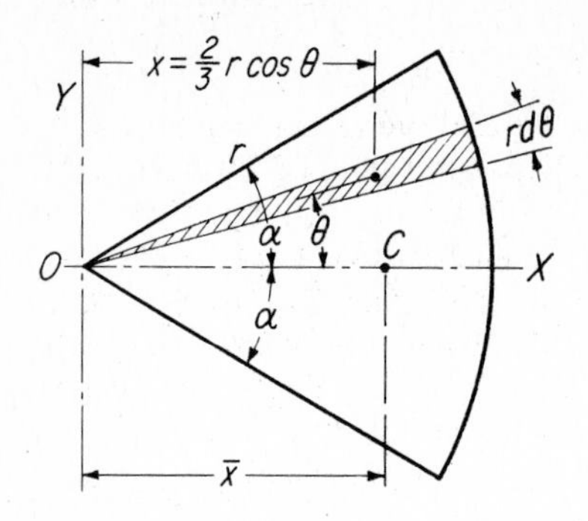

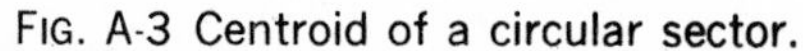
FIG. A-3 Centroid of a circular sector.

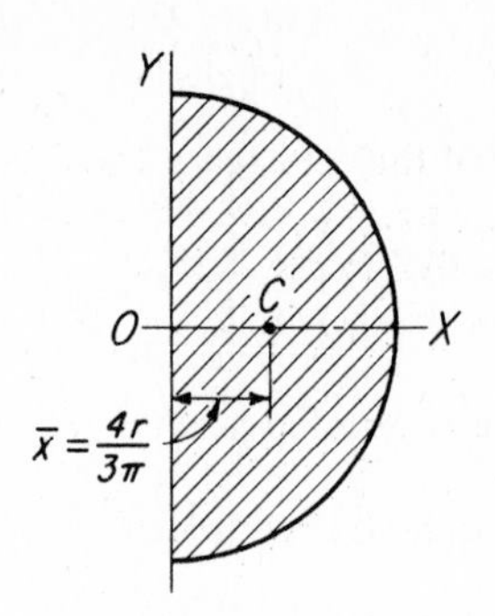

FIG. A-4 Centroid of a semicircle.

or

$$\frac{1}{3}ka^3 \cdot \bar{y} = \frac{k^2a^5}{10} \qquad \text{and} \qquad \bar{y} = \frac{3}{10}ka^2 = \frac{3}{10}b \qquad \textit{Ans.}$$

A-3. Locate the centroid of the area of the sector of a circle shown in Fig. A-3.

Solution: Let the area of the sector be symmetrical about the X axis. The radius is r and the subtended angle is 2α. The differential area dA here selected is a triangle whose area is $\frac{1}{2}r^2\,d\theta$. The centroid of this triangle lies at a distance $x = \frac{2}{3}r\cos\theta$ from the Y axis. The area of the sector is

$$\left[A = \int dA\right] \qquad A = \int_{-\alpha}^{+\alpha} \frac{1}{2}r^2\,d\theta = r^2\alpha \tag{h}$$

Then from Eq. (a)

$$\left[A\bar{x} = \int x\,dA\right] \qquad r^2\alpha \cdot \bar{x} = \int_{-\alpha}^{+\alpha} \frac{2}{3}r\cos\theta \cdot \frac{1}{2}r^2\,d\theta \tag{i}$$

or

$$r^2\alpha \cdot \bar{x} = \frac{1}{3}r^3 \int_{-\alpha}^{+\alpha} \cos\theta\,d\theta = \frac{2}{3}r^3\sin\alpha \tag{j}$$

and

$$\bar{x} = \frac{2}{3} \cdot \frac{r\sin\alpha}{\alpha} \qquad \textit{Ans.}$$

When $\alpha = 90°$, as in the semicircle in Fig. A-4, $\bar{x} = 4r/3\pi$.

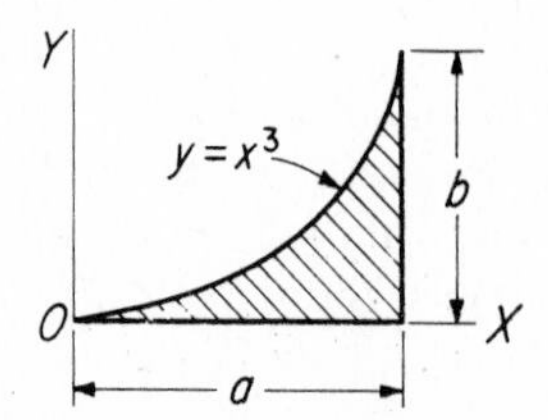

FIG. A-5 Prob. A-5.

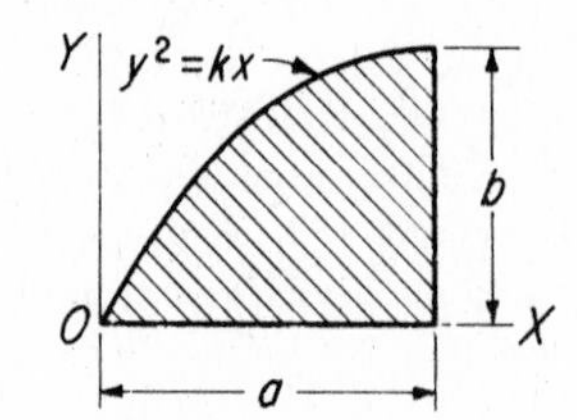

FIG. A-6 Prob. A-6.

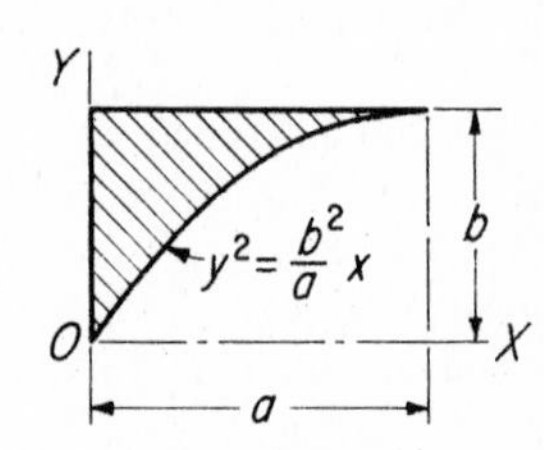

FIG. A-7 Prob. A-7.

PROBLEMS

A-4. Find the distance $\bar{y}$ from the X axis to the centroid of the triangle shown in Fig. A-1 using (*a*) Rule 2 and a strip parallel to the Y axis and (*b*) Rule 1 and a strip parallel to the X axis.
Ans. $\bar{y} = \frac{1}{3}a$

A-5. Locate the position of the centroid of the shaded area shown in Fig. A-5.
Ans. $\bar{x} = \frac{4}{5}a$; $\bar{y} = \frac{2}{7}b$

A-6. Determine the position of the centroid of the shaded area shown in Fig. A-6.
Ans. $\bar{x} = \frac{3}{5}a$; $\bar{y} = \frac{3}{8}b$

A-7. Locate the centroid of the shaded area shown in Fig. A-7.
Ans. $\bar{x} = \frac{3}{10}a$; $\bar{y} = \frac{3}{4}b$

APPENDIX B

Moments of Inertia of Plane Areas

Rectangular Moment of Inertia. As stated in Art. 11-1, the moment of inertia of an area with respect to an axis in the plane of the area is required in the design of beams and columns. The exact moments of inertia of simple plane areas, such as a rectangle, a triangle, or a circle, must be determined by integration. The basic expression for the moment of inertia I of any plane area with respect to an axis in its plane, as was mentioned in Art. 11-1, is

$$I = \int y^2 \, dA \qquad (a)$$

According to this equation, *the moment of inertia of a plane area with respect to an axis in the plane is the sum of the products of the elemental areas, each multiplied by the square of the distance from the axis to its centroid.*

Rules. In selecting the elemental area, the following rules must be observed:

1. All points in the elemental area must be the same distance from the axis.
2. The moment of inertia of the elemental area with respect to the inertia axis must be known. The moment of inertia of the entire area is then the summation of the moments of inertia of all the elements.

ILLUSTRATIVE PROBLEMS

B-1. Find the moment of inertia of the rectangular area shown in Fig. B-1 with respect to (*a*) a centroidal axis parallel to the base b and (*b*) an axis coinciding with the base.

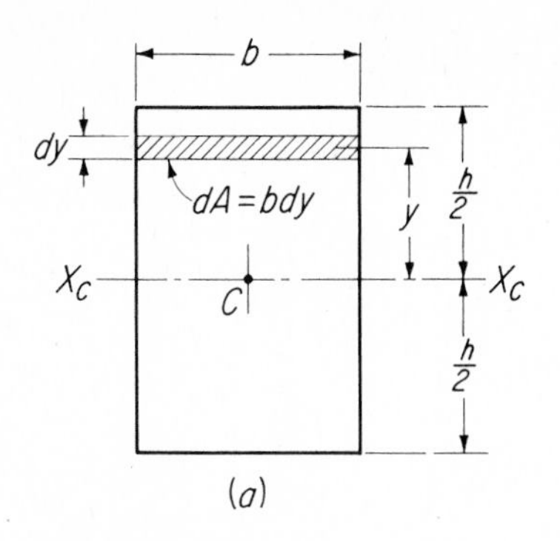

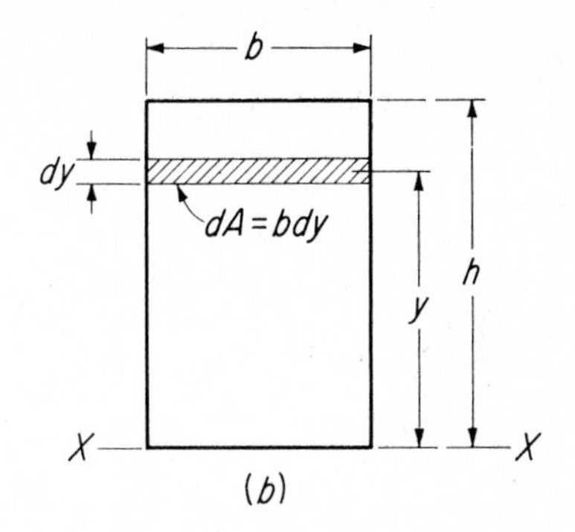

FIG. B-1 Moment of inertia of a rectangular area. (*a*) Axis through centroid. (*b*) Axis coinciding with base.

Solution: (*a*) *Axis through centroid.* Let the width of the area be b and its depth h, and let the elemental area be $b\,dy$, a strip parallel to the inertia axis X-X. The moment of inertia of a single strip with respect to this axis is $y^2\,dA$ and of the entire area between the limits of $h/2$ and $-h/2$ then is

$$\left[I = \int y^2\,dA\right] \qquad I_{X_C} = \int_{-h/2}^{+h/2} y^2 b\,dy = b\left[\frac{y^3}{3}\right]_{h/2}^{+h/2}$$

or

$$I_{X_C} = \frac{b}{3}\left(\frac{h^3}{8} + \frac{h^3}{8}\right) = \frac{bh^3}{12} \qquad \textit{Ans.} \qquad (b)$$

(*b*) *Axis coinciding with base.* The summation is now made between the limits of 0 and h. That is,

$$\left[I = \int y^2\,dA\right] \qquad I_X = \int_0^h y^2 b\,dy = b\left[\frac{y^3}{3}\right]_0^h = \frac{bh^3}{3} \qquad \textit{Ans.}$$

B-2. Determine the moment of inertia of the triangular area shown in Fig. B-2

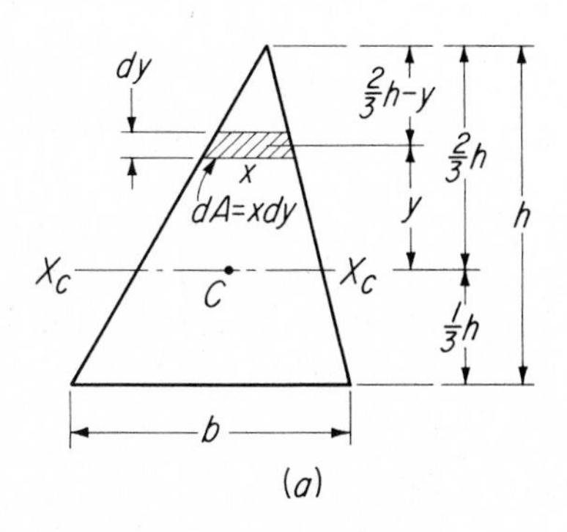

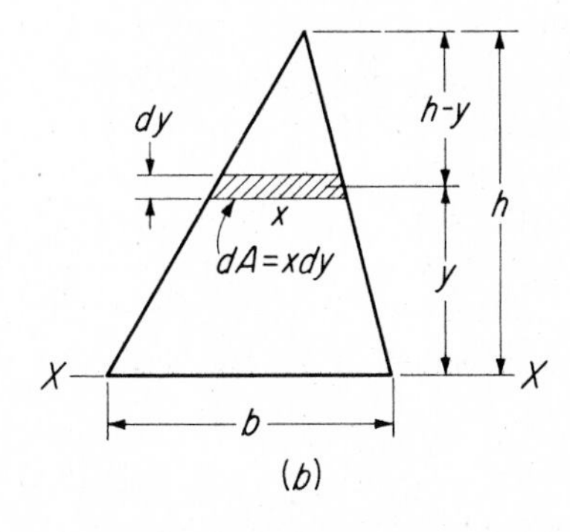

FIG. B-2 Moment of inertia of a triangular area. (*a*) Axis through centroid. (*b*) Axis coinciding with base.

with respect to (*a*) a centroidal axis parallel to the base b and (*b*) an axis coinciding with the base.

Solution: (*a*) *Axis through centroid.* The elemental area selected is a strip parallel to the centroidal inertia axis X_C. The moment of inertia of this strip with respect to axis X_C is $y^2\,dA$, or $y^2x\,dy$, and of the entire triangle then is

$$\left[I = \int y^2\,dA\right] \qquad I_{X_C} = \int_{-h/3}^{+2h/3} y^2 x\,dy \tag{c}$$

By similar triangles

$$\frac{x}{\frac{2}{3}h - y} = \frac{b}{h} \qquad \text{or} \qquad x = \frac{b}{h}\left(\frac{2}{3}h - y\right) \tag{d}$$

The moment of inertia of the entire area then is

$$\left[I = \int y^2\,dA\right] \qquad I_{X_C} = \int_{-h/3}^{+2h/3} y^2 \cdot \frac{b}{h}\left(\frac{2}{3}h - y\right) dy$$

$$= \frac{2}{3}b\int_{-h/3}^{+2h/3} y^2\,dy - \frac{b}{h}\int_{-h/3}^{+2h/3} y^3\,dy$$

$$= \frac{2}{3}b\left[\frac{y^3}{3}\right]_{-h/3}^{+2h/3} - \frac{b}{h}\left[\frac{y^4}{4}\right]_{-h/3}^{+2h/3}$$

$$= \frac{2}{9}b\left(\frac{8}{27}h^3 + \frac{1}{27}h^3\right) - \frac{b}{4h}\left(\frac{16}{81}h^4 - \frac{1}{81}h^4\right)$$

or

$$I_{X_C} = \frac{bh^3}{36} \qquad \textit{Ans.} \tag{e}$$

(*b*) *Axis coinciding with base.* When the same elemental strip is used,

$$\left[I = \int y^2\,dA\right] \qquad I_X = \int_0^h y^2 x\,dy \tag{f}$$

But by similar triangles

$$\frac{x}{h - y} = \frac{b}{h} \qquad \text{and} \qquad x = \frac{b}{h}(h - y) \tag{g}$$

The moment of inertia of the entire area then is

$$\left[I = \int y^2\,dA\right] \qquad I_X = \int_0^h y^2 \cdot \frac{b}{h}(h - y)\,dy$$

$$= b\int_0^h y^2\,dy - \frac{b}{h}\int_0^h y^3\,dy$$

$$= b\left[\frac{y^3}{3}\right]_0^h - \frac{b}{h}\left[\frac{y^4}{4}\right]_0^h$$

$$= \frac{bh^3}{3} - \frac{bh^3}{4} = \frac{bh^3}{12} \qquad \textit{Ans.} \tag{h}$$

B-3. Find the rectangular moment of inertia of the circular area shown in Fig. B-3, with respect to a diameter.

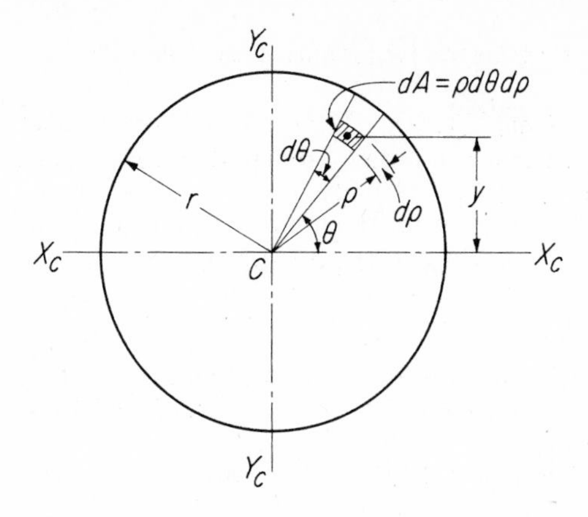

FIG. B-3 Rectangular moment of inertia of a circular area.

Solution: The elemental area is selected as shown in Fig. B-3. Its moment of inertia with respect to the X_C axis is $y^2\,dA$, or $\rho^2 \sin^2\theta \cdot \rho\,d\theta\,d\rho$, since $y = \rho \sin\theta$. A double integration is now necessary. The moment of inertia of the elemental sector is $\int_0^r \rho^2 \sin^2\theta \cdot \rho\,d\theta\,d\rho$. When this quantity is then again integrated between the limits of $\theta = 0$ and $\theta = 2\pi$ radians, we obtain the moment of inertia of the entire circle. That is,

$$\left[I = \int y^2\,dA\right] \qquad I_{X_C} = \int_0^r \int_0^{2\pi} \rho^3\,d\rho \cdot \sin^2\theta\,d\theta$$

$$= \frac{r^4}{4}\int_0^{2\pi} \sin^2\theta\,d\theta = \frac{\pi r^4}{4} \qquad \textit{Ans.} \qquad (i)$$

The Transfer Formula for Parallel Axes. In Art. 11-4 was described the method of computing the moment of inertia of an area with respect to a noncentroidal axis in the plane of the area. This was accomplished by use of Eq. (11-2), the so-called **transfer formula,** the derivation of which is shown below.

The moment of inertia of the elemental area dA in Fig. B-4 with respect to axis XX is

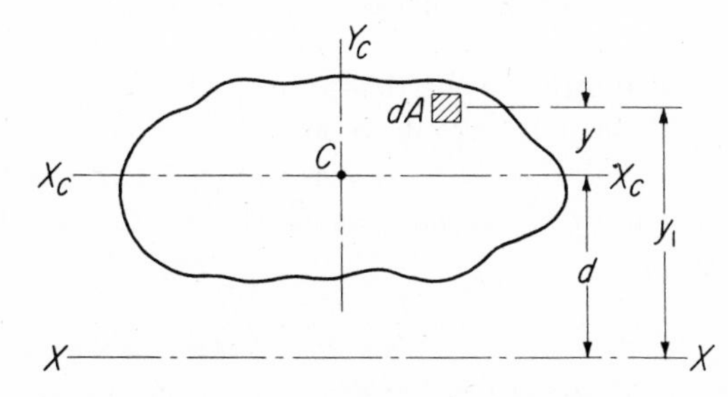

FIG. B-4 Moment of inertia of an area with respect to a noncentroidal axis.

$$I_X = \int y_1^2 \, dA = \int (y + d)^2 \, dA \qquad (j)$$

Expanding this, we obtain

$$I_X = \int y^2 \, dA + 2d \int y \, dA + d^2 \int dA \qquad (k)$$

The first term is recognized as the moment of inertia of the area with respect to its own centroidal axis X_C. The second term when integrated becomes $\bar{y}A$, the first moment of the area with respect to axis X_C. When this axis passes through the centroid of the area as it does, $\bar{y}$ becomes zero, since the first moment of an area with respect to a centroidal axis is zero. Hence the second term becomes zero. The third term when integrated becomes Ad^2. Accordingly

$$\boldsymbol{I_X = I_{X_C} + Ad^2} \qquad \textbf{(11-2)}$$

As an example of the use of this equation, let us obtain the moment of inertia of the triangle in Prob. B-2 with respect to the centroidal axis X_C, knowing its moment of inertia with respect to the X axis to be $bh^3/12$. Now d, the distance between the two parallel axes, is $\frac{1}{3}h$. Hence

$$\frac{bh^3}{12} = I_{X_C} + \left(\frac{bh}{2}\right)\left(\frac{h}{3}\right)^2 \qquad (l)$$

and

$$I_{X_C} = \frac{bh^3}{12} - \frac{bh^3}{18} = \frac{bh^3}{36} \qquad (m)$$

B-4. In Prob. B-3 the moment of inertia of the circular area with respect to a diameter was found to be $\frac{1}{4}\pi r^4$. Consequently the moment of inertia of a semicircle with respect to the same axis would be one-half of that, which is $\frac{1}{8}\pi r^4$, or $0.3927R^4$, as listed in Table 11-1, Art. 11-3. By use of the transfer formula, find the moment of inertia of the semicircular area with respect to the centroidal axis X_C, which is given as $0.110R^4$ $(r = R)$.

Solution: Since $A = \pi R^2/2$ and $d = 0.424R$ (Art. 10-4),

$$[I_X = I_{X_C} + Ad^2] \qquad 0.3927R^4 = I_{X_C} + \frac{1}{2}\pi R^2(0.424R)^2$$

or

$$I_{X_C} = 0.3927R^4 - 0.2827R^4 = 0.1100R^4 \qquad \textit{Ans.} \qquad (n)$$

Polar Moment of Inertia. In the deisgn of shafts and other members subjected to twisting, we need to know the moment of inertia of the cross-sectional area of the member with respect to an axis *perpendicular* to the plane of the area. This is called the *polar moment of inertia* of the area and is denoted by the symbol J.

B-5. Find the polar moment of inertia of the circular area shown in Fig. B-5 with respect to an axies perpendicular to the plane of the area and passing through its centroid C.

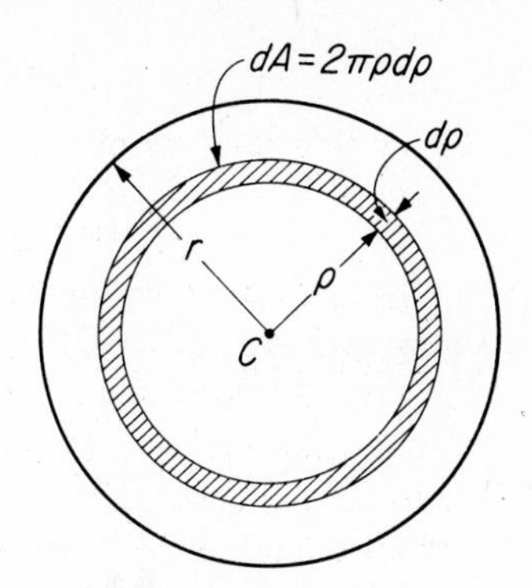

FIG. B-5 Polar moment of inertia of a circular area.

Solution: The elemental area selected is circular, as shown in Fig. B-5. Its moment of inertia with respect to an axis through C is $\rho^2\,dA$. Integrating this quantity between the limits of $\rho = 0$ and $\rho = r$ gives the moment of inertia of the entire area. That is, since $dA = 2\pi\rho\,d\rho$,

$$\left[J = \int \rho^2\,dA\right] \qquad J = \int_0^r \rho^2 \cdot 2\pi\rho\,d\rho$$

$$= 2\pi \int_0^r \rho^3\,d\rho = 2\pi \left[\frac{\rho^4}{4}\right]_0^r = \frac{\pi r^4}{2} \qquad \textit{Ans.} \qquad (o)$$

APPENDIX C

Moments of Inertia of Solid Bodies

In Art. 18-4, the term $\Sigma wr^2/g$ was defined as the *moment of inertia* I of a solid body about its centroidal axis of rotation. When the weight of each particle is infinitesimally small, or dw, and r is a variable radius ρ, this term becomes

$$I = \int \frac{\rho^2\,dw}{g} \qquad (a)$$

From this general expression we may derive the separate expressions for the moments of inertia of various solids of revolution, such as a cylinder, a cone, a sphere, and others, as is illustrated below.

ILLUSTRATIVE PROBLEMS

C-1. Cylinder. Derive the expression for the moment of inertia, about its geometric axis C, of the homogeneous right circular cylinder shown in Fig. C-1.

Solution: The elemental weight dw is the product of the density δ and the elemental volume dV, or $dw = \delta\,dV$. From Fig. C-1, the elemental area is $dA = (\rho\,d\theta)(d\rho)$. The elemental volume is $dV = (dA)(dx)$. Hence, the elemental weight $dw = \delta\,dV = (\delta)(\rho\,d\theta)(d\rho)(dx)$, and the moment of inertia then is

$$I = \int \frac{\rho^2\,dw}{g} = \frac{\delta}{g}\int \rho^2\,dV$$

$$= \frac{\delta}{g}\int_0^l \int_0^{2\pi} \int_0^R (\rho^3\,d\rho)(d\theta)(dx)$$

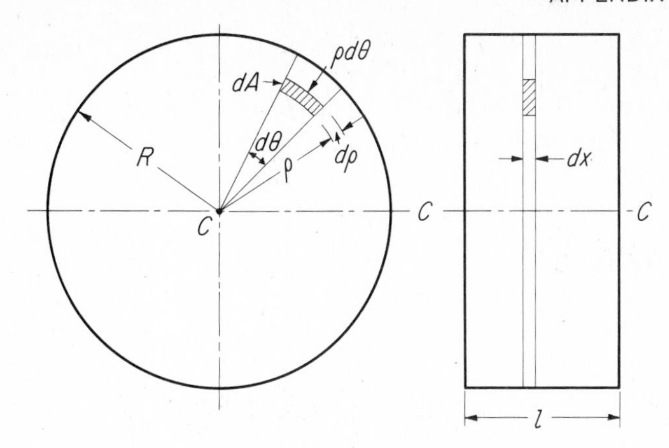

FIG. C-1 Moment of inertia of a cylinder.

$$= \frac{\delta}{g}\left[\frac{\rho^4}{4}\right]_0^R \cdot \left[\theta\right]_0^{2\pi} \cdot \left[x\right]_0^l$$

$$= \frac{\delta}{g}\left(\frac{R^4}{4} - 0\right)(2\pi - 0)(l - 0) = \left(\frac{\delta}{2g}\right)(\pi R^4 l) \qquad (b)$$

But the weight W of the cylinder is $(\delta)(\pi R^2 l)$, the product of its density δ and volume V, which is $\pi R^2 l$. Therefore, the moment of inertia of a solid cylinder is

$$I = \frac{WR^2}{2g} \qquad (c)$$

C-2. Sphere. To obtain the moment of inertia of a sphere with respect to a centroidal axis, we shall consider the shaded circular element shown in Fig. C-2. Since dx is infinitesimal, this element is, in effect, a cylinder of radius $R = y$ and length $l = dx$, whose moment of inertia then is $I = (\delta/2g)(\pi y^4\, dx)$, according to Prob. C.1. We obtain the moment of inertia of the entire sphere by summing up the moments of inertia of all these elemental cylinders between the limits of $x = + R$

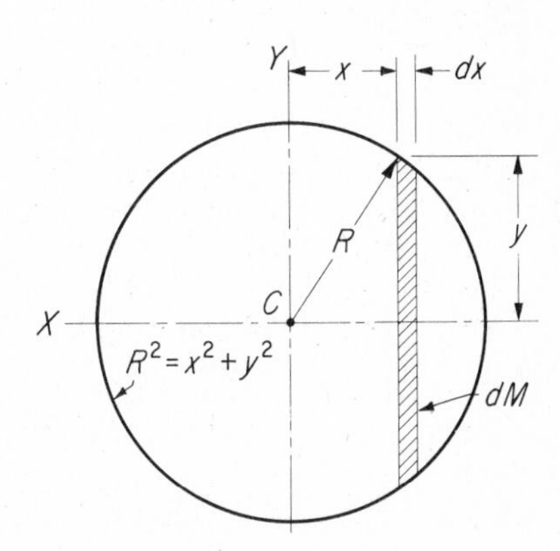

FIG. C-2 Moment of inertia of a sphere.

and $x = -R$. But note that $y^2 = R^2 - x^2$. Therefore $y^4 = (R^2 - x^2)^2$. Hence

$$I = \frac{\pi\delta}{2g}\int_{x=-R}^{x=+R} y^4\,dx$$

$$= \frac{\pi\delta}{2g}\int_{x=-R}^{x=+R} (R^2 - x^2)^2\,dx \qquad (d)$$

If now we let the lower limit of x equal 0 instead of $-R$, we obtain I for the right half of the sphere only, which when multiplied by 2 gives I for the entire sphere. That is,

$$I = 2\frac{\pi\delta}{2g}\int_{x=0}^{x=R} (R^4 - 2R^2x^2 + x^4)\,dx$$

$$= \frac{8}{15}\delta\pi R^5 = \frac{2}{5}R^2\left(\delta\cdot\frac{4}{3}\pi R^3\right) = \frac{2}{5}\frac{WR^2}{g} \qquad (e)$$

since the weight of the sphere $W = \delta V = \delta(4/3 \cdot \pi R^3)$.

No specific problems are given here, but any of the terms listed in Table 18-1, Art. 18-5, may be derived, as time permits.

The Transfer Formula. The moment of inertia of a body about an axis of rotation A which does not pass through its center of gravity is $I_A = I_C + Md^2$, or, since $M = W/g$,

$$I_A = I_C + \frac{W}{g}\cdot d^2 \qquad (f)$$

in which I_A = moment of inertia with respect to any axis of rotation A
I_C = moment of inertia with respect to a parallel centroidal axis
W = weight of body
d = perpendicular distance between the parallel axes

As proof of this, let us consider a section of a body cut perpendicular to its centroidal axis which pierces the section at C in Fig. C-3. We desire to obtain the moment of inertia of this body about some other parallel axis AA which pierces the section at A.

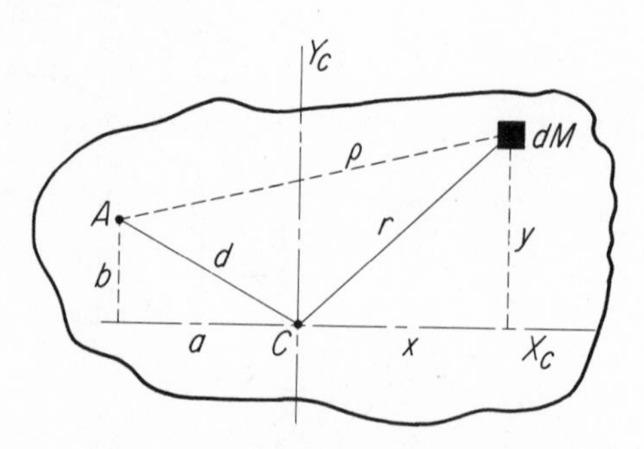

FIG. C-3 The transfer formula.

The fundamental equation is

$$I = \int \frac{\rho^2 \, dw}{g} = \int \rho^2 \, dM \tag{g}$$

when the *elemental mass is* $dM = dw/g$.

From Fig. C-3, we see that

$$\begin{aligned}\rho^2 &= (x + a)^2 + (y - b)^2 \\ &= x^2 + y^2 + a^2 + b^2 + 2ax - 2by\end{aligned}$$

But $x^2 + y^2 = r^2$, and $a^2 + b^2 = d^2$. Hence,

$$\rho^2 = r^2 + d^2 + 2ax - 2by$$

When this value of ρ^2 is substituted in Eq. (g), we obtain

$$I = \int r^2 \, dM + d^2 \int dM + 2a \int x \, dM - 2b \int y \, dM \tag{h}$$

Now, $2a\int x \, dM = 2a\bar{x}M$ and $2b\int y \, dM = 2b\bar{y}M$. But these terms both equal zero, because $\bar{x}$ and $\bar{y}$ are zero when x and y are measured from the *centroidal* axis.

The first term in Eq. (h) we recognize as the moment of inertia I_C of the body with respect to its centroidal axis. The second term becomes Md^2. Therefore,

$$I_A = I_C + Md^2 \tag{i}$$

which is the required proof. A word statement of this relationship is: *The moment of inertia of a body with respect to any noncentroidal axis equals its moment of inertia with respect to a parallel centroidal axis plus the product of its mass and the square of the distance between the two axes.*

Radius of Gyration. From Eq. (18-3) of Art. 18-8, $I = Wk^2/g = Mk^2$. Hence, Eq. (i) may also be written

$$Mk_A{}^2 = Mk_C{}^2 + Md^2$$

from which

$$k_A{}^2 = k_C{}^2 + d^2 \tag{j}$$

thus giving a simple relationship between the centroidal and the noncentroidal radii of gyration.

APPENDIX D

Variable Acceleration

Rectilinear Motion. The velocity of a body was defined (Art. 13-6) as its rate of change of displacement with respect to time. The fundamental differential equation expressing this relationship is

$$v = \frac{ds}{dt} \tag{a}$$

Similarly, acceleration is the rate of change of velocity with respect to time, for which the differential equation is

$$a = \frac{dv}{dt} = \frac{d^2s}{dt^2} \tag{b}$$

These relationships are readily applied to the solution of problems involving variable acceleration.

ILLUSTRATIVE PROBLEM

D-1. The velocity of a body starting from rest is expressed by the equation $v = 8t + 6t^2$. Calculate its acceleration a, velocity v, and total displacement s at the end of 3 sec.

Solution: The equation expressing the acceleration of the body is obtained by differentiation of the equation $v = 8t + 6t^2$. We then obtain

$$a = \frac{dv}{dt} = 8 + 12t$$

To obtain a and v at the end of 3 sec,

$[a = 8 + 12t]$ $\quad a = 8 + (12)(3) = 44$ ft per sec^2 *Ans.*
$[v = 8t + 6t^2]$ $\quad v = (8)(3) + (6)(3)^2 = 78$ ft per sec *Ans.*

The equation expressing the displacement s is obtained by integrating the velocity equation $v = 8t + 6t^2$ between the limits of 0 and 3 sec. Then

$$s = \int_0^t v\,dt = \int_0^3 (8t + 6t^2)\,dt$$

$$= \int_0^3 8t\,dt + \int_0^3 6t^2\,dt = \left[\frac{8t^2}{2}\right]_0^3 + \left[\frac{6t^3}{3}\right]_0^3$$

from which $\quad s = [(4)(3)^2 - 0] + [(2)(3)^3 - 0] = 90$ ft *Ans.*

PROBLEMS

D-2. The velocity of a body starting from rest is expressed by the equation $v = 6t^2$. Determine its acceleration, velocity, and displacement at the end of 4 sec.

Ans. $a = 48$ ft per sec^2; $v = 96$ ft per sec; $s = 128$ ft

D-3. The displacement of a body starting from rest is expressed by the equation $s = 9t^2 - t^3$. Calculate its initial acceleration a_0, and its acceleration, velocity, and displacement at the end of 3 sec.

Ans. $a_0 = 18$ ft per sec^2; $a = 0$; $v = 27$ ft per sec; $s = 54$ ft

D-4. A car passes point A with a velocity expressed by the equation $v = 8 + 2t + 1.2t^2$, and 5 sec later reaches point B. Determine its acceleration and velocity at B, and the distance it has traveled from A to B.

Ans. $a = 14$ ft per sec^2; $v = 48$ ft per sec; $s = 115$ ft

D-5. The instant the brakes are applied to bring a locomotive to a stop, its velocity is expressed by the equation $v = 15 - 0.6t^2$. Calculate the time elapsed before it stops, its deceleration, and the distance traveled before it stops.

Ans. $t = 5$ sec; $a = -6$ ft per sec^2; $s = 50$ ft

Rotation. Variable accelerations of angular motion are treated the same as those of rectilinear motion. The fundamental differential equations are

$$\omega = \frac{d\theta}{dt} \quad (c) \qquad \text{and} \qquad \alpha = \frac{d\omega}{dt} = \frac{d^2\theta}{dt^2} \qquad (d)$$

ILLUSTRATIVE PROBLEM

D-6. The angular velocity of a flywheel is expressed by the equation $\omega = 8 + 6t^2$, where ω is in radians per second and t in seconds. Calculate its angular acceleration α, its velocity ω, and its displacement θ at the end of 4 sec.

Solution: The equation expressing the acceleration α is obtained by differentiation of the velocity equation $\omega = 8 + 6t^2$. That is,

$$\alpha = \frac{d\omega}{dt} = 12t$$

Then, when $t = 4$ sec,

$$\alpha = 12t = (12)(4) = 48 \text{ rad per sec}^2$$

and $\omega = 8 + 6t^2 = 8 + (6)(4)^2 = 8 + 96 = 104$ rad per sec

The displacement θ is obtained by integration of the velocity equation $\omega = 8 + 6t^2$ between the limits of 0 and 4. That is,

$$\theta = \int_0^t \omega\, dt = \int_0^4 (8 + 6t^2)\, dt$$

or

$$\theta = \int_0^4 8\, dt + \int_0^4 6t^2\, dt = [8t]_0^4 + \left[\frac{6t^3}{3}\right]_0^4$$

whence $\theta = [(8)(4) - 0] + [(2)(4)^3 - 0] = 32 + 128 = 160$ rad *Ans.*

PROBLEMS

D-7. The angular displacement of a pulley is expressed by the equation $\theta = 10t + 2t^3$, where θ is in radians and t in seconds. Write the equations for ω and α. Determine the initial angular velocity ω_0 of the pulley and the angular acceleration α, velocity ω, and displacement θ, at the end of 3 sec.

Ans. $\omega = 10 + 6t^2$; $\alpha = 12t$; $\omega_0 = 10$ rad per sec; $\alpha = 36$ rad per sec^2; $\omega = 64$ rad per sec; $\theta = 84$ rad

D-8. During a certain period of 4 sec, a drum rotates with an angular velocity expressed by the equation $\omega = 48 - 24t + 3t^2$, in which ω is in revolutions per second and t is in seconds. Write the equations for α and θ, and compute the angular displacement θ at the end of 4 sec.

Ans. $\alpha = -24 + 6t$; $\theta = 48t - 12t^2 + t^3$; $\theta = 64$ rev

D-9. A high-speed turbine rotating with an angular velocity ω_0 of 54 rev per sec is brought to a stop in 6 sec by a deceleration expressed by the equation $\alpha = -3t$, where α is in revolutions per second per second and t is in seconds. Write the equations for ω and θ, and compute the total displacement θ (rev) during the 6 sec.

Ans. $\omega = 54 - 1.5t^2$; $\theta = 54t - 0.5t^3$; $\theta = 216$ rev

APPENDIX E

Derivation of the Laws of Motion Diagrams

The laws of motion diagrams, stated in Art. 14-5, are as follows:

First Law. The slope of the curve at any point in any diagram equals the length of the ordinate at the corresponding point in the next higher diagram.

Second Law. The difference in length of any two ordinates in any diagram equals the area between corresponding ordinates in the next higher diagram.

The proofs of these statements are shown in Fig. E-1 and in the following discussion.

Acceleration is defined as the time rate of change of velocity.

The differential equation expressing this relationship is

$$a = \frac{dv}{dt} = \frac{d^2s}{dt^2}$$

Velocity is the time rate of change of displacement, for which the differential equation is $v = ds/dt$. In Fig. E-1 are shown sections of the acceleration, velocity, and displacement curves expressing the motion of a particle. Together with their respective reference axes and explanatory matter, they will be referred to as the a, v, and s diagrams.

The ordinates a, v, and s represent, respectively, the magnitudes of the acceleration, velocity, and displacement of the particle at a given time t. The two ordinates enclosing the shaded area in each diagram may be considered to be equal, since dt is infinitesimal.

Proof of First Law: In the v diagram, the slope of the curve at any point

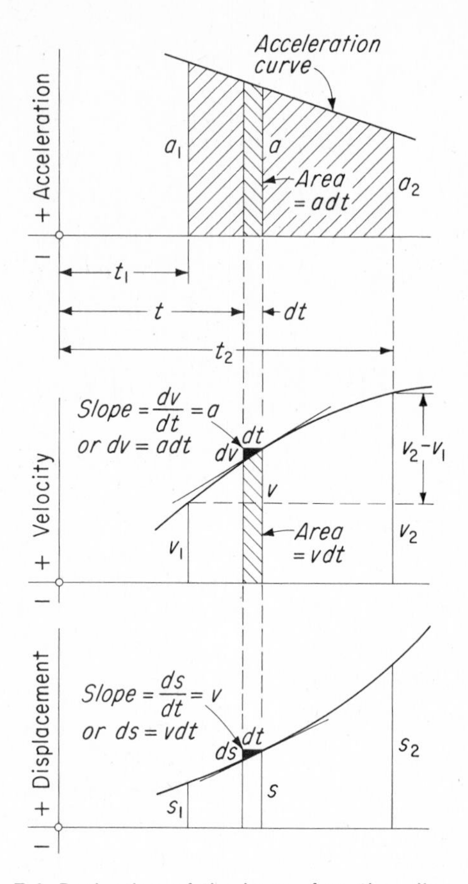

FIG. E-1 Derivation of the laws of motion diagrams.

is dv/dt, which equals a, the ordinate at the corresponding point in the next higher diagram. A similar relationship exists between the s and the v diagrams.

Proof of Second Law: In the v diagram, the difference between the two velocity ordinates enclosing the shaded area is dv, which equals $a\ dt$, the area between corresponding ordinates in the next higher diagram. A similar relationship exists between the s and the v diagrams. Between definite time limits t_1 and t_2, the difference in velocity ordinates, Δv or $v_2 - v_1$, is equal to the entire shaded area in the a diagram, which is

$$v_2 - v_1 = \int_{t_1}^{t} a\ dt$$

The mathematical relationships stated above, and illustrated in Fig. E-1, prove the correctness of the word statements of the laws of motion diagrams, as they are given in Art. 14-5.

Index